Green Photo-active Nanomaterials
Sustainable Energy and Environmental Remediation

RSC Green Chemistry

Editor-in-Chief:
Professor James Clark, *Department of Chemistry, University of York, UK*

Series Editors:
Professor George A. Kraus, *Department of Chemistry, Iowa State University, Ames, Iowa, USA*
Professor Andrzej Stankiewicz, *Delft University of Technology, The Netherlands*
Professor Peter Siedl, *Federal University of Rio de Janeiro, Brazil*

How to obtain future titles on publication:
A standing order plan is available for this series. A standing order will bring delivery of each new volume immediately on publication.

For further information please contact:
Book Sales Department, Royal Society of Chemistry, Thomas Graham House, Science Park, Milton Road, Cambridge, CB4 0WF, UK
Telephone: +44 (0)1223 420066, Fax: +44 (0)1223 420247
Email: booksales@rsc.org
Visit our website at www.rsc.org/books

Green Photo-active Nanomaterials
Sustainable Energy and Environmental Remediation

Edited by

Nurxat Nuraje
Texas Tech University, Lubbock TX, USA
Email: nurxat.nuraje@ttu.edu

Ramazan Asmatulu
Wichita State University, Wichita KS, USA
Email: ramazan.asmatulu@wichita.edu

Guido Mul
University of Twente, Enschede, The Netherlands
Email: G.mul@utwente.nl

THE QUEEN'S AWARDS
FOR ENTERPRISE:
INTERNATIONAL TRADE
2013

RSC Green Chemistry No. 42

Print ISBN: 978-1-84973-959-7
PDF eISBN: 978-1-78262-264-2
ISSN: 1757-7039

A catalogue record for this book is available from the British Library

Published by The Royal Society of Chemistry,
Thomas Graham House, Science Park, Milton Road,
Cambridge CB4 0WF, UK

Registered Charity Number 207890

For further information see our web site at www.rsc.org

Printed in the United Kingdom by CPI Group (UK) Ltd, Croydon, CR0 4YY, UK

Preface

Energy and environmental issues are of great concerns for the public and will keep increasing in the next few decades. The demand for clean energy sources in our current society also increases with large-scale economic developments and population growth. It is crucial to build clean energy systems in order to solve both the environmental issues and the energy demands. Alternative energy sources to replace fossil and mineral-based fuels have been actively searched for to meet the clean energy demands. Among the renewable energy sources defined as clean energy from natural sources, such as solar, rain, tides, wind, waves, biomass, and geothermal heat, solar energy is one of the greatest sources of renewable energy for meeting the above demands.

Along with developments of related solar technologies, storage of this energy as chemical energy in the form of hydrogen is a promising method to add to the solar cell technology, due to its sporadic nature. At present, solar hydrogen production from water has been achieved by the following several methods:

(1) electrolysis of water using a solar cell
(2) reforming of biomass
(3) photocatalytic or photoelectrochemical water splitting.

Photocatalytic water splitting is an artificial photosynthesis technique and contributes to a definitive green sustainable chemistry to solve energy and environmental issues.

This book, entitled *Green Photo-active Nanomaterials: Sustainable Energy and Environmental Remediation*, is an advanced book about the fundamentals of solar energy conversion, natural and artificial photosynthetic systems, nanotechnology and nanoscience, and the application of

RSC Green Chemistry No. 42
Green Photo-active Nanomaterials: Sustainable Energy and Environmental Remediation
Edited by Nurxat Nuraje, Ramazan Asmatulu and Guido Mul
© The Royal Society of Chemistry 2016
Published by the Royal Society of Chemistry, www.rsc.org

nanoscience and nanotechnology in energy and environmental remediation, as well as educational and training purposes. Furthermore, nanotechnology has a great potential to design artificial photosynthesis systems to store solar energy, produce fuels from biomass, reduce organic contaminants from the environment, and convert carbon dioxide to useful hydrocarbon fuels because of the outstanding mechanical, electrical (conductive and semi-conductive), optical, magnetic, quantum mechanics, and thermal properties of nanomaterials. These unique properties of nanoscale materials, such as nanoparticles, nanotubes, nanowires, nanofibers, nanocomposites, nano-pores, and nanofilms, allow them to design the next generation of photosynthetic devices in energy and environmental applications.

Recent publications in green photoactive nanostructured materials for energy and environment have shown that increased and sophisticated progress has been made using innovatively designed nanostructured materials in various devices. It is very important for us to provide an advanced book which can provide the basic science of nanomaterial and solar spectrum interactions, green synthesis of nanomaterials, and descriptions of natural photosynthetic systems which will inspire us to design more efficient photoelectrochemical devices. This book will detail recent developments in green photo-active nanostructures materials in water splitting, biomass, and environmental remediation. It also emphasizes the recent development of nanostructured materials for carbon dioxide conversion, degradation of pollutants in environment, and green chemistry. The book also discusses the safety and risk assessments of the nanostructured materials used for various energy production systems. Therefore, this book will be informative for researchers in photoactive nanomaterials in energy and environment application, and also will be an excellent text book for advanced study in the Universities from fundamental points of views.

Thus, the editors are very pleased to present the recent progress in photo-active nanomaterials in energy and environment remediation in the publication of this great book for engineers, scientists and other readers, policy makers, and scientific communities. We are also thankful for the authors' hard work and contributions, and reviewers' comments and suggestions. During the editing process, we also have received tremendous support from the editorial team of the RSC, including Dr. Merlin Fox and Dr. Mina Roussenova. We also specially thank Dr. Sindee Simon (Texas Tech) for her kind advice and support. We acknowledge all support from Texas Tech University, MIT, and Wichita State University. Without all the above support, it would not have been possible for us to publish this book.

Dr Nurxat Nuraje, Texas Tech University, USA
Dr Ramazan Asmatulu, Wichita State University, USA
Dr Guido Mul, University of Twente, The Netherlands

Contents

RSC Green Chemistry No. 42
Green Photo-active Nanomaterials: Sustainable Energy and Environmental Remediation
Edited by Nurxat Nuraje, Ramazan Asmatulu and Guido Mul
© The Royal Society of Chemistry 2016
Published by the Royal Society of Chemistry, www.rsc.org

Chapter 10 Hybrid Inorganic and Organic Assembly System for Photocatalytic Conversion of Carbon Dioxide   **240**
Xin Zhang, Yu Lei and Nurxat Nuraje

Chapter 11 Biological Systems for Carbon Dioxide Reductions and Biofuel Production   **274**
E. Asmatulu

CHAPTER 1

Introduction to Green Nanostructured Photocatalysts

R. ASMATULU,*[a] N. NURAJE*[b] AND G. MUL*[c]

[a] Department of Mechanical Engineering, Wichita State University, 1845 Fairmount, Wichita, KS 67260, USA; [b] Department of Chemical Engineering, Texas Tech University, P.O. Box 43121, Lubbock, TX 79409, USA; [c] Faculty of Science & Technology, University of Twente, PO Box 217, Meander 225, 7500 AE, Enschede, The Netherlands
*Email: ramazan.asmatulu@wichita.edu; nurxat.nuraje@ttu.edu; G.mul@utwente.nl

1.1 Introduction

1.1.1 General Background

Fossil fuel-based sources of energy, such as coal, oil, and natural gas, have been used to meet the world's energy demands for centuries; however, overproduction and overconsumption of these fuels have created many known and unknown concerns. Knowledge about the sources of mineral fuel, including nuclear energy, are also inadequate in terms of long-term waste disposal and lack of technology.[1] Fossil fuel-based energy systems have a huge impact on the environment and are considered to be the major cause of global warming as well as air, soil, and water contamination and pollution. Because of dramatic economic development, population growth, environmental and health concerns, and increasing demands on clean energy sources, many countries have been seeking to find alternative energy sources to replace fossil and mineral-based fuels.[1-3] These new sources of

RSC Green Chemistry No. 42
Green Photo-active Nanomaterials: Sustainable Energy and Environmental Remediation
Edited by Nurxat Nuraje, Ramazan Asmatulu and Guido Mul
© The Royal Society of Chemistry 2016
Published by the Royal Society of Chemistry, www.rsc.org

energy should be renewable, minimize/eliminate concerns, and, at the same time, be inexpensive and affordable by many nations of the world.

Renewable energy is usually defined as clean energy, which mainly comes from natural sources such as sunlight, rain, tides, wind, waves, biomass, and geothermal heat, and can be naturally replenished in a shorter period of time without harming the Earth. Solar energy is one of the greatest sources of renewable energy for meeting the world's demand because of its enormous magnitude – approximately 10^5 terawatts.[2] The current energy consumption of the world is about 12 terawatts; this represents only 0.01% of the total amount of the Sun's energy that reaches the Earth's surface. This energy could be generated from an area 10^5 km^2 in size that is installed with solar cells working at 10% efficiency. However, today, many energy conversion systems can easily pass the 10% energy conversion levels.[2–4]

Even though energy from the Sun is one of the most widely considered renewable energy sources, new studies need to be conducted to address some concerns with solar energy, such as harnessing incident photons, lowering production costs, enhancing efficiency, storing energy, eliminating waste materials, eliminating health and environmental risks, dealing with seasonal changes, addressing the lack of technology, and so on.[4–8] Nanotechnology is an emerging technology that could address these concerns by using innovative strategies.

1.1.2 Nanotechnology in Energy Systems

Nanotechnology is the development of materials, components, devices, and/or systems at the near-atomic level or nanometer scale. One of the dimensions of nanotechnology is between 1 and 100 nm.[9] This technology mainly involves fabricating, measuring, modeling, imaging, and manipulating matter at the nanoscale. Nanotechnology consists of highly multidisciplinary fields, including chemistry, biology, physics, engineering, and some other disciplines. For more than two decades, significant progress has been made in designing, analyzing, and fabricating nanoscale materials and devices, and this trend will continue for a few more decades in various fundamental studies and in research and development fields.[10]

Nanomaterials are the major building blocks of solar energy conversion devices and have been applied in the following three ways:[2]

(a) the assembly of molecular and clusters of donors–acceptors mimicking photosynthesis
(b) the production of solar fuel using semiconductor-assisted photocatalysis
(c) the use of nanostructured semiconductor materials in solar cells.

Among the nanostructured solar energy conversion systems and devices, binary and ternary metal oxides are the most widely used and have a promising future in this field.[2–4]

Even though several books have been published on renewable energy, solar cells, solar conversion, and solar fuels, very few books have been published on green photo-active nanomaterials and their major applications. Most books cover a broad spectrum of photocatalysts, including metal oxides and non-metal oxides. However, this book introduces and summarizes the fundamentals of harnessing solar energy using nanomaterials, synthetic approaches to green photo-active nanomaterials and their applications in designing artificial photochemical systems for solar energy conversion, and microorganisms found in solar energy conversion up until the present time. It describes the natural photosynthetic system in plants, the mechanisms involved in photosynthesis, and how components contribute to this sophisticated orchestration. Relevant cell biology as well as variations of the process used by plants in hot and dry environments are also discussed. The potential for biomass to contribute to meeting humanity's growing need for sources of energy is described, and a context is provided to frame efforts in mimicking natural photosynthesis in order to generate energy.

This book also focuses on applications of organic and inorganic nanomaterials utilized for fuel production from carbon dioxide and biomass, removal of contamination, water splitting, modeling, and health and environmental aspects of these green photo-active nanomaterials.

1.1.3 Environmental Considerations

Industrialization has significantly increased gas emissions and suspended particulate concentrations, and these concerns will likely continue for the next few decades, in turn further worsening the quality of air, soil, and water in the world and jeopardizing human life over the long term. Methane, carbon dioxide (CO_2), and nitrogen oxide (NOx) are the primary greenhouse gas sources involved in global warming and climate change, so reducing these emissions is now a worldwide challenge. Microorganisms (*e.g.*, microalgae, bacteria, viruses, fungi, and molds) can be an effective way of addressing some of these concerns. Nanomaterials can also offer structural features for reducing CO_2 and other emissions in an environmentally friendly manner.[11]

Combining microorganisms with nanomaterials can effectively capture greenhouse gasses from the atmosphere and convert them into carbon sources for the production of biomass and biofuels for industrial and household heating, transportation, agriculture practices, and many other uses. Also, plants can naturally absorb CO_2 emissions and other contamination for their growth media and reduce toxicity levels. As an outcome of this cycle, concentrations of specific pollutants in the air, soil, and water can be significantly decreased. Carbon dioxide contains an abundant source of carbon, which supports the growth of microbial species and plants in the environment, and can be biochemically transformed into biomass and renewable energy sources to meet the world's demands.[12–14]

1.2   Photo-active Nanomaterials

Some binary and ternary metal oxides are photoactive and are used for photocatalytic activities in solar cells, water splitting, and other solar-driven reactions. Synthetic methods for binary and ternary metal oxide photo-catalysts emphasize green reaction processes. The advent of green, facile, and benign methods of producing these nanomaterials is necessary to comply with modern environmental concerns. An important aspect for such green methods is low temperature, fast reaction rate, and reduced toxic agents. The second chapter of this book highlights new techniques to produce photo-active nanomaterials in order to minimize the use and generation of hazardous substances during the manufacturing process. Such techniques include hydrothermal approaches along with the polymer gel method, chemical precipitation technique, solvothermal method, ultrasound sonication, and hybrid synthesis method. For example, even though several methods are currently available, such as solid state reactions, the polymerizable complex method, and the hydrothermal method, titanium dioxide (TiO_2) is usually synthesized *via* sol–gel methods. Typically, particles synthesized by soft methods, including the polymerizable complex and sol–gel methods, provide higher performance than those synthesized using a solid state reaction because of the small particle size, shape, and good crystallinity.[2]

The band gaps of metal oxides with d^0 metal ions are usually formed from O 2p orbitals and nd orbitals from a metal cation, which are more negative than the zero potential of hydrogen ions. The band gaps of metal oxides are usually in the ultraviolet (UV) range. Powdered titania photocatalysts cannot split water without modification, such as a platinum (Pt) cocatalyst.[2–4] Hydrogen production experiments have been conducted using a TiO_2 photocatalyst with a band gap of 3.2 eV under different conditions, including pure water, vapor, and an aqueous solution including an electron donor with the assistance of a cocatalyst. Sodium hydroxide (NaOH) or sodium carbonate (Na_2CO_3) have been used to split water with a loaded Pt. Under UV irradiation, the efficiency of titania doped with other metal ions is considerably improved.[3]

Zirconium dioxide (ZrO_2) with a band gap of 5.0 eV is a photocatalyst that can split water without a cocatalyst under UV irradiation owing to the position of its high conduction band. Photocatalytic activity of ZrO_2 decreased when it was loaded with cocatalysts, such as Pt, copper (Cu), gold (Au), and ruthenium oxide (RuO_2). It is likely that the height of the electronic barrier of the semiconductor band metal impeded electron transport and stopped further molecular water-splitting reactions. Nevertheless, photocatalytic activity improved with the addition of Na_2CO_3.[2]

Niobium pentoxide (Nb_2O_5) with a band gap of 3.4 eV is not active without any modification under UV irradiation. It decomposes water efficiently in a mixture of water and methanol after being loaded with a Pt cocatalyst. Its higher photocatalytic activity under UV irradiation was observed as assembled mesoporous Nb_2O_5. Tantalum pentoxide (Ta_2O_5) with a band gap of

4.0 eV is also a well-known photocatalyst. It can produce a small amount of hydrogen and no oxygen without any modification. Ta_2O_5 loaded with nickel oxide (NiO) and RuO_2 shows great photocatalytic activity for generating both hydrogen and oxygen. The addition of Na_2CO_3 and a mesoporous structure of the catalyst showed enhanced photocatalytic activity. Nanostructured vanadium dioxide (VO_2) with a body-centered cubic (BCC) structure and a large optical band gap of 2.7 eV demonstrated excellent photocatalytic activity in hydrogen production from a solution of water and ethanol under UV irradiation. It also exhibited a high quantum efficiency of 38.7%.[2] Additionally, all of the metal oxides with d^{10} metal ions (Zn^{2+}, In^{3+}, Ga^{3+}, Ge^{4+}, Sn^{4+}, and Sb^{5+}) are effective photochemical water-splitting catalysts under UV irradiation.[3]

Even though binary metal oxides with d^0, d^{10}, and f^0 metal ions show efficient photocatalytic activity, their ternary oxides have been widely studied and proven to have the same photocatalytic effects. For instance, strontium titanate ($SrTiO_3$) with a band gap of 3.2 eV and potassium tantalite ($KTaO_3$) with band gap of 3.6 eV photoelectrodes can be photoactive without an external bias because of their high conduction bands. These materials can be employed as powder photocatalysts for solar cells and water splitting. Domen and co-workers studied the photocatalytic performance of NiO-loaded $SrTiO_3$ powder for water splitting. A reduction in hydrogen gas (H_2) is responsible for the activation of the NiO cocatalyst for H_2 evolution. Then, subsequent oxygen gas (O_2) oxidation to form an NiO/Ni double-layer structure provides a further path for the electron migration from a photocatalyst substrate to a cocatalyst surface. The NiO cocatalyst prevents the back reaction between H_2 and O_2, which is totally different for Pt.[2–4] The enhanced photocatalytic activity of $SrTiO_3$ was also reported using a new modified preparation method or suitable metal cation doping (*e.g.*, La^{3+}, Ga^{3+}, and Na^+).

Many ternary titanates are efficient photocatalysts for water splitting under UV irradiation. The H_2 evolution of photocatalysts of sodium titanate $Na_2Ti_3O_7$ (layered crystal structure), potassium titanate $K_2Ti_2O_5$ (layered crystal structure), and potassium titanate $K_2Ti_4O_9$ (layered crystal structure) from aqueous methanol solutions in the absence of a Pt cocatalyst was reported. The quantum yield of materials studied for H^+-exchanged $K_2Ti_2O_5$ reaches 10%. The method of catalyst preparation also shows a different activity. Barium titanate ($BaTiO_3$) with a band gap energy of 3.22 eV and perovskite crystal structure prepared using a polymerized complex method has high photocatalytic activity in comparison with materials prepared by the traditional method because of the smaller size and larger surface area.[2]

Calcium titanate ($CaTiO_3$) with a band gap energy of 3.5 eV and perovskite crystal structure loaded with Pt showed good photocatalytic activity under UV irradiation. The activity of $CaTiO_3$ doped with a zirconium ion (Zr^{4+}) solid solution was further increased. Quantum yields were reported to be up to 1.91% and 13.3% for H_2 evolution from pure water and an aqueous ethanol solution, respectively. A number of lanthanum titanate perovskites

(including La$_2$TiO$_5$, La$_2$Ti$_3$O$_9$, and La$_2$Ti$_2$O$_7$) with layered structures were reported with much higher photocatalytic activities under UV irradiation than bulk LaTiO$_3$. The photoactivities of La$_2$Ti$_2$O$_7$ doped with barium (Ba), strontium (Sr), and (calcium) Ca was improved sufficiently. The lanthanum titanate perovskite La$_2$Ti$_2$O$_7$ (band gap energy of 3.8 eV) prepared using a polymerized approach showed higher photoactivity than when the traditional solid-state method was used.[2]

Biological materials used as templates, such as bacteriophages, offer environmentally friendly synthesis and organization of functional materials at the nanoscale, where there is an efficiency of energy transfer by increasing the probability of the energy transfer groups being precisely positioned. A biological system such as M13 viruses presents a rational design and assembly of nanoscale catalysts based on biological principles (which are required for the water-splitting reaction) for the production of oxygen and hydrogen gas driven by light.

1.3 Microorganisms in Energy Mitigations

Recent studies have indicated that nanotechnology materials and processes could be applied to microorganism growth processes to potentially improve biological biomass production from the atmosphere. This technology can significantly enhance biodiesel production and biomass conversion rates. It can also improve enzyme immobilization, lipid accumulation and extraction, enzyme loading capacity, nanoscale catalysis activity, storage capacity, separation and purification rates of liquid from other liquids and solids, and bioreactor design and applications.[11-14]

Microorganisms such as bacteria, viruses, algae, molds, and fungi are living creatures and have survived in extreme environmental conditions for millions of years. They usually deposit fat, lipids/oil, glucose, starch, and other hydrocarbons and organic substances in their bodies that can be extracted and converted into useful products.

Bacteria are a single-cell form of life, and each individual cell is unique. They often grow into different colonies; however, each bacteria cell has its own independent life. New bacteria are reproduced by a process known as cell division. It is estimated that more than 3000 species of bacteria are living in totally different environments and conditions. Nevertheless, some of them are found only in a very specific environment, thus requiring specialized types of food, temperature, and light.[11]

A virus is a small infectious organism that can only replicate inside living cells of other cells and organisms. They can infect all kinds of animals, bacteria, plants, and so on. Unlike bacteria, viral populations do not grow through cell division since they are acellular, instead they use the machinery and metabolism of a host cell to produce multiple copies of themselves and then assemble inside those cells. To date, approximately 5000 viruses have been scientifically described in detail. A group of scientists has recently announced that they can successfully modify a virus to split water molecules, which can be an efficient and non-energy-intensive method of producing H$_2$.

These scientists genetically modified a commonly known, harmless bacterial virus in order to assemble the components for separating water molecules into H_2 and O_2 molecules, in turn yielding a fourfold boost in production efficiency. This novel process mimics plants that use the power of sunlight to make chemical fuel for their growth. In this research, the scientists engineered the virus as a kind of biological scaffold to split a water molecule.[12]

Algae comprise several different species (2800) of relatively simple living organisms that are found all over the world, capturing light energy through photosynthesis and converting inorganic/organic substances into simple sugars and other substances using photon energy. Algae can be considered the early stage of simple plants, and some are closely related to more complex plants as well. Some algae also appear to represent different protist groups (large and diverse groups of eukaryotic microorganisms), alongside other organisms that are traditionally considered more animal-like (*e.g.*, protozoa). Therefore, algae do not represent a single evolutionary direction but rather a level of organization that may have developed several times in the early history of microorganism life on the earth's crust.[13]

Some microorganisms usually require the following conditions for their growth:

- pH of 5–9 (lower pH may be seen)
- presence of organic substances (waste water, city waste, leaves)
- temperature between 4 °C and 40 °C; sulfur-, iron-, copper-, zinc-, cobalt-, and manganese-rich conditions;
- the presence of carbon and CO_2, nitrogen, phosphoros, oxygen, hydrogen, and sunlight (more sunlight, less UV rays).

As an example, the growth conditions of microorganisms found in Yellowstone National Park, which is an extremely hot, and mineral- and ion-rich environment, are totally different than the growth conditions of similar species that live in coastal and lake areas.[14]

Botryococcus braunii, a green algae with a pyramid-shaped planktonic structure, is one of the most important algae in biotechnology. These algae colonies are held together by a lipid biofilm and are usually found in tropical lakes, rivers, and creeks. They will bloom in the presence of dissolved inorganic phosphorus and other nutrients in a growth condition. *B. braunii* has great potential for algae farming because it produces hydrocarbons, which can be chemically converted into different fuels. It has been estimated that up to 86% of the dry weight of this alga can be composed of long-chain hydrocarbons, and some of its useful hydrocarbon oils can be found outside of the cell. *B. braunii* can convert 61% of its biomass into oil, which drops to only 31% under different conditions. It grows best between 22 °C and 25 °C, and is a great choice for biofuel production.[12]

Recently, nanotechnology-associated studies have been conducted on microorganisms to increase the efficiency of their growth and rates of fuel conversion. Nanotechnology probes tap into algae and bacteria cells to extract electrical energy. It has been postulated that *Chlamydomonas reinhardtii*, a

single-cell alga, might be an ideal cell for energy harvesting. This new study demonstrated the feasibility of collecting high-energy electrons in steps of the photosynthetic electron transport chain prior to the downstream process. However, cells usually die after a period of time because of leaks in the membrane where the nanoscale electrodes penetrate into the body of the cell.[14]

1.4   Environmental Health and Safety

Nanomaterials have outstanding electrical, optical, mechanical, magnetic, quantum mechanical, and thermal properties. Because of these unique properties, a variety of nanoscale materials, such as nanoparticles, nanofibers, nanocomposites, nanotubes, and nanofilms, all of which are considered to be the next generation of materials, have been utilized in several different in-dustries worldwide.[8] It has been stated that nanomaterials are already found in more than 1500 different products/processes, including solar cells, water-splitting reactions, bacteria-free cloths, concrete, sunscreens, car bumpers, tennis rackets, toothpastes, polymeric coatings, wrinkle-resistant clothes, and various electronic, optical, diagnostic, and sensing devices.

Recently, several studies have focused on photo-active nanomaterials for renewable energy generation from solar energy. However, some exposure of workers during large-scale nanomaterial production is inevitable, even when incidents of unusual release do not occur. It is necessary to determine the degree of exposure that can cause major health impacts. The toxicities of photo-active nanomaterials are among the most studied in nanotechnology because of their numerous applications in both energy and medical fields. The risk of exposure is greatest to workers in the nanotechnology field, but others would also be vulnerable, such as following the environmental release of nanomaterials during consumer product use, transportation, storage, *etc.*[1]

Nanomaterials are more reactive and potentially more damaging than bulk-scale materials of the same composition because of their high surface area-to-volume ratio. Testing with a variety of organisms to determine the impact of nanoparticle characteristics on toxicity has yielded inconsistent results, but some generalizations can be made. Smaller nanomaterials tend to be more toxic than larger nanomaterials. Fibrous or rod-like nanoparticles of any composition tend to be more hazardous than spherical or agglom-erated nanoparticles. For TiO_2, the crystalline phase may influence the degree or mode of toxicity, but both rutile and anatase nanoparticles can cause toxic reactions at nanoscale. As the nanoparticle concentration increases, toxicity tends to increase, unless particles aggregate, which tends to reduce toxicity significantly. Nevertheless, aggregations may reduce the efficiency of solar energy conversion systems. Some tests indicate that the size of nanoparticles has less effect on toxicity than the material itself. Nanoparticle surface charge can impact toxicity and may prevent contact with cells if they share a surface charge.

The variables that affect nanoparticle toxicity continue to be refined, but impacts remain difficult to predict. The risk-control method involving

nanomaterials entails a chain of assessment, planning, implementation, and corrective action, which is repeated in order to minimize employee exposure to unwanted nanoparticles, thus allowing them to work in acceptable environments.[15] The three main categories of risk control are engineering techniques, administrative measures, and personal protection.[8] Table 1.1 shows this hierarchy applied to nanoparticle assessment.

Engineering controls, such as design, elimination/substitution, isolation/confinement, and ventilation seem to be more effective than the administrative measures (*e.g.*, information/training, work procedures, cleaning and equipment, personal hygiene, and work periods). Personal protective equipment, such as respiratory protection, and skin and eye protection, is less effective than other elements of the hierarchy.[8] Once the conditions of the engineering techniques are satisfied, then workers can reduce the possibility of contact and potential risk. This also shows the effectiveness of these three categories relative to the overall risk-control hierarchy.[15,16] Table 1.2 summarizes some properties of nanoparticles currently used by

Table 1.1 Risk control hierarchy applied to nanoparticles.

Risk control hierarchy	Actions	Effectiveness
Engineering techniques	Design Elimination/substitution Isolation/confinement Ventilation	+
Administrative measure	Information/training Work procedures Cleaning and equipment Personal hygiene Work periods	
Personal protective equipment	Respiratory protection Skin, eye, and other protections	−

Table 1.2 Some concerns capable of influencing nanomaterials toxicity.

Primary concerns	Secondary concerns
Number of particles	Surface shape and geometry
Specific surface area	Solubility/dissolution
Size and granulometric distribution	Clustering/agglomeration
Concentration	Crystalline structures
Chemical compositions	Surface oxidation
Surface properties	Surface hydrophobicity/hydrophilicity
Zeta potential	Manufacturing techniques
Functional groups	Inertness/reactivity
Oxidative stresses	Biocompatibility/biodegradability
Free radicals	Metal and alloy, ceramic, polymer, composite
Cell viability	Dispersion/settlement
Surface coverage	Impurities and defects

various industries. Overall, more effort needs to be conducted in the field to minimize/eliminate the toxicity of nanomaterials used for photocatalyst purposes.

1.5 Contents of This Book

This book summarizes the most recent developments in the field of photo-active green nanomaterials, as well as their recent applications in solar energy conversion, mitigation of contamination, water splitting, health and environmental aspects, modeling, CO_2 emissions reduction, and nano and biological systems, in order to address some concerns of climate change and global warming. The following chapters will be included in this book, all of which were written by world-renowned authors in their fields:

Chapter 1:   Introduction to Green Nanostructured Photocatalysts
Chapter 2:   Fundamentals of Sunlight–Materials Interactions
Chapter 3:   Green Nanomaterials Preparation: Sustainable Methods and Approaches
Chapter 4:   Natural Photosynthesis System
Chapter 5:   Bioinspired Photocatalytic Nanomaterials
Chapter 6:   Hybrid Molecular–Nanomaterial Assemblies for Water-Splitting Catalysis
Chapter 7:   Hierarchical Nanoheterostructures for Water Splitting
Chapter 8:   Nanophotocatalysis in Selective Transformations of Lignocellulose-derived Molecules: A Green Approach for the Synthesis of Fuels, Fine Chemicals, and Pharmaceuticals
Chapter 9:   Photocatalytic CO_2 Conversion to Fuels by Novel Green Photocatalytic Materials
Chapter 10: Hybrid Inorganic and Organic Assembly System for Photocatalytic Conversion of Carbon Dioxide
Chapter 11: Biological Systems for Carbon Dioxide Reduction and Biofuel Production
Chapter 12: Organic Reactions using Green Photo-active Nanomaterials
Chapter 13: Hierarchical Nanoheterostructures: Double Layer Hydroxide-based Photocatalytic Materials
Chapter 14: Health and Environmental Aspects of Green Photo-active Nanomaterials
Chapter 15: Risk Assessments of Green Photo-active Nanomaterials
Chapter 16: Energy Harvesting from Solar Energy Using Nanoscale Pyroelectric Effects

1.6 Conclusions

Fossil fuels, such as coal, oil, and natural gas, have been heavily utilized by many developed and developing countries. Because of their overproduction

and consumption, fossil fuel-based energy systems have had a dramatic impact on the environment and on health, causing global warming as well as air, soil, and water contamination and pollution. Inspiration from the natural environment can be an option for solving these problems. By mimicking photo-active green nanomaterials found in nature, we can create light-harvesting assemblies, devise new methods for synthesizing fuels, and develop tools to synthesize novel functional materials for solar cells, water-splitting units, pollution control devices, and so on. Photo-active green nanomaterials can be more reactive, and potentially more damaging, than bulk materials of the same composition because of their high surface area-to-volume ratio. The properties of these nanomaterials should be determined prior to any photocatalytic applications.

Acknowledgements

The authors gratefully acknowledge the Kansas NSF EPSCoR (#R51243/ 700333), Wichita State University, and Texas Tech University for the financial and technical support of this work.

References

1. A. V. Rosa, *Fundamentals of Renewable Energy Processes*, Academic Press, New York, 3rd edn, 2012.
2. N. Nuraje, R. Asmatulu and S. Kudaibergenov, *Curr. Inorg. Chem.*, 2012, **2**, 124–146.
3. Y. Lei, R. Asmatulu and N. Nuraje, *ScienceJet*, 2015, **4**, 169–173.
4. N. Nuraje, S. Kudaibergenov and R. Asmatulu, in *Production of Fuels Using Nanomaterials*, ed. R. Luque and A. M. Balu, Taylor and Francis, 2013, pp. 95–117.
5. N. Nuraje, W. S. Khan, M. Ceylan, Y. Lie and R. Asmatulu, *J. Mater. Chem. A*, 2013, **1**, 1929–1946.
6. R. Asmatulu, M. Ceylan and N. Nuraje, *Langmuir*, 2011, **27**(2), 504–507.
7. R. Asmatulu, H. Haynes, M. Shinde, Y. H. Lin, Y. Y. Chen and J. C. Ho, *J. Nanomater.*, 2010, 715282, 3 pages.
8. R. Asmatulu, *Nanotechnology Safety*, Elsevier, Amsterdam, The Netherlands, August, 2013.
9. G. Cao and Y. Wang, *Nanostructures and Nanomaterials: Synthesis, Properties, and Applications*, World Scientific, 2nd edn, 2011.
10. B. Rogers, J. Adams and S. Pennathur, *Nanotechnology Understanding Small Systems*, CRC Press, 2nd edn, 2011.
11. V. Babu, A. Thapliyal and G. K. Patel, *Biofuels Production*, John Wiley & Sons, 2013.
12. X. L. Zhang, S. Yana, R. D. Tyagia and R. Y. Surampalli, *Renewable Sustainable Energy Rev.*, 2013, **26**, 216–223.

13. S. T. Yang, H. El-Ensashy and N. Thongchul, *Bioprocessing Technologies in Biorefinery for Sustainable Production of Fuels, Chemicals, and Polymers*, John Wiley & Sons, 2013.
14. S. Siva and C. Marimuthu, *Int. J. ChemTech Res.*, 2015, 7(4), 2112–2116.
15. R. Asmatulu, O. Nguyen and E. Asmatulu, in *Nanotechnology Safety*, ed. R. Asmatulu, Elsevier, 2013, pp. 57–72.
16. R. Asmatulu, in *Bronchitis*, ed. I. Martin-Loeches, InTec, 2011, pp. 95–108.

CHAPTER 2

Fundamentals of Sunlight–Materials Interactions

YUNHUI ZHU[a] AND HANG YU*[b,c]

[a] Department of Mechanical Engineering, Massachusetts Institute of Technology, Cambridge, MA 20139, USA; [b] Department of Materials Science and Engineering, Massachusetts Institute of Technology, Cambridge, MA 20139, USA; [c] Department of Materials Science and Engineering, Virginia Tech, Blacksburg, VA 24061, USA
*Email: hangyu@vt.edu

2.1 Introduction

Green photo-active nanomaterials are promising for sustainable energy and environmental remediation technologies, especially for those involving sunlight–materials interactions. Owing to the extremely high surface area to volume ratio and the unique materials structures on the nanoscale, the interactions with sunlight can be significantly enhanced in green nanostructured materials as compared to bulk materials, rendering the former an ideal candidate for a wide variety of emerging energy and environmental applications that are based on photocatalysis, photosynthesis, and photovoltaics.[1–9] Materials preparation and characterization for these applications will be discussed in detail in the following chapters. In this chapter, our focus is on the fundamentals of sunlight–materials interactions, with the aim to provide a comprehensive understanding of the underlying physics in light harvesting and energy conversion. This will include discussions of the generation and propagation of sunlight, the physical processes involved in

RSC Green Chemistry No. 42
Green Photo-active Nanomaterials: Sustainable Energy and Environmental Remediation
Edited by Nurxat Nuraje, Ramazan Asmatulu and Guido Mul
© The Royal Society of Chemistry 2016
Published by the Royal Society of Chemistry, www.rsc.org

light–matter interactions, and the approaches for evaluating the efficiency of light–materials interactions.

2.2 Solar Energy and the Solar Spectrum

2.2.1 The Sun as the Ultimate Energy Source

The Sun is the ultimate energy source for human beings. Solar energy is transmitted onto the Earth's surface through a constant flow of electromagnetic radiation waves. Among the total radiated energy in the solar system, only a very small portion is captured by the Earth. Some of the captured energy is ultimately converted into wind and ocean flows, and some will be converted and stored by plants. The latter type of solar energy utilization, namely photosynthesis, has been the dominant approach for solar energy utilization and is essential to the life of human beings *via* agriculture and forestry. During the Industrial Revolution, a more convenient energy source in the form of fossil fuels was actively exploited. Raw fossil fuel materials are nothing but a time capsule of ancient solar energy stored in plants and animals. As the industrialized age proceeds, the cost of fossil fuels, as well as the barrier to access fossil fuel minerals, is persistently increasing. The trend has almost reached a point that direct utilization of solar radiation becomes economically and environmentally favorable as compared to utilization of fossil fuels. The world is thus moving toward the 'new energy' era, where biofuel energy, wind power, and, of course, solar energy are recognized as alternative, renewable, and sustainable energy sources for the future.[10]

The total power of solar radiation is 384.6 yotta watts (3.846×10^{26} W). This huge amount of energy flow is uniformly radiated into all directions from the Sun. At the distance of 150 million kilometers away, the exposed area of the Earth receives approximately 1368 W m^{-2}. As 30% of the irradiated solar energy is reflected by the atmosphere, approximately 1000 W m^{-2} of solar energy is received everywhere on the surface of the Earth. While this amount of total energy provided by the Sun is far more than sufficient, the real challenge is about how to efficiently collect and use the solar energy.

2.2.2 The Spectrum of Sunlight

Light is an electromagnetic wave which propagates at the speed of light, c (3×10^8 m s^{-1}), in a vacuum. As a wave, it can be described by its oscillation frequency, ν, or equivalently by its wavelength, λ, in a vacuum. A light wave is also recognized as quantized energy chunks known as photons. The energy of a single photon is equal to the frequency multiplied by the Plank's constant h. The higher the frequency, the more energy a single photon possesses. The interaction between light and materials is critically dependent on the frequency (and hence, energy) of the light. For all the

applications involving light–matter interactions, it is thus important to identify the energy distribution of light as a function of light wave frequency. This frequency range is known as the spectrum of light. Sunlight has a complex spectrum extending from the ultraviolet range into the infrared range, originating from the generation and propagation processes of sunlight.

2.2.2.1 Fusion in the Core

The Sun is powered by the process of nuclear fusion. In the center of the Sun, the so-called core area (Figure 2.1), the extremely high temperature and pressure lead to a spontaneous fusion process. The heat and pressure generated by the fusion gases in turn balance the pressure associated with gravitational forces, preventing the Sun from collapsing by its own weight. The primary fusion process in the Sun is associated with the formation of a helium nucleus from four free hydrogen nuclei (protons). The amount of energy released during this process is proportional to the mass difference between a helium nucleus, which consists of two protons and two neutrons, and four free hydrogen nuclei, *i.e.* four protons. The radiation output of such a fusion process consists of both hadrons (with high kinetic energy) and high-energy photons. Fusion radiation is blocked by the outer layers of the Sun, and is therefore unobservable.

2.2.2.2 Thermal Radiation from the Radiation Zone

Outside the core area, the physical conditions no longer support nuclear fusion. As a result, energy is not produced, but is just redistributed. The redistributed radiation frequently interacts with the dense matter (mainly ionized hydrogen) outside the core area of the Sun. During the long journey from the core to the surface of the Sun, most (98%) of the fusion radiation is

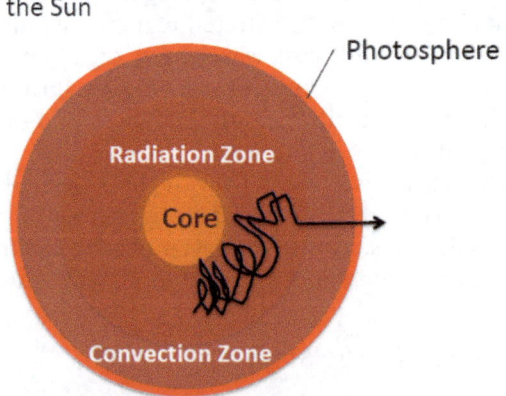

Figure 2.1 The conceptual structure of the Sun.

scattered by nuclei in the radiation zone (see Figure 2.1). As the fusion energy slowly diffuses out, it heats up the dense matter and converts itself to thermal radiation. When the radiation reaches the surface of the Sun, it has been cooled down to an equilibrium temperature of 5778 K, at which the radiation becomes visible light.

In thermodynamic equilibrium with the environment, the solar radiation can be described by the simple model of blackbody radiation. The thermal radiation of an ensemble at thermodynamic equilibrium can be precisely described using a single parameter, the temperature T, by solving for the distribution that maximizes the entropy for Boson particles. Plank's law gives the spectral radiance $I(\nu)$ by[11]

$$I(\nu) = \frac{2h\nu^3}{c^2} \frac{1}{e^{h\nu/k_B T} - 1} \tag{2.1}$$

where $h = 6.63 \times 10^{-34}$ J s (the Planck constant); ν is the frequency of the light; $c = 3 \times 10^8$ m s^{-1} (the speed of light in vacuum); $k_B = 1.38 \times 10^{-23}$ J K^{-1} (the Boltzmann constant).

The profile of the blackbody radiation spectrum is a smooth curve with a single peak at a characteristic frequency, ν_{max}, given by

$$\nu_{max} = T \times 58.8 \text{ GHz K}^{-1} \tag{2.2}$$

When expressed in terms of wavelength, the spectrum peak at λ_{max} is given by

$$\lambda_{max} = \frac{b}{T} \tag{2.3}$$

where $b = 2.90 \times 10^{-3}$ K m. Eqn (2.3) is called Wien's displacement law, and was actually discovered before Plank's law.

At the surface temperature of the Sun, $T = 5778$ K, the peak of radiation is found at a greenish wavelength $\lambda_{max} = 502$ nm. On the shorter wavelength side, the spectrum extends to the visible purple range and falls sharply within the ultraviolet (UV) range. On the longer wavelength side, the spectrum extends much more into the infrared region; the radiation falls to half of its peak value at the infrared wavelength of 1 μm, with a long tail extending well beyond 3 μm. Our eyes are only sensitive to a small region around the peak wavelength in the sunlight spectrum – we call this small region the visible range.

The sunlight spectrum is compared with the blackbody radiation spectrum in Figure 2.2. The spectrum generally agrees well with the prediction from blackbody radiations. The small deviations are mainly caused by surface layer absorption of the Sun, as will be discussed next.

2.2.2.3 Absorption by the Surface of the Sun

Before final reaching out to the vacuum of space, thermal radiation is filtered by the surface layer of the Sun. As the density of matter and the

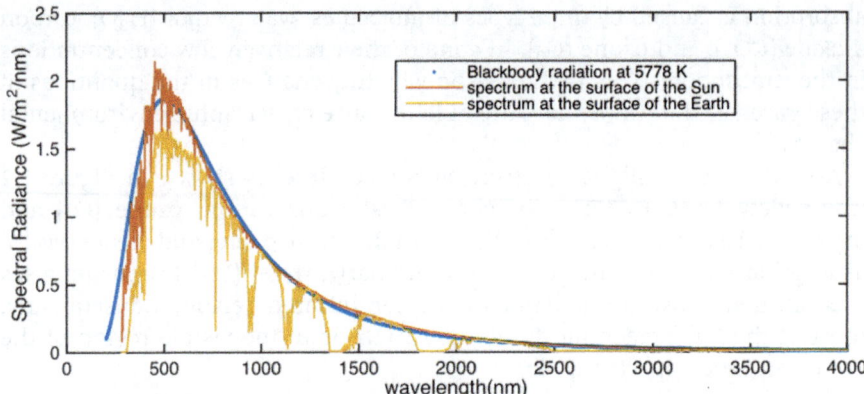

Figure 2.2 The solar spectrum. The blue dotted line shows the theoretical predication of blackbody radiation spectrum at 5778 K. The red curve shows the measured solar spectrum in space, and the yellow curve shows the measured solar spectrum on the Earth at sea level.

temperature decrease, atoms (mainly hydrogen and helium atoms) start to form within the surface layer of the Sun. This dramatically reduces the scattering probability, allowing the radiation to escape to space. Because the surface layer is no longer opaque to visible light, it is also called the photosphere. Specific wavelengths of the radiation are absorbed by the atoms in the photosphere, resulting in absorption lines in the spectrum of the output light, as shown in Figure 2.2. In addition, owing to the continuous absorption band of hydrogen atoms in the photosphere, the whole spectrum is slightly attenuated in the UV range.

2.2.3 Interaction of Sunlight with the Atmosphere on Earth

The sunlight emitted from the photosphere is still not in the final form for human observation on the Earth. The photons have to go through the Earth's atmosphere, which strongly reflects and absorbs the sunlight and alters the spectrum of light through wavelength-sensitive scattering processes. The interaction between the atmosphere and sunlight is far more complicated than that between the photosphere and sunlight, owing to the presence of high density atoms and molecules in the atmosphere that result from the much lower temperature on the Earth as compared to the photosphere.

2.2.3.1 *Characteristic Absorption of Molecules*

Similar to the absorption lines generated by atoms in the photosphere, the atoms and molecules in the atmosphere also absorb light at characteristic wavelengths. The absorption lines of the Earth's atmosphere are much more abundant due to the large density and variety of molecule types. Most of the

absorption is caused by three types of molecules, water vapor (H_2O), carbon dioxide (CO_2), and ozone (O_3), in spite of their relatively low concentrations in the atmosphere. This is the reason why tiny changes in the quantities of these gases in the atmosphere are able to cause catastrophic environmental changes.

A plot of the sunlight spectrum at the sea level is shown in Figure 2.2 (the yellow line). In terms of characteristic absorption, ozone (O_3) and oxygen (O_2) gases block X-ray radiation in the atmosphere, and create several absorption bands in infrared region. Similarly, water (H_2O) vapor imposes a continuous absorption band in the far infrared region. Consequently, most of the light reaching the Earth surface is in the visible region of the electromagnetic spectrum.

2.2.3.2   *Light Scattering in the Atmosphere*

The overall electric polarization of the molecules and atoms leads to elastic scattering of the light in the atmosphere, in which the energy and wavelength of light remain unchanged, and only the propagation direction is redirected. Two elastic scattering processes dominate in the atmosphere, Rayleigh scattering and Mie scattering.

Rayleigh scattering[12] is used to describe light–particle interactions when the particle size is much smaller than the wavelength of the incident light. In the atmosphere, the constituent molecules of air (Ar, N_2, O_2, *etc.*) scatter the sunlight by this process. The scattered intensity of light can be precisely expressed for randomly localized particles with probability α,

$$I = I_0 \frac{8\pi^4 \alpha^2}{\lambda^4 r^2} \left(1 + \cos^2 \phi\right) \tag{2.4}$$

where I_0 is the intensity of the incident light, r is the particle radius, λ is the wavelength and ϕ is the angle between the scattered light and incident light. It can be seen that the scattering intensity is very sensitive to the wavelength; the intensity scales with the 4th power of the inverse of wavelength. As a result, Rayleigh scattering is much stronger for blue light than for red light. Rayleigh scattering is responsible for the blue color of the sky and yellowish color of the Sun. This process also attenuates the UV light more than the longer wavelength part of the spectrum. It should also be noted that Rayleigh scattering is anisotropic, with a strong preference in the forward direction. Rayleigh scattering is the main scattering process on a clear day, in which about 30% of the incident sunlight is re-directed from the main beam.

Mie scattering describes the re-emitted radiation from spherical particles with sizes that are comparable to the wavelength of the incident light. The exact solution of the scattered intensity is expressed in the form of infinite series;[12] here we only discuss the general features of the process. Compared to Rayleigh scattering, Mie scattering has a much higher scattering probability with a much lower wavelength sensitivity, resulting in the white

appearance of a Mie scattering-dominated medium such as milk. Mie scattering process is almost isotropic, so the sunlight is scattered in all directions. Examples of Mie scattering in the atmosphere include the interactions of sunlight with cloud and haze, where the scattering particles are water droplets or pollution particles. The isotropic nature of Mie scattering is responsible for the large loss of sunlight in a cloudy day. The solar energy production is therefore vitally dependent on weather conditions.

2.2.4 The Geographic Intensity Distribution of Sunlight

In the previous section, both the light absorption and scattering by the atmosphere are discussed for the case of normal incidence. As shown in Figure 2.3, a tilted incident angle would further reduce the intensity of sunlight received on the ground. As compared to normal incidence, the tilted light beam projects the same amount of energy onto a larger area on the ground, reducing the received solar energy per unit area. For example, a flat device placed parallel to the ground receives an intensity reduced by a factor of cos θ. In addition, a tilted incident light wave has a substantially longer propagation distance in the atmosphere, with an increased probability of being absorbed or scattered.

The angle of sunlight with respect to the horizon is called the elevation angle. It is 0° at sunrise and 90° when the Sun is right overhead. The incident angle of sunlight is determined both by the location on the Earth's sphere, the time of the year, and in the time in the day. When averaged throughout the year, the sunlight power decreases with latitude. This sunlight radiation variation is a result of the curved surface of the Earth. As seen from Figure 2.3, when the sunlight is normally incident on the Equator, the sunlight angle with respect to the horizon decreases as the latitude increases. An efficient solar energy application should consider optimization of sunlight incident angles. For example, solar panel plants should

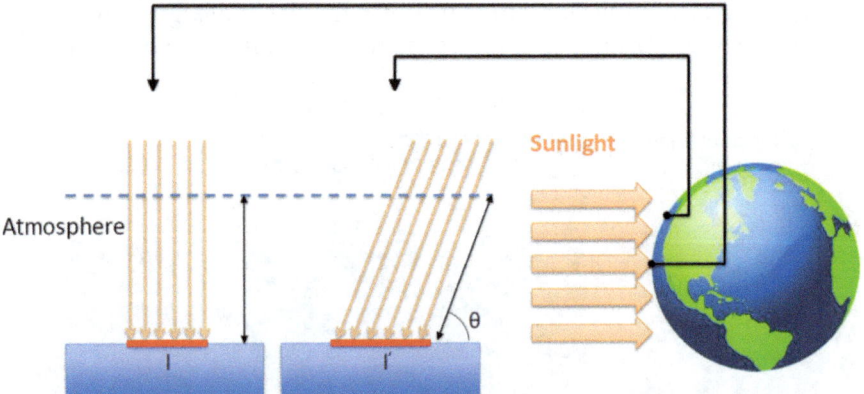

Figure 2.3   Illustration of the geographic intensity distribution of the sunlight.

preferentially be built in areas of low latitude. In addition, to ensure normal incidence of sunlight, some solar panels are made to rotate during the day while some are built with multiple facets at different directions.

2.3   Interactions of Light and Photo-active Materials

In Section 2.1, we have already discussed several types of light–matter interactions when the solar radiation propagates through the outer layers of the Sun and through the atmosphere of the Earth. In this section, we will systematically discuss the physical processes involved in interactions between the sunlight and photo-active materials, as illustrated in Figure 2.4. In particular, we will examine the interband absorption that is characterized by the electronic band structure of materials. We will also discuss several processes that do not lead to changes in the electronic states in the materials, such as reflection, scattering, and transmission.[13]

2.3.1   Reflection

Reflection occurs at the interface of two materials with different refractive index n. The reflectivity R, which describes the fraction of energy being reflected, is sensitive to the incidence angle. At normal incidence, R is given by

$$R = \frac{n_1 - n_2}{n_1 + n_2} \tag{2.5}$$

where n_1 and n_2 are refractive indices of the materials at the two sides of the interface. Here, refractive index n is defined as the ratio of speed of light in vacuum *versus* in the material. Reflectivity increases when the incident angle increases, which is another reason why the solar energy applications are better to be placed facing normal to the incident direction of light. Because reflection occurs at all interfaces with an abrupt change of refractive index, sharp interfaces should be avoided in solar energy applications, for instance, by reducing the number of protection layers, using materials with matching refractive index, employing structures of refractive index gradients, or designing textured surfaces that direct the reflected light onto other faces.[14]

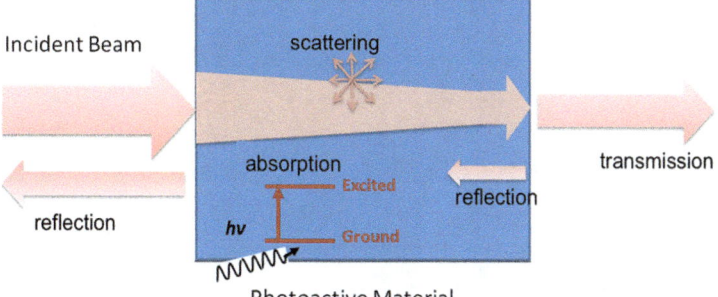

Figure 2.4   Illustration of light–matter interaction processes.

2.3.2 Scattering

Scattering refers to the random change of the light direction as the light propagates and interacts with the medium. It is a macroscale concept including many possible microscale mechanisms. There are two types of scattering, known as elastic scattering and inelastic scattering.

Elastic scattering refers to the scattering process where the energy of the photon and the medium is conserved before and after the scattering. Most common elastic scattering processes include Rayleigh scattering and Mie scattering; these processes have been discussed in detail in Section 2.2.3.2.

Elastic scattering occurs when the oscillating electrical field of incident light induces electromagnetic dipole radiation at the same frequency. The dipole radiation is strongest along the original incident direction of the optical field, while it re-directs a small amount of radiation in all directions in the space.

Inelastic scattering changes both the momentum and energy of photons by transferring energy between the photons and the medium. If the energy of scattered photons is reduced, the scattering is called Stokes scattering; whereas in rare cases of anti-Stokes scattering, energy is transferred from the medium to the photons. Inelastic scattering involves interactions of photons with excitation states in the materials. The most common inelastic scattering is mediated by the vibrational and rotational excitation of materials. The elementary excitations in the materials are known as phonons. Inelastic scattering includes Raman scattering and Brillouin scattering, which are distinguished by the amount of energy transfer in the scattering process. The high frequency phonon involved in Raman scattering is called the optical phonon, while the low energy phonon involved in the Brillouin scattering is called the acoustic phonon.

Essentially, scattering does not reduce the number of photons, but reduces the number of photons going in the forward direction. Scattering is thus considered as a source of attenuation. As discussed in Section 2.2.3.2, the scattering processes in the atmosphere significantly reduce the intensity of sunlight. Elastic scattering only changes the direction of the incident light, so it does not affect the absorption of photons. In fact, elastic scattering inside the active materials of solar energy applications could enhance absorption by effectively increasing the optical path length of the light in the materials. In contrast, inelastic scattering changes the spectrum of the light by transferring energy between the photon and medium, which could substantially influence the absorption efficiency by pulling the photons in or out of resonance.

2.3.3 Absorption

Absorption occurs when the frequency of the light matches the transition frequency of the electrons in materials. Absorption by isolated atoms or molecules often gives rise to sharp characteristic absorption lines, owing to

the quantized electronic energy states. For example, the hydrogen and helium atoms in the photosphere lead to sharp characteristic absorption lines
in the sunlight spectrum, as shown in Figure 2.2. On the other hand,
absorption by solid materials with electronic band structure manifests a
continuous interband absorption spectrum with a cut-off frequency matching the band gap energy. This is because interband absorption only occurs
when the energy of a photon is high enough to excite the electron transition
across the band gap between the valence band and the conduction band.[15]
When an electron in the valence band jumps to the conduction band, the
original state of the electron is left as an unoccupied hole. The excitation of a
free electron and a hole is known as the generation of an electron–hole pair.
Absorption eliminates the photon, and therefore attenuates the light beam.
In the absorption process, the radiation energy of light is converted into the
internal energy of excited states. As a key step in photosynthesis, photocatalysis, and photovoltaics, absorption is essential for most solar energy
applications.

There are two types of band gaps, direct band gaps and indirect band gaps
(Figure 2.5). Note that the mass-less particle (photon) does carry a (relativ-
istic) momentum that is equal to the photon energy divided by the speed of
light c. For direct band gaps, the transition occurs at the two states with the
same momentum, whereas for indirect band gaps a momentum change is
required for the transition to occur. The direct band gap transition only
involves exchange of energy between the photon and the electron, whereas
the indirect band gap transition involves transfer of both energies of momentum. Because the momentum of a photon is negligible compared to that
of an electron, the indirect band gap transition cannot be completed without
extra momentum provided by phonons. Because extra phonons are required
for the interaction, the indirect band gap transition is a second-order
process, and is much weaker than direct band gap absorption.

The broadband sunlight spectrum results in multiple electron transitions
in the materials and multiple channels for light absorption. Such phenomena are evident from the complex characteristic absorption lines caused

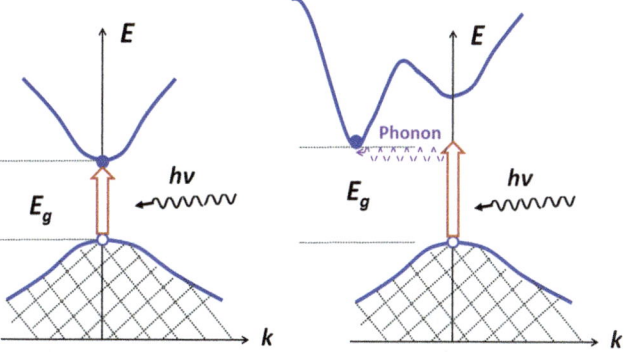

Figure 2.5   Concepts of direct (left) and indirect (right) band gaps.

by the atmosphere of the Earth. The mismatch between the sunlight spectrum and the applicable light range in a material seriously affects the efficiency of solar energy applications. For example, the infrared spectrum cannot provide enough energy to overcome the band gap in Si-based solar energy applications and therefore cannot create electrical work in a single step. Absorption of the infrared light is converted into vibrational or rotational kinetic energy of the lattice, which is then dissipated as heat. In most cases, thermal absorption of the sunlight is not useful for solar energy applications and may even harm their normal operation.

2.3.4 Luminescence

Luminescence refers to the re-emission of light caused by the spontaneous decay of atoms from excited states to their ground state. It can be viewed as the reversed process of absorption. It may occur after the absorption of a photon, which leads to the electron transition from the ground state to the excited states. This is called photoluminescence (see Figure 2.6). Luminescence can also be caused by other excitation methods, such as electrical excitations.

It takes a characteristic lifetime for an electron to decay after its excitation *via* radiative spontaneous decay. The luminescence process is efficient when the lifetime of radiative decay is shorter than non-radiative decays. The spectrum of emitted photons reflects the characteristics of the relaxation mechanisms of the electron, rather than characteristics of incident light or how the electron is excited. The direction of luminescence is also randomized into all directions. As a reverse process of interband absorption, interband luminescence occurs when the excited electron falls back to the valence band and simultaneously emits a photon. It is also known as the recombination of an electron–hole pair. Direct band gap materials have strong interband absorption as well as strong interband luminescence.

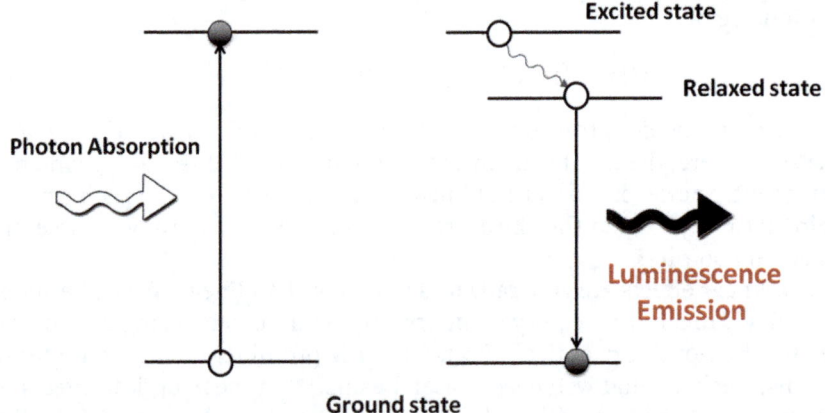

Figure 2.6 Illustration of the luminescence process.

Physical defects and impurities in crystalline materials may localize free electrons (and specific electron states) and therefore can act as luminescence centers, featuring a strong coupling between the vibrational excitations and the electron states. For example, electron excitations are significantly broadened by the vibrational coupling when they are localized near the defects, resulting in a larger absorption as well as a larger emission bandwidth.

In the luminescence process, the excited electron states first relax to a lower energy vibrational level, and then decay to the ground states. The emission spectrum is red shifted (or Stokes shifted) due to the energy transfer to the vibrational relaxation. Luminescence is often used to generate light from electricity in light-emitting diodes (LEDs). However, it is not desired in solar energy applications, as it converts the internal energy of excited states back to radiation energy of light. The energy of photons is often reduced in this process, and becomes no longer capable of exciting another electron–hole pair. For example, in solar energy applications, luminescence occurs when the excited electron–hole pair recombines. Recombination of electron–hole pairs reduces the photocurrent and decreases the efficiency. As a result, collection of the charge carriers before any recombination is of paramount importance in the design of solar energy applications.

2.3.5 Nonlinear Effects

The light–matter interactions discussed so far only involve individual photons. Such interactions are known as linear optical interactions, where the interaction probability is proportional to the magnitude of the electric field. It is also possible to introduce interactions involving multiple photons, which are known as nonlinear effects.[16] The interaction probability is proportional to the N-th order of the electric field, where N is the number of photons involved. The material response is expressed in terms of polarization vector P,

$$P(t) = \varepsilon_0(\chi^{(1)}E(t) + \chi^{(2)}E^2(t) + \chi^{(3)}E^3(t) + \cdots) \tag{2.6}$$

where ε_0 is the dielectric constant, $\chi^{(n)}$ is the n-th order susceptibility of the material. Here, the first term represents the linear response, the second term represents the second order nonlinearity that involves two photons, and the third term represents the third order nonlinearity that involves three photons, and so on.

Nonlinear effects are generally much weaker than linear effects, because it is unlikely to simultaneously gather multiple photons in a single event. As a result, the nonlinear optical effects are only prominent with high-intensity beams, such as highly focused laser beams. Nonlinear optical effects are generally not considered in solar energy applications, because the sunlight intensity is too weak to trigger them.

2.3.6 Transmission

When none of the aforementioned processes occur, the light wave propagates through the material medium without atomic- or electronic-level interactions. This is the transmission process. As the light propagates through the interface of two materials with different refractive indices, n_1 and n_2, the propagation direction will change, with the incidence angle θ_1 and refraction angle θ_2 satisfying the following equation:

$$n_1 \sin \theta_1 = n_2 \sin \theta_2 \tag{2.7}$$

This is commonly known as Snell's law.

Transmission and refraction do not change the intensity of the light; they only lead to a phase shift of the light wave. To reduce optical losses due to transmission in solar energy applications, it is important to enhance light–matter interactions before the light leaves photo-active materials, *e.g.* by employing highly reflective interfaces that reflect otherwise transmitted light back into materials, effectively enhance the total propagation distance, and increase absorption probability.

2.4 Quantum Yield in Solar Energy Conversion

2.4.1 Definition of Quantum Yield

As explained in Section 2.2.1, only a small fraction of the sunlight completes the long journey from the Sun to the surface of solar devices on the Earth. Once photons are absorbed by the device, again, only a fraction of the absorbed light contributes to the function of the device. Here, we introduce an important figure of merit, quantum yield, to describe the efficiency of the photon-energy utilization. For a radiation-induced process, quantum yield is defined as the ratio between the number of photon-induced events and number of absorbed photons, *i.e.*

$$\Theta = \frac{\text{number of photo-induced events}}{\text{number of photons absorbed}} \tag{2.8}$$

The quantum yield, Θ, takes values from 0, in which case no photon absorption contributes to the desired process, to 1, in which case all the absorbed photons contribute to the desired process. In rare cases, the quantum yield can be larger than 1. This occurs when a chain reaction is involved where a high-energy single photon induces multiple reactions with lower energy photons, thus with more than one quanta of yield.

Sunlight is absorbed as quantized energy packages of photons, so the light–matter interactions often involve the participation of a single photon with a single molecule at the micro-level. For a given event, a photon either participates or not. Quantum yield measures the probability of participation. In order to enable a photon-induced event, the energy of the photon needs to be larger than the energy gap required for the interaction. However, the

excess energy of a photon cannot be saved for another event. On the other hand, the probability is substantially lowered when multiple photons and molecules must be involved in the interaction, so it is difficult to enable a nonlinear process by the accumulated energy from multiple photons.

Quantum yield is measured differently for different devices and applications. In terms of photocatalysis or photovoltaics, quantum yield is defined as the number of electron–hole pairs generated per photon absorbed by the device. In these circumstances, the electron–hole pair is responsible for the free radicals that enable secondary reactions in photocatalysis, and for electric current generation in photovoltaics. More examples of quantum yield in light–materials interactions are discussed in more detail in the next section.

2.4.2 Quantum Yield in Light–Materials Interactions

2.4.2.1 Photocatalysis

Photocatalysis refers to the acceleration of a light-induced reaction in the presence of a catalyst. One typical example in photocatalysis is water splitting (Figure 2.7),[17–21] in which the interactions between sunlight and nanoscale catalysts (*e.g.* TiO_2 nanoparticles) split the water molecules into oxygen and hydrogen atoms, producing hydrogen gas as a clean energy resource. Because water is transparent within the sunlight spectrum, the efficiency of this process crucially relies on the nanoscale catalysts, which absorb the sunlight and convert the photon energy into chemical energy for water molecules.

The water-splitting reaction is simply given by

$$H_2O + \textit{light energy} \rightarrow \frac{1}{2}O_2 + H_2$$

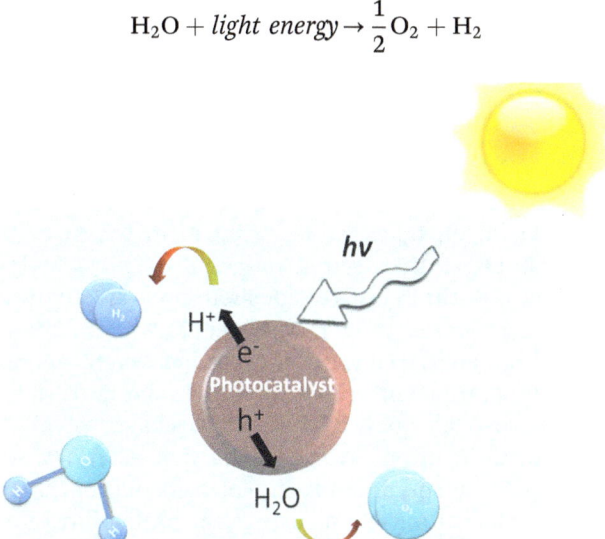

Figure 2.7   Illustration of the water-splitting processes.

The required light energy for water splitting is 1.23 eV, corresponding to a wavelength of about 1 μm. In principle, most photons within the sunlight spectrum could trigger water splitting. However, water cannot directly absorb the sunlight, because it is transparent over the entire sunlight spectrum. To introduce the photon energy to the water molecules, a catalyst is needed that first absorbs the sunlight and then transfers the energy to the water. This is similar to the functions of pigments and reaction centers in plants, as will be discussed in the process of photosynthesis in Section 2.4.2.2.

The quantum yield of photocatalytic water splitting is given by the ratio between the number of generated hydrogen molecules and the number of absorbed photons, *i.e.*,

$$\Theta = \frac{\text{number of hydrogen molecules generated}}{\text{number of photons absorbed}} \tag{2.9}$$

Note that the definition of quantum yield has excluded optical loss by only accounting for photons that are absorbed by the system. Like all other radiation-enabled applications, photocatalysis faces all types of issues that decrease the quantum yield efficiency. For photocatalytic water splitting, relevant problems include the following:

- Losses due to absorption of photons without electron state excitation. For example, catalysts such as TiO_2 are efficient for water splitting, because they have a band gap larger than 1.23 eV. However, these types of catalysts can only be excited with higher energy photons in the UV range. Absorption of sunlight in the visible range is not able to induce an excitation in such catalysts.
- Sometimes the excited catalysts are not able to interact with a water molecule, but instead fall back to the ground state. In other words, the catalysts fail to react with the water molecules before recombination of the electron–hole pair.
- Decay of the catalyst. Water, especially seawater, is corrosive for many materials. Many catalysts will decay and form defects in the humid environment. These defects will further enhance the recombination rate and reduce the quantum yield.

2.4.2.2 Photosynthesis

Photosynthesis is a process that converts energy from sunlight into chemical energy that is stored in certain types of chemical materials. Plants use the photosynthesis processes to synthesize carbohydrate ($C_6H_{12}O_6$) and oxygen (O_2) *via* the degradation of carbon dioxide (CO_2) and water (H_2O):

$$6CO_2 + 12H_2O + light\ energy \rightarrow C_6H_{12}O_6 + 6O_2 + 6H_2O$$

The overall chemical reactions are carried out in two steps, known as the light reaction and the dark reaction.

In the light reaction, chlorophyll and other pigments with different colors absorb sunlight at different parts of the spectrum, primarily at the red and the blue parts. The energy of the absorbed photon is first converted into an excited electron state of the pigment, which then travels through other pigments as excitons. The excitons, if lucky, are then trapped by a special protein. The excited protein acts as a reaction center to provide the energy to transfer an electron from a donor molecule to an acceptor molecule. Once the primary photochemical reaction of photosynthesis is initiated, electron transfer interactions will follow subsequently. In the end of the light reaction, the photon energy breaks water molecules and produces ATP (adenosine triphosphate), a compound universally used as fuel storage by living cells.

In the dark reaction, the energy from the ATP is used to convert CO_2 to sugar in the photosynthetic carbon reduction cycle.

The quantum yield of a photosynthesis system is defined as the ratio between the number of carbon dioxide molecules consumed and the number of absorbed photons, *i.e.*

$$\Theta = \frac{\text{number of } CO_2 \text{ molecules consumed}}{\text{number of photons absorbed}} \tag{2.10}$$

Not all the absorbed photons are used in reactions with CO_2 molecules, and the quantum yield of a plant is usually between 3% and 6%. Most of the absorbed photons are converted and dissipated as heat. A small portion of the absorbed photons can also be re-emitted later at a longer wavelength *via* the process of fluorescence.

2.4.2.3 Photovoltaics

As shown in Figure 2.8, a photovoltaic device converts sunlight to an electric current through interband absorption[22] in a P–N junction, which is an interface between p-type and n-type of semiconductor materials. The interband photon absorption leads to electron excitation from the valence band into the conduction band, where the electrons are free to move. Meanwhile, as an electron leaves the original bonded position in the atom, a vacancy is

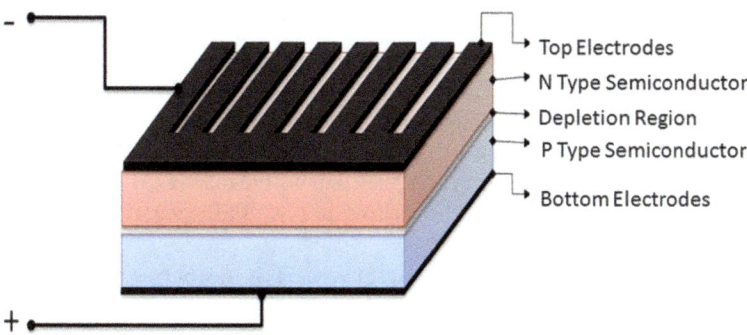

Figure 2.8 Illustration of a photovoltaic device.

left as a hole with net positive charge. As adjacent electrons can easily fall in and fill the position sequentially, this results in the net effect of the hole moving. The positive and negative charge carriers are then separated under an electric field within the deletion region in the P–N junction. An open circuit voltage is therefore developed between the P-side and N-side of the junction. When connected externally in a loop, an electric current will be generated. This current is called the photocurrent.

The quantum yield of a photovoltaic device is defined as the ratio between the number of electron–hole pairs that are separated and collected and the number of absorbed photons. It is called the internal quantum efficiency (IQE) of the solar cell:

$$\mathrm{IQE} = \frac{\text{number of electrons collected}}{\text{number of photons absorbed}} \tag{2.11}$$

Quantum efficiency can also be expressed in units of ampere per watt as the ratio of generated current to the absorbed photon energy:

$$\mathrm{IQE} = \frac{\text{number of electrons collected per second} \times \text{charge of electron}}{\text{number of photons absorbed per second} \times \text{energy of the photon}}$$
$$\tag{2.12}$$

It is convenient for engineers to incorporate reflection and transmission loss of the incident light and discuss the overall efficiency of the solar device. The corresponding external quantum efficiency (EQE) is defined as the ratio of the number of electrons collected to the number of photons incident on the device:

$$\mathrm{EQE} = \frac{\text{number of electrons collected}}{\text{number of incident photons}} \tag{2.13}$$

Various processes in photocurrent generation keep the quantum efficiency far lower than 1. Possible issues include:

- Light might not be able to get to the photovoltaic active material. The incident light might get reflected or absorbed by the connecting anodes or supporting materials of the device.
- Long wavelength photons with the energy below the band gap are not able to excite electron–hole pairs. The quantum efficiency at such wavelengths is therefore zero.
- The absorption of a photon might not excite an electron–hole pair due to energy dissipation into other channels. Instead, the absorption of a photon sometimes lead to fluorescence or heat.
- The generated electron–hole pair can recombine before being collected. Recombination takes place if the generated electron–hole pair does not have enough time to diffuse to the depletion region with the electric field, as is often exacerbated by material defects. Recombined electron–hole pairs do not contribute to the photocurrent.

Another commonly used figure of merit is known as the solar cell efficiency, which is defined as the ratio of output electric power to the input photon power, *i.e.*,

$$\eta = \frac{P_e}{IA} \tag{2.14}$$

where P_e is the power of the generated electricity, I is the intensity of the incident light, and A is the area of the active solar panel. This quantity describes the net energy conversion efficiency. Unlike quantum yield efficiency, which can approach 100% for appropriate wavelengths, solar cell efficiency is fundamentally bounded by the thermodynamic efficiency limit. Even for a theoretically 'perfect' Carnot engine, the efficiency is limited to 85%,[23,24] based on the temperature of the photons from the surface of the Sun and the temperature on the Earth. Such a theoretical limit cannot be achieved in practice. In fact, the solar cell efficiency is further reduced when the whole spectrum of sunlight is treated as a whole: when the photon energy is lower than the band gap energy, the electron–hole pair cannot be generated; whereas when the photon energy is higher than the band gap energy, the excess absorbed energy can only be dissipated as heat. As a consequence, the best efficiency of a commercial silicon solar cell is only about 25.6%.[25]

2.4.2.4 Fluorescence

A form of luminescence, fluorescence refers to the re-emission of light from a material that has absorbed light of a short wavelength such as X-rays or ultraviolet light. In this process, a photon is first absorbed to initiate an excited state in the material, which can be used as the energy source for subsequent events. However, excited states are inherently unstable; there is a finite probability of spontaneous decay from the excited state back to the stable ground state or lower level excited states. The decay of the excited state is often associated with the emission of another photon. Owing to energy dissipation into other channels in the conversion process, the re-emitted photon often has less energy than the absorbed photon, and thus has a longer wavelength. On the other hand, it is also possible for the excited state to decay without the emission of a photon. Such occasions are known as the non-radiative decay processes, which involve conversion of the radiation energy to kinetic energy, vibration and rotation energy of the molecules, and energy transfer to another molecule.

The efficiency of a fluorescence process is influenced by both the lifetime of the spontaneous decay for excited states and the rates of non-radiative decays. Quantum yield of a fluorescence process is defined as the ratio of the number of re-emitted photons to the number of absorbed photons:

$$\Theta = \frac{\text{number of photons re-emitted}}{\text{number of photons absorbed}} \tag{2.15}$$

Since spontaneous decay naturally occurs together with other processes, it is often used as an approach to probing the primary process. The energy changes of the absorbed and emitted photons in fluorescence reflect the internal energy levels of the material; fluorescence spectroscopy[26] has been established as a powerful tool for materials characterization.

2.4.3 Beyond Quantum Yield: Eco and Environmental Sustainability

Quantum yield is the most important figure of merit for measuring the efficiency of solar energy applications. However, other figures of merit, such as economic efficiency, whole life cycle assessment, and environmental sustainability, are also important in evaluating the overall functionality of the applications.

2.4.3.1 Cost Efficiency

To create real impact, solar energy applications need to be cost competitive as compared to petroleum-based energy generation approaches. For example, it is important to measure the economic efficiency of an application in terms of cost per unit energy generation, *e.g.* in dollars per watt. The solar energy cost has been dramatically decreased owing to the industrialized mass production of silicon-based solar cell panels. Because traditional energy products do not include the cost of pollution in their price tag, it is difficult to make a fair comparison among different energy sources. Nevertheless, with the tax on pollution and CO_2 emission, solar energy production has been increasing its market share recently (although it is only about 0.5% in United States in 2014).[27]

2.4.3.2 Life Cycle Assessment

A solar energy application can be extremely efficient and environmentally friendly when it is in operation. However, the conclusion can be quite different when taking into account the entire life cycle of the system. For example, manufacturing of solar cell panels is energy intensive, and the maintenance cost might be expensive too. Moreover, disposal of used solar energy devices can be pollutive. This whole 'green energy' concept would fail if the energy generated using the solar energy devices cannot compensate for the energy used for manufacturing and disposal.

To compare environmental effects including everything associated with a product from cradle to grave, a so-called life cycle assessment (LCA)[28] is now becoming the standard to evaluate green technology suitability. The term 'life cycle' refers to assessments that include the potential impact of all intervening transportation steps necessary for the product, such as raw material production, manufacturing, distribution, utilization, and disposal.

This standard provides the whole picture for green technology assessment, and is becoming widely used in both public and private sectors, such as in business, education, and policymaking processes.

References

1. M. C Beard, J. M. Luther and A J. Nozik, *Nat. Nanotechnol.*, 2014, **9**, 951.
2. T. Soga, *Nanostructured Materials for Solar Energy Conversion*, Books24×7 Inc, Elsevier, Amsterdam, Boston, 2006.
3. P. Spinelli, V. E. Ferry, J. van de Groep, M. van Lare, M. A. Verschuuren, R. E. I. Schropp, H. A. Atwater and A. Polman, *J. Optics*, 2012, 14.
4. D. V. Bavykin, J. M Friedrich and F. C. Walsh, *Adv. Mater.*, 2006, **18**, 2807.
5. A. J. Frank, N. Kopidakis and van de Lagemaat, *J. Coord. Chem. Rev.*, 2004, **248**, 1165.
6. D. P. Hagberg, T. Edvinsson, T. Marinado, G. Boschloo, A. Hagfeldt and L. C. Sun, *Chem. Commun.*, 2006, 2245.
7. A. Kay, I. Cesar and M. Gratzel, *J. Am. Chem. Soc.*, 2006, **128**, 15714.
8. C. I. Yeo, E. K Kang, S. K Lee, Y. M. Song and Y. T. Lee, *IEEE Photonics J.*, 2014, 6.
9. L. K. Verma, M. Sakhuja, J. Son, A. J. Danner, H. Yang, H. C. Zeng and C. S. Bhatia, *Renew. Energ.*, 2011, **36**, 2489.
10. M. Kaltschmitt, W. Streicher and A. Wiese, *Renewable Energy: Technology, Economics, and Environment*, Springer, Berlin, New York, 2007.
11. R. Loudon, *The Quantum Theory of Light*, Oxford University Press, Oxford, New York, 2000.
12. J. D. Jackson, *Classical Electrodynamics*, Wiley, New York, 1999.
13. M. Fox, *Optical Properties of Solids*, Oxford University Press, Oxford, New York, 2010.
14. J. Poortmans and V. Arkhipov, *Thin Film Solar Cells: Fabrication, Characterization and Applications*, Wiley, Chichester, England, Hoboken, NJ, 2006.
15. N. W. Ashcroft and N. D. Mermin, *Solid State Physics*, Saunders College Pub., Fort Worth, 1976.
16. R. W. Boyd, *Nonlinear Optics*, Books24×7 Inc, Academic Press, Burlington, Mass., 3rd edn, 2008.
17. I. Cesar, K. Sivula, A. Kay, R. Zboril and M. Graetzel, *J. Phys. Chem. C*, 2009, **113**, 772.
18. J. Z. Su, L. J Guo, N. Z. Bao and C. A. Grimes, *Nano Lett.*, 2011, **11**, 1928.
19. R. van de Krol, Y. Q. Liang and J. Schoonman, *J. Mater. Chem.*, 2008, **18**, 2311.
20. A. Wolcott, W. A. Smith, T. R. Kuykendall, Y. P. Zhao and J. Z. Zhang, *Adv. Funct. Mater.*, 2009, **19**, 1849.
21. J. F. Zhu and M. Zach, *Curr. Opin. Colloid Interface Sci.*, 2009, **14**, 260.
22. R. F. Pierret, *Semiconductor Device Fundamentals*, Addison-Wesley, Reading, Mass, 1996.

23. R. Brendel, J. H. Werner and H. J. Queisser, *Sol. Energy Mater. Sol. Cells*, 1996, **41–42**, 419.
24. J. Nelson, *The Physics of Solar Cells*, Imperial College Press, London, 2003.
25. M. A. Green, K. Emery, Y. Hishikawa, W. Warta and E. D. Dunlop, *Prog. Photovoltaics*, 2015, **23**, 1.
26. J. R. Albani, *Principles and Applications of Fluorescence Spectroscopy*, Wiley InterScience (Online service), Blackwell Science, Oxford, Ames, Iowa, 2007.
27. V. Smil, *Sci. Am.*, 2014, **310**, 52.
28. W. Klopffer, *Environ. Sci. Pollut. Res.*, 1997, **4**, 223.

CHAPTER 3

Green Nanomaterials Preparation: Sustainable Methods and Approaches

XIAOYANG XU,* QIAN LYU, JOSHUA BADER AND
MEAGAN ACCORDINO

Department of Chemical, Biological and Pharmaceutical Engineering, New
Jersey Institute of Technology, 161 Warren Street, Newark, NJ 07103, USA
*Email: xiaoyang@njit.edu

3.1 Introduction

Photo-active nanomaterials are extensively used in energy production and environmental applications[1] – such as sensors,[2] degradation of organic pollutants,[3] photoelectric conversion,[4] electrochromic displays,[5] and likely numerous undiscovered applications. This chapter explores the problems of conventional synthetic methods, which usually consume more chemicals, generate more waste, and require more energy for reaction than current environmental standards allow for in commercial and industrial applications. In order to minimize the use and generation of hazardous substances during the manufacturing process of photo-active nanomaterials, environmentally benign and sustainable synthetic techniques must be developed. This chapter will highlight the photocatalytic ability of nanomaterials. The properties of photo-active nanomaterials give great advantages over other nanomaterials, notably in the application of degradation of

RSC Green Chemistry No. 42
Green Photo-active Nanomaterials: Sustainable Energy and Environmental Remediation
Edited by Nurxat Nuraje, Ramazan Asmatulu and Guido Mul
© The Royal Society of Chemistry 2016
Published by the Royal Society of Chemistry, www.rsc.org

organic pollutants[3] and the splitting of water,[6] using benign and facile methods. In addition, green synthesis methods for the most widely used inorganic, inorganic–inorganic, and inorganic–organic hybrid photo-active nanomaterials will be summarized, including TiO_2, CdS, ZnO, TiO_2-ZnO, TiO_2-CdS and graphene-based TiO_2 nanomaterials. Characterization methods used to verify the successful synthesis of these photo-active nanomaterials and the demonstration of the improvement of their photo-activity will also be introduced.

3.2 Titanium Dioxide

Titanium dioxide is one of the most widely used photo-active nanomaterials. The photo-active property of TiO_2 motivates the study of its application in: photocatalysts,[7] splitting water,[6] chemical and biosensors,[2] electrochromic displays,[5] photoelectric conversion,[4] and solar cells.[1] TiO_2 can exist in nature in three different forms, each of which can either be crystalline or amorphous; they include anatase, rutile, brookite, or a mixture of the three. It is important to note that pure crystalline anatase is the most photo-active phase of TiO_2. The high stability of photoelectric chemistry, and high photoelectric conversion efficiency, provides TiO_2 with its special properties. The photocatalytic properties of TiO_2 nanoparticles bestow the potential to rapidly degrade pollutants in wastewater, and split water into hydrogen and oxygen. Organic pollutants can be degraded in the presence of TiO_2, along with an energetic light source, and an oxidizing agent.[8]

Several different synthetic methods are used to prepare TiO_2 photo-active nanoparticles including: sol–gel processing,[9] reverse microemulsion,[10] dialysis hydrolysis,[11] microwave-assisted emulsion polymerization,[12] alcohothermal method,[13] hydrothermal,[14] combustion synthesis,[15] and gas-phase methods.[16] Among them, the hydrothermal method is a very useful technique to synthesize TiO_2. The reaction can be initiated at relatively low temperature (100 °C). The particle size, morphology, and composition are tunable by controlling the reaction conditions such as temperature and heating time. Importantly, extremely small TiO_2 nanoparticles, from 6 nm to 10 nm, can be synthesized, which improves the photo-activity due to the large surface area.[17]

Two green and simplified methods are often used to prepare the photo-active TiO_2 nanomaterials.

- The microwave-assisted hydrothermal method (MW-HT) to synthesize TiO_2 from $TiCl_4$ (eqn (3.1))

$$TiCl_4 + O_2 \rightarrow TiO_2 + 4HCl \tag{3.1}$$

provides multiple benefits such as fast reaction, energy savings, and elimination of metastable phases.
- The polymer-gel technique is a simple and cost-effective method of producing pure metal oxides.

3.2.1 Microwave-assisted Hydrothermal Method

In general, the hydrothermal synthetic method uses an oxidant and a reducing agent in aqueous solution. Komarneni *et al.*[18] compared the conventional hydrothermal method with the microwave-assisted hydrothermal method. In the microwave-assisted hydrothermal method, TiO_2 nanoparticles were synthesized from metal alkoxides and hydrogen peroxide with a molar ratio of $0.05:0.12$ ($Ti:H_2O_2$). Table 3.1 shows the synthesis conditions including temperature, time, phase, and crystallite size, prepared from different titanium alkoxides under conventional-hydrothermal and microwave-hydrothermal conditions.

The results demonstrate that the desired nanophase, anatase, can be obtained using either method. These methods can lead to the formation of very small nanomaterials (6–10 nm). However, the conventional hydrothermal method requires more time (up to 24 hours) compared to the microwave-assisted hydrothermal method, which was synthesized in only 1 hour, due to rapid and homogenous heating.

Sridhars' research also shows that anatase synthesized by either method results in similar NO_x destruction at all wavelengths. However, the NO_x destruction by nanophase anatase synthesized under both conventional hydrothermal method (C-H) and the microwave-assisted hydrothermal

Table 3.1 Synthesis conditions of TiO_2 under conventional-hydrothermal (C-H) and microwave-hydrothermal (MW-HT) conditions, including preparation time, phase, and crystallite size of TiO_2 by using different titanium alkoxides as Ti source and different temperatures. Copyright permission from ref. 18.

	Run no.	Ti source	Crystallization T (°C)	t (h)	Phase	Crystallite size (nm) XRD	TEM	Surface area (m² g⁻¹)
C-H	1	Ti ethoxide	100	24	Anatase	6.9	4–10	212
	2	Ti ethoxide	125	24	Anatase	9	8–12	177
	3	Ti ethoxide	150	24	Anatase	9.3	10–15	174
	4	Ti ethoxide	175	24	Anatase	10	10–20	157
	5	Ti isopropoxide	100	24	Anatase	7.3	4–8	193
	6	Ti isopropoxide	125	24	Anatase	7.7	8–15	174
	7	Ti isopropoxide	150	24	Anatase	8.9		155
	8	Ti isopropoxide	175	24	Anatase	9.5		156
	9	Ti butoxide	100	24	Anatase	7.7	4–8	181
	10	Ti butoxide	125	24	Anatase	8.2	10–15	166
	11	Ti butoxide	150	24	Anatase	8.9		163
	12	Ti butoxide	175	24	Anatase	10.1		147
M-H	Ex. 15	Ti ethoxide	100	1	Anatase	6.1	5–8	232
	Ex. 17	Ti ethoxide	125	1	Anatase	6.3		215
	Ex. 19	Ti ethoxide	150	1	Anatase	6.9	8–15	201
	Ex. 23	Ti ethoxide	175	1	Anatase	7.5		187
	Ex. 14	Ti isopoproxide	100	1	Anatase	5.5	3–8	298
	Ex. 16	Ti isopoproxide	125	1	Anatase	6.6		212
	Ex. 18	Ti isopoproxide	150	1	Anatase	6.7	5–12	209
	Ex. 22	Ti isopoproxide	175	1	Anatase	7.6		194

method is superior to the standard commercial TiO_2. Thus, such green and sustainable synthesis methods not only benefits the environment but also improves anatase ability to decompose NO_x.

3.2.2   Polymer-gel Method

The production of nanosize anatase through the polymer-gel method is far less intricate than the hydrothermal method. In general, the polymer-gel method begins by mixing a polymer in a metal added solution to obtain metal–polymer mixtures. The resulting nanoparticle size and purity are similar to the products obtained using the hydrothermal method.

There are several disadvantages with the hydrothermal technique compared to the polymer-gel method. For example, the hydrothermal technique has difficulty controlling the reaction time and requires the use of an autoclave. The final products must then be purified, unlike the polymer-gel method which produces pure products.

In Karthick's[19] experiments; poly-(vinylpyrrolidone) (PVP) was added to pure ethanol to produce a homogeneous solution. Titanium(IV) isopropoxide was slowly added until the solution changed visually from clear to a light yellow color, while being vigorously stirred to maintain a homogenous solution. The ratio of titanium(IV) isopropoxide, PVP, and ethanol in solution is $1:1:10$. HCl was then added until the pH of the solution reached 2.0. The final mixture was then separated and heated to 300 °C, 400 °C, and 500 °C respectively to obtain the final product, TiO_2 anatase nanoparticles.

The characterization of anatase TiO_2 nanocrystals was determined by X-ray diffraction (XRD). Figure 3.1 shows the typical X-ray diffraction patterns, the peaks shown for the sample at 400 °C corresponds to the X-ray diffraction patterns of standard anatase (JSPDS 89-4921). The XRD pattern also shows that there is negligible rutile and brookite TiO_2 structures formed from the absence of diffraction peaks at 27° and 31°. The absence of a diffraction peak at 30°, which represents either the crystal defects or the absence of long-range ordering in the XRD pattern, indicates that the only phase present in appreciable quantities is pure tetragonal anatase formed at 400 °C, compared to samples calcined at 300 °C and 500 °C which do not produce pure anatase, shown in Figure 3.1.

High-resolution transmission electron microscopy (HR-TEM) as a characterization method was used to provide direct observations on the atomic scale. The calcined product was analyzed by HR-TEM (Figure 3.2), which showed the TiO_2 nanomaterial obtained at 400 °C to measure approximately 12 nm. UV absorption displays a peak below 400 nm (Figure 3.3a), and the band gap energy was calculated to be 3.55 eV (Figure 3.3b). The UV absorption and photoluminescence peaks at 391 nm and 496 nm, imply that the anatase TiO_2 nanoparticles, formed by the simple polymer-gel technique, are comparable with the structure and morphology of the product obtained by the tedious hydrothermal technique.

In summary, anatase TiO_2 nanoparticles can be prepared successfully from the simple polymer-gel procedure. The hydrothermal method is more

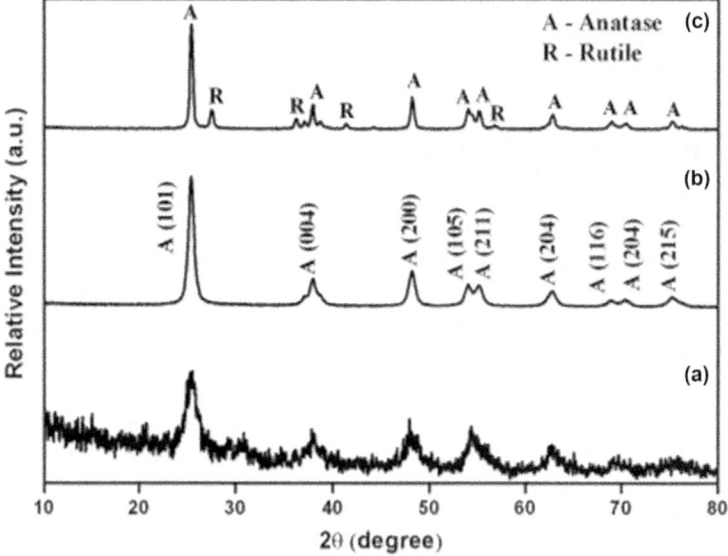

Figure 3.1 X-ray diffraction patterns of TiO₂ morphologies calcined at 300 °C, 400 °C, and 500 °C.
Copyright permission from ref. 19.

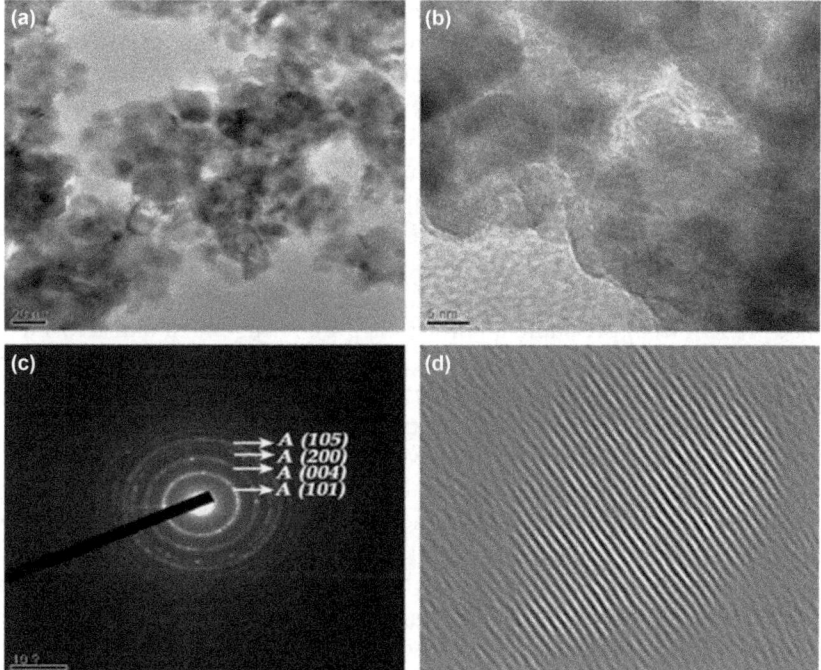

Figure 3.2 High-resolution transmission electron microscopy characterization of TiO₂ nanoparticles calcined at 400 °C.
Copyright permission from ref. 19.

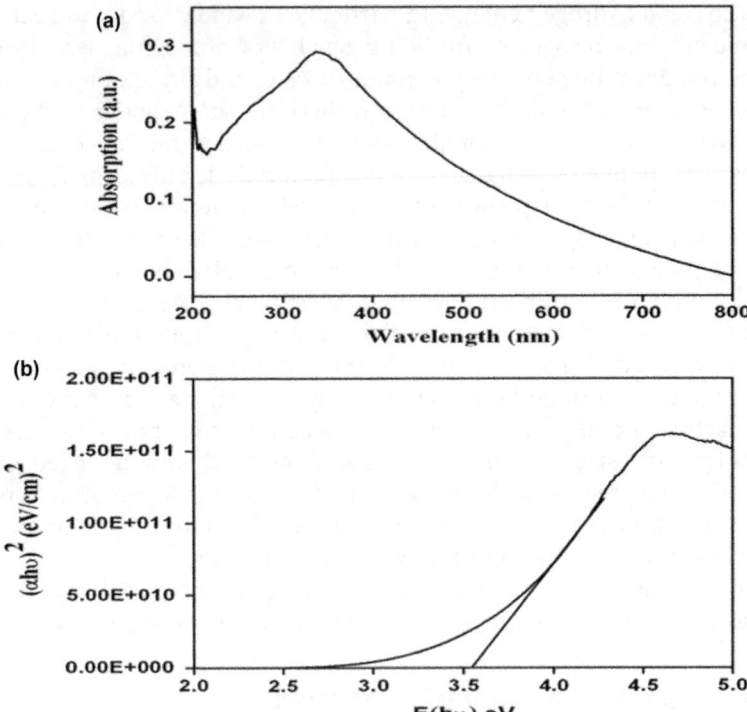

Figure 3.3   (a) Absorption *versus* wavelength spectrum; (b) plots of $(\alpha h\nu)^2$ *versus* $E(h\nu)$ to show band gap energy hν for the TiO$_2$ nanoparticles calcined at 400 °C.
Copyright permission from ref. 19.

time consuming, is temperature sensitive, and is costly. Due to the lower temperature and economic procedure, the polymer-gel method is not only a simple method to synthesize the pure metal oxides, but also corresponds with today's environmental concerns.

3.3   Cadmium Sulfide

The worldwide dependence on fossil fuels drives researchers to discover new alternative energy sources. Hydrogen (H$_2$), because it is clean and can be easily stored, has become the focus as an excellent alternative energy source[20] H$_2$ can be produced by spitting water (see eqn (3.2)),[6]

$$2H_2O \xrightarrow{\text{Photon energy} > 1.23 \text{ eV}} 2H_2 + O_2 \qquad (3.2)$$

which usually requires a photocatalyst and solar energy to stimulate the decomposition of water into hydrogen and oxygen. The earliest mention of this synthesis method was in 1972 by Fujishima and Honda, who used a semiconductor as a photocatalyst.[6] Once this method was successfully developed, researchers began to focus on developing a better photocatalyst to

produce clean hydrogen energy. In particular, a widely used photocatalyst is a semiconductor recognized for being capable of producing clean hydrogen energy and degrading organic pollutants in water and air. Although numerous semiconductor materials have been synthesized, the majority of these photocatalysts are only capable of absorbing ultraviolet light. This characteristic makes semiconductors unfavorable considering UV is only a small portion of solar energy. CdS is a type of photocatalyst with a narrow band gap, 2.30 eV, compared with TiO_2, 3.0–3.2 eV, which therefore allows CdS to absorb the visible light which accounts for 43% of the total solar energy.[21]

Many methods have been developed to prepare CdS nanomaterials; such as ion exchange,[22] hydrothermal,[23] chemical precipitation,[24] microwave-assisted methods,[25] solvothermal, thermal evaporation, and hot colloidal.[26] Furthermore, solution-based methods are exceptional for nanomaterials manufacturing due to the relatively low reaction temperature requirement, flexibility with size control of the nanomaterials, and no need to pre-synthesize the precursors. Various morphologies of CdS photocatalysts have also been produced, such as nanosheets,[27] nanowires,[28] nanorods,[29] nanosheet flowers,[22] and multi-armed CdS nanorods.[30]

Two synthetic methods are currently widely used for CdS nanoparticle synthesis – the chemical precipitation method and the solvothermal method.

3.3.1 Chemical Precipitation Method

The chemical precipitation method has the ability to produce various sizes of CdS nanostructures by adjusting the quantities of capping agents at low cost. This dynamic method can react rapidly at room temperature,[31] while also producing nanomaterials within a narrow size range.[32]

Eskandari *et al.*[31] developed a CdS nanoparticle preparation approach by using mercaptoethylamine hydrochloride (MEA) as the capping agent. This capping agent causes the nanostructure surface to carry a positive charge, which will provide strong interactions with polar molecules and prevent agglomeration through charge repulsion. In this approach, cadmium acetate and sodium solution were first prepared by dissolving the respective salts in deionized water. The capping agent was dissolved in the cadmium solution and then the Na_2S solution was added to the mixture to form CdS nanoparticle at room temperature (see eqn (3.3)). The precipitates were centrifuged and dried. The CdS nanoparticles with various sizes were successfully synthesized with varying contents of 0, 0.05, 0.5, 1.0, 1.5, and 2.0 g of MEA in the solution, accompanied with a color change from orange to light yellow of the nanoparticle suspension.

$$Cd(CH_3COO)_2 + Na_2S \rightarrow CdS + 2CH_3COONa \qquad (3.3)$$

UV-vis spectroscopy can be used to determine the band gap of the synthesized nanomaterial, calculated by eqn (3.4),

$$E = \frac{hc}{\lambda} \qquad (3.4)$$

along with the particle size obtained by the effective mass approximation.[33] The final results gave the values 2.5, 2.69, 2.9, 3.05, 3.07, and 3.1 eV for the band gap and the size values were estimated accordingly at 6.95, 4.49, 3.27, 3.25, 3.21, and 3.61 nm. X-ray diffraction measurement was used to determine the CdS nanostructure phase and composition. The peaks at 26.5°, 44.0°, and 52.1° confirmed the cubic zinc blende structure of CdS. In addition, XRD demonstrated that the crystallinity can be improved by increasing the quantities of the capping agent. Diffuse reflectance spectrum (DRS) measurement further confirmed that the CdS nanostructure size changed with the relative amount of capping agent. The results showed that as the quantity of capping agent increased, the particle size decreased and band gap increased.

To prove that the prepared CdS nanomaterial could absorb visible light with a wavelength below 510 nm (2.4 eV), a blue LED lamp was chosen as the light source. By using this light source the photocatalytic ability of CdS nanoparticles was studied. The smaller nanoparticles held a higher degradation yield of methylene blue (MB) than the larger particles,[31] both within the same reaction time. The results also showed that the photoactivity of CdS nanoparticles in terms of MB degradation was increased under sunlight irradiation compared to LED irradiation. CdS photoactivity proved better than the commercially available Dugussa P25 TiO_2 nanoparticle under blue LED irradiation.

3.3.2   Solvothermal Method

The solvothermal method is another simple method for CdS nanoparticle synthesis. Wang *et al.* used this solvothermal method to successfully synthesize the CdS multipods in dodecylamine solvent (Figure 3.4).[34]

Chen *et al.*[35] and Gao *et al.*[36] also applied the same method in ethylenediamine (en) solvent and successfully prepared CdS multi-arm nanorods (see eqn (3.5)).

$$H_2O \xrightarrow{\text{irradiated}} e_{aq}^-, H, {}^{\bullet}OH, H_3O^+, \ldots$$

$$S_2O_3{}^{2-} + e_{aq}^- \rightarrow S^{2-}$$

$$\left[Cd(en)_2\right]^{2+} + S^{2-} \rightarrow CdS\downarrow + 2en$$

$$(3.5)$$

$${}^{\bullet}OH + C_2H_8N_2 \rightarrow C_2H_7N_2 + H_2O$$

Yu and co-workers reported a different solvothermal route to synthesize CdS multi-arm nanorods.[30] The synthesized nanorods were applied to H_2 production.

In Yu's[30] method, $Cd(OAc)_2 \cdot 2H_2O$ is dissolved in dodecylamine. Thioacetamide was then added to the $Cd(OAc)_2$ solution and the mixture was sent to an autoclave and heated to 100 °C, 140 °C, and 180 °C for 12 hours. The mixtures were then cooled down to room temperature and the yellow precipitate was collected by centrifugation and dried in an oven.

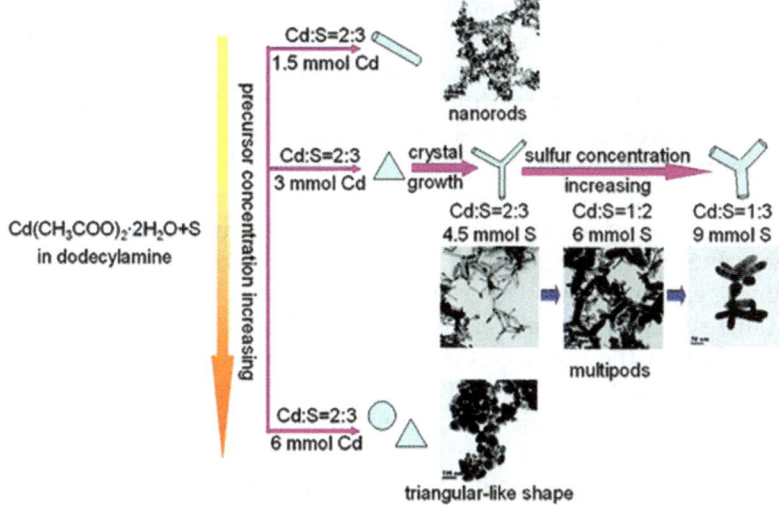

Figure 3.4 A method to produce CdS multipods in dodecylamine solvent using the solvothermal method.
Copyright permission from ref. 34.

The nanoparticle samples were labeled as R100, R140, and R180 respectively.

For comparison, CdS nanoparticles can be prepared under the same conditions using distilled water instead of dodecylamine. These nanoparticles obtained at 100 °C, 140 °C, and 180 °C were labeled as P100, P140, and P180 respectively. These parallel experiments were run to study the effects of CdS morphology on water splitting ability in terms of H_2 production rate.

XRD patterns show that all diffraction peaks of R100, R140, and R180 belong to pure hexagonal CdS nanostructure, while CdS nanoparticles (P100, P140, and P180) prepared without the addition of dodecylamine show cubic and hexagonal mixture phases. As the autoclave temperature increased, the intensity of the diffraction peaks increased and peak width decreased, which shows the formation of longer CdS nanorods and improved crystallization. TEM was also used to prove the successful synthesis of multi-armed CdS nanostructures.

UV-vis was used to study the band gap of the synthesized sample. The results showed that there was a large absorption at 520 nm from the R140 sample, which represents the intrinsic band gap absorption at 2.38 eV. In contrast with P140, R140 has a larger band gap resulting in a larger reduction and oxidation power. This makes the electron–hole difficult to recombine, which can improve the photocatalytic ability.

Additionally, Yu *et al.* measured photocatalytic activity under visible light by adding lactic acid as a scavenger agent and Pt as a cocatalyst. By comparing the photocatalytic ability of the CdS multi-armed nanorods and CdS

nanoparticles under visible light irradiation, nanorods were found to be more pragmatic to use due to their rapid H_2 generation rate. This accelerated rate is associated with the morphology of nanorods, which includes a large surface area and pore volume, higher photocatalytic activity of hexagonal nanocrystals, and a short bulk-to-surface diffusion distance. It was also determined that the H_2 generation rate was dependent on the Pt cocatalyst content. This was proved by sample R140 that had almost no H_2 generation, but was significantly increased by adding only a minute amount of Pt (0.08 wt%). However, the addition of too much Pt (higher than 0.23 wt%) can lead to suppressed photocatalytic activity. In conclusion, the visible change of the solution turning opaque may contribute to a decrease in light absorption and thus the decrease in photocatalytic activity.

3.4 Zinc Oxide

Due to the unique benign environmental impacts and wide applications in various fields, zinc oxide (ZnO) is another widely used semiconductor material. ZnO has a wide band gap of 3.37 eV and a large excitonic binding energy of 60 meV.[37] As a result of zinc oxide's attractive physical and chemical features, it has many applications in biomedicine, UV absorption, photocatalysis, solar cells, and photonic materials.[38]

By taking the advantage of the photocatalytic activity of ZnO, ZnO nanoparticles can be used to degrade phenol[39] and chlorinated phenols, such as 2,4,6-trichlorophenol.[40] It is also used for the degradation of methylene blue,[41] direct dyes,[42] acid red,[43] and ethyl violet,[44] which are common organic pollutants in waste water. There have been a great deal of synthetic methods designed for the preparation of ZnO nanoparticles including the solvothermal method, sol–gel, vapor-liquid-solid technique, pulsed-laser process, thermal evaporation method, template-assisted process, molecular beam epitaxy method, chemical vapor and deposition, electrochemical process, and reverse micellar.[41–46] However, these synthetic methods have certain limitations, such as high temperature (850–925 °C),[45] high pressure (100 MPa),[46] long reaction time (30 days),[46] and toxic reagents (O_3).[47] To address those challenges, the development of green, sustainable, and highly efficient preparation methods are in great demand. Among the aforementioned ZnO nanoparticle preparation methods, microwave-assisted hydrothermal and ultrasound sonication are two green synthesis methods that fit this description and will be further explored in this chapter.

3.4.1 Microwave-assisted Hydrothermal Technique

The microwave-assisted hydrothermal technique utilizes an oxidant and a reducing agent in an aqueous solution to synthesize nanosize ZnO in a green and facile manner. Al-Hazmi and co-workers successfully applied the microwave-assisted hydrothermal technique[48] for ZnO nanoparticle

preparation. Al-Hazmi *et al.* dissolved 0.6 mol of zinc acetate dehydrate, and 0.4 mol urea in 50 mL distilled water. The solution was then transferred to an autoclave to react at 200 °C, the solid product was then dried and calcined at 400 °C, producing nanosize ZnO (see eqn (3.6)).

$$CO(NH_2)_2 + H_2O \rightarrow CO_2 + 2NH_3$$

$$\uparrow\downarrow + H_2O$$

$$NH_4^+ + OH^-$$

$$\downarrow + Zn^{2+} \tag{3.6}$$

$$Zn(OH)_4^{2-}$$

$$\downarrow \text{Microwave Hydrothermal Method}$$

$$ZnO(s)$$

The prepared ZnO nanosheets crystal was analyzed using X-ray diffraction patterns, Fourier transform infrared spectroscopy (FTIR), and photo-luminescence spectrum (PL). The XRD pattern (Figure 3.5) does not show any characteristic peaks related to impurities, which demonstrates that pure ZnO nanomaterials were prepared.

The final product was also analyzed by FTIR spectra (Figure 3.6); the wavenumber 487 cm^{-1} indicates the ZnO vibrations.[38a,38b] The SEM characterization shows that the ZnO nanosheets average 70 nm lateral dimensions and 12 nm thickness. The PL spectrum graph (Figure 3.7) for the

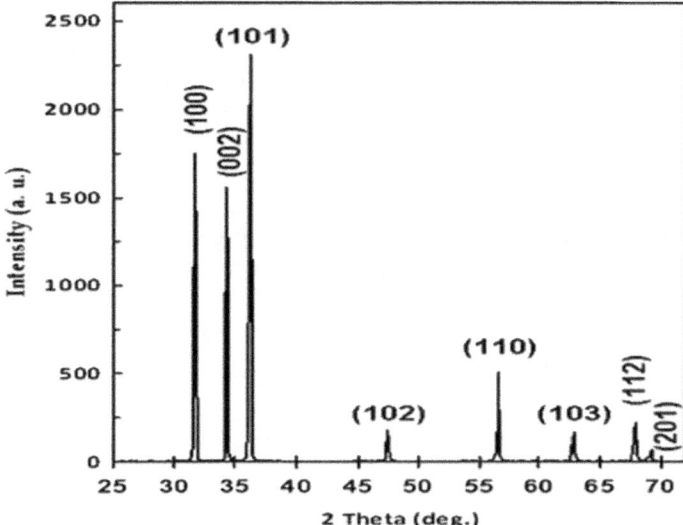

Figure 3.5 X-ray powder diffraction pattern of ZnO nanosheets prepared through microwave-assisted hydrothermal method, showing no impurities. Copyright permission from ref. 48.

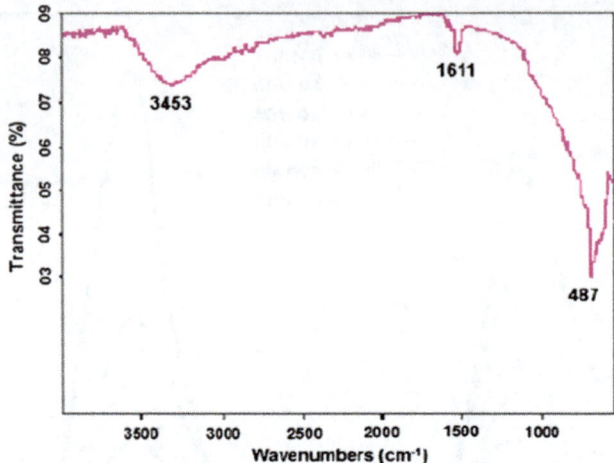

Figure 3.6 FTIR spectra of ZnO nanosheets synthesized through microwave-assisted hydrothermal method, showing high purity ZnO.
Copyright permission from ref. 48.

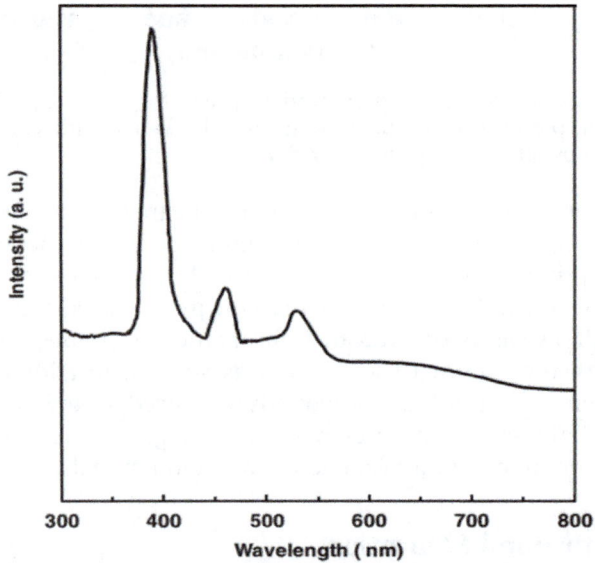

Figure 3.7 Room temperature PL spectra of ZnO nanosheets (excited at 389 nm) synthesized through the microwave-assisted hydrothermal method, showing high purity ZnO.
Copyright permission from ref. 48.

prepared nanosize ZnO shows that the 389 nm ultraviolet band, blue-green band at 460 nm, and green band at 531 nm confirm the high purity of ZnO nanosheets.

To analyze the photoactivity, Al-Hazmi *et al.* used methylene blue as a model pollutant substance to be photo-decomposed. Figure 3.8 displays the

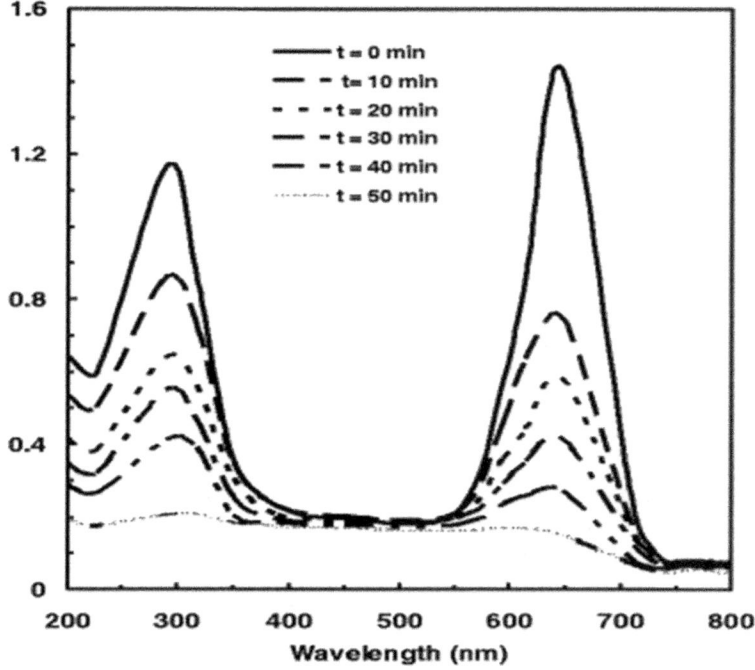

Figure 3.8 Variation of the UV-vis absorption spectra for methylene blue solution in the presence of ZnO nanosheets under UV light at different time intervals. Copyright permission from ref. 48.

UV-vis absorption spectra of methylene blue. The methylene blue absorption intensity changes with time: after 50 minutes of UV irradiation, the maximum absorption intensity disappeared, which indicates that MB dye was completely degraded by the synthesized ZnO photo-active nanomaterial.

This microwave-assisted hydrothermal method is simple, fast, and cost-effective, with the minimal use of water as solvent. In addition, no other organic solvents or hazardous compounds were used during the preparation process, and therefore this presents a model green chemistry synthetic method for the preparation of photo-active nanomaterials.

3.4.2 Ultrasound Sonication

Another green synthesis method that produces nanosize ZnO is the ultrasound sonication approach. Sonication is a method which applies sound energy in order to generate acoustic cavitations in the liquid phase. During this process, cavities form and grow into bubbles which immediately collapse into the liquid phase, in which high pressure (1000–2000 atm) and temperature (5000 K) are generated; formation of these cavitations promotes the synthesis of the final product.[49] The ability of a substance to form cavitations is dependent on the solvent used for the synthesis, which is based on its vapor pressure. A solvent with a low vapor pressure will have a high boiling point, thus hydrocarbons with a low vapor pressure are vital to

the sonication method to ensure a high temperature bubble during the synthesis process. Bhatte *et al.* used 1,3-propanediol as the solvent, which is environmentally benign and has a boiling point of 200 °C,[50] to prepare ZnO nanoparticles through the ultrasound sonication method (see eqn (3.7)).

$$Zn(CH_3COO)_2 + CH_2(CH_2OH)_2$$

$$\downarrow \text{Sonication} \quad CH_3OCO-CH_2-CH_2-CH_2-OCOCH_3 + HO-Zn-OH$$

$$\downarrow \text{Sonication}$$

$$ZnO \tag{3.7}$$

Bhatte and co-workers[51] successfully demonstrated the ultrasound sonication technique in a typical procedure, in which 0.5 g of $Zn(CH_3COO)_2$ was added to 10 mL of 1,3-propanediol, the mixture was then sonicated under a 22 kHz frequency ultrasound horn for 2 hours. The reaction mixture was then centrifuged to separate out the product, finally the ZnO produced was subjected to vacuum drying at 100 °C for 10 hours.

XRD, transmission electron microscopy (TEM), and elemental detection spectroscopy analysis (EDS) were used to characterize the ZnO nanoparticles. The TEM image (Figure 3.9) shows that the polygonal shaped ZnO particles were formed and the product was a type of crystalline material. The TEM

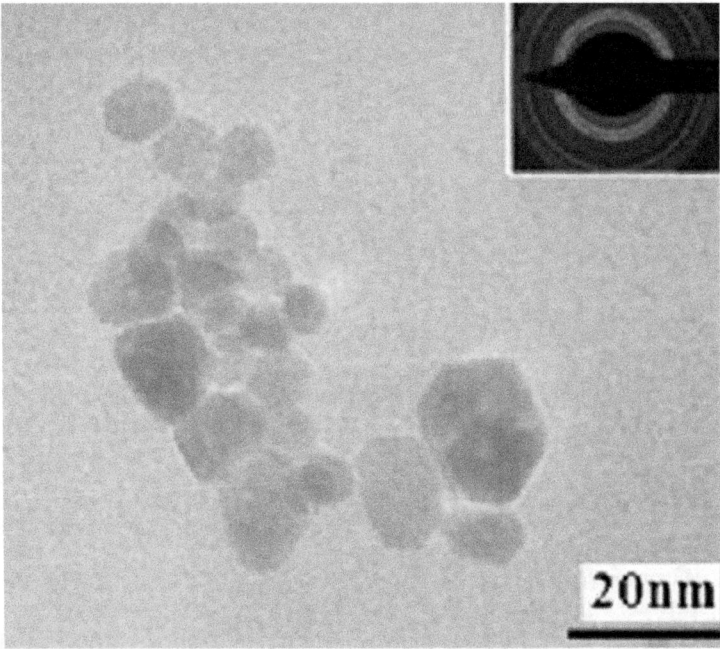

Figure 3.9 Transmission electron microscopy image displaying polygonal shaped ZnO particles.
Copyright permission from ref. 50.

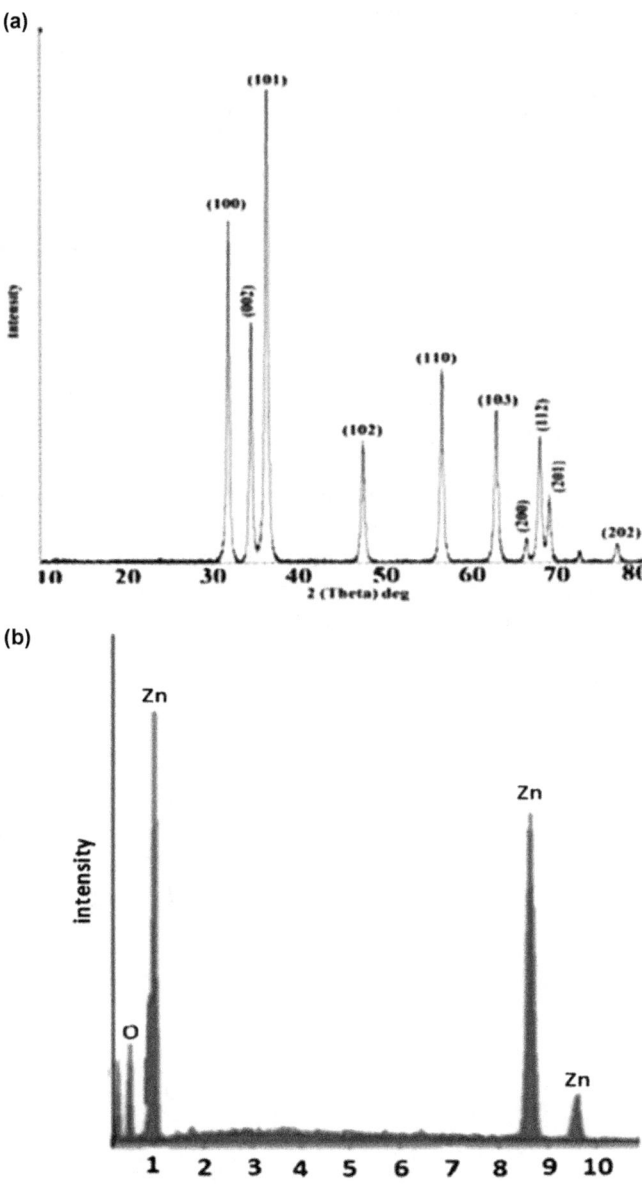

Figure 3.10   (a) XRD pattern of nanosize zinc oxide. (b) EDAX spectrum of nanosize
zinc oxide, which verifies synthesis of pure ZnO using ultrasound
sonication.
Copyright permission from ref. 50.

image combined with the XRD pattern (Figure 3.10) shows (100), (002), (101),
(102), (110), (103), (200), (112), (201), and (202) X-ray diffraction peaks, which
verifies the formation of ZnO.

A parallel experiment was carried out without utilizing sonication to prepare ZnO nanoparticles. However, there was no detectable ZnO synthesized with prolonged reaction time. Ultrasound sonication is vital for the reaction to occur.

The ultrasound sonication method is a simple, additive and calcination-free, rapid, green, and an energy-efficient synthetic method to prepare ZnO nanoparticles. It also promotes the development of wet chemical techniques which can be expanded for industrial and commercial use.

3.5 Hybrid Nanomaterial

Currently, series of green preparation methods for inorganic TiO_2, CdS, and ZnO photo-active nanomaterials are being developed. Earlier in this chapter, TiO_2 nanomaterials were discussed as a photo-active material. They gained a great deal of interest due to their distinguished properties – stable, non-toxic, and low cost. However, the inherited characteristics of its wide band gap (~ 3.2 eV) limit TiO_2 to absorbing only UV photons. Because of this limited absorption boundary, the interaction results in about 3–5% utilization of solar energy.[52] In addition, TiO_2 exhibits fast recombination of photo-induced electrons and holes leading to a decrease in photocatalytic activity.[53] Given these limitations, there have been numerous synthetic methods developed to solve these problems, such as non-metal element doping,[54] coupling with semiconductors,[55] noble metal loading,[56] and especially the recombination of TiO_2 with newly emerging carbon-based nanomaterials.[53c,57]

Each of these methods delineates and deciphers the drawbacks of the TiO_2 nanomaterials. TiO_2 coupling with metal oxides or sulfides, such as ZnO/TiO_2,[55b] CuO/TiO_2,[55a] SnO_2/TiO_2,[55c] CdS/TiO_2,[55d] and ZnS/TiO_2[58] will fulfill the electron–hole pair separation whether under UV or visible light illumination. A larger contact interface between titanium nanotubes (TNT) and graphene is expected to promote the photo-induced charge separation which will improve the photocatalytic activity. Researchers have been using hydrogenation,[59] metal and non-metal doping,[54a,56a] and sensitizing with a low band gap semiconductor material to reduce band gaps and improve the efficiency of the use of sunlight. All of these methods have shown improvements in the photocatalytic activities. Three types of hybrid synthetic photo-active nanomaterials have been recently developed using green chemistry. They can be prepared respectively by using metal oxides or metal sulfides (ZnO, CdS) to achieve photo-active ZnO-TNT and $CdS-TiO_2$ composites, and using graphene oxide and graphene to synthesize the graphene-based nanomaterial.

3.5.1 ZnO/TiO$_2$ Composite

3.5.1.1 Microwave-assisted Hydrothermal Method

The potential to achieve a more efficient electron–hole pair separation under UV-visible light can be accomplished by coupling TiO_2 with other metal

oxides, thereby increasing the reaction rate at the surface of the hybrid nanomaterial composite. Coupled composites such as ZnO/TiO_2, CuO/TiO_2, and SnO_2/TiO_2 have shown enhanced performance in photocatalytic activities either for splitting of water or degrading pollutants, as compared to pure TiO_2.[55a–c]

The preparation of ZnO/TiO_2 composites usually requires a high reaction temperature, a long reaction time, and more reagents. Sustainable synthesis techniques are under development that need fewer reagents, a lower temperature, and a relatively short reaction time.

Cho *et al.*[60] reported a microwave-assisted method for ZnO-TNT composites preparation. TiO_2 and ZnO powder were first mixed in a NaOH and H_2O aqueous solution in varying amounts of ZnO from 3 to 40 wt%. Microwave conditions for the synthesis of ZnO-TNT were set to 150 °C and 195 W. Under these conditions, pure TiO_2 nanotubes, 3 wt%, 5 wt%, 10 wt%, 20 wt%, and 40 wt% ZnO-TNT were synthesized and labeled, TNTz3, TNTz5, TNTz10, TNTz20, and TNTz40, respectively.

SEM micrographs reveal that below 20 wt% ZnO, Zn ions are either diffused inside or doped into the TiO_2 framework. At higher concentrations, *e.g.* 40 wt% ZnO, the saturation limit of ion exchange of Zn^{2+} and Na^+ is exceeded during the hydrothermal process and fine ZnO particles aggregate on the surface of the TiO_2 nanotube. The lengths and diameters of the TiO_2 nanotube can also be varied with different amounts of ZnO.

The XRD pattern was also used to characterize the various products synthesized from NaOH solution. For example, XRD demonstrates that TNTz3 is composed of sodium-titanium oxide ($Na_2Ti_9O_{19}$), and further analysis revealed that TNTz3 possessed a mixture of sodium titanate (Na_2TiO_3) and titanium hydrogen oxide hydrate ($H_2Ti_4O_9 \cdot H_2O$).

By using UV-vis DRS spectra, Cho *et al.*[60] calculated the band gap energy for different samples using a modified Kubelka–Munk method,[61] which showed that the band gap energy for TNTz40 is less than that of TNTz3. The photocatalytic activities for the synthesized material were also evaluated, using RhB as the degradation substances (see eqn (3.8)).

$$TiO_2/RGO \rightarrow TiO_2(h^+) - RGO(e^-)$$

$$RGO(e^-) + O_2 \rightarrow RGO + O_2^-$$

$$TiO_2(h^+) + H_2O/OH^- \rightarrow TiO_2 + {}^\bullet OH \qquad (3.8)$$

$${}^\bullet OH + RhB \rightarrow \text{degradation products}$$

The results show that under 2 hour UV light, TNTz3 has 14.7% decolorization of RhB and TNTz40 has 18.6% decolorization of RhB. For the sample analyzed under 2 hour visible light, there were 2.9%, 19.9%, 20.4%, 25.8%, and 30% decolorization for TNTz3, TNTz5, TNTz10, TNTz20, and TNTz40, respectively.

3.5.1.2 One-pot Hydrothermal Method

The synthesis of ZnO/TNT in Cho's microwave hydrothermal approach requires three hours of reaction time followed by 48 hours of freeze drying. The one-pot hydrothermal method, a convenient preparation method, has been extensively studied given its reaction at a low temperature in a short period of time. Kim *et al.*[62] reported a one-pot hydrothermal method requiring a relatively low reaction temperature (140 °C), shorter reaction time (2 hours) and drying time (12 hours at 60 °C) to synthesize TiO_2/ZnO nanocomposites.

When compared to Bahadur's study using ethanol and NH_4OH as the reagents,[63] Kim *et al.* worked with only zinc nitrate hexahydrate, bishexamethylene triamine, and TiO_2 NPs using water as solvent and successfully prepared TiO_2/ZnO nanocomposites. FE-SEM, TEM, and XRD were used to demonstrate that the TiO_2 nanoparticle is doped on the surface of ZnO nanoflower. Photodegradation ability of the nanocomposite was also evaluated through methylene blue (MB) photodecomposition. The results showed a greater catalytic ability compared with pure ZnO nanoparticle or the commercially available P25 TiO_2 nanoparticle. This hydrothermal process is convenient to synthesize TiO_2/ZnO nanocomposites without the use of any organic solvent. The preparation is environmental friendly and low cost.

3.5.2 CdS/TiO_2 Nanocomposite

The photocatalytic ability of TiO_2 can be improved by coupling with a low band gap semiconductor such as CdS, which is another widely studied nanomaterial used for degradation of wastewater and splitting of water based on its photo-active properties. Coupling CdS with TiO_2 will also increase the photocatalytic ability of CdS. CdS is unstable in aqueous media during photocatalytic reactions because of photocorrosion, it can be readily oxidized by the photogenerated holes. By coupling CdS nanoparticles to other wide band gap semiconductors such as TiO_2, the stability of CdS can be improved while the photocorrosion effects are suppressed.[64]

There have been quite a few methods that have been developed to prepare CdS/TiO_2 composites, such as photodeposition,[65] co-precipitation,[66] chemical vapor deposition,[67] sonochemical method,[68] and hydrothermal method.[69] However, most of these methods require high temperature calcinations (200–400 °C) leading to the decrease of the specific surface area and an increase of the crystallite size[68] which therefore limits the photocatalytic activity. These methods also require the presence of various surfactants and long reaction times, at least 20–24 hours. Correspondingly, CdS/TiO_2 composite can be oxidized during the long thermal treatment. Furthermore, the presence of various surfactants can lead to blockages at the active sites on the surface and hence decrease their photocatalytic activities. In order to progress from these setbacks it is desired to prepare CdS/TiO_2 composites with highly exposed surface area at a low reaction temperature and in surfactant-free synthesis conditions.

To prepare CdS/TiO$_2$ hybrid nanomaterial, Li *et al.*[70] first synthesized pure CdS by mixing Na$_2$S·9H$_2$O with Cd(NO$_3$)$_2$·4H$_2$O with a molar ratio of 1:1. Titanium isopropoxide was then added into the prepared CdS solution followed by 1 hour of stirring at room temperature. A series of nanostructured CdS/TiO$_2$ composites were successfully synthesized with varying molar ratios of CdS:TiO$_2$.[70] TiO$_2$ nanomaterials were prepared in water with a pH ranging from 9.0 to 13.0. The nanostructure and band gaps have been characterized and the photocatalytic activities of the CdS/TiO$_2$ were also studied by Li *et al.*

As the molar ratio of CdS:TiO$_2$ varies from 2:1 to 1:20, the XRD patterns for the CdS/TiO$_2$ nanostructure shows that the intensity of the CdS becomes weaker. This shows that the lowered intensity of the nanostructured CdS is due to an increased thickness of amorphous TiO$_2$.

The UV-vis diffuse reflectance spectra of the as-prepared CdS samples, with different molar ratios of Na$_2$S:Cd(NO$_3$)$_2$, show that as the molar ratio of Na$_2$S to Cd(NO$_3$)$_2$ increases from 0.04 to 2.0, the band gap of the resulting nanostructured CdS decreases from 2.46 eV to 2.21 eV. The results also display that as the pH decreases from 13.0 to 9.0, as the band gaps of TiO$_2$ decreases from 3.30 eV to 3.14 eV almost linearly. In addition, 9.1% of CdS in the CdS/TiO$_2$ composite will greatly improve the absorption in the visible light region, which indicates that small amounts of CdS can be used to improve the absorbing capability of the amorphous TiO$_2$ for the visible spectrum of sunlight.

To study the photodegradation of methylene blue (see eqn (3.9)), CdS was prepared using Na$_2$S and Cd(NO$_3$)$_2$ with a molar ratio of 1:1. TiO$_2$ nanomaterials were prepared by adding titanium isopropoxide into a NaOH solution with a pH ranging from 9.0 to 13.0. The TiO$_2$ nanoparticle prepared at a pH of 13.0 shows the highest photocatalytic activities in the series. As the CdS prepared with 1:1 molar ratio of Na$_2$S to Cd(NO$_3$)$_2$ and the TiO$_2$ prepared under a pH of 13.0 shows the most photocatalytic activity, Li *et al.* synthesized the CdS/TiO$_2$ composite under those conditions and only varied the molar ratio. It was determined that the composite shows the highest photocatalytic activity when the molar ratio of CdS:TiO$_2$ is 1:50.

$$TiO_2/CdS + h\nu \rightarrow e_{CB}^- + h_{VB}^+$$

$$e_{CB}^- + O_2 \rightarrow O_2 \cdot^-$$

$$2O_2^- + 2H^+ \rightarrow O_2 + H_2O_2$$

$$O_2 \cdot^- + 2H^+ + 2e_{CB}^- \rightarrow H_2O_2 \qquad (3.9)$$

$$H_2O_2 + e_{CB}^- \rightarrow {}^\bullet OH + OH^-$$

$$H_2O + h_{VB}^+ \rightarrow {}^\bullet OH + H^+$$

In conclusion, using this surfactant-free, room temperature, and facile preparation method, a series of nanostructured CdS/TiO$_2$ photocatalysts can

be successfully synthesized and show superior photocatalytic activity compared to CdS or TiO_2 nanoparticles.

3.5.3 Graphene/TiO$_2$ and Graphene/TNT Nanocomposite

Graphene oxide and graphene are carbonaceous materials which are well suited to act as catalyst supports. Graphene oxide can increase the dispersion of TiO_2 catalyst and decrease the tendency of electron–hole pair recombination. Both graphene oxide and graphene can be hybridized with TiO_2 to form a graphene-TiO_2 (TiO_2/GR) or graphene oxide-TiO_2 (TiO_2/GO) nanocomposite, which can be utilized to split water and degrade wastewater pollutants.

Zhou *et al.* recently synthesized TiO_2/GR nanocomposites, and found that their photoactivity is improved when compared to the commercially available TNPs, P25, when photo-degrading methylene.[71] Zhang *et al.*[72] also reported a type of TiO_2/GR nanomaterial which shows better photoactivity than P25, by using a one-pot hydrothermal preparation method.

3.5.3.1 One-step Hydrothermal Method

Commonly used methods to prepare graphene-based TiO_2 nanocomposites are known as anchoring methods.[73] Anchoring methods provide graphene nanosheets with functional groups which can facilitate the formation of nanocrystallites, or immobilize pre-synthesized nanomaterials with fixed dimensions to the graphene nanosheets. The drawback of this method is that TiO_2 tends to agglomerate leading to a decrease of effective surface area for graphene, TiO_2, and phase separation.

Park *et al.*[74] successfully synthesized graphene-wrapped TiO_2 core/shell nanoparticles through a one-step hydrothermal GO reduction approach. By using this highly efficient hydrothermal method, GO was successfully reduced and TiO_2 converted from amorphous to a crystalline form. In this method, 3-aminopropyl-trimethoxysilane (APTMS) must be used as an agent to graft an amine functional group to the TiO_2 nanoparticles surface. However, APTMS is highly toxic and has a destructive quality to the mucous membranes and the upper respiratory track, specifically targeting the nerve system and organs, such as the liver and kidney. Developing a direct synthesis method for TiO_2/GO nanomaterials preparation without the use of modification agents (*i.e.*, APTMS) is necessary for today's green and environmental protection requirements.

Zhang *et al.* successfully prepared TiO_2/GR nanocomposites using ethanol as a solvent,[75] without the use of highly toxic APTMS. Jiang[72] and co-workers proposed a one-step hydrothermal synthesis method to prepare aforementioned TiO_2/GR nanomaterials without the use of either APTMS or ethanol.[76]

This TiO_2/GR nanomaterial can be easily synthesized. $TiCl_4$ was gradually added into graphene oxide aqueous solution followed by 1 hour of stirring.

The suspension was then transferred into an autoclave and heated at 180 °C for 6 hours. The composite was collected, centrifuged, and washed with deionized (DI) water. The composite was then dried at 50 °C for 48 hours. A series of composites were successfully synthesized with GR content of 0.8, 5.0, and 10.0 wt%. GO was obtained conveniently by ultrasonic stirring graphite oxide, which can be prepared through the Hummer method.[77]

XRD pattern was used to characterize the composition and size of TiO_2/GR composite. It shows that TiO_2/GS composites exhibit mainly anatase TiO_2 crystal phase with low graphene content, while TiO_2/GR composites are composed of both anatase and rutile TiO_2 crystal phases with higher graphene content. The average crystal sizes are 15, 16, 6, and 7 nm for the as-synthesized TiO_2/GR composites with GS contents as 0.8, 2.0, 5.0, and 10.0 wt%, respectively, suggesting that the existence of graphene can suppress the crystal growth of anatase TiO_2. Typical TEM images of TiO_2/GR (2.0 wt% GR) and TiO_2/GR (5.0 wt% GR) are shown in Figure 3.11, demonstrating that TiO_2 nanoparticles were successfully loaded onto the GR planes.

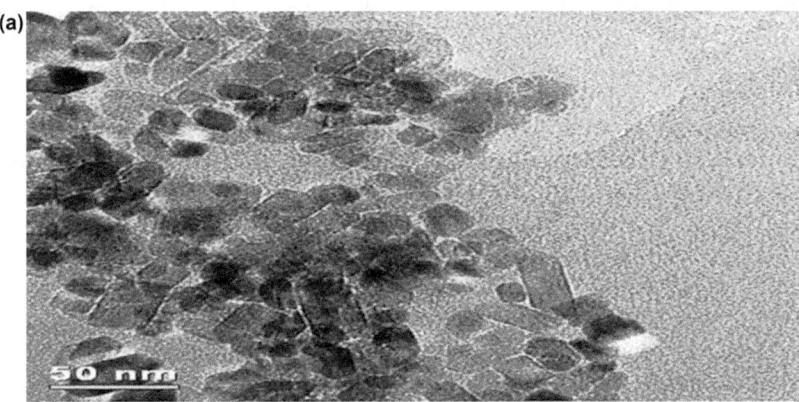

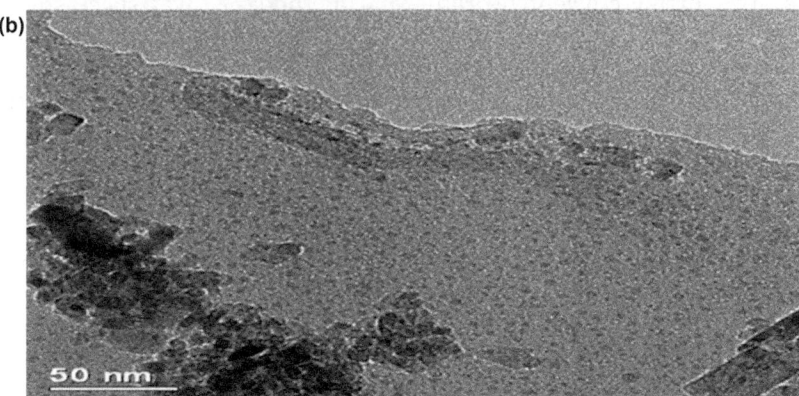

Figure 3.11 TEM images of GR composites (a) TiO_2/GR 2.0 wt%GR and (b) TiO_2/GR 5.0 wt% demonstrates that TiO_2 nanoparticles were successfully loaded onto the GR planes.
Copyright permission from ref. 72.

Jiang *et al.* also evaluated the photocatalytic activity of the synthesized TiO_2/GR nanocomposites through a hydrogen evolution experiment. An enhancement of photocatalytic activity can be observed compared with that of P25. For example, the average H_2 evolution rate increases to 4.5 mmol/hour (TiO_2/GR 0.8 wt%) and 5.4 mmol/hour (TiO_2/GR 2.0 wt%) from 3.4 mmol/hour (P25). The enhancement of photocatalytic activity for TiO_2/GR composites is mainly attributed to the excellent electron conductivity of GR which can suppress the recombination of photo-generated electron–holes, and the formation of chemically bonded TiO_2/GR enhance light absorption in the visible range.[23]

A green and facile preparation of TiO_2/GR nanocomposites with high photoactivity performance has been achieved *via* a one-step hydrothermal method. This convenient preparation method is environmentally friendly and features a direct synthesis by a chemical reaction and has potential applications for generating clean and renewable energy.

3.5.3.2 One-pot Facile Hydrothermal Method (TNT/GR)

TiO_2 nanotubes with increased catalytic activity can be prepared facilely under specific conditions. TNT combines the properties of conventional TiO_2 nanoparticles with its own unique properties, such as high conductivity, large specific surface area, good mechanical and electrical properties, high aspect ratio, and a large contact interface.[78] To improve TNT activity, it is preferred to improve the adsorption of reactant molecules and light due to the large number of active sites. In addition, the high aspect ratio of TNT nanostructure and a larger contact interface between TNT and GR are expected to promote the photo-induced charge separation. All of the aforementioned properties contribute to the improvement of the photocatalytic activity of the TiO_2 nanomaterial.

In addition, the photocatalytic performance of TNT can be further improved by hybridizing with graphene (Figure 3.12). Researchers have developed many synthetic methods to prepare TNT/GR nanocomposites, which have high electrical and heat conductivity, large specific surface area, and great optical transparency. Most synthetic processes utilize hydrazine as a reductant to reduce graphene oxide into graphene.

For example, Zhai *et al.*[79] used a one-pot hydrothermal method to synthesize TNT/GR photocatalysts by using hydrazine as the reductant, and GR and TiO_2 as the reactants. This type of nanomaterial can degrade Rhodamine-B under visible light, and the photocatalytic ability is superior to P25 and pure TNT. However, hydrazine is associated with significant toxicity and corrosion issues, which brings safety concerns as well as environmental protection issues and limits the manufacturing of graphene-based nanocatalyst. Therefore, researchers need to develop a green synthesis method to prepare GR-based TNT catalysts without using hydrazine.

One-pot facile hydrothermal synthesis is based on the ability of an alkali to act as a reductant to reduce graphene oxide to graphene, while

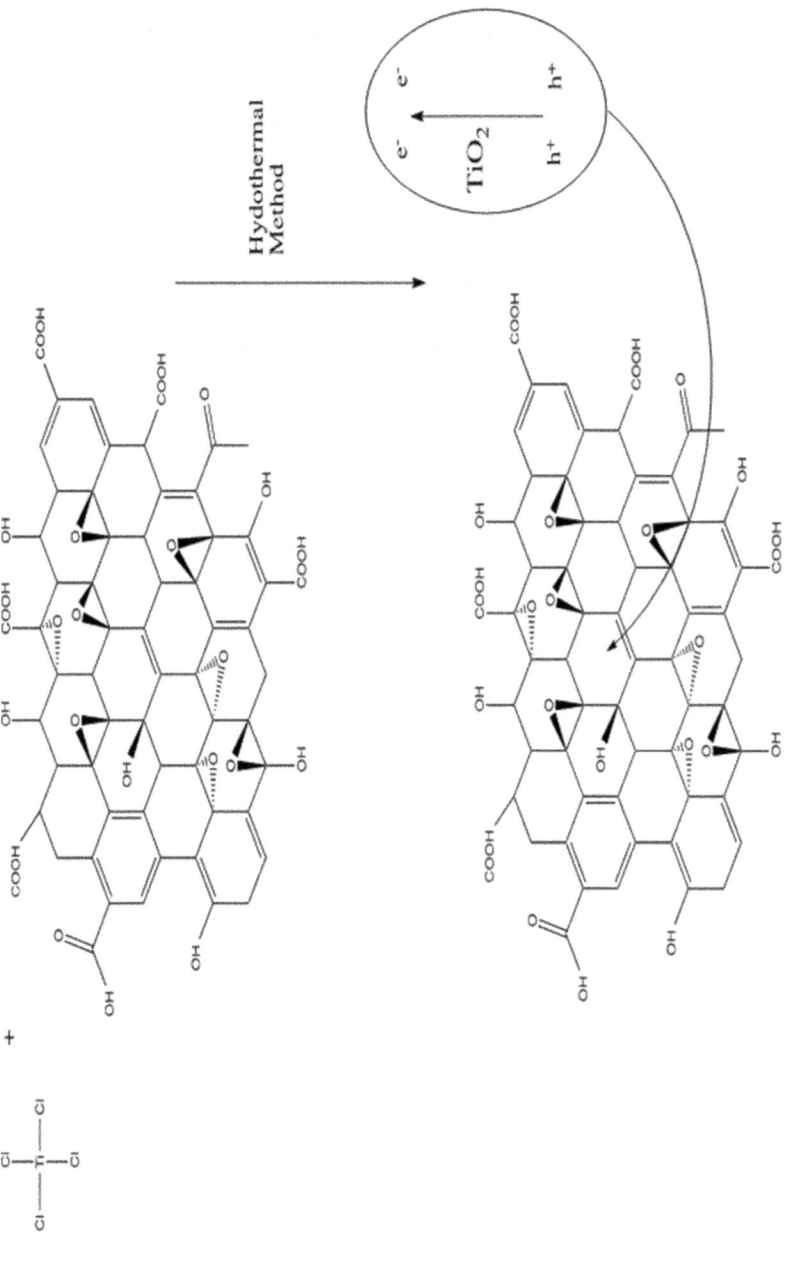

Figure 3.12 TiO$_2$ nanotubes hybridized with graphene to form a TNT/GR nanocomposite.

simultaneously converting the titanium dioxide nanoparticles into nano-tubes. The well-defined TNTs grow directly and distribute uniformly on the graphene sheets, forming an efficient photocatalyst nanocomposite. This synthetic method does not use or produce any hazardous agents during the preparation process.

Dang *et al.*[80] used this method to prepare TNT/GR photo-active nano-materials in NaOH solution. P25 and GO were added to the solution and the reactants were heated to 150 °C for 24 hours in an autoclave. The final products were washed and dried at 60 °C overnight, followed by calcination at 400 °C with argon atmosphere protection.

The final product was characterized by XRD and the results shows that the 10.8° 2θ peak, which belongs the pure GO, had disappeared. A broad 25.0° 2θ diffraction peak appeared which represents the formation of GR. The reactant solution changes from golden brown (GO) to dark brown (GR) (Figure 3.13). This observed phenomenon is similar to the previous experi-ment, which used hydrazine as the reducing agent.

The TEM image (Figure 3.14b) shows the morphology of the final product and proves the formation of GR sheets, which are wrinkled and covered with the tubular-like TNTs. In addition, the TEM image does not show any of the original TNP, which qualitatively demonstrates that all TNP reactant were reacted to form TNTs.

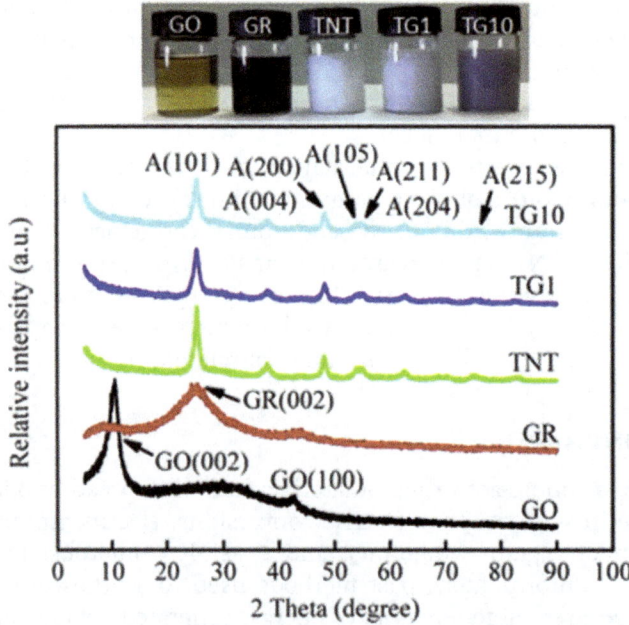

Figure 3.13 Samples and the color of each sample XRD patterns of GO, GR, pure TNT, TG1, and TG10.
Copyright permission from ref. 80.

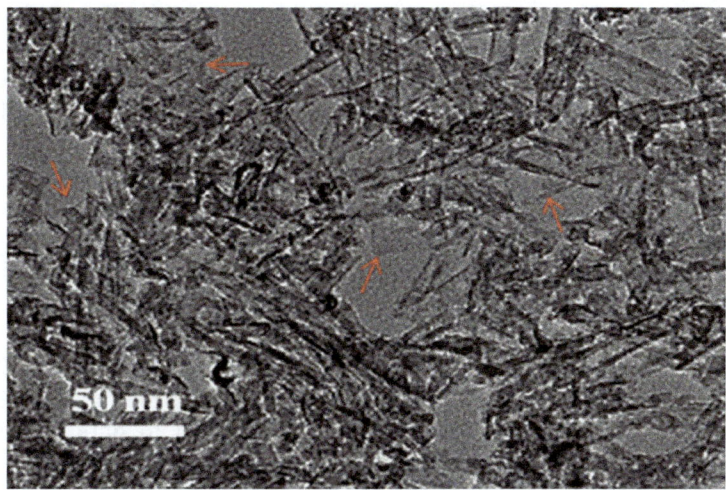

Figure 3.14   TEM image of TNT/GR composite synthesized under the one-pot facile
hydrothermal method with 1.0 wt% graphene, TG1.
Copyright permission from ref. 80.

The contribution of different graphene material contents (0.5, 1, 2, 5, 10 wt%) to the photocatalytic ability of splitting water was also studied using a 300 W lamp to irradiate the photocatalyst containing aqueous solution with 25% methanol. The nanocomposite samples with different graphene contents were marked as TG0.5, TG1, TG2, TG5, and TG10 respectively. The results show that the H_2 evolution rates were increased significantly when compared with pure TNT, especially for TG1 in which H_2 evolution rates increased more than 300%.

Overall, a facile, green hydrothermal synthetic method was developed for TNT/GR nano photocatalyst preparation without the use of any hazardous materials. Importantly, due to the addition of carbonaceous material, the photoactivity of TNT was increased in terms of hydrogen evolution rate for water. This environmentally friendly one-pot hydrothermal method and newly synthesized TNT/GR material will have many potential applications in addressing energy as well as environmental-related issues.

3.6   Conclusion

The advent of photo-active nanomaterials has widespread application, including use in sensors, photoelectric conversions, electrochromic displays, degradation of organic pollutants, and possibly numerous undiscovered applications. Unfortunately, past methods used to synthesize these nanomaterials required large amounts of energy, generated copious amounts of waste, and required the use of hazardous substances – which prevented the manufacture of photo-active nanomaterials on a commercial scale. Over the past several decades the public has become increasingly aware of the effects

of pollution on the environment; this has driven researchers to find ways to synthesize photo-active nanomaterials using green chemistry. Green chemistry pursues methods to reduce the environmental impact of chemicals by using less natural resources and reducing pollutants.

Studies of a series of green and sustainable methods to prepare photoactive nanomaterials have been demonstrated. These methods successfully synthesized nanomaterials with considerations such as low temperature, fast reaction rate, and reduced toxic reagents. While these methods have considerably decreased the environmental impacts of producing photoactive nanomaterials on a commercial scale, the cost of producing these nanomaterials is still very high. Researchers have not yet found a way to use absolute green chemistry, and the photo-active efficiencies are still low.

There is a promising future for photo-active nanomaterials, as their ability to split water to produce hydrogen may help alleviate our growing demand for energy. The ability to degrade organic pollutants is of great interest as we attempt to reduce global pollution. Researchers will continue to find new ways to synthesize photo-active nanomaterials that will reduce cost and maintain green chemistry.

References

1. X. Chen and S. S. Mao, *Chem. Rev.*, 2007, **107**, 2891–2959.
2. Y. Zhu, J. Shi, Z. Zhang, C. Zhang and X. Zhang, *Anal. Chem.*, 2002, **74**, 120–124.
3. B. Serrano and H. de Lasa, *Ind. Eng. Chem. Res.*, 1997, **36**, 4705–4711.
4. Z. S. Wang, F. Y. Li, C. H. Huang, L. Wang, M. Wei, L. P. Jin and N. Q. Li, *J. Phys. Chem. B*, 2000, **104**, 9676–9682.
5. Y. J. Kim, H. K. Jeong, J. K. Seo, S. Y. Chai, Y. S. Kim, G. I. Lim, M. H. Cho, I. M. Lee, Y. S. Choi and W. I. Lee, *J. Nanosci. Nanotechnol.*, 2007, 7, 4106–4110.
6. A. Fujishima and K. Honda, *Nature*, 1972, **238**, 37–38.
7. U. G. Akpan and B. H. Hameed, *J. Hazard. Mater.*, 2009, **170**, 520–529.
8. S. Ahmed, M. G. Rasul, W. N. Martens, R. Brown and M. A. Hashib, *Desalination*, 2010, **261**, 3–18.
9. C. Rath, P. Mohanty, A. C. Pandey and N. C. Mishra, *J. Phys. D: Appl. Phys.*, 2009, **42**, 205101.
10. Y. Wang, A. Zhou and Z. Yang, *Mater. Lett.*, 2008, **62**, 1930–1932.
11. Y. Chen, A. Lin and F. Gan, *Powder Technol.*, 2006, **167**, 109–116.
12. H. L. Luo, J. Sheng and Y. Z. Wan, *Mater. Lett.*, 2008, **62**, 37–40.
13. Y. Li, X. Sun, H. Li, S. Wang and Y. Wei, *Powder Technol.*, 2009, **194**, 149–152.
14. J. Beusen, M. K. Van Bael, H. Van den Rul, J. D'Haen and J. Mullens, *J. Eur. Ceram. Soc.*, 2007, **27**, 4529–4535.
15. G. Sivalingam, K. Nagaveni, M. S. Hegde and G. Madras, *Appl. Catal., B*, 2003, **45**, 23–38.

16. S. Y. Lu and S. W. Chen, *J. Am. Ceram. Soc.*, 2000, **83**, 709–712.

17. M. Anpo, T. Shima, S. Kodama and Y. Kubokawa, *J. Phys. Chem.*, 1987, **91**, 4305–4310.

18. S. Komarneni, S. Esquivel, Y. D. Noh, S. Sitthisang, J. Tantirungrotechai, H. Li, S. Yin, T. Sato and H. Katsuki, *Ceram. Int.*, 2014, **40**, 2097–2102.

19. S. N. Karthick, K. Prabakar, A. Subramania, J.-T. Hong, J.-J. Jang and H.-J. Kim, *Powder Technol.*, 2011, **205**, 36–41.

20. (a) M. Gratzel, *Nature*, 2001, **414**, 338–344; (b) R. D. Cortright, R. R. Davda and J. A. Dumesic, *Nature*, 2002, **418**, 964–967.

21. Q. Xiang, J. Yu, W. Wang and M. Jaroniec, *Chem. Commun.*, 2011, **47**, 6906–6908.

22. Q. Xiang, B. Cheng and J. Yu, *Appl. Catal., B*, 2013, **138–139**, 299–303.

23. T. Zhai, X. Fang, L. Li, Y. Bando and D. Golberg, *Nanoscale*, 2010, **2**, 168–187.

24. J. Ran, J. Yu and M. Jaroniec, *Green Chem.*, 2011, **13**, 2708–2713.

25. H. Yang, C. Huang, X. Li, R. Shi and K. Zhang, *Mater. Chem. Phys.*, 2005, **90**, 155–158.

26. (a) M. Zhang, T. Zhai, X. Wang, Y. Ma and J. Yao, *Cryst. Growth Des.*, 2010, **10**, 1201–1206; (b) Y.-w. Jun, S.-M. Lee, N.-J. Kang and J. Cheon, *J. Am. Chem. Soc.*, 2001, **123**, 5150–5151; (c) Y. Li, X. Li, C. Yang and Y. Li, *J. Mater. Chem.*, 2003, **13**, 2641–2648; (d) J. Joo, H. B. Na, T. Yu, J. H. Yu, Y. W. Kim, F. Wu, J. Z. Zhang and T. Hyeon, *J. Am. Chem. Soc.*, 2003, **125**, 11100–11105.

27. N. Bao, L. Shen, T. Takata and K. Domen, *Chem. Mater.*, 2008, **20**, 110–117.

28. N. Bao, L. Shen, T. Takata, D. Lu and K. Domen, *Chem. Lett.*, 2006, **35**, 318–319.

29. C. M. Janet and R. P. Viswanath, *Nanotechnology*, 2006, **17**, 5271–5277.

30. J. Yu, Y. Yu and B. Cheng, *RSC Adv.*, 2012, **2**, 11829–11835.

31. P. Eskandari, F. Kazemi and Y. Azizian-Kalandaragh, *Sep. Purif. Technol.*, 2013, **120**, 180–185.

32. Y. Nosaka, N. Ohta, T. Fukuyama and N. Fujii, *J. Colloid Interface Sci.*, 1993, **155**, 23–29.

33. (a) S. Kolahi, S. Farjami-Shayesteh and Y. Azizian-Kalandaragh, *Mater. Sci. Semicond. Process.*, 2011, **14**, 294–301; (b) P. E. Lippens and M. Lannoo, *Phys. Rev. B: Condens. Matter Mater. Phys.*, 1989, **39**, 10935–10942; (c) Y. Kayanuma, *Solid State Commun.*, 1986, **59**, 405–408.

34. X. Wang, Z. Feng, D. Fan, F. Fan and C. Li, *Cryst. Growth Des.*, 2010, **10**, 5312–5318.

35. M. Chen, Y. Xie, J. Lu, Y. Xiong, S. Zhang, Y. Qian and X. Liu, *J. Mater. Chem.*, 2002, **12**, 748–753.

36. F. Gao, Q. Lu, S. Xie and D. Zhao, *Adv. Mater.*, 2002, **14**, 1537–1540.

37. Z. Deng, M. Chen, G. Gu and L. Wu, *J. Phys. Chem. B*, 2008, **112**, 16–22.

38. (a) R. M. Al-Tuwirqi, A. A. Al-Ghamdi, F. Al-Hazmi, F. Alnowaiser, A. A. Al-Ghamdi, N. A. Aal and F. El-Tantawy, *Superlattices Microstruct.*, 2011, **50**, 437–448; (b) A. A. Al-Ghamdi, F. Al-Hazmi, O. Al-Hartomy,

F. El-Tantawy and F. Yakuphanoglu, *J. Sol-Gel Sci. Technol.*, 2012, **63**, 187–193; (c) A. Umar, M. S. Chauhan, S. Chauhan, R. Kumar, G. Kumar, S. A. Al-Sayari, S. W. Hwang and A. Al-Hajry, *J. Colloid Interface Sci.*, 2011, **363**, 521–528; (d) Y.-z. Lv, C.-r. Li, L. Guo, F.-c. Wang, Y. Xu and X.-f. Chu, *Sens. Actuators, B*, 2009, **141**, 85–88; (e) Y.-x. Han, Y.-z. Ding, W.-z. Yin and Z.-x. Ma, *Trans. Nonferrous Met. Soc. China*, 2006, **16**, 1205–1212; (f) P. Sharma, M. Kumar and A. Pandey, *J. Nanopart. Res.*, 2011, **13**, 1629–1637; (g) M. A. Majeed Khan, M. Wasi Khan, M. Alhoshan, M. S. AlSalhi and A. S. Aldwayyan, *Appl. Phys. A: Mater. Sci. Process.*, 2010, **100**, 45–51; (h) K. Chandrappa, T. Venkatesha, K. Vathsala and C. Shivakumara, *J. Nanopart. Res.*, 2010, **12**, 2667–2678.

39. L. Xu, Y.-L. Hu, C. Pelligra, C.-H. Chen, L. Jin, H. Huang, S. Sithambaram, M. Aindow, R. Joesten and S. L. Suib, *Chem. Mater.*, 2009, **21**, 2875–2885.

40. W.-J. Huang, G.-C. Fang and C.-C. Wang, *Colloids Surf., A*, 2005, **260**, 45–51.

41. Y. J. Jang, C. Simer and T. Ohm, *Mater. Res. Bull.*, 2006, **41**, 67–77.

42. N. Sobana and M. Swaminathan, *Sol. Energy Mater. Sol. Cells*, 2007, **91**, 727–734.

43. N. Sobana and M. Swaminathan, *Sep. Purif. Technol.*, 2007, **56**, 101–107.

44. C.-C. Chen, *J. Mol. Catal. A: Chem.*, 2007, **264**, 82–92.

45. Y. Dai, Y. Zhang, Q. K. Li and C. W. Nan, *Chem. Phys. Lett.*, 2002, **358**, 83–86.

46. C.-L. Zhang, W.-N. Zhou, Y. Hang, Z. Lü, H.-D. Hou, Y.-B. Zuo, S.-J. Qin, F.-H. Lu and S.-L. Gu, *J. Cryst. Grow.*, 2008, **310**, 1819–1822.

47. Y. W. Heo, V. Varadarajan, M. Kaufman, K. Kim, D. P. Norton, F. Ren and P. H. Fleming, *Appl. Phys. Lett.*, 2002, **81**, 3046–3048.

48. F. Al-Hazmi, N. Aal, A. Al-Ghamdi, F. Alnowaiser, Z. Gafer, A. Al-Sehemi, F. El-Tantawy and F. Yakuphanoglu, *J. Electroceram.*, 2013, **31**, 324–330.

49. E. B. Flint and K. S. Suslick, *Science*, 1991, **253**, 1397–1399.

50. K. D. Bhatte, D. N. Sawant, R. A. Watile and B. M. Bhanage, *Mater. Lett.*, 2012, **69**, 66–68.

51. K. D. Bhatte, D. N. Sawant, D. V. Pinjari, A. B. Pandit and B. M. Bhanage, *Mater. Lett.*, 2012, 77, 93–95.

52. G. Liu, Y. Zhao, C. Sun, F. Li, H. M. Cheng and G. Q. Lu, *Angew. Chem., Int. Ed.*, 2008, **47**, 4516–4520.

53. (a) P. Dong, Y. Wang, L. Guo, B. Liu, S. Xin, J. Zhang, Y. Shi, W. Zeng and S. Yin, *Nanoscale*, 2012, **4**, 4641–4649; (b) X. Pan, Y. Zhao, S. Liu, C. L. Korzeniewski, S. Wang and Z. Fan, *ACS Appl. Mater. Interfaces*, 2012, **4**, 3944–3950; (c) L.-L. Tan, S.-P. Chai and A. R. Mohamed, *ChemSusChem*, 2012, **5**, 1868–1882.

54. (a) B. K. Vijayan, N. M. Dimitrijevic, D. Finkelstein-Shapiro, J. Wu and K. A. Gray, *ACS Catal.*, 2012, **2**, 223–229; (b) Y.-P. Peng, S.-L. Lo, H.-H. Ou and S.-W. Lai, *J. Hazard. Mater.*, 2010, **183**, 754–758.

55. (a) S. Xu, A. J. Du, J. Liu, J. Ng and D. D. Sun, *Int. J. Hydrogen Energy*, 2011, **36**, 6560–6568; (b) N. Wang, X. Li, Y. Wang, Y. Hou, X. Zou and G. Chen, *Mater. Lett.*, 2008, **62**, 3691–3693; (c) K. R. Gopidas,

M. Bohorquez and P. V. Kamat, *J. Phys. Chem.*, 1990, **94**, 6435–6440; (d) N. Kakuta, K. H. Park, M. F. Finlayson, A. Ueno, A. J. Bard, A. Campion, M. A. Fox, S. E. Webber and J. M. White, *J. Phys. Chem.*, 1985, **89**, 732–734.

56. (a) Y. Su and Y. Deng, *Appl. Surf. Sci.*, 2011, **257**, 9791–9795; (b) Q. Zhao, M. Li, J. Chu, T. Jiang and H. Yin, *Appl. Surf. Sci.*, 2009, **255**, 3773–3778.

57. (a) Y. Park, S. H. Kang and W. Choi, *Phys. Chem. Chem. Phys.*, 2011, **13**, 9425–9431; (b) H.-i. Kim, G.-h. Moon, D. Monllor-Satoca, Y. Park and W. Choi, *J. Phys. Chem. C*, 2012, **116**, 1535–1543.

58. H. Li, B. Zhu, Y. Feng, S. Wang, S. Zhang and W. Huang, *J. Solid State Chem.*, 2007, **180**, 2136–2142.

59. Y. W. Chen, J. D. Prange, S. Dühnen, Y. Park, M. Gunji, C. E. D. Chidsey and P. C. McIntyre, *Nat. Mater.*, 2011, **10**, 539–544.

60. S. H. Cho, G. Gyawali, R. Adhikari, T. H. Kim and S. W. Lee, *Mater. Chem. Phys.*, 2014, **145**, 297–303.

61. A. A. Christy, O. M. Kvalheim and R. A. Velapoldi, *Vib. Spectrosc.*, 1995, **9**, 19–27.

62. H. R. Pant, C. H. Park, B. Pant, L. D. Tijing, H. Y. Kim and C. S. Kim, *Ceram. Int.*, 2012, **38**, 2943–2950.

63. N. M. Bahadur, T. Furusawa, M. Sato, F. Kurayama and N. Suzuki, *Mater. Res. Bull.*, 2010, **45**, 1383–1388.

64. (a) Y. Bessekhouad, D. Robert and J. V. Weber, *J. Photochem. Photobiol., A*, 2004, **163**, 569–580; (b) L. Wu, J. C. Yu and X. Fu, *J. Mol. Catal. A: Chem.*, 2006, **244**, 25–32; (c) X. Wang, G. Liu, F. Li, H. M. Cheng, Z. G. Chen, L. Wang and G. Q. Lu, *Chem. Commun.*, 2009, 3452–3454.

65. Y. Jin-nouchi, S.-i. Naya and H. Tada, *J. Phys. Chem. C*, 2010, **114**, 16837–16842.

66. J. S. Jang, H. Gyu Kim, P. H. Borse and J. S. Lee, *Int. J. Hydrogen Energy*, 2007, **32**, 4786–4791.

67. J. Luo, L. Ma, T. He, C. F. Ng, S. Wang, H. Sun and H. J. Fan, *J. Phys. Chem. C*, 2012, **116**, 11956–11963.

68. N. Ghows and M. H. Entezari, *Ultrason. Sonochem.*, 2012, **19**, 1070–1078.

69. D. He, M. Chen, F. Teng, G. Li, H. Shi, J. Wang, M. Xu, T. Lu, X. Ji, Y. Lv and Y. Zhu, *Superlattices Microstruct.*, 2012, **51**, 799–808.

70. X. Li, T. Xia, C. Xu, J. Murowchick and X. Chen, *Catal. Today*, 2014, **225**, 64–73.

71. K. Zhou, Y. Zhu, X. Yang, X. Jiang and C. Li, *New J. Chem.*, 2011, **35**, 353–359.

72. X. Zhang, Y. Sun, X. Cui and Z. Jiang, *Int. J. Hydrogen Energy*, 2012, **37**, 811–815.

73. Y. Cong, M. Long, Z. Cui, X. Li, Z. Dong, G. Yuan and J. Zhang, *Appl. Surf. Sci.*, 2013, **282**, 400–407.

74. J. S. Lee, K. H. You and C. B. Park, *Adv. Mater.*, 2012, **24**, 1084–1088.

75. H. Zhang, X. Lv, Y. Li, Y. Wang and J. Li, *ACS Nano*, 2010, **4**, 380–386.

76. H. Liu, X. Dong, X. Wang, C. Sun, J. Li and Z. Zhu, *Chem. Eng. J.*, 2013, **230**, 279–285.
77. W. S. Hummers and R. E. Offeman, *J. Am. Chem. Soc.*, 1958, **80**, 1339.
78. D. V. Bavykin, J. M. Friedrich and F. C. Walsh, *Adv. Mater.*, 2006, **18**, 2807–2824.
79. Q. Zhai, B. Tang and G. Hu, *J. Hazard. Mater.*, 2011, **198**, 78–86.
80. H. Dang, X. Dong, Y. Dong and J. Huang, *Int. J. Hydrogen Energy*, 2013, **38**, 9178–9185.

CHAPTER 4

Natural Photosynthesis System

DAVID S. GRAY*[a] AND NULAZHI YEERXIATI*[b]

[a] Massachusetts Institute of Technology, Cambridge, MA 02139, USA;
[b] Princeton University, Princeton, NJ 08544, USA
*Email: dsgray@alum.mit.edu; erxat318@126.com

4.1 The Wonder and Impact of Photosynthesis

What is the pathway for photons from the Sun to contribute to fuel propelling a rocket towards the stars, a flower's fragrant molecules, and the bee's buzzing ability? Amazingly, the basic resources of sunlight, water, and carbon dioxide are the main ingredients powering so much motion and activity in the world. The machinery enabling this wonder of nature consists of components that operate in a sophisticated orchestration of steps. A more thorough understanding of the fundamentals of photosynthesis not only increases our awe of nature's methods, but it also has helped in trying to meet humanity's rapidly increasing energy needs.

The development of oxygenic photosynthesis resulted in a profound change in the ecosystem as the atmosphere then became able to sustain aerobic metabolism and more complex life. Although it is considered a waste product in the context of photosynthesis, most organisms – including photosynthetic organisms – use oxygen in cellular respiration. While molecular oxygen is essential for that purpose, the gas is actually highly reactive and must be continually replenished in the atmosphere due to this reactivity.

Interestingly, oxygenic photosynthesis is not the only sort used by photoautotrophs (any organism that uses light to make food and uses

RSC Green Chemistry No. 42
Green Photo-active Nanomaterials: Sustainable Energy and Environmental Remediation
Edited by Nurxat Nuraje, Ramazan Asmatulu and Guido Mul
© The Royal Society of Chemistry 2016
Published by the Royal Society of Chemistry, www.rsc.org

carbon dioxide as a source of carbon). The use of water as a source of electrons is described below, but some organisms have been able to use other substrates as electron donors. Some examples include hydrogen gas and small organic molecules, including formate, acetate, or methanol, and even currently green sulfur bacteria use hydrogen sulfide. At the end of this chapter is a summary of how oxygenic photosynthesis is thought to have arisen over billions of years past.

The face of the world has been altered as life can thrive wherever sunlight reaches moisture and nutrients. With photons produced from fusion inside the Sun, photoautotrophs turn water and carbon dioxide into stores of energy. Most of the energetic actions on the Earth are made possible by harvesting the Sun's oscillating electromagnetic waves, which enables this erstwhile sterile planet to brim with motion.

Much of this book is dedicated to attempts and potential approaches to learn from nature's method of harnessing the Sun's energy. With solid-state solar cells and organic thin film arrays already surpassing the efficiency of plants in the percentage conversion of solar energy into a more useful energy, it may seem at this point that little could be gleaned from nature's approach in photosynthesis, but further understanding of this process can continue to improve the conversion of sunlight to electricity. Some features of photosynthesis not replicated in synthetic systems include self-assembly, repair, and self-replication.

4.2   The Method and Products of Photosynthesis

The position of electrons in molecules is responsible for much of the phenomena in chemistry. All chemical reactions result from the flow of electrons, so studying their movement is essential to understanding chemical processes. In photosynthesis, a series of transfers of excited electrons is core to the conversion of sunlight to energy storage in molecules. This movement is described in terms of oxidation and reduction reactions. The term 'oxidizing' refers to the taking away of a molecule's electrons, and 'reducing' refers to the giving of electrons.

Chemical bonds within molecules store energy in cells, and the formation of bonds is achieved through electron movement. The following reaction showing the formation of carbohydrates broadly summarizes photosynthesis, but it only conveys mass balance aspects and does not show additional products that can be made:

$$CO_2 + H_2O \rightarrow (CH_2O)_x + O_2 \qquad (4.1)$$

In the process of photosynthesis, cells store energy temporarily in two molecules, ATP and NADPH. In later reactions, ATP and NADPH are involved to achieve long-term energy storage by synthesizing carbohydrates.

Light energy is the input used to form the 'high energy bond' between ADP and P_i (inorganic phosphate) in the formation of ATP. This molecule is thought of as 'currency' in this context as it serves as an investment for later

reactions that otherwise would not proceed. More specifically, it is able to reverse the equilibrium of reactions by coupling them with ATP hydrolysis. For instance, the equilibrium of the reaction adding phosphate to glucose is shifted to the product, glucose-6-P.

$$ATP + H_2O \rightarrow ADP + P_i \ (-31 \text{ kJ mol}^{-1}) \tag{4.2}$$

$$P_i + glucose \rightarrow glucose\text{-}6\text{-}P + H_2O \ (+14 \text{ kJ mol}^{-1}) \tag{4.3}$$

$$glucose + phosphate \rightarrow glucose\text{-}6\text{-}P \ (-17 \text{ kJ mol}^{-1}) \tag{4.4}$$

The movement of electrons to reduce $NADP^+$ also enables the transfer of protons across a membrane. This helps set up a proton concentration gradient, which produces the proton motive force (pmf). The movement of protons through a channel in the membrane from the side of high concentration to low concentration generates a mechanical force that produces a conformational change in an enzyme, which in turn enables it to catalyze the production of ATP.

The set of reactions producing ATP and NADPH is known collectively as the light-dependent reactions and is summarized as:

$$2H_2O + 2NADP^+ + 3ADP \rightarrow 3P_iO_2 + 2NADPH + 3ATP \tag{4.5}$$

where P_i is inorganic phosphate, which is a free phosphate ion that is negatively charged.

The portability of ATP and NADP allows for this set of reactions to occur in a separate place from where energy from light is collected. This is analogous to how electricity generated in one place can be stored in batteries or fuel for use elsewhere.

In summary, the production of carbohydrates by photosynthesis occurs by two sets of reactions: (1) those yielding ATP and NADPH and (2) reactions that use these molecules to generate organic products.

4.3 Harvesting Light Energy

When an electron leaves its location within a bond in a molecule, this leaves behind a positive charge and results in charge separation. Since opposite charges attract, their separation requires an input of energy, which comes from light. The captured photonic energy promotes electrons within pigment molecules of proteins to higher energetic levels.

Ultimately, at least eight photons – but more likely nine or ten – must be captured to fix each molecule of CO_2. These photons journey epically from space before participating in photosynthesis but then are very quickly utilized. Photons produced by nuclear reactions in the Sun's center travel a tortuous path lasting 10 000 years to escape the Sun, cover a distance of 150 000 000 km in about 8 minutes to reach the Earth, and within 100 picoseconds are captured and channeled into reaction centers by photoautotrophs.[1]

light-harvesting components in photosynthesis absorb photons of
ain wavelength ranges very effectively. In fact, one of the protein com-
.exes in photosynthesis performs almost perfectly – an electron is excited
almost every time a photon is absorbed within the proper wavelength range.
This wonder of science is accomplished by a complex system – there are
more than 200 pigments incorporated in the protein in addition to many
other cofactors. Excited electrons routinely pass through these pigments
before reaching their destination.[2]

The efficiency of photosynthesis is limited, though, by its utilization
of only a subset of visible light. Furthermore, the visible spectrum,
400–800 nm, represents just half of the solar energy imparted to the Earth.
In photosynthesis, most of the light harvested is in the red and blue regions
of light. The range of wavelengths available for use in each photoautotroph
is determined by its particular pigment molecules. The green color of plants
is due to reflection and transmission of that portion of the spectrum
(Figure 4.1).

Electrons excited by the absorption of light flow through a series of small
molecules and proteins to other regions for reactions.

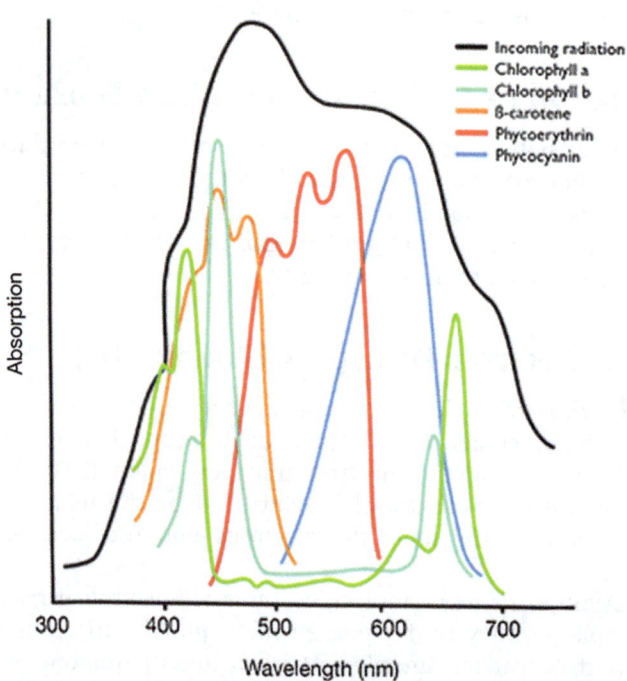

Figure 4.1 The spectrum of sunlight in the visible region and absorption of various
pigment molecules. Most of the sunlight utilized is in the blue and red
regions.
(Reproduced from ref. 11 with permission from The Royal Society of
Chemistry. © Royal Society of Chemistry 2010.)

4.4 Water Splitting for Supply of Protons and Electrons

The continual use of electrons requires an external source to resupply them, which comes from water. The capability in an organism to oxidize water – thereby generating protons, electrons, and molecular oxygen – was a necessary innovation that helped bring about oxygenic photosynthesis.

$$2H_2O \rightarrow O_2 + 4H^+ + 4e^- \tag{4.6}$$

Part of the wonder of the photosynthetic system is the redox power of its machinery – among the components are the strongest oxidant and the strongest reductant in biology. This oxidation power is put to use to produce molecular oxygen from water.

In addition, protons are produced from water. As mentioned, they are not able to permeate the membrane except through channels, and they contribute therefore to a higher concentration of protons on one side of a membrane. The membrane protein, ATP synthase, contains a proton channel within it, and by the movement of protons (a form of chemiosmosis) through it, ATP generation is made possible through the addition of phosphate (P_i) to adenosine diphosphate (ADP).

4.5 Fixation of Carbon from Carbon Dioxide

A source of carbon is needed to form the main end product of photosynthesis, carbohydrates. Carbon dioxide is the source of carbon in photosynthesis. This gas is obtained from the atmosphere, and there are some unique aspects in plants that retrieve gas in hot and dry environments, which are discussed later in this chapter.

4.6 The Electron Transport Chain (ETC)

After photon capture, excited electrons move from one molecule to another to reach a reaction center. Each step in this transport happens by means of a redox reaction along the electron transport chain (ETC) (Figure 4.2). The pathway the electrons travel is controlled by the relative location of the carriers able to transport the electrons and the local environment (Figure 4.3).

In photosynthesis, small molecules serve as electron carriers between protein complexes. They bind to pockets in the proteins to transfer electrons. Examples of these carriers are plastoquinone and ubiquinone. Proteins can also function as electron carriers by means of their internal metal ion complexes and aromatic groups in their polypeptide chains. Excited electrons can move through these structures. For instance, cytochrome is a protein which contains iron pigments that carry electrons. The iron atom in the protein serves to transfer the electron by accepting an electron

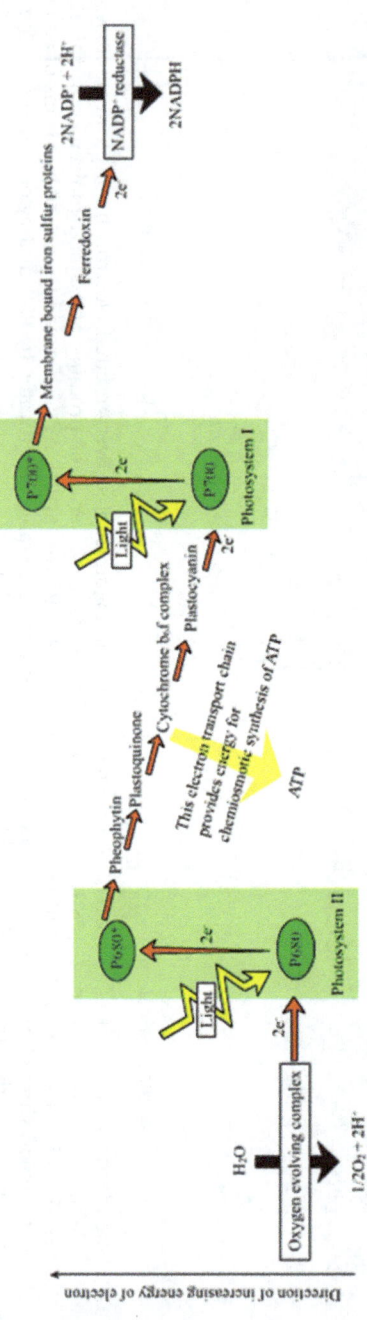

Figure 4.2 A simplified scheme of the light reactions of photosynthesis taken from http://en.wikipedia.org/wiki/Photosynthesis. Two photons must be absorbed per electron extracted from water and transferred to CO_2. The production of O_2 requires splitting two water molecules, so the overall process involves the absorption of four photons per PSII and PSI reaction center. The first photon is absorbed by photosystem II (PSII), which is the enzyme responsible for splitting water to generate protons and electrons that replenish those that flow to reduce $NADP^+$ by means of photosystem I (PSI) producing the strong reducing species, NADPH. In the production of O_2, two water molecules are split. The diagram indicates the rise of the energy of electrons from the energy of photons. The flow of higher energy electrons also enables the proton pumping action of cytochrome b_6f (Cyt b_6f), and these protons help to convert ADP to ATP. Both ATP and NADPH are used in fixing CO_2. (Reproduced from ref. 1 with permission from The Royal Society of Chemistry. © Royal Society of Chemistry 2008.)

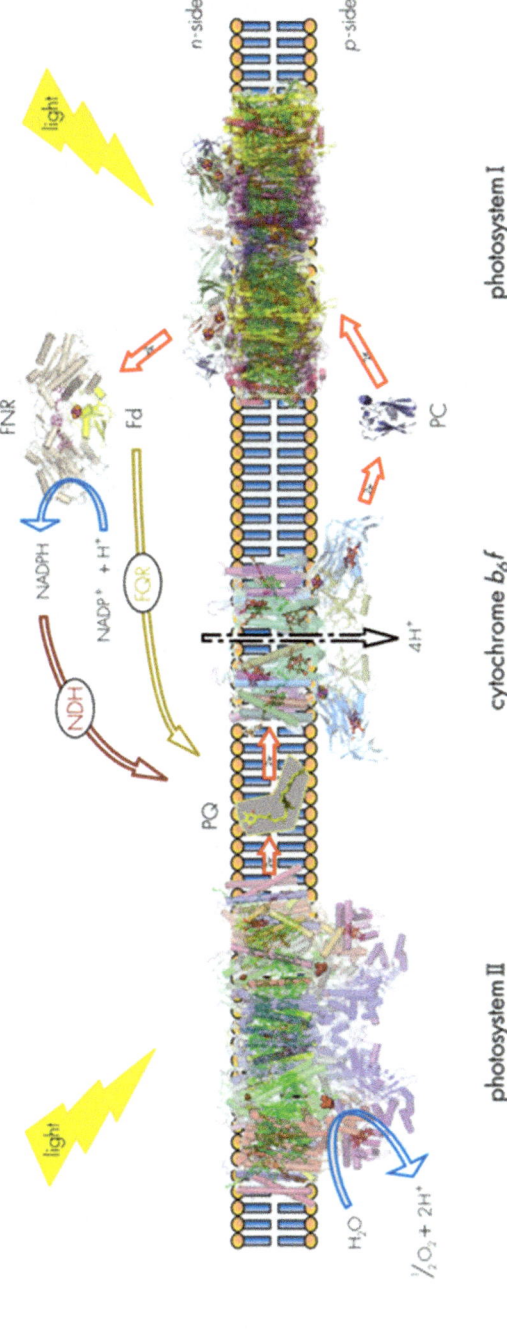

Figure 4.3 The electron transport chain (ETC) in oxygen-evolving photoautotrophs. Every excited electron involved in photosynthesis originates in photosystem II (PSII). The electrons flow through a series of proteins and small molecules in the ETC, as described in this chapter. In addition to the reduction of NADP$^+$ to NADPH, ATP is formed from ADP by means of an enzyme (not shown) through which protons flow. These protons are generated from water by PSII and cytochrome b$_6$f. The *n*-side corresponds to the stroma, and the *p*-side corresponds to the lumen.
(Reproduced from ref. 10 with permission from The Royal Society of Chemistry. © Royal Society of Chemistry 2008. Original by Professor W.W. Cramer and modified by S. Renger and D. Gray.)

(becoming reduced) and then giving that electron (becoming oxidized) to the next carrier.

All electrons originate at a reaction center through the splitting of water. They are excited twice by photons along the *ETC*. The end point for the flow of electrons in photosynthesis is a reaction center where $NADP^+$ becomes NADPH through reduction and the addition of a proton.

4.7 Chloroplasts: Containers for Photosynthesis

In plants and algae, the apparatus for photosynthesis is contained within chloroplasts, which are a spheroid shape approximately 5–7 μm by 1–2 μm.[3] Membranes are inside, surrounded by a colorless stroma (Figure 4.4). The proteins that help produce ATP and NADPH are located on or near these membranes. The number of chloroplasts in cells range from one in some algae to 100–200 in leaf cells with a high level of photosynthetic activity.

Lamellae are the sheets running throughout the chloroplasts. They are each roughly 10–15 nm thick. The lamellae are what make up the thylakoids, which are hollow disks. Within the thylakoid membranes are the protein complexes that make up the photosynthetic machinery. Additional components serve as electron transfer agents and are soluble or positioned on the sides of the membrane (Figure 4.5).

In chloroplasts, the grana are regions in which these sac-like structures are very tightly stacked. The thylakoids can extend from one grana into another grana with the thylakoids connected.

4.8 The Photosynthetic Machinery

In summary, through the capture of photons, the transfer of excited electrons *via* redox reactions, and the fixation of carbon captured in CO_2, photosynthesis yields organic products. The following steps occur in photosynthesis:

(1) charge separation through light harvesting
(2) transfer of electronic excited states to reaction centers
(3) the generation of ATP and NADPH
(4) the production of organic molecules in the latter stage of photosynthesis by means of electrons, hydrogen ions, and those energy-carrying molecules, ATP and NADPH.

With knowledge about the process and where photosynthesis occurs, let us turn to specifics of the components.

There are two excitation events in the first stage of photosynthesis, known as the light-dependent reactions. These excitations happen in photosystem I (PSI) and photosystem II (PSII) using light-harvested energy. There is an incongruity in the labeling of PSII and PSI due to the order of discovery of the

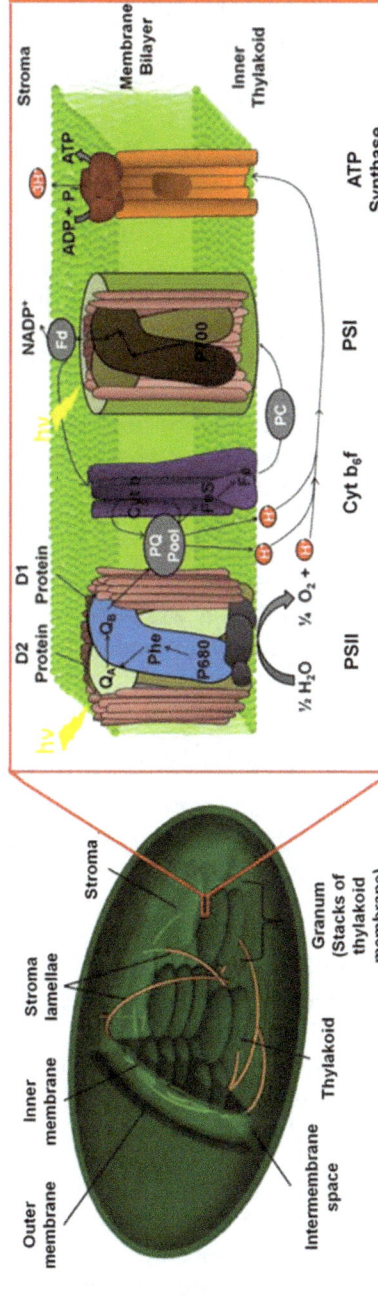

Figure 4.4 A schematic of the main features of chloroplasts as well as the arrangement of components of the photosynthetic machinery within the thylakoid membrane. Within the thylakoid membranes are embedded protein complexes that constitute the photosynthetic machinery, and additional components, including protons and electron carriers, are positioned on both sides. The generation of ATP and NADPH by the processes of photosynthesis occurs in the stroma. (Reproduced from ref. 8 with permission from The Royal Society of Chemistry. © Royal Society of Chemistry 2011.)

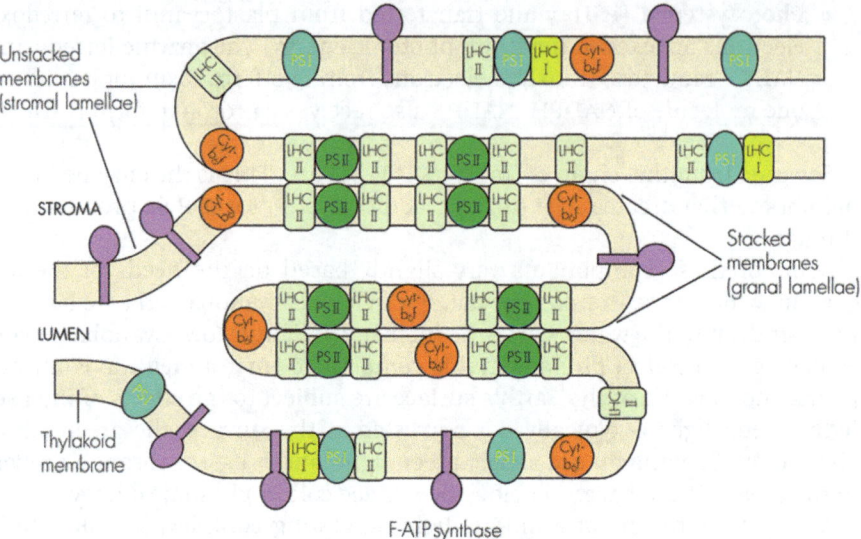

Figure 4.5 Schematic of the distribution of the main complexes of the photosynthetic apparatus in the thylakoid membrane of chloroplasts in higher plants.
(Reproduced from ref. 10 with permission from The Royal Society of Chemistry. © Royal Society of Chemistry, 2008.)

two photosystems. The first photosystem, PSII, splits water to produce oxygen and protons. The electrons from water also provide a continual supply flowing from PSII to $NADP^+$, which occurs at the second photosystem, PSI.

The two photosystems, PSII and PSI, absorb photons of different wavelengths based on the pigments they contain. PSII absorbs maximally at about 680 nm, while PSI absorbs photons of 700 nm maximally (slightly lower energy). As mentioned above, by capturing photons of different portions of the spectrum, more of the energy offered by solar light can be utilized.

Excited electrons flow from PSII to $NADP^+$ in this order (proteins in bold):

PSII → **plastoquinone** → cyctochrome b_6f → plastocyanin → PSI → ferredoxin → **FNR**

In brief, these membrane protein complexes perform the following functions:

- **Photosystem II (PSII):** all electrons in photosynthesis originate in PSII, where photons are captured to excite electrons to higher energy states. Water is oxidized to produce protons and electrons. Electrons from PSII reduce plastoquinone (PQ), which shuttles the electrons to the cytochrome b_6f complex.
- **Cytochrome-b_6f:** mediates transfer of electrons from PSII to PSI and pumps protons across the thylakoid membrane.

- **Photosystem I (PSI)**: while transferred from plastocyanin to ferredox, electrons are excited here with photonic energy. The enzyme ferredoxin-$NADP^+$ reductase transfers electrons from two ferrodoxin molecules to one molecule of NADPH. $NADP^+$ also receives a proton to form NADPH.

Separate from this chain of transfers is F-ATPase. This is the enzyme in the membrane that creates ATP through addition of P_i to ADP as protons flow through it.

Some of these components vary slightly based on the needs of the organism, which arise from its habitat. For instance, cyanobacteria are large in number deep in the water where the intensity of light is low. Cyanobacteria's challenge is to collect photons in this competitive environment. In contrast, plants and bacteria at the Earth's surface are subject to the effects of intense light. Green light is typically not harvested at the surface; the green algae above cyanobacteria do not collect green light, and it passes through deeper in the ocean. Cyanobacteria below the surface collect photonic energy of this wavelength by means of a unique light-harvesting complex, the phycobilisome. This is one of several unique light-harvesting complexes in cyanobacteria. Also, cyanobacteria utilize a larger version of PSI (trimeric) to capture more light.

Although there are some small differences in the photoautotrophs' machinery for photosynthesis, the core process is similar. Details on how the parts work in concert to create ATP and NADPH are provided below, and the reactions that lead to the generation of carbohydrates are discussed afterwards.

4.8.1   PSII

PSII is a particularly special component, for all of the excited electrons in photosynthesis flow from PSII. Of all enzymes PSII is the only one known to oxidize water. It produces the strongest oxidizing potential in any known biological system.[2] Electrons acquired in the oxidation of water replenish those that are excited in PSII and flow away from the pigment molecules.

Molecular oxygen and hydrogen ions are also generated from water, and the protons yield a gradient of higher H^+ concentration in the thylakoid lumen relative to the chloroplast stroma. The resulting pmf is needed for ATP generation by ATP synthase with protons flowing through that enzyme.

PSII contains a large number of cofactors, including 20 protein subunits with 35 chlorophyll a molecules, 12 carotenoid molecules, and 25 integral lipids, one chloride ion per monomer, and the Mn_4CaO_5 cluster, which is discussed below.[4]

4.8.1.1   Photon Absorption

For the light reactions performed in the photosystems, a transfer of one electron requires the absorption of one photon. The oxidization of each

water molecule supplies two electrons. In the process of forming oxygen gas, O_2, two molecules of H_2O are oxidized, thus providing four electrons. Since each electron is excited twice in photosynthesis eight photons is considered at least the minimum needed in using two water molecules in photosynthesis. In fact, the quantum requirement for the use of water in photosynthesis is nine or ten since ATP formation requires additional light absorption in a separate pathway, the cyclic photophosphorylation pathway.

Although pigments in the proteins responsible for photon absorption can trap light at a wide range of wavelengths, energy from photons higher in energy than that of the red region is not fully utilized; rather, the excess energy is degraded. It forms heat within the light-harvesting pigments.

A few calculations are involved in determining the theoretical efficiency. First is the energy of a molecule of glucose, a typical product of carbon fixation. This energy is 672 kcal mol^{-1} (2810 kJ mol^{-1}). Six moles of CO_2 are needed to make each mole of glucose. For every CO_2 molecule fixed, the equivalent of eight 'red' photons must be absorbed. The energy from 48 moles of 680 nm light is 2016 kcal (8435 kJ mol^{-1}). Therefore, the theoretical efficiency is about 30%.[1] Later in this chapter is a discussion about the actual efficiency achieved in photosynthesis, which is much lower.

To maximize capture of incoming light and excite electrons, both photosystems, PSII and PSI, contain an antenna complex that consists of light-absorbing molecules. The electrons in these molecules are excited by light and are subsequently directed to a reaction center. The absorption of light by the antenna complex within the visible spectrum is due to the characteristics of the pigments, and the number of each varies by organism. For instance, there can be from 25 to 120 chlorophyll molecules per P700 reaction center.[5]

Supplementing chlorophyll in plants are additional light-absorbing molecules – carotenes and carotenoids. One of the roles of carotenoids is to protect against the effects of excess energy by scavenging free radicals. More directly relevant to the process of converting light to biological energy, they pass on photonic energy to chlorophyll. This happens in the pathway to the reaction center.

A set of cofactors in PSII makes up the PSII primary donor, P680. An electron is excited in P680 by either absorbing a photon or by receiving energy transferred from nearby chlorophylls in PSII. P680 consists of four weakly interacting chlorophyll monomers. Its name derives from being a pigment ('P') and from the wavelength of maximum absorption being close to 680 nm (it is actually 684 nm). The chlorophyll monomers that constitute P680 are not arranged in pairs in PSII, and as a result, $P680^+$, the positively charged form of P680 that results after its excited electron is taken by a donor, is highly oxidized with a redox potential of 1.17 V. $P680^+$ is able to provide the necessary redox potential for the splitting of water since an oxidizing potential of just 1 V is needed to split two molecules of water at physiological pH to form dioxygen and four reducing equivalents. In the splitting of two water molecules, four electrons are taken one by one by an as of yet unidentified carrier.

4.8.1.2 Active Site

In the process of oxidizing water to molecular oxygen, four electrons are extracted from two water molecules yielding molecular oxygen and four protons:

$$2H_2O \xrightarrow{h\nu} O_2 + 4e^- + 4H^+ \qquad (4.7)$$

In total, PSII needs to absorb four photons to drive the water-splitting reaction. After an electron is excited in a pigment as mentioned above, the resonance energy is transferred to the P680 site to create electron–hole pairs. A hole is not an actual particle; rather, it is the lack of an electron where one could exist. At the oxygen-evolving complex (OEC), the hole remains for use in the oxidation of water by PSII. The OEC is known as the metallo-oxo cluster (Figure 4.6).

Excited electrons within PSII are transferred first from P680 to pheophytin (Phe) and then to quinone Q_A, which is tightly bound, reducing Q_A. Next, the

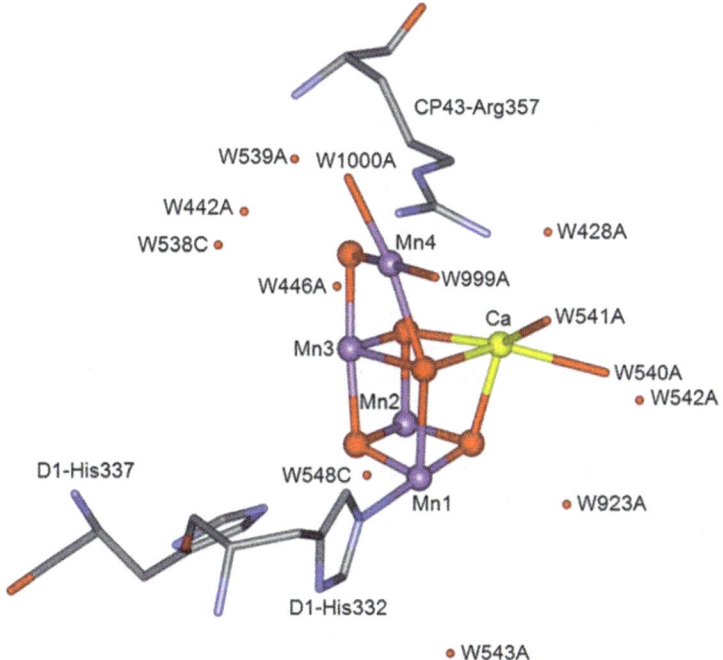

Figure 4.6 The structure of the Mn4Ca-oxo cluster as observed in the X-ray crystal structure of PSII. The four protons of four water molecules are shown coordinated with the manganese (W1000A and W999A with Mn4) and calcium ions (W541A and W540A with Ca). Three residues on PSII close to the Mn4Ca-oxo cluster are also shown (D1-His332, D1-His337, and CP43-Arg357), and nine water molecules are indicated in the diagram. (Reproduced from ref. 12 with permission from The Royal Society of Chemistry. © Royal Society of Chemistry, 2013.)

potential difference between Q_A and the oxidized Q_B results in movement of the electron to mobile Q_B. The transfer is aided by a non-heme iron (Fe) positioned between them. Finally, after reduction of Q_B a second time, Q_B^{2-} accepts two protons to become Q_BH_2 and moves into the lipid bilayer. This is the reaction with Q_BH_2 in sum:

$$4H^+ + 4e^- + Q \rightarrow O_2 + 2Q_BH_2 \qquad (4.8)$$

The electrons from here move along the ETC for use in generating the pmf and reducing compounds.

The catalytic center, OEC, consists of a cooperative Mn_4CaO_5 cluster and acts concertedly with a redox-active tyrosine residue (D1-Tyr161). After two additional photochemical cycles, the Mn_4CaO_5 cluster is reset to a reduced state, and oxidized quinone from the membrane replaces the lost, reduced Q_BH_2. Models to propose the splitting of water have been proposed including at least two mechanisms to describe the formation of dioxygen. Three of the manganese atoms in Mn_4CaO_5 are positioned in a cubic formation with Ca^{2+}, and an Mn ion is positioned adjacent to Ca^{2+} and outside of cubane. It is suggested that these two ions provide the 'catalytic' surface to bind two water molecules and subsequently oxidize them.[1] One mechanism suggests that the substrate water is deprotonated and converted to an oxo, which would necessitate Mn_4 being converted to a high oxidation state prior to the formation of the O–O bond. It is thought then that the other three Mn ions transition to high valency states while the reactive oxo is electron deficient. As a result, it is a susceptible to nucleophile attack by oxygen in a second bound water molecule.[1] Alternatively, based on calculations from depth density function theory (DFT), it has been suggested that the deprotonated water molecule located on Mn-4 attacks an oxo-ligand of the Mn_3CaO_4-cubane after becoming an oxyl radical.[1]

Notwithstanding the astounding ability in biology to oxidize water and its significance in the biosphere, not all the details in the mechanism are fully known yet. The challenge is based on the difficulty of probing with conventional methods the complex where this occurs, and one of the main challenges has been to learn more about the water molecules that are involved.

4.8.1.3 Replacement of Electrons

As mentioned above, the absorption of photons results in charge separation and the flow of electrons from pigment molecules to the reaction center. The molecules oxidized in this process must be reduced in order for them to be able again to provide energized electrons. In the case of oxygenic photosynthesis, the source of external electrons is water. The electrons yielded from the four related charge-separation reactions are first moved to a tyrosine residue, which reduces P680, thereby restoring it.

4.8.2   Plastoquinol

Three agents in series subsequently transfer the electrons energized by photons captured within PSII before reaching PSI. The first of these transfer agents is a small molecule, plastoquinone (PQ), which is reduced to plastoquinol. PQ shuttles both protons and electrons. It accepts two protons from the stroma and two electrons from PSII and then transports the protons to the lumen and the electrons to the cytochrome b_6f protein complex. It can diffuse through the thylakoid membrane since it is lipid soluble.

Generally, quinones are a class of compounds found in bacteria, fungi, and plants as pigment molecules consisting of an oxidized aromatic ring with two carbonyl groups attached. PQ is the type of quinone that is part of the photosynthetic process, and it is found in high number in the lamallae.

4.8.3   Cytochrome b_6f Complex

The cytochrome b_6f complex transfers electrons from plastoquinol to PSI, which facilitates the movement of plastiquinol's two protons from the stroma to the lumen.

$$QH_2 + 2Pc(Cu^{2+}) + 2H^+(stroma) \rightarrow Q + 2Pc(Cu^{2+}) + 4H^+(lumen)$$
$$(4.9)$$

where QH_2 is plastiquinone and Pc is plastocyanain (see Section 4.8.4).

This process moves up to two-thirds of the protons to establish the pmf in photosynthesis.

4.8.4   Plastocyanin

Plastocyanin (Pc) is the third electron transfer agent moving electrons donated from cytochrome f of the cytochrome b_6f complex to $P700^+$ of PSI. Plastocyanin in most plants is most often 99 amino acids in length and contains copper, which is involved in the first step of electron transfer/reduction of plastocyanin by cytochrome b_6f:

$$Cu^{2+}Pc + e^- \rightarrow Cu^+Pc \qquad (4.10)$$

After reduction, plastocyanin dissociates from cytochrome f, diffuses in the lumen, and subsequently reduces PSI's $P700^+$ after binding to it.

4.8.5   PSI

PSI is another photosystem, a magnificent protein complex with 110 cofactors. It is also known as plastocyanin:ferredoxin oxidoreductase. PSI is the photosystem that provides strong reducing equivalents. With energy

harvested by PSI from the absorption of light, an electron is excited and then transferred from a chlorophyll molecule and passed through electron acceptors to lower and lower energy states. The energy created through those sequential losses is utilized to move protons into the lumen across the thylakoid membrane. In addition, the electron is used in the reduction of $NADP^+$ to NADPH (with the addition also of a proton). Some fundamental features of PSI are described below.

4.8.5.1 *Photosytem I Primary Donor (P700)*

P700 is a reaction center in PSI and is made up of two chlorophyll molecules (thought to be one chlorophyll a and one chlorophyll a′ molecule). P700 absorbs light best at 700 nm, can receive energy from the antenna complex, and can raise an electron to a higher energy level. Excited electrons in PSI are transferred to electron acceptors.

4.8.5.2 *Ferredoxin and Ferredoxin NADP$^+$ Reductase*

Ferrodoxin, like plastocyanin, is a small protein that acts as an electron transfer agent. The specific role of ferrodoxin is to transfer electrons to $NADP^+$ through the catalytic action of ferredoxin:$NADP^+$ oxidoreductase (FNR). Ferrodoxin contains iron and sulfur atoms in iron–sulfur clusters, which can accept or discharge electrons through the change of iron's oxidation state. The enzyme, FNR, reduces $NADP^+$ through the oxidation of ferredoxin:

$$2 \text{ reduced ferredoxin} + NADP^+ + H^+ \rightarrow 2 \text{ oxidized ferredoxin} + NADPH$$
$$(4.11)$$

The transfer of electrons from ferrodoxin to NADPH happens only in the presence of light, and ferredoxin-$NADP^+$ reductase is the final destination from which electrons are moved from PSI to NADPH.

FNR is soluble as a monomer in the stroma and can bind to the thylakoid membrane and exist as a dimer. This binding increases in acidic conditions, which occurs in the dark, thus providing a method to store and stabilize FNR in the dark.

4.8.6 **Comparison and Contrast of PSII and PSI**

Based on the structure of PSII and PSI, the two photosystems are thought to have evolved from a common ancestor, but considering that the efficiency of PSI is approximately 100% and that the maximum thermodynamic potential of PSII is about 70%, there are some significant differences (Figure 4.7). The similarities in their architecture include the same helical arrangement of the initial electron-transfer components. The N termini of two domains within PSI, PsaA and PsaB, are similar in sequence, structure, and pigment content

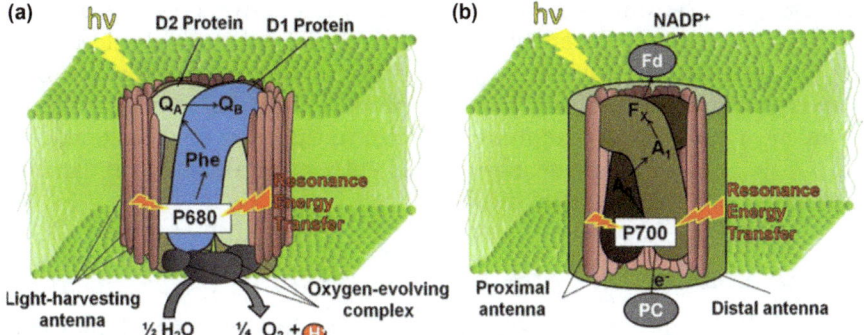

Figure 4.7   Structure of PSII and PSI. (a) The hole generated by photon absorption in PSII is used in the oxidation of water to produce oxygen. The high energy electron transfers to the Phe and QA sites of the complex and then to the QB site, where it is subsequently removed to enter the ETC in cytochrome b_6f. (b) Electrons that reach PSI are excited by photon absorption at the P700 site. The electron is subsequently extracted to the ferredoxin (Fd) site and then used to reduce $NADP^+$ to NADPH. Both PSII and PSI contain a surrounding antenna that enhances the absorption of light associated with a wider range of wavelengths throughout the solar spectrum.
(Reproduced from ref. 8 with permission from The Royal Society of Chemistry. © Royal Society of Chemistry 2013.)

to the subunits, CP43 and CP47, in PSII.[2] Further, the electron-transfer components are arranged in two branches in both photosystems, and the pairs of cofactors show a pseudo-twofold symmetry.[2]

The contrasting functions of the two photosystems, though, demands differences in the environment for the cofactors within each. For instance, in PSI there are about 50 chlorophylls only 20–30 Å away from the electron transport chain able to provide P700 with excitation energy.[2] On the other hand, there are just two light-harvesting chlorophylls in the central domain of PSII.[2] Further, there are no protective carotenoids near P680, perhaps due to its high oxidizing ability that could potentially oxidize nearby pigments.[2] With regard to the similarly structured twin branches, only one of the branches is active in PSII to reduce the risk of back transfer to P680 from quinone reduced just once.[2]

4.8.7   F-ATPase

F-ATPase is a form of ATPase located in the chloroplast thylakoid membrane. Like other ATPases, F-ATPase is an enzyme that produces ATP by joining ADP with P_i in an elegant, mechanical fashion (Figure 4.8). The pmf that is established between the two sides of the thylakoid membrane enables the flow of protons through F-ATPase, and the proton's passage results in the rotation of a portion of the enzyme. This internal movement results in the catalytic sites passing through conformational changes leading to the release of newly formed ATP.

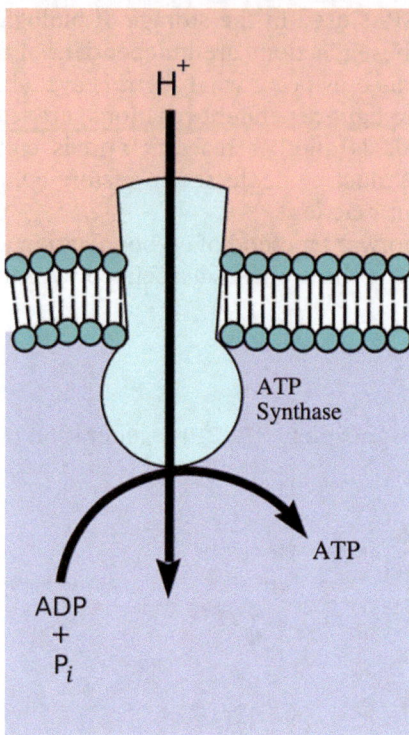

Figure 4.8 The movement of protons across a membrane in chloroplasts results in the generation of ATP from ADP and P_i by the action of F-ATPase. (Reproduced from http://en.wikipedia.org/wiki/ATP_synthase – image in public domain.)

4.9 Cyclic and Non-cyclic Electron Flow

The components described above perform the steps in the first of two stages in photosynthesis, which is called collectively the light-dependent reactions. The reactions are either cyclic or non-cyclic. In the cyclic form of the light-dependent reactions, only ATP is produced, and the electrons from PSI that would be used to reduce NADPH are instead transferred again to plastoquinol. This results in a cyclic flow. In the non-cyclic form, ATP and NADPH are both synthesized. To review for purposes of comparison, the transfer of electrons from PSI to FNR and finally to reduce $NADP^+$ results in non-cyclic electron flow.

Depending on the cell's physiological conditions, cyclic or non-cyclic photosynthesis occurs. The flexibility to perform the cyclic form allows for the production of an appropriate ratio of three molecules NADPH to two molecules ATP, which is needed in the fixation of carbon.

4.9.1 Light-independent Reactions

The light-independent reactions occur in the stroma of chloroplasts and yield glyceraldehyde 3-phosphate (G3P). The process is also known as the

Calvin cycle, and G3P is used in the storage of biological energy. Although the name suggests these reactions are independent of the presence of light, in fact, these reactions only occur when light is available. NADPH and ATP, which come from the light-dependent reactions, are reactants in the Calvin cycle. Carbon dioxide is another reactant and is collected from the air. A percentage of G3P must be dedicated to restore components involved in the light-independent reactions.

The most straightforward method of carbon fixation is the C3 method and is described below. The three phases associated with the light-independent reactions are:

(1) carbon fixation
(2) reduction
(3) ribulose 1,5-bisphosphate (RuBP) regeneration (Figure 4.9).

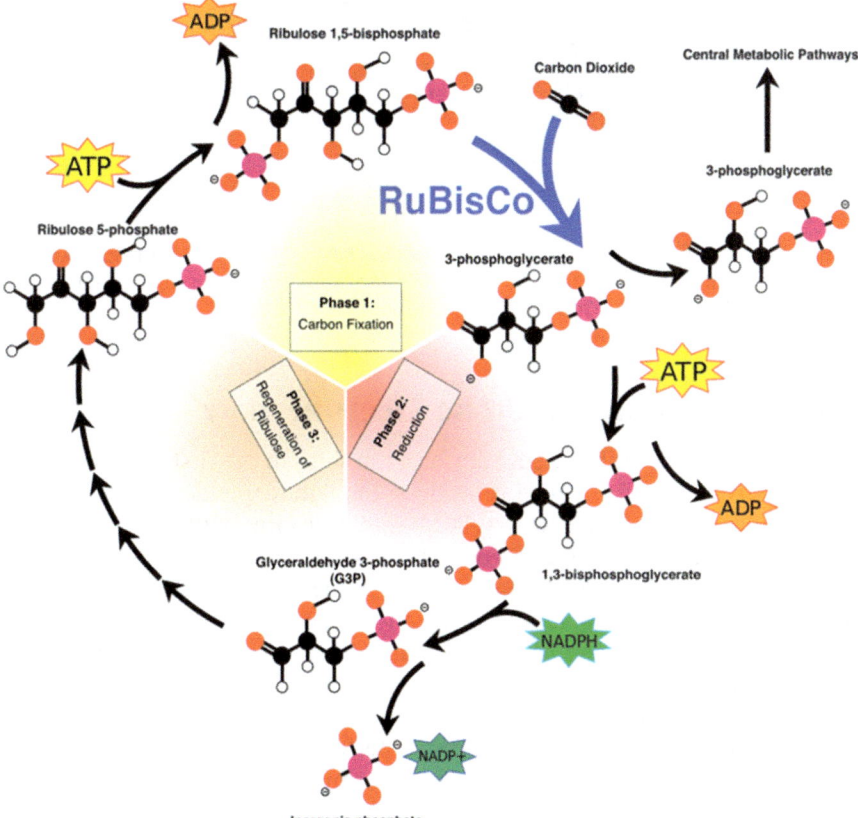

Figure 4.9 A depiction of the Calvin cycle showing the three phases: (1) carbon fixation, (2) reduction, and (3) regeneration of ribulose.
(Taken from http://en.wikipedia.org/wiki/Light-independent_reactions with permission provided by CC BY-SA 3.0 – http://creativecommons.org/licenses/by-sa/3.0/).

Plants in hot and dry environments have developed more elaborate processes, more well-suited to their specific needs for fixing carbon. A short description of these methods follows the summary of the C3 method.

4.9.1.1   Phase 1: Carbon Fixation

In the first phase of the Calvin cycle, CO_2 is fixed by the enzyme RuBisCO into 3-phosphoglycerate. This is achieved by RuBisCo catalyzing the carboxylation of RuBP which forms a six-carbon intermediate that immediately splits into two molecules of 3-phosphoglycerate (3-PGA).

$$RuBP + CO_2 \xrightarrow{\text{RuBisCO}} \text{unstable intermediate} \rightarrow 2\,3 - PGA \qquad (4.12)$$

4.9.1.2   Phase 2: Reduction Reactions

Next, phosphoglycerate kinase catalyzes the phosphorylation of 3-PGA to form 1,3-biphosphoglycerate (1,3BPGA, glycerate 1,3-bisphosphate). Another enzyme, glyceraldehyde-3-phosphate dehydrogenase, then catalyzes the reduction of 1,3BPGA by NADPH forming G3P.

$$2\,3 - PGA + 2ATP + 2NADPH \rightarrow 2G3P + 2ADP + P_i + NADP^+$$

4.9.1.3   Phase 3: Ribulose 1,5-bisphosphate (RuBP)

The last stage in the Calvin cycle results in regeneration of RuBP. This requires several additional enzymes, another molecule of ATP, and three more molecules of CO_2.

4.9.1.4   Summary of the Calvin Cycle

In summary, two molecules of G3P are produced in a complete turn of the Calvin cycle. But five out of six carbons in those two G3P are dedicated to regenerating RuBP. Thus, only one carbon is fixed per turn of the Calvin cycle, so three turns are needed to make a molecule of G3P for purposes beyond regeneration. Three turns require nine ATP, six NADPH, and three CO_2 molecules.

$$3CO_2 + 6NADPH + 5H_2O + 9ATP \rightarrow \text{glyceraldehyde 3-phosphate (G3P)}$$
$$+ 2H^+ + 6NADP^+ + 9ADP + 8P_i \qquad (4.13)$$

G3P is the starting point for the synthesis of other carbohydrates, such as starch and cellulose.

4.9.2   Photorespiration (C2 Cycle)

RuBisCo can bind oxygen in addition to carbon dioxide, and imperfect specificity provides the potential for oxygenase activity by RuBisCo. When

this occurs it is wasteful in the sense that sugars are not produced despite an input of energy. Oxygenase activity becomes more likely as the concentration of carbon dioxide lowers relative to oxygen. Since carbon dioxide is fixed and oxygen is evolved in chloroplasts, the concentration of oxygen can be higher than atmospheric (20%), and, contrastingly, the concentration of carbon dioxide can be lower than atmospheric (0.039%). The relative difference can increase when, for instance, the plant pores (stomata) are closed in C3 plants in which case CO_2 is not collected and the concentration of oxygen gas within the chloroplast increases. Also, higher temperatures can cause the relative concentration of CO_2 to decrease due to reduced solubility. Further, RuBisCo is less able to discriminate between O_2 and CO_2 at higher temperatures.

Oxygenase activity of RuBisCo involves the addition of oxygen gas to RuBP. This results in 3-phosphoglycerate (PGA) and 2-phosphoglycolate (2PG, or PG):

$$RuBP + O_2 \rightarrow PG + PGA + 2H^+ \tag{4.14}$$

PGA is normally produced in the carboxylation of RuBP and does re-enter the Calvin cycle when oxygenase activity occurs. But the formation of PG in this reaction results in a number of disadvantages. PG inhibits some enzymes that play a role in carbon fixation and is thought of as an 'inhibitor of photosynthesis'. Also, glycolate is formed by the quick metabolization of PG and is toxic, and salvaging it is expensive energetically. Carbon is lost, and ammonia is produced in the glycolate pathway.

The salvaging pathway, known as photorespiration and, alternatively, the C2 cycle, acts on the products of RuBisCo oxygenase activity. The simplified sum reaction in photorespiration is:

$$2 \text{ glycolate} + ATP \rightarrow 3\text{-phosphoglycerate} + \text{carbon dioxide} + ADP + NH_3 \tag{4.15}$$

Remarkably, there are variations on the light-independent reactions that satisfy the needs of plants located in arid conditions to maintain their moisture but also capture CO_2.

4.9.3 Two Additional Methods of Carbon Fixation

To maintain moisture and reduce photorespiration, plants have adapted methods, known commonly as carbon-concentrating mechanisms (CCMs), to increase the concentration of CO_2 near RuBisCo. This reduces the likelihood of O_2 reacting with RuPB and leading to the unfavorable products described above. In both C4 plants and CAM plants, the enzyme, phosphoenolpyruvate carboxylase (PEPC), is used to capture CO_2. Compared to RuBisCo, it is more selective for CO_2 and results in a faster reaction.

4.9.3.1 Crassulacean Acid Metabolism (CAM) Plants

In order to collect CO_2 from the surrounding air and release oxygen, plants open their pores, known as stomata. While the pores are open, water vapor is also released from plants (a process termed transpiration), and the fact that water is inevitably lost in the collection of CO_2 poses a potential conundrum to plants in hot, dry environments. This challenge is addressed by the Crassulacean acid metabolism (CAM) process. To thrive in arid conditions, CAM plants collect CO_2 only at night by storing it in the form of malic acid in their vacuoles. During the day, the leaves' stomata are shut, and the malate is moved to the chloroplasts and converted to CO_2 for use in photosynthesis.

4.9.3.2 C4 Carbon Fixation

Special structures and biochemistry within the leaves of C4 plants help prevent the level of photorespiration found in C3 plants. In the mesophyll cells, CO_2 is added to phosphoenolpyruvate (PEP), a three-carbon molecule, to create a four-carbon molecule, oxaloacetic acid, by PEP carboxylase. Also known as malate, this molecule is then translocated to the bundle sheath cells to release CO_2 to be then fixed by RuBisCo and utilized to generate carbohydrates through the C3 pathway. This is possible because chloroplasts exist in C4 plants in both mesophyll cells and bundle sheath cells. By separating the sequestration of CO_2 from its fixation into three-carbon 3-phosphoglyceric acids, the concentration of CO_2 is boosted and the need for photorespiration reduced.

4.10 Actual Efficiency of Photosynthesis

In considering the efficiency of photoautotroph's conversion of light energy to chemical energy, it is informative to consider the actual efficiency in context of the nominal efficiency and the theoretical maximum. As calculated above by considering the energy necessary to convert CO_2 to glucose and the energy of 'red' photons, with a minimum of eight photons of light assumed to fix a molecule of CO_2, the nominal efficiency in converting those photons is about 30%. Since only about 45% of the light in sunlight is in the range that is available for photosynthesis, the maximum theoretical efficiency is about 11%. But the overall photosynthetic efficiency varies from about 0.1% to 2% of the total solar radiation. This is due to inefficiencies associated with not absorbing all the light available and with all energy harvested not funneled to create biomass and variables such as availability of nutrients and water.[1] The following calculations provided by Hall and Rao demonstrate how the overall efficiency is quickly reduced by a series of inefficiencies:[6]

(1) 53% of solar light is in the 400–700 nm range
(2) 30% is lost to incomplete absorption
(3) 24% is lost to wavelength-mismatch degradation

(4) 32% efficient conversion of ATP and NADPH to D-glucose
(5) of the 9% that remains and which is collected as sugar, 35–40% is recycled or consumed by the leaf in the dark and photo-respiration
(6) this leaves 5.4% net efficiency.

As we will see below, others have calculated the theoretical maximum of converting light to chemical energy to be merely 4.5%.[1]

4.11   Recovering from Damage from Light

On the one hand, photoautotrophs have had to adapt to environments with very low light and compete for the relatively few photons that reach their chloroplasts. On the other hand, the intensity spectrum changes by orders of magnitude during the day. Photoautotrophs in these conditions need to handle the higher level of energies from solar irradiance. Under bright sunlight, light is harvested and produces excitations faster than the rate at which reaction centers are ready to accept them. If not properly handled, the excess excitations are capable of forming triplet states and then singlet oxygen that harms biological molecules.[7] Amazingly, if the self-repair processes were not in operation, plants would produce at least twenty times less and would cease functioning after only a few minutes of exposure to intense illumination.[8] The toxic oxygen radicals cleave and denature proteins, bleach light-harvesting antennas, and damage machinery for gene expression.[8]

There are a variety of mechanisms that help prevent damage that derives from the singlets and triplets. Radical scavengers, such as xanthophylls, carotenes, and some enzymes, help to deactivate oxygen radicals.[8] Also, as mentioned, carotenoids dissipate excess energy to avoid damage. Triplets that are formed on the chlorophyll are quenched by carotenoids.[7] Further, carotenoids participate in the process of non-photochemical quenching under conditions of high light intensity.[7] The light-harvesting antennae operate in this case in a state during which most of the excitation energy then dissipates as heat.[7]

The mechanism of non-photochemical quenching operates when the machinery for the transfer of electrons in photosynthesis is saturated and the chance that charge-separated states recombines increases.[7] Damage can result from the recombination events due to the formation of triplet states.[7] Intense light can cause a build-up of a trans-membrane proton gradient in just tens of seconds, and this results in switching on non-photochemical quenching.[7] Two mechanisms that both might operate in parallel to enable this switch between light harvesting and quenching are based on conformational changes that affect the interaction between the carotenoid and the chlorophyll.[7]

4.12   Biomass

Oil is the product of millions of years of biological energy being converted by intense heat and pressure (Figure 4.10). A critical question in terms of

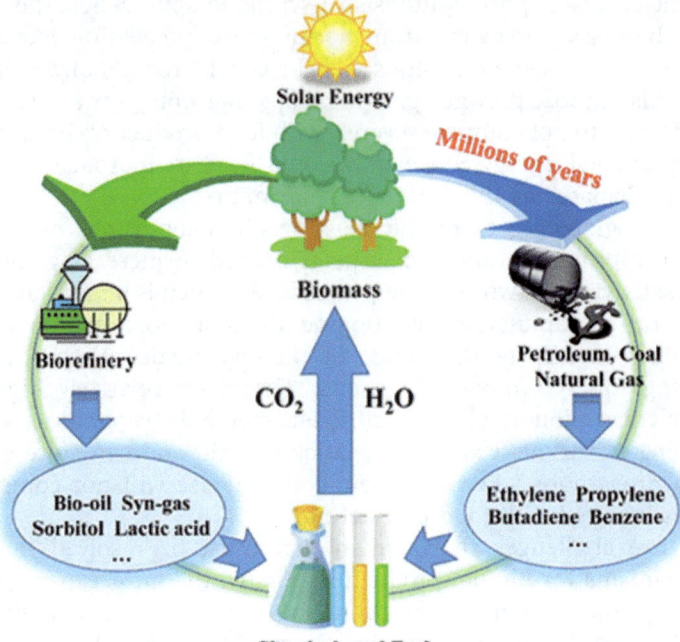

Figure 4.10 Cycle of biomass utilization towards chemical and fuel production and regeneration of biomass by photosynthesis with the by-products. The ingredients of CO_2, H_2O, and sunlight are combined towards the generation of biomass, which over millions of years at high temperature and pressure is converted to petroleum, coal, and natural gas. This ancient biomass is converted to petrochemicals, including primarily olefins (such as ethylene, propylene, and butadiene) and aromatics (benzene, for instance). Olefins and aromatics are used as reagents towards the production of chemicals, and the hydrocarbons found in crude oil function as fuels. On the other hand, biorefineries acting on biomass bypass the millions of years needed to convert it to useful materials towards generating chemicals and fuels. Over time, by-products of CO_2 and H_2O are generated through the use of chemicals and fuels, which are incorporated in new biomass.
(Reproduced from ref. 13 with permission from The Royal Society of Chemistry. © Royal Society of Chemistry 2013.)

the world's energetic needs is how soon this resource will be depleted. Considering the total amount of energy that photoautotrophs utilize from the Sun every year and that the Earth receives from the Sun enough energy in 1 hour to fulfill humanity's energetic needs for the period of a year, it is little wonder that the prospect of tapping biomass for electricity would be turned to in hopes of reducing reliance on oil deposits for energy needs.

Yet, some scientists see the conversion of plant matter to electricity as prohibitively inefficient. The world demand for energy is expected to reach 20 TW by 2030, which would require three times the amount of cultivatable land that is now used for agriculture globally.[1] In addition to the already

lower efficiencies of photosynthesis to synthetic approaches, the energetic costs of biomass harvesting, transporting to a processing location, and converting into a usable form must also be considered. Finally, using fertile land for this purpose prevents growing crops that would otherwise help feed humanity, and to not compete severely with food production by farming, the efficiency of solar energy conversion would need to approach 4.5%, which has been calculated as the theoretical maximum.[1]

Genetic engineering to modify existing photoautotrophs offers the potential to improve a number of aspects related to increasing their utility towards satisfying the world's energy needs. Research is already underway to increase the accumulation of compounds that store energy in photoautotrophs.[9] Modifying the genes of photosynthetic microorganisms is compelling for a number of reasons. They have relatively high photosynthetic conversion efficiencies, different metabolic capabilities providing for thriving in various ecosystems, quick growth, and can store or secrete useful hydrocarbons.[9] Further, their potential use will not compete with resources involved with food production.[9]

There are challenges that remain that must be resolved in utilizing microorganisms for biofuel production. Firstly, the most expensive part of processing downstream is thought to be the harvesting and dewatering steps.[9] One possibility, given the slow speed of settling and flocculating, the higher cost of centrifugation and filtration, and the very tough outer cells wall of many microalgal species, is to engineer the microalgal cells to directly secrete fuels or feedstocks.[9]

It has been predicted that the overall efficiency of solar energy conversion will not typically be above 1% and often will be significantly lower.[1] While the overall efficiency may be quite low in terms of biomass production, the earlier photochemical and chemical reactions in photosynthesis are much more efficient. With this in mind, a more realistic approach to effectively utilizing photosynthesis is perhaps to mimic processes and components involved in water splitting with a synthetic, molecular-based technology.

4.13   Biomimicry

Much of the work to achieve artificial photosynthesis has been towards constructing systems of molecules that donate and accept electrons towards charge separation *via* light. With this has come much analysis on how electrons are transferred efficiently based on physical and chemical properties, such as the distance between donors and acceptors and their respective orientation, the free energy of the reaction, and interactions of electrons.[1] This, in turn, has yielded information guiding how to make these systems and the benefits of multi-step charge separation.[1]

Figure 4.11 serves as summary of many of the processes discussed in this chapter. The purpose here is to show how the various components fit in context and for consideration of what aspects might be mimicked. Examples of elements currently under investigation for biomimetic systems are

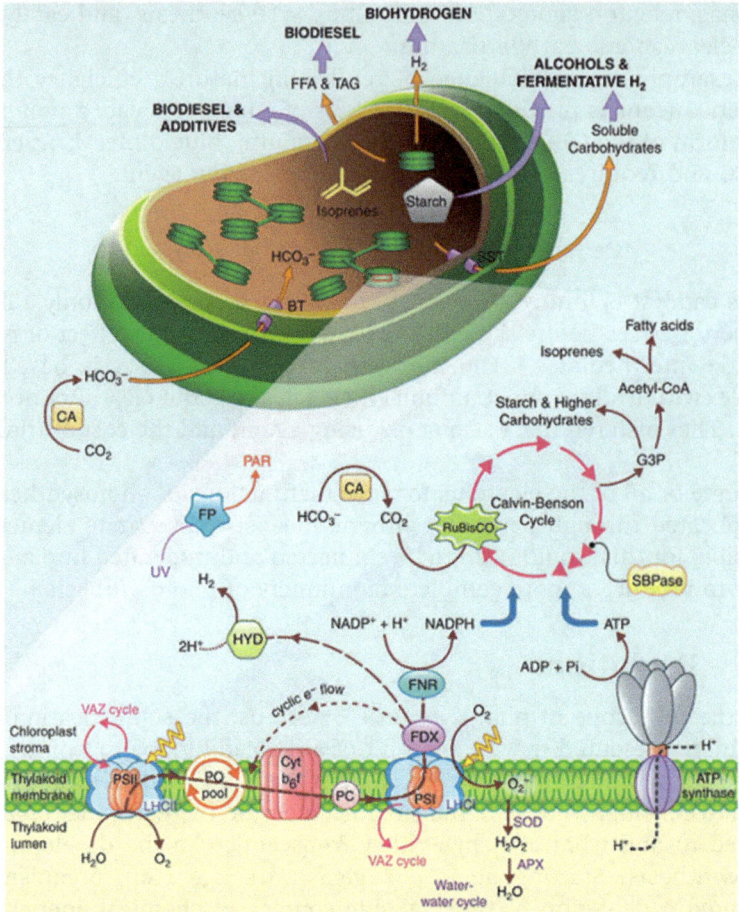

Figure 4.11 A chloroplast of a green alga with the ETC along the thylakoid membrane and metabolites towards bioenergy production. The two yellow, squiggly arrows represent the absorption of photonic energy by PSII and PSI, which is used to excite electrons. The flow of ATP and NADPH to be used in the Calvin–Benson cycle is shown, and in this cycle G3P is synthesized through the fixation of CO_2 for creation of units of longer-term energy storage. Also shown is the flow of electrons in the cyclic form of the light-dependent reactions with the dotted arrow labeled 'cyclic e- flow'. The various biofuels are listed that can be generated through photosynthesis and potentially used to meet humanity's increasing needs for energy. (APX: ascorbate peroxidase, BT: bicarbonate transporter, CA: carbonic anhydrase, Cyt b_6f: cytochrome b_6f, FDX: ferredoxin, FFA: free fatty acids, FNR: ferredoxin–$NADP^+$ reductase, FP: fluorescent protein, G3P: glyceraldehyde 3-phosphate, HCO_3: bicarbonate, HYD: hydrogenase, LHC: light-harvesting complex, PAR: photosynthetically active radiation, PC: plastocyanin, PS: photosystem, PQ pool: plastoquinone pool, SBPase: sedoheptulose-1,7-bisphosphatase, SOD: superoxide dismutase, SST: soluble sugar transporter, TAG: triacylglycerol, UV: ultraviolet light, VAZ: xanthophyll cycle.)
(Reproduced from ref. 14 with permission from The Royal Society of Chemistry. © Royal Society of Chemistry 2014. Original by Professor W. W. Cramer and modified by S. Renger and D. Gray.)

antennas, reaction centers, proton pumps, ATP synthesis, and catalysts for water cleavage and carbon fixation.

An example of the challenge of replicating nature's efficiency through synthetic means is provided by the results of efforts to create a non-protein based form of the OEC near PSII. The ruthenium 'blue dimer' is as efficient as OEC and requires activation by a strong oxidizing agent:

$$4Ce(\text{IV}) + 2H_2O \xrightarrow{\substack{\text{Ruthenium} \\ \text{'bluedimer'}}} 4Ce(\text{III}) + O_2 + 4H^+ \qquad (4.12)$$

This catalyst is limited, though, in catalyzing reactions for only a limited number of cycles before the efficiencies decrease from the effect of reactive catalytic intermediates.[8] The next generation of catalysts is based on a complex with iridium and maintain efficiencies of about 66% after over 2000 cycles. They also require a strong oxidizing agent, and the cost of iridium is high.[8]

If some or all of the elements for specific functions of photosynthesis can be replicated through partially synthetic means, the separate elements responsible for this would need to be connected and integrated into a unified whole to produce a more complete biomimicry of photosynthesis.

4.14   Evolution

With the full scope of photosynthesis before us, the sophistication of the machinery presented may instill a bit of awe, leading one to wonder about the means by which it came to be. Within the framework of evolution, it is useful to consider what were the circumstances on the Earth that organisms adapted to and what challenges they transcended in the development of photosynthesis. Starting about 3.7 giga years ago (Ga), organisms encountered a depletion of the available sources of chemical energy at the time, including geothermal, reducing compounds, and hydrogren.[10]

Initially, a basic form of respiration probably developed involving cytochromes and iron–sulfur centers, and anoxygenic bacteria first engaged in these types of processes, potentially using chlorophylls and bacteriochlorophylls. It is possible that PSI evolved 3.5 Ga with a homodimeric reaction center first resembling that found in currently in green bacteria.[2] Eventually, a heterodimeric reaction center evolved as a more complex reaction center and proteins to harvest light were developed. By about 2.2–2.8 Ga, cyanobacteria could split water – thus providing hydrogen, which is necessary for the reduction of carbon dioxide.[10] Finally, the three dominant domains of multicellular organisms developed: animals, plants, and fungi.

4.15   Conclusion

Considering that the continuation of our existence depends on plants harnessing the Sun's energy, it would be difficult to overstate the significance of photosynthesis. Surely its value is evident after scores of years of use of the

massive stores of biological energy underground formed over eons. That worthiness and continued relevance is shown in humanity's increasingly urgent need to identify an alternative source of energy before the this energy is fully depleted. Within the past decade, about 85% of energy use was derived from burning fossil fuels.[1]

It is fascinating to think that all the elements involved in photosynthesis required about a billion years to develop. Perhaps even more intriguing is our attempt to mimic this functionality in the coming years to meet new needs of energy brought about by the evolution of technological advance. It is certainly worth our efforts to develop a sustainable approach to energy production through our own ingenuity and nature's example. Hopefully, the remaining chapters in this volume will guide and inspire those on the path to develop effective methods of synthetic photosynthesis.

References

1. J. Barber, *Chem. Soc. Rev.*, 2008, **38**, 185.
2. N. Nelson and A. Ben-Shem, *Nat. Rev. Mol. Cell Biol.*, 2004, **5**, 971.
3. Encyclopædia Britannica Online, s. v. 'chloroplast,' accessed April 03, 2015, http://academic.eb.com/EBchecked/topic/113761/chloroplast.
4. A. Guskov *et al.*, *Nat. Struct. Mol. Biol.*, 2009, **16**, 334.
5. V. Shubin, N. Karapetyan and A. Krasnovsky, *Photosynth. Res.*, 1986, **9**, 3.
6. D. Hall and K. Rao, *Photosynthesis*. Cambridge University Press, Cambridge, UK, 1999.
7. G. Scholes, G. Fleming, A. Olaya-Castro and R. Grondelle, *Nat. Chem.*, 2011, **3**, 763.
8. A. Boghossian, M.-H. Ham, J. Choi and M. Strano, *Energy Environ. Sci.*, 2011, **4**, 3834.
9. R. Radakovits, R. E. Jinkerson, A. Darzins and M. C. Posewitz, *Eukaryotic Cell*, 2010, **9**, 486.
10. (a) G. Renger, *Primary Processes of Photosynthesis, Part 1: Principles and Apparatus*, ed. G. Renger, RSC Publishing, Cambridge, UK, 2007; (b) *Angew. Chem., Int. Ed.*, 2008, 47, 6944–6945.
11. P. Williams and L. Laurens, *Energy Environ. Sci.*, 2010, **3**, 554.
12. S. Milikisiyants *et al.*, *Energy Environ. Sci.*, 2012, **5**, 7747.
13. X. Liu *et al.*, *RSC Adv.*, 2014, DOI: 10.1039/C4RA06466F.
14. Y. Shen, *RSC Adv.*, 2014, **4**, 49672–49722.

CHAPTER 5

Bioinspired Photocatalytic Nanomaterials

MD NASIM HYDER* AND ZAKIA SULTANA

Massachusetts Institute of Technology, 77 Massachusetts Avenue, Cambridge, MA 02139, USA
*Email: nasim@mit.edu

5.1 Introduction

The demands on our energy system are growing rapidly due to the limited supply of fossil fuels. In recent years, unprecedented motivation and research efforts are being made for the emergence of clean, renewable, and green energy sources and technologies. Solar energy is one of the few alternative renewable energy sources that could meet the increased future demand for energy, as the yearly average power intensity of solar radiation is 1367 W m^{-2}.[1] The total solar energy irradiating the Earth is 120 000 TW per year, of which less than 0.02% is sufficient to entirely replace the total present energy demands from the fossil fuels and nuclear power.[2] However, for the solar energy revolution, it is essential to convert solar energy efficiently in a way that can be economically competitive with the energy produced from fossil fuels, nuclear power, or other renewable energy resources.

Among various renewable energy options, hydrogen – as a synthetic combustible fuel – appears to be a viable option for energy sources for stationary, portable, and automotive applications. For example, hydrogen can be used to power vehicles or heating using internal combustion engines with little modification to the current technology due to its compatibility with other fuel cell systems. It can be used within the entire energy sector as a fuel

RSC Green Chemistry No. 42
Green Photo-active Nanomaterials: Sustainable Energy and Environmental Remediation
Edited by Nurxat Nuraje, Ramazan Asmatulu and Guido Mul
© The Royal Society of Chemistry 2016
Published by the Royal Society of Chemistry, www.rsc.org

if hydrogen gas can be produced economically in a sustainable manner. However, one of the critical challenges appears to be there is no natural pure hydrogen resource on the Earth even though it is the most abundant element in the universe; hydrogen is always chemically bonded to other elements.

As hydrogen is an efficient energy carrier, it needs be produced from other energy sources that are abundant on the Earth, *e.g.*, from the electrolysis of water. In addition, hydrogen is environmentally clean when produced from solar energy – by mimicking natural photosynthesis. Since the last oil crisis, extensive research efforts have been made worldwide to developed functioning artificial photosynthetic technology; however, synthetic mimics are often too complex when incorporating both the antenna system and reaction center of its natural counterpart.

Plants use light energy to split water into oxygen and a form of hydrogen in an enzyme complex called photosystem II; while this hydrogen powers all life on earth, the other product of photosynthesis is oxygen, which is critical to sustain life. Hydrogen can be produced by various bacterial methods such as direct biophotolysis, indirect biophotolysis, photo-fermentation, dark fermentation, and the water-gas shift reaction of photoheterotrophic bacteria. Cyanobacteria (commonly known as blue-green algae) and green algae are both capable of converting solar energy to hydrogen gas.[3] The process is catalyzed by hydrogenases: proteins that can consume or produce hydrogen, or both. However, challenges remain to overcome the limitation of the oxygen sensitivity of the hydrogen-evolving enzymes in order to increase both the efficiency and gas purity. An alternative way to produce hydrogen is *via* combining efficient silicon solar cells with electrolysis, which is prone to generate waste heat from electrical energy. In recent years, researchers around the world have focused on developing photoelectrochemical cells (PEC) where the electrode materials are selected such that they also act as a light-harvesting component – hydrogen production occurs directly in the cell to avoid any energy loss.[4,5]

Among various nanoscale assemblies, biological templates offer environmentally friendly synthesis and organization of hybrid materials at the nanoscale where efficiency can be improved by increasing the probability of the energy transfer groups being precisely positioned.[6–8] New properties can be obtained by utilizing functional materials in the nanometer scale that arise due to the small dimensions and large surface areas, short charge carrier diffusion lengths, and low reflectivity.[9] Rationally designed nanostructures, including metals, alloys, nitrides, carbides, phosphides, oxides, and organic–inorganic hybrids, can be employed to achieve a photoconversion system with high efficiency and ultra-stability.[10,11] Bioinspired design of nanostructures, a low energy intensive technique, employs biological templates of desired morphology for direct organization, nucleation, or growth of advanced materials into a functional matrix.[12] It offers the opportunity to design novel nanomaterials *via* genetic engineering to design hierarchical nanostructures. For example, the M13 virus has been studied

extensively and can be genetically engineered to organize nanoscale organic/inorganic materials into nanowire (NW) or nanotube (NT) networks.[13,14] Similarly, genetically modified tobacco mosaic virus (TMV 1-cys) has also been used as a template to develop nanoscale photocatalytically active metal oxides for water splitting.[15] Based on biological principles and peptide engineering, functional materials such as metals, perovskite, carbon nanotubes, conducting polymers, and semiconductors can be incorporated in nanostructured architecture for the precise placement of catalysts with biotemplated design and assembly for efficient photoelectrochemical devices.

This chapter will cover some of the current research efforts towards biomimetic or biological templates to design nanostructured electrodes decorated with catalysts and artificial structures mimicking biological processes that can be introduced into solar water-splitting production schemes.

5.2   Mechanism of Photoelectrochemical Water Splitting

In photosynthesis, the key reaction is where electrons are extracted from water using energy from the Sun. The electromotive force (E_o) required for the water dissociation reaction is 1.23 V (at pH $= 0$ and standard temperature and pressure, STP) which is defined as

$$E_o = -\frac{\Delta G^o}{nF}$$

where n is the number of electrons transferred and ΔG^o, the Gibbs free energy, is defined as $\Delta G^o = \Delta H^o - T\Delta S^o$ and can be calculated from the chemical reaction:

$$H_2O \text{ (liquid)} \rightarrow H_2 \text{ (gaseous)} + \frac{1}{2}O_2 \text{ (gaseous)}$$

Among various challenges, the catalysts must also absorb radiant light with enough photon energies for the so-called overpotentials beside the surface stability of the light absorbing material and the stability of the photocatalysts.[16] In particular, the overpotential at the anode side, where four electrons should be transferred *via* the electrode–electrolyte interface to oxidize two water molecules to obtain one O_2 molecule, is typically rather large, reaching values between 0.4 and 0.6 V depending on the type of catalyst.[17] Because these overpotentials are an inherent limiting parameter of the process, the overall efficiency is affected by the efficiency of the components of the cell.

5.3   Materials Design and Synthesis

PEC cells use photoactive electrodes immersed in an aqueous electrolyte, illuminated by sunlight, and have the ability to split water into hydrogen and

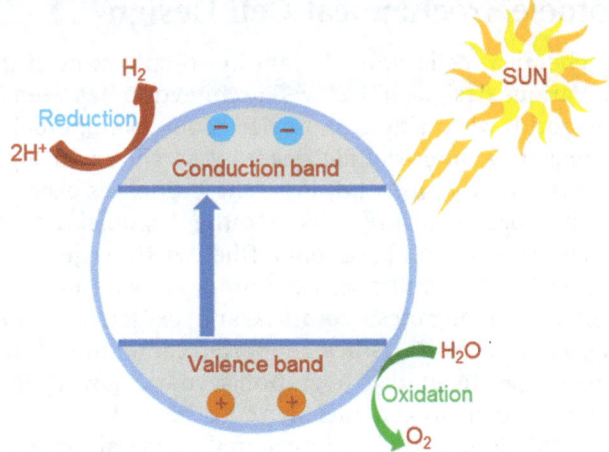

Figure 5.1 Schematic representation of water-splitting reaction on a photocatalyst.

oxygen (Figure 5.1). In the water electrolysis case, the cations are hydrogen ions and the anions are the oxygen atoms and the process depends on the supply of electricity. This technology offers huge potential but materials challenges still remains in term of their cost and lifespan. For example, alkaline electrolyzers can have a hydrogen output of above 99% purity, but a short lifespan due to the highly corrosive electrolyte, especially at high temperatures.

On the other hand, a proton-exchange membrane (PEM) electrolyzer that operates the 'in reverse' principle of a PEM fuel cell can produce 99.999% pure hydrogen from highly pure water using expensive noble metal catalysts (Pt, Ir, Ru). Most of the current photovoltaic and photoelectrochemical technologies employ a planar junction design, requiring the use of expensive, high-purity semiconducting materials for efficient and stable operation. With artificial photosynthesis, these shortcomings can be circumvented by mimicking plant photosynthesis processes to only utilize sunlight and water as the inputs and so will produce hydrogen and oxygen as the outputs. Photochemical splitting of water requires both a photosensitizer (to capture the photon that provides the energy to split the water molecule) and a catalyst. For efficient movement of electrons and holes, the photosensitizer and catalyst need to be correctly positioned with respect to each other. With the recent developments in nanotechnology for materials design and assembly, it is possible to achieve the precise positioning of the sensitizer and catalyst, which were previously challenging and constrained. This is possible due to heterogeneous covalent linkage, surface immobilization, or amorphous entrapment of materials. Owing to their small dimensions, photo-generated carriers in nanocrystals are always created near a surface, where water conversion takes place.

5.4   Photoelectrochemical Cell Design

Photoelectrochemical cells have the ability to split water into hydrogen and oxygen. Various designs of PEC cell architecture has been investigated around the world with single-photoelectrode, two-photoelectrode and hybrid-photoelectrode-photovoltaic systems. As shown in Figure 5.2, the cell assembly is rather simple; however, the properties expected from the materials of hydrogen-evolving cells is rather complex and demanding. The cell encloses two compartments filled with water but separated from each other by a membrane. On the anode side of the membrane, water is split by a manganese complex and oxygen is evolved. On the cathode side, a cocatalyst (*i.e.*, silica-supported cadmium sulfide, $CdS\text{-}SiO_2$) evolves hydrogen gas. In an ideal operation scenario, it will generate oxygen gas at one end and hydrogen fuel in the other cell. The hydrogen can be used immediately in a fuel cell or stored in metal organic framework (MOF) materials or metal hydrides.[18] A modular cell design approach would allow the system to be self-sufficient, requiring only water to be replenished.[19]

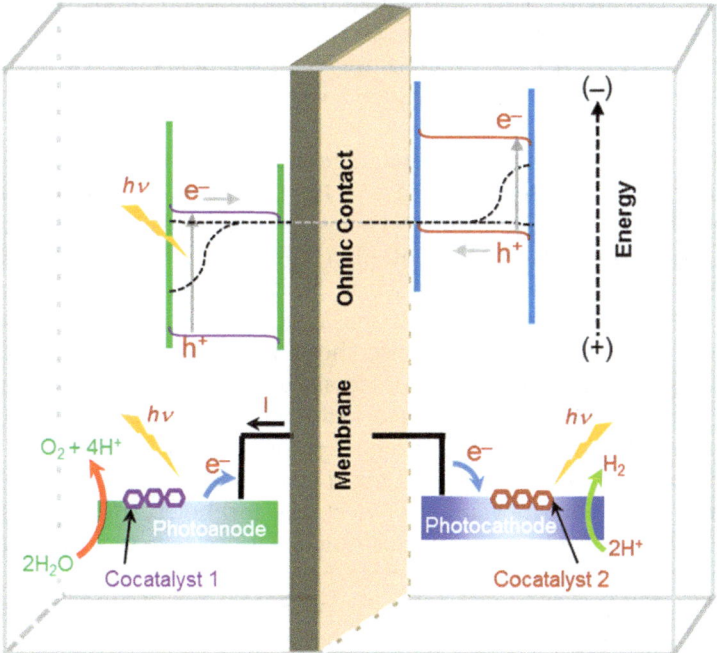

Figure 5.2   Schematic of a photoelectrochemical cell arrangement for a complete water-splitting cell assembly with photoanode, photocathode, and a membrane.
(Adapted from ref. 19 with permission from The Royal Society of Chemistry.)

In most PEC cells, four main steps occur in the water-splitting process:

(1) generation of an electronic charge at the surface of the photoanode due to incident solar radiation, generating electron–hole pairs
(2) oxidation of water by the holes at the photoanode, generating oxygen gas
(3) transport of the protons and electrons from the photoanode to the cathode by electrolyte and external circuit, respectively
(4) reduction of protons at the cathode by electrons to generate hydrogen molecules.

It is crucial to select suitable photoanode materials (n-type semi-conductor) with the existence of a 'perfect' bandgap (equivalent to 1.23 eV, the minimum theoretical electromotive force needed to split a water molecule) and the ability to resist recombination of charge carriers. In contrast to the process of photosynthesis, in which non-noble metal catalysts (Mn- and Fe-Ni-clusters) convert CO_2 and water into O_2 and hydrocarbons, current artificial PEC cells (combining a solar cell with an electrolyzer) employ platinum and ruthenium oxide in the process of water splitting.[20]

For example, a schematic design of a standalone photoelectrochemical solar water-splitting system is shown in Figure 5.3 (from the Center for Chemical Innovation, CCI, Solar Fuels).[17,21] It consists of an artificial photosynthetic system that will only utilize sunlight and water as the inputs and will produce hydrogen and oxygen as the outputs. The cell consists of three principal components: the photoanode, the photocathode, and the product-separating but ion-conducting membrane. The cell design consists of two separate, photosensitive semiconductor/liquid junctions that will collectively generate the 1.7–1.9 V at open circuit necessary to support both the oxidation

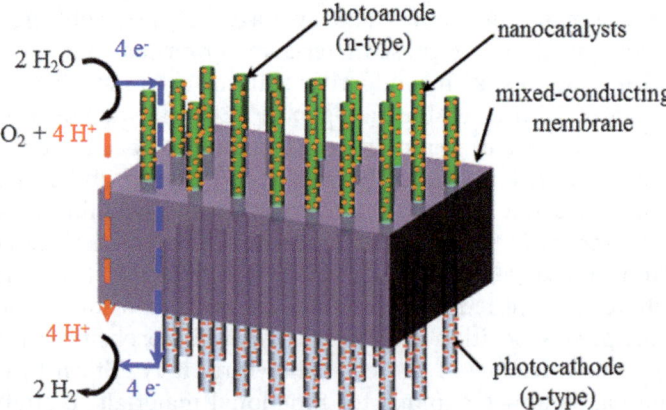

Figure 5.3 Schematic of a photoelectrochemical cell arrangement for a complete water-splitting cell design.
(Adapted from ref. 21 with permission from The Royal Society of Chemistry.)

of H_2O (or OH^-) and the reduction of H^+ (or H_2O). To drive the oxidation or reduction reactions at low overpotentials, heterogeneous multi-electron transfer catalysts are embedded in the rod-like semiconductor photoanode and photocathode partially embedded in the separator membrane. The low-cost high aspect ratio electrodes allow orthogonalization of light absorption for efficient charge carrier collection and lower the flux of charge carriers over the rod array surface. This cell design allows lowering the photocurrent density at the solid/liquid junction to maximize the activity of catalysts. The cell also requires a polymeric membrane with electron- and ion-conducting properties that also acts as a separator to prevent the mixing of the gases.

The overall efficiency of photoelectrolysis is highly dependent on the precise semiconductor materials design and development for water splitting into H_2 and O_2 using sunlight. The design could encompass either semiconductor liquid junctions (SCLJ) or photovoltaic cells (PV) or a combination of the two (PV/SCLJ) for optimum performance.[22–24] Therefore, it is necessary to design and develop novel electrode materials to enhance the overall energy efficiency of the water photoelectrolysis assembly. The design and synthesis of complex molecules, macromolecules, and nanoparticles are keys for the development of efficient photoelectrochemical systems. However, one of the pressing issues is the assembly of these nanoscale building blocks into functional assemblies for functional PEC systems.

5.4.1   Design of Photocatalytic Electrodes

For the photoanode, titanium dioxide (TiO_2) is the most widely investigated material in PEC cells since Fujishima and Honda tested it in 1970.[25] It has generated widespread interest in the solar photoelectrolysis of water due to the applicability of TiO_2 in the paint/pigment industry. Since then, researchers has investigated tungsten trioxide (WO_3) and hematite (Fe_2O_3); however, these materials require a bias voltage for water splitting since the cell photovoltage applied to these materials is lower than 1.23 eV.[26] Beyond metal oxides materials, semiconducting gallium arsenide (GaAs) has been also investigated with favorable bandgap in PEC cells, demonstrating high efficiency.[27,28] In recent years, indium, niobium, or tantalum particles exhibit fast reaction kinetics for hydrogen and oxygen under visible light irradiation.[29] In a photochemical cell, the key in designing ideal photosensitizer materials is to capture the 'photon' that provides the energy for the 'electron' to escape that is balanced by the 'electron' from the catalyst and the 'hole' transferred through the catalyst that oxidizes water. Additionally, the precise position and orientation of photosensitizer and catalyst determines the efficiency of the cell. Therefore, it is critical to develop a suitable nanoassembly technique for functional materials to attain the desired structure and properties of the device. Mesoporous semiconducting nanostructures with a high active surface area are beneficial to maximize catalytic or photo-active sites and loading of functional materials. Additionally, continuous pathways within the semiconducting structures are

critical to transport the photo-excited carriers to the current collectors efficiently.[14] Therefore, optimizing semiconducting architectures for practical photoelectrochemical applications requires a high degree of control over the spatial organization of nanomaterials within the mesoporous structures.

Nature exhibits a paradigm for designing a solar-driven energy system efficiently, which is also thermodynamically viable, through biotemplated functional assemblies for energy carriers and energy storage in a dynamic network. Inspiration from natural photosynthesis can provide the sole starting point, but it might not necessarily be the ideal system from an operational perspective. Functional assemblies with precise decoration of photocatalysts can be developed through biotemplated inspired self-assembly and organization. Among various bottom-up nanofabrications, biotemplating offers specific molecular recognition patterns that can be manipulated through genetic control – this results in ordering of nanoscale materials with exotic physicochemical properties leading to novel engineering systems.[30] Biotemplating can either replicate the morphology or guide the assembly of nanomaterials by rendering unprecedented molecular control over nucleation, growth, and stabilization of inorganic materials similar to biomineralization (Figure 5.4).

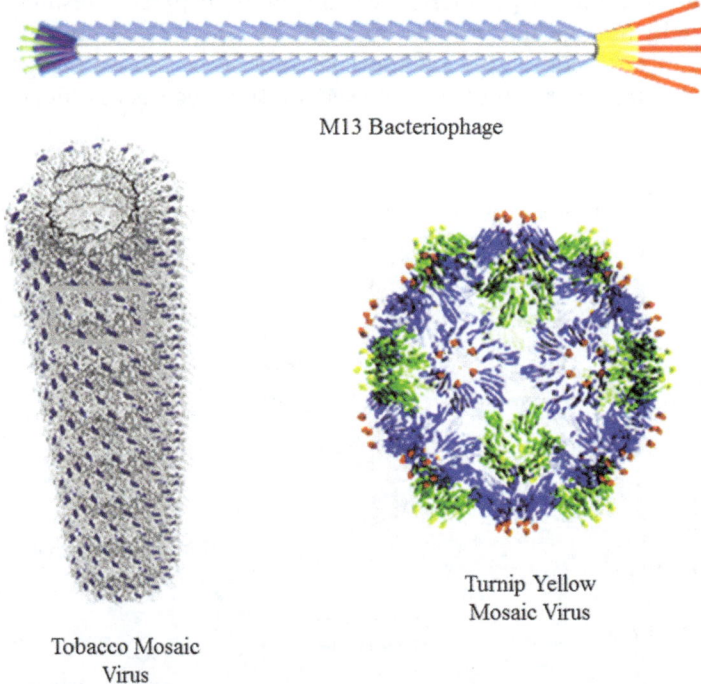

M13 Bacteriophage

Turnip Yellow
Mosaic Virus

Tobacco Mosaic
Virus

Figure 5.4 Representative virions used as biotemplates for the assembly of functional nanostructures.
(Adapted from ref. 32 with permission from The Royal Society of Chemistry.)

The process is governed by covalent/ionic interactions and molecular recognition along with the inherent ability of biological molecules to self-assemble at multiple length scales. Similarly, the replicating process results in either a negative, positive, or exact copy of the template of different biological species: virus, bacteria, DNA, proteins, spider silk, diatoms, insect wings, and textiles.[31–33] The templated nanostructure can develop different features. Examples are: nanopores from diatoms, nanotubes/nanowires from virus, or hierarchical architecture from insect wings. The combination of biotemplated nanomaterials with biological self-assembly enables an unprecedented opportunity to design and create new hierarchical ordered structures and functional assemblies.

For example, M13 bacteriophage, shown in Figure 5.5, is a virus that infects bacteria, and is made up of single-stranded DNA encapsulated by approximately 2700 copies of the major coat proteins (P8) and four minor coat proteins including P3 located on one end of the phage.[13] Also, the surface of the phage can be modified both genetically and chemically to introduce specific peptides. The filamentous phage ($\sim 880\,nm$ long and $\sim 6.8\,nm$ wide), composed of relatively few proteins and genes, can be used as a template for producing functional nanomaterials upon which the components, the catalyst, and photosensitizer can be assembled.[34] The highly anisotropic shape, along with the monodispersity, results from its high-fidelity biological reproduction of M13 in suspension. In step 1, metal oxides can be incorporated on the coat protein to fabricate high surface area nanowires or nanotubes, followed by the electrostatic deposition of

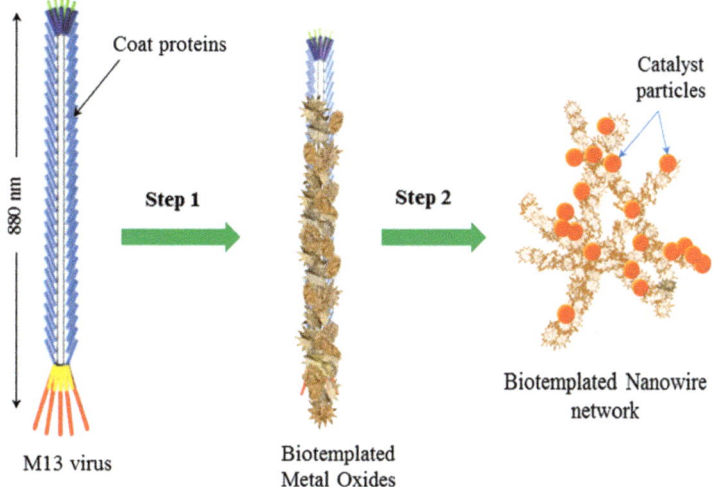

Figure 5.5   Schematic representation of biotemplated assembly of nanowire electrodes loaded with the photocatalysts.
(Adapted by permission from Macmillan Publishers Ltd: Nam *et al.*, *Nature Nanotechnology*, 2010, **5**, 340, Copyright 2010, and Oh *et al.*, *Nature Communications*, 2013, **4**, Copyright 2013.)

nanocatalysts in step 2. Researchers have chemically attached a photo-sensitizer, zinc porphyrin (ZnDPEG) to the major coat protein of the phage; the protein coat can be also employed to nucleate a catalyst, iridium oxide (IrO_2), through the peptide motif.[6] The biotemplate with dual labeling allowed the precise incorporation of photosensitizers and catalysts, including controlling the thickness of the catalyst. The nanowire network consisting of the ZnDPEG and IrO_2 were precisely spaced to trigger the water-splitting reaction; the pigment acts like an antenna to capture the light and transfer the energy along the nanowires, where the catalyst promote the water-splitting reaction. The resulting nanocatalysts demonstrated the oxidation of water driven by light with high quantum efficiency and turnover rate of oxygen production nearly fourfold. To overcome the issue of bio-templated nanowire aggregation, researchers fabricated porous microgels using microfluidic technique as an immobilization matrix, as shown in Figure 5.6, maintaining the structural integrity of the IrO_2–ZnDPEG nano-structures. The monodisperse virus encapsulated microgels developed without any phase segregation (Figure 5.6b) were later used as templates for assembling IrO_2 hydrosol clusters and photosensitizers. It was observed that the microgels with a lower IrO_2:ZnDPEG ratio showed significantly more oxygen evolution between the two different IrO_2–ZnDPEG microgels with $r = 35$ and 109. However, the catalytic activity of the nanowire was decreasing consistently, which was attributed to the photochemical degradation of the photosensitizers attached to the virus.[6,35] M13 viruses has also been used as a template to develop TiO_2 nanowires of mesoporous semiconducting network; the technique allows to tune the diameter of the nanowire, the size of the crystallites, and the film thickness, uniformity, and porosity.[36]

Tobacco mosaic virus (TMV 1-cys), another of the widely studied biotem-plates, is cylindrical in shape, 300 nm long and 18 nm in diameter, with a 4 nm open inner channel.[31] The genetically modified virus expresses coat proteins containing regularly spaced cysteine residues at the 1 position. The thiol groups (−SH) on each cysteine are capable of forming strong metal–thiol bonds, which were utilized in previous works as nucleation sites for noble metal particle growth. The properties of the biotemplated TMV 1-cys will be used to exercise two dimensions of control over the size and number of particles formed. By tuning the surface coverage density of the substrate, the number of particles capable of forming can be controlled. The size of the particle can then be controlled by the number of repetitions of the adsorption–hydrolysis reactions executed.

Similarly, the formation of hydrogen molecules from the protons and electrons in the second half of the reaction requires a reduction of the protons. In PEC water-splitting devices, a catalyst is needed for the hydrogen evolution reaction (HER) and the free energy of hydrogen adsorption, ΔG_H, is a good descriptor for identifying HER catalysts.[37] An ideal HER catalyst needs to produce hydrogen at low overpotentials and be optically transpar-ent to form a stable electrical contact with the photo-absorbing material. Platinum (Pt), the best HER catalyst followed by ruthenium oxide (RuO_2), is

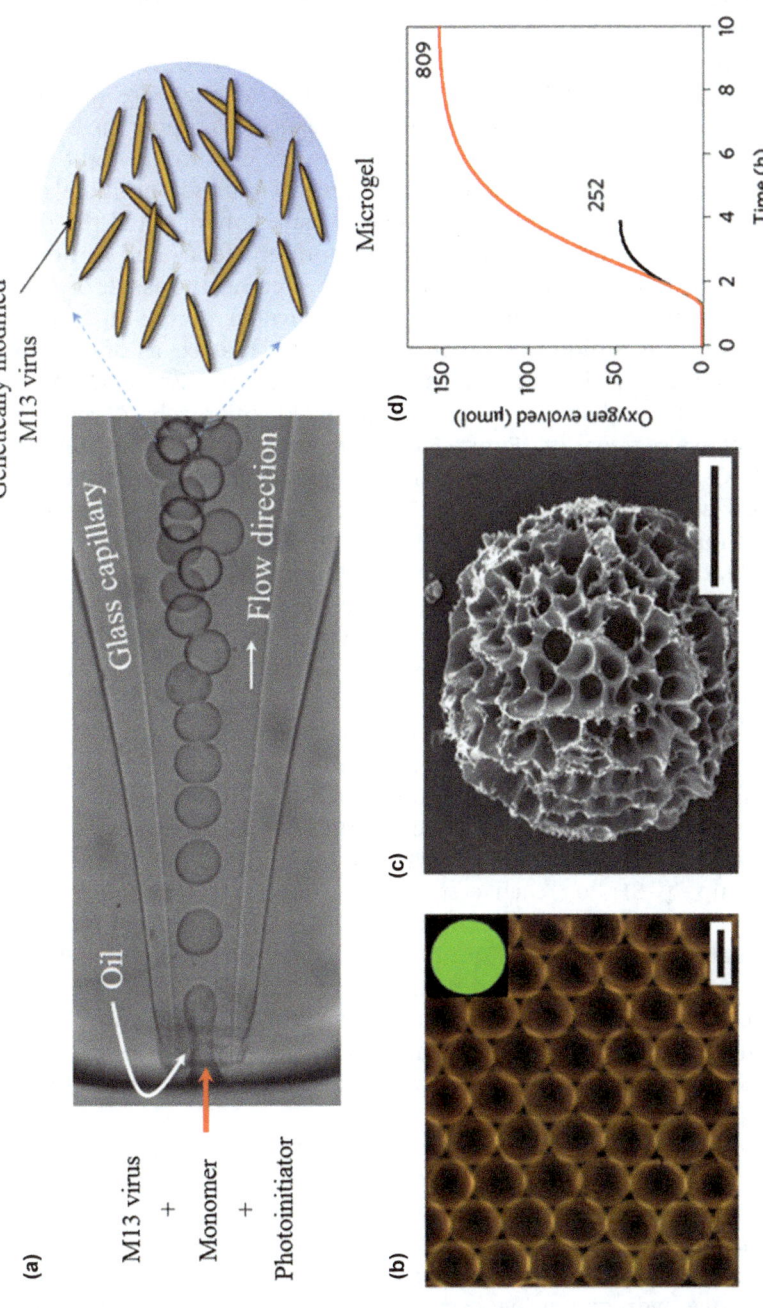

Figure 5.6 Fabrication and performance of biotemplated catalytic materials: (a) fabrication of virus-loaded microgels; (b) optical microscopy image of monodisperse microgels (scale bar = 100 μm); (c) microstructure of a freeze-dried microgel (scale bar = 20 μm); (d) oxygen evolution profiles from IrO$_2$-ZnDPEG microgels with $r = 35$ (black) and $r = 109$ (red). The numbers above the curve in part (d) indicate TON values. (Adapted by permission from Macmillan Publishers Ltd: Nam *et al*, *Nature Nanotechnology*, 2010, 5, 340, Copyright 2010.)

often deposited on a photocathode to promote PEC hydrogen production.[38] However, Pt is a precious noble metal and not suitable for a large-scale application. Surface-protected Cu_2O is one of the most promising low-cost materials (a p-type oxide) for PEC hydrogen production.[39] The direct band gap of Cu_2O is suitable for capturing a large portion of the visible spectrum with maximum solar-to-hydrogen (STH) efficiency. The transition metal sulfides in nitrogenases and hydrogenases are excellent HER catalysts with close to optimum ΔG_H values. In recent years, amorphous molybdenum sulfide films have been used as HER catalysts under acidic and neutral conditions.[40]

Recently, M13-virus templated perovskite nanowires have been fabricated and demonstrated to show hydrogen production and photovoltaic behavior under solar irradiation that nicely match the energy requirements for the oxidation and reduction potentials of water.[7,8] The genetically engineered M13 phage with three glutamates and one aspartate expressed at the N-terminus of each pVIII subunit from the coat protein of the filamentous body was employed as a bioscaffold; the negatively charged viruses electrostatically interact with cationic metal precursors and serve as nucleation centers for strontium titanate ($SrTiO_3$) nanomaterials.[7] The visible light irradiation energy was sufficient enough to excite the electrons to the conduction band of nitrogen-treated strontium titanate to reduce positive hydrogen ions to hydrogen gas with help of a platinum cocatalyst. Under UV irradiation, perovskite materials (including strontium titanate nanowire) demonstrated excellent photocatalytic properties although they lack performance under visible light due to a large band gap.

5.4.2 Membrane Separator

The membrane in an electrochemical cell not only acts as a separator for the gases but it should be also ionically and electronically conducting to work as a standalone device. Compartmentalization of the redox processes in a photochemical cell requires spatial distribution of the reactive sites on the photocatalysts while maintaining adequate chemical transport throughout the system. In addition, the thin film membranes in a photoelectrochemical water-splitting device need to have the characteristics of providing electrical contact between the anode and the cathode, while managing the proton transport to maintain a low pH gradient, along with low absorption of light in the visible range.[17] The flexible separator membrane should also have low permeability for hydrogen and oxygen gases that are generated during the operation of the device and be sufficiently strong to support the device, while maintaining the arrangement of microrod electrodes in a robust polymer matrix (as shown in Figure 5.3). The distance between the anode and the cathode reaction media increases the electrical resistance between the photoanode and photocathode, thus it can drastically affect the cell efficiency. Efforts to reduce the distance between these reaction media bring about the catalyst-coated membranes which are known as a membrane electrode assembly (MEA).[41,42] Applying catalyst on both sides of the

membrane minimizes the distance between the anode and cathode electrodes; in fact, the only media between anode and cathode reaction is the
membrane itself. By combining biotemplated catalysis deposition with layer-
by-layer (LbL) assembly, dual conducting thin-film membranes can be developed with the desired structural and physicochemical properties. The
design minimizes the electrical resistances, since the reaction occurs on
both surface of the membrane due to the presence of the catalyst. For example, Nafion, a sulfonated tetrafluoroethylene-based proton exchange
membrane relevant for solar water-splitting devices as both a proton conductor and gas separation membrane, as well as OH^- anion exchange
membrane, has been developed for use in alkaline conditions.[43]

Conjugated polymers such as PEDOT:PSS that are promising for use
in solar energy conversion applications such as poly(3,4-ethylenedioxy-
thiophene) PEDOT are also of interest when combined with ionically
conducting polymers such as poly (styrene sulfonate) (PSS).[44] The photoelectrodes can be embedded in the above polymers to study the influence of
bonding character on mechanical behavior that can be improved through
side-group chemistry, surface energy, and polymer physical properties. The
modular system design approach allows each piece to be independently
fabricated, modified, tested, and improved, as future advances in semiconductor, polymeric, and catalytic materials are made. These 'smart'
membranes can be sturdy and be based on inorganic, organic, or hybrid
systems. There are many features of biological system assembly that are
important to emulate to organize materials on multiple length scales and to
create advanced materials despite any presence of disorder and defects. This
will allow developing advanced photoelectrochemical cells with artificial
membranes with template function for self-repair and fault tolerance
including low cost and long lifetime.

In recent years, researchers have significantly advanced the field of
structural biology and operational mechanisms of the photosynthetic assemblies that permitted the synthesis and control of nanostructures, the use
of self-organization, and the control of materials assembly for cell development. One of the major technological challenges in artificial photosynthesis
is then to learn how to exploit these designs and assemblies that make
photosynthesis affordable and viable in the days ahead. This requires improved understanding of the organic/inorganic soft interface, including the
ability to tie multiple interactions to create and optimize functional assemblies. Another important challenge will be the design of biomimetic
catalysts inspired by the active sites of natural enzymes. Fast and long-lived
charge separation need to be established in a bio-inspired structure, coupling the charge separating device to a photosensitizer and coupling proton to
electron transfer. The electrons or holes needed for the catalytic reaction
must be accumulated and stored long enough to allow the completion of
the catalytic cycle. It is also expected that a catalyst should allow a high
conversion rate, high turnover number, and be composed of cheap and
environmental friendly materials.

5.5 Summary

Advanced materials design and self-assembly play key roles in the design and functionality of photochemical cell development for efficient reactions to control the movement of electrons, protons, and holes. Using nature's concept, phage templated design of organic/inorganic nanomaterials has showed tremendous potential for the development of superior electrode architecture for energy conversion and storage technologies. Rational design and advanced materials selection criteria are necessary to find a match between anode and cathode with ohmic resistance minimization for the standalone system integration of a PEC cell. Biotemplating provides a new way of controlling the incorporation and precise deposition of catalytic materials in the electrode assembly. While challenges remain concerning the development of high-efficiency, corrosion-resistant photoelectrodes, biotemplating-based self-assembly provides a remarkable opportunity for creating novel photocatalysts with exotic physicochemical properties for photoelectrochemical cell development for solar water splitting to produce hydrogen fuel.

References

1. W. C. Livingston, L. Wallace and O. R. White, *Science*, 1988, **240**, 1765.
2. N. S. Lewis and D. G. Nocera, *Proc. Natl. Acad. Sci.*, 2006, **103**, 15729–15735.
3. D. Dutta, D. De, S. Chaudhuri and S. K. Bhattacharya, *Microb. Cell Fact.*, 2005, **4**, 36.
4. J. J. Concepcion, R. L. House, J. M. Papanikolas and T. J. Meyer, *Proc. Natl. Acad. Sci.*, 2012, **109**, 15560–15564.
5. R. E. Blankenship, D. M. Tiede, J. Barber, G. W. Brudvig, G. Fleming, M. Ghirardi, M. R. Gunner, W. Junge, D. M. Kramer, A. Melis, T. A. Moore, C. C. Moser, D. G. Nocera, A. J. Nozik, D. R. Ort, W. W. Parson, R. C. Prince and R. T. Sayre, *Science*, 2011, **332**, 805–809.
6. Y. S. Nam, A. P. Magyar, D. Lee, J.-W. Kim, D. S. Yun, H. Park, T. S. Pollom, D. A. Weitz and A. M. Belcher, *Nat. Nanotechnol.*, 2010, **5**, 340–344.
7. N. Nuraje, Y. Lei and A. Belcher, *Catal. Commun.*, 2014, **44**, 68–72.
8. N. Nuraje, X. Dang, J. Qi, M. A. Allen, Y. Lei and A. M. Belcher, *Adv. Mater.*, 2012, **24**, 2885–2889.
9. K. T. Nam, D.-W. Kim, P. J. Yoo, C.-Y. Chiang, N. Meethong, P. T. Hammond, Y.-M. Chiang and A. M. Belcher, *Science*, 2006, **312**, 885–888.
10. G. Whitesides, J. Mathias and C. Seto, *Science*, 1991, **254**, 1312–1319.
11. B. Pokroy, A. K. Epstein, M. C. M. Persson-Gulda and J. Aizenberg, *Adv. Mater.*, 2009, **21**, 463–469.
12. R. Nagao, A. Moriguchi, T. Tomo, A. Niikura, S. Nakajima, T. Suzuki, A. Okumura, M. Iwai, J.-R. Shen, M. Ikeuchi and I. Enami, *J. Biol. Chem.*, 2010, **285**, 29191–29199.

13. C. Mao, D. J. Solis, B. D. Reiss, S. T. Kottmann, R. Y. Sweeney, A. Hayhurst, G. Georgiou, B. Iverson and A. M. Belcher, *Science*, 2004, **303**, 213–217.

14. P.-Y. Chen, M. N. Hyder, D. Mackanic, N.-M. D. Courchesne, J. Qi, M. T. Klug, A. M. Belcher and P. T. Hammond, *Adv. Mater.*, 2014, **26**, 5101–5107.

15. C. Yang, A. K. Manocchi, B. Lee and H. Yi, *Appl. Catal., B*, 2010, **93**, 282–291.

16. J. P. McEvoy, J. A. Gascon, V. S. Batista and G. W. Brudvig, *Photochem. Photobiol. Sci.*, 2005, **4**, 940–949.

17. J. M. Spurgeon, S. W. Boettcher, M. D. Kelzenberg, B. S. Brunschwig, H. A. Atwater and N. S. Lewis, *Adv. Mater.*, 2010, **22**, 3277–3281.

18. J. Yang, A. Sudik, C. Wolverton and D. J. Siegel, *Chem. Soc. Rev.*, 2010, **39**, 656–675.

19. K. Sivula and M. Gratzel, in *Photoelectrochemical Water Splitting: Materials, Processes and Architectures*, The Royal Society of Chemistry, 2013, pp. 83–108.

20. J. Ihssen, A. Braun, G. Faccio, K. Gajda-Schrantz and L. Thöny-Meyer, *Curr. Protein Pept. Sci.*, 2014, **15**, 374–384.

21. J. McKone and N. Lewis, in *Photoelectrochemical Water Splitting: Materials, Processes and Architectures*, The Royal Society of Chemistry, 2013, pp. 52–82.

22. S. Licht, *J. Phys. Chem. B*, 2003, **107**, 4253–4260.

23. T. Inoue and T. Yamase, *Chem. Lett.*, 1985, **14**, 869–872.

24. E. L. Miller, B. Marsen, D. Paluselli and R. Rocheleau, *Electrochem. Solid-State Lett.*, 2005, **8**, A247–A249.

25. A. Fujishima and K. Honda, *Nature*, 1972, **238**, 37–38.

26. A. Mao, J. K. Kim, K. Shin, D. H. Wang, P. J. Yoo, G. Y. Han and J. H. Park, *J. Power Sources*, 2012, **210**, 32–37.

27. Z. Mazouz, L. Beji, J. Meddeb and H. Ben Ouada, *Arabian J. Chem.*, 2011, **4**, 473–479.

28. S. Hu, C.-Y. Chi, K. T. Fountaine, M. Yao, H. A. Atwater, P. D. Dapkus, N. S. Lewis and C. Zhou, *Energy Environ. Sci.*, 2013, **6**, 1879–1890.

29. J. R. Swierk and T. E. Mallouk, *Chem. Soc. Rev.*, 2013, **42**, 2357–2387.

30. S. H. Yang, W.-J. Chung, S. McFarland and S.-W. Lee, *Chem. Rec.*, 2013, **13**, 43–59.

31. L. A. Lee, Z. Niu and Q. Wang, *Nano Res.*, 2009, **2**, 349–364.

32. L. A. Lee, H. G. Nguyen and Q. Wang, *Org. Biomol. Chem.*, 2011, **9**, 6189–6195.

33. S. Oh, E.-A. Kwak, S. Jeon, S. Ahn, J.-M. Kim and J. Jaworski, *Adv. Mater.*, 2014, **26**, 5217–5222.

34. D. Oh, J. Qi, Y.-C. Lu, Y. Zhang, Y. Shao-Horn and A. M. Belcher, *Nat. Commun.*, 2013, **4**.

35. G. R. Soja and D. F. Watson, *Langmuir*, 2009, **25**, 5398–5403.

36. Z. Hosseinidoust, A. L. J. Olsson and N. Tufenkji, *Colloids Surf., B*, 2014, **124**, 2–16.

37. Y. Hou, B. L. Abrams, P. C. K. Vesborg, M. E. Björketun, K. Herbst, L. Bech, A. M. Setti, C. D. Damsgaard, T. Pedersen, O. Hansen, J. Rossmeisl, S. Dahl, J. K. Nørskov and I. Chorkendorff, *Nat. Mater.*, 2011, **10**, 434–438.

38. J. Parrondo, T. Han, E. Niangar, C. Wang, N. Dale, K. Adjemian and V. Ramani, *Proc. Natl. Acad. Sci. U. S. A.*, 2014, **111**, 45–50.

39. C. G. Morales-Guio, L. Liardet, M. T. Mayer, S. D. Tilley, M. Grätzel and X. Hu, *Angew. Chem., Int. Ed.*, 2015, **54**, 664–667.

40. J. K. Norskov, T. Bligaard, J. Rossmeisl and C. H. Christensen, *Nat. Chem.*, 2009, **1**, 37–46.

41. J. Ronge, D. Nijs, S. Kerkhofs, K. Masschaele and J. A. Martens, *Phys. Chem. Chem. Phys.*, 2013, **15**, 9315–9325.

42. W. Hui, W. Jian-Tao, O. Xue-Mei, L. Fan and Z. Xiao-Hong, *Nanotechnology*, 2014, **25**, 265401.

43. X. Shen, B. Sun, F. Yan, J. Zhao, F. Zhang, S. Wang, X. Zhu and S. Lee, *Acs Nano*, 2010, **4**, 5869–5876.

44. W. Lu and Z. Jun, Symposium on Photonics and Optoelectronics (SOPO), 2012.

CHAPTER 6

Hybrid Molecular– Nanomaterial Assemblies for Water Splitting Catalysis

NAN JIANG, MEILI SHENG AND YUJIE SUN*

Department of Chemistry and Biochemistry, Utah State University, Logan, UT 84322, United States
*Email: yujie.sun@usu.edu

6.1 Introduction

Water is an earth-abundant and renewable chemical utilized by nature in photosynthesis which converts CO_2 to hydrocarbons and stores solar energy in chemical forms. Mimicking nature's approach to produce chemical fuels from water with renewable energy input, such as solar energy, has been a research focus for decades. Among various strategies, light-driven water splitting to generate H_2 and O_2 is widely considered as an appealing approach to capture and store solar power in chemical bonds.[2] Water splitting consists of two half reactions, H_2 evolution reaction (HER) and O_2 evolution reaction (OER), both of which are kinetically slow under ambient conditions and require competent catalysts.

Owing to their well-defined active sites and tunable properties *via* structural/electronic substituents, molecular complexes have attracted much attention in developing catalytic systems for water splitting. Although molecular catalysts might possess high intrinsic activity, their stability is usually inferior compared to solid-state heterogeneous catalysts, potentially due to undesirable inter-molecular collision of high-energy catalytic

RSC Green Chemistry No. 42
Green Photo-active Nanomaterials: Sustainable Energy and Environmental Remediation
Edited by Nurxat Nuraje, Ramazan Asmatulu and Guido Mul
© The Royal Society of Chemistry 2016
Published by the Royal Society of Chemistry, www.rsc.org

intermediates.[1] Along with the rapid development of nanomaterials science over the last three decades, an ever increasing research focus has been shifted towards assembling hybrid molecular-nanomaterial systems for water splitting catalysis, taking the advantages of both homogeneous and heterogeneous catalytic reactions. There are many apparent benefits of hybrid molecular-nanomaterial assemblies: (i) ligand modification controlling the catalytic function at the molecular level; (ii) defined active sites allowing detailed mechanistic studies; (ii) well-controlled micro-environment prohibiting undesirable collision of catalytic intermediates; (iii) high density of catalyst loading leading to increased catalytic activity per geometric area; (iv) improved electronic communication between catalyst and electrode if the hybrid assemblies are directly decorated on electrode. This chapter intends to highlight recent development of hybrid molecular-nanomaterial assemblies for both H_2 and O_2 evolution reactions, with an emphasis on the construction of hybrid assemblies and improved performance compared to those of molecular counterparts.

6.2 Hydrogen Evolution Reaction

Billion years of evolution provided Nature with hydrogenases to perform the H^+/H_2 inter-conversion with a remarkably high efficiency under ambient conditions. Hydrogenases are able to catalyze H_2 evolution near the thermodynamic potential and a turnover frequency (TOF) of ~ 9000 s^{-1} can be achieved at room temperature.[2,3] In order to mimic the catalytic fashion of hydrogenases, a large number of bioinspired molecular complexes containing the core structures of hydrogenase cofactors, including [Fe–Fe], [Ni–Fe], and [Fe], have been explored extensively.[4–10] Many delicate models have been synthesized and investigated. Those studies led to our deeper understanding in the catalytic mechanisms of hydrogenases, paving the way for designing improved biomimetic HER catalysts. On the other hand, molecular HER catalysts containing non-hydrogenase metals, such as molybdenum[11,12] and cobalt,[13–20] were also found to be active for electro- and photocatalytic H_2 evolution.

Traditionally, molecular HER catalysts were studied in organic solvents with the addition of an organic or inorganic proton source, owing to the limited solubility and/or stability of those catalysts in aqueous media.[21–26] However, in order to develop a catalytic system for water splitting at an industrial scale, it is more desirable to directly utilize water as the reaction medium. Indeed, recent years have witnessed the emergence of robust molecular HER catalysts which are able to function in pure aqueous media, such as molybdenum-oxo and cobalt-qua species coordinated with pentadentate pyridyl ligands, albeit with relatively high overpotentials.[27–30]

Another parallel strategy is to develop solid-state heterogeneous HER catalysts, which tend to possess better stability in aqueous media. Although platinum has been well-known as an excellent HER catalyst, its scarce and thus high cost limit its large-scale application. Therefore, many research groups are

devoted to developing solid-state HER catalysts only composed of earth-abundant elements. Some of those non-noble HER catalysts, such as transition metal chalcogenides,[31–41] carbides,[42,43] phosphides,[44–51] and alloys,[52] start to rival the performance of platinum. Being lack of well-defined active sites, solid-state HER catalysts is a challenging system for detailed mechanistic study.

In order to take the advantages of both molecular and solid-state HER catalysts, hybrid molecular-nanomaterial assemblies have attracted much attention in recently years. Multiple integration strategies, including covalent attachment and π-π stacking, have been established to immobilize molecular HER catalysts onto nanostructured substrates. A variety of nanomaterials, particularly those with high surface area, superior electrical conductivity, and versatile functional groups for grafting molecular catalysts, have been intensively employed as ideal platforms for the formation of hybrid molecular-nanomaterial assemblies of catalytic H_2 evolution. Similar scenario was applied to hybrid catalytic systems for the O_2 evolution reaction as well. The following paragraphs highlight several representative hybrid assemblies for H_2 evolution first, followed by hybrid systems for water oxidation to produce O_2.

6.2.1   Hybrid Assemblies Containing Iron Catalysts

Bioinspired diiron complexes that mimic the active sites of [FeFe]-hydrogenase have been extensively studied, with the aim of elucidating the catalytic mechanism in H_2 uptake/evolution in natural systems as well as exploring competent HER catalysts in artificial systems. Although it is challenging to incorporate a synthetic analogue of [FeFe]-hydrogenase active sites within a solid-state substrate, it could be highly rewarding if this strategy can afford new electrode materials for H_2 evolution. It is particularly appealing if catalysis can be matched to the conducting regime of the supporting substrate and fast electron relays are also incorporated. Following their previous work on iron-sulfur cluster analogues built into cysteinyl functionalized poly(pyrroles),[53,54] Pickett *et al.* reported the incorporation of a subsite of the [FeFe]-hydrogenase within a poly(pyrrole) framework for electrocatalytic H_2 evolution.[55] As shown in Figure 6.1, the pentafluorophenolate 'active ester' monomer C was co-electropolymerised with the cationic tetra-alkylammonium monomer E on glassy carbon or platinum electrodes. The resulting co-polymer films (thickness: ~100 nm) were further reacted with the 2-hydroxyethane derivatised diiron unit B to produce the {2Fe3S}-carbonyl/cyanide assemblies (GC|H_{poly}). Indeed, the subsite-loaded polymeric films exhibited proton reduction at −1.22 V *vs.* Ag/AgCl, roughly 120 mV positive of that of the catalyst-free polymer. Although this electrocatalysis occurred at a quite high overpotential, this report established a feasible approach to assemble molecular [FeFe]-hydrogenase mimics to solid-state substrates.

Besides electrocatalytic H_2 evolution, it is more attractive to directly produce H_2 from water upon light irradiation. Since the first attempt by Sun *et al.* to construct an artificial photocatalytic system for H_2 evolution, a large number of synthetic model complexes have been pursued to mimic the

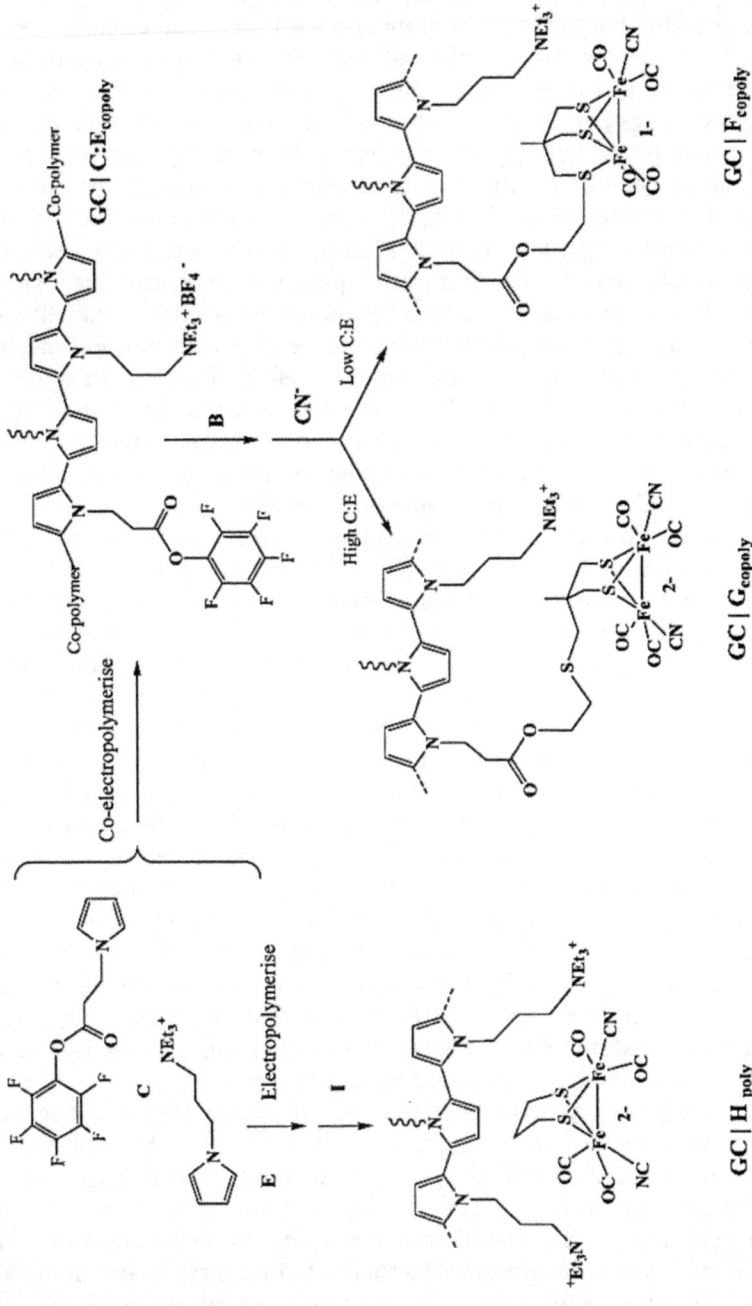

Figure 6.1 Routes to assemble molecular diiron complexes in electrode bound electropolymer films for electrocatalytic H_2 evolution. Reproduced from ref. 55 with permission from The Royal Society of Chemistry.

structure and functionality of the diiron cofactor of the [FeFe]-hydrogenase.[56] Common approaches involve the utilization of molecular photosensitizers, such as $[Ru(bpy)_3)]^{2+}$ (bpy = 2,2'-bipydiyl), $[Ir(ppy)_3]^+$ (ppy = 2-(2-pyridyl)-phenyl), and zinc porphyrin, in conjunction with sacrificial electron donors. Partially due to the intrinsic reactivity of those organic N-donor ligands of the photosensitizers at excited states, these systems usually suffer from poor stability over long-term operation. However, solid-state semiconductors and quantum dots (QDs) typically show better stability as light harvesting materials. Compared with organic and organometallic chromophores, QDs also offer unique size-dependent absorption, large absorption cross sections over a broad spectral range, long exciton lifetime, and superior photostability. QDs can simultaneously absorb multiple photons, or continuously absorb multiple photons even after electrons or holes are accumulated, thus enabling the coupling of single-photon/electron events with multiple-electron redox reactions necessary for photocatalytic H_2 production. In addition, the surface of QDs can be readily modified for specific functional targets and/or reaction environment, therefore it can form molecular assemblies with catalysts and/or sacrificial electron donors. These characteristics make QDs superb candidates for photocatalytic generation of H_2.

Since iron-sulfur carbonyl assemblies have been shown to electrocatalyze proton reduction to H_2 and their sulfide bridges can potentially bind to indium, Pickett's group reported a photocathode by assembling InP QDs and $[Fe_2S_2(CO)_6]$ for photoelectrocatalytic H_2 evolution.[57] The photocathode was constructed on a gold substrate. A monolayer of 1,4-dibenzenedithiol was first assembled on the gold surface, which was able to bind a primary layer of InP nanocrystals. Layer-by-layer build-up of the nanoparticle assembly was achieved by alternating exposure to the dithiol and nanocrystal solutions. The modified electrode was later immersed into a $[Fe_2S_2(CO)_6]$ solution to load the HER catalysts. A carton of the modified electrode and the iron-sulfur carbonyl cluster is shown in Figure 6.2. A stable photocurrent was obtained for at least one hour at a bias potential of −400 mV under a 395 nm LED irradiation. A current yield of approximately 60% was achieved. The speculated mechanism involves the excitation of electrons into the conduction band of InP under light irradiation. Since the conduction band of InP lies at ∼ −1.0 V *vs.* Ag/AgCl, the excited electrons were energetically favourable to transfer into the LUMO (−0.90 V *vs.* Ag/AgCl) of the iron-sulfur carbonyl catalysts, which thereby reduced proton to H_2.

Similar integrated systems of QDs and molecular [FeFe]-hydrogenase mimics were developed in Wu's group as well. In 2011, Wu and her colleagues reported a photocatalytic system composed of CdTe QDs as the photosensitizer, an artificial [FeFe]-hydrogenase mimic as the catalyst, and ascorbic acid as both the proton source and the sacrificial electron donor (Figure 6.3a).[58] A cyanide group was tethered to link three hydrophilic ether chains to the active site of the [FeFe]-hydrogenase mimic to increase the water solubility of the catalyst. CdTe QDs stabilized by 3-mercapatopropionic acid was selected as the photosensitizer owing to its broad visible light absorption and aqueous dispersion. In pH 4 aqueous media, such a

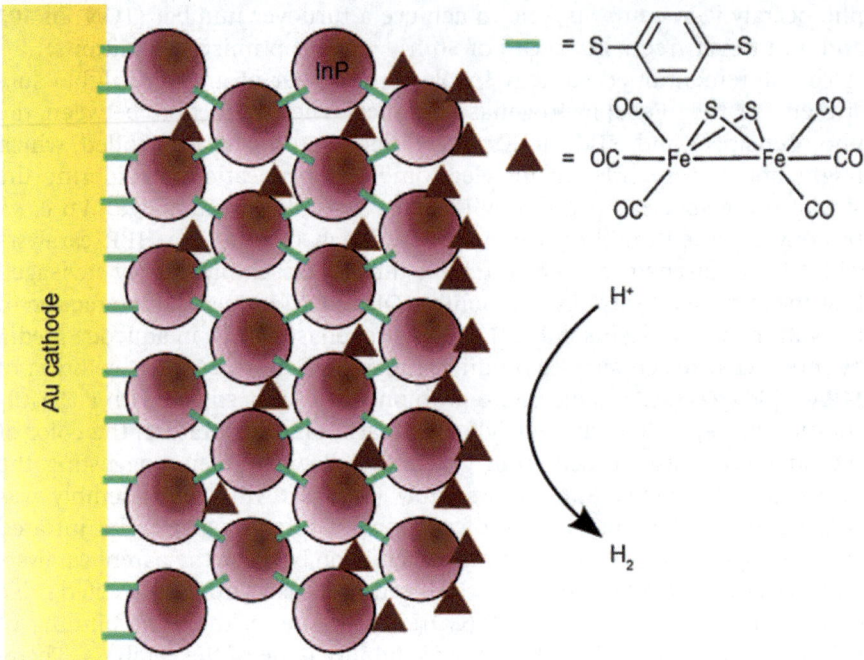

Figure 6.2 Cross-section of an InP nanocrystal-modified gold electrode with adsorbed/intercalated [Fe$_2$S$_2$(CO)$_6$] subsite analogue for photoelectrocatalytic H$_2$ generation.
Reproduced from ref. 57 with permission from John Wiley & Sons, Inc.

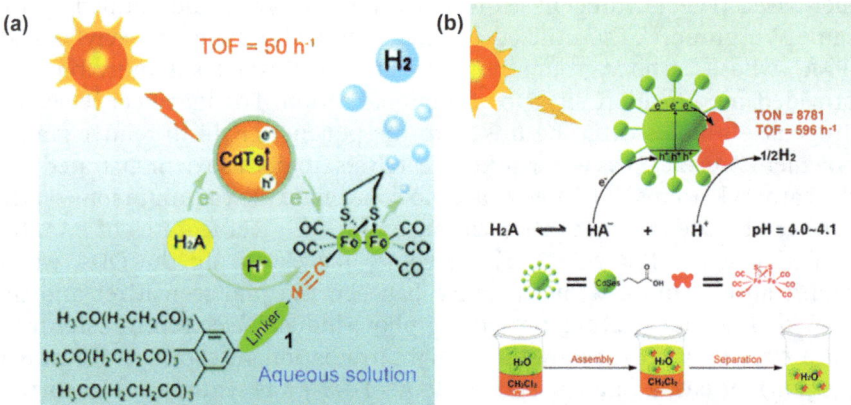

Figure 6.3 (a) Schematic representation of photocatalytic system with [FeFe]–H$_2$ases mimic as a catalyst and MPA-CdTe QDs as a photosensitizer. Reproduced from ref. 58 with permission from John Wiley & Sons, Inc. (b) Schematic representation of the interface-directed assembly of CdSe/Fe$_2$S$_2$(CO)$_6$ and its H$_2$ photogeneration. Reproduced from ref. 59 with permission from The Royal Society of Chemistry.

photocatalytic system was able to achieve a turnover number (TON) of 505 and a turnover frequency (TOF) of 50 h^{-1} under optimized conditions.

The aforementioned strategy involves the tether of a water-soluble substituent to the [FeFe]-hydrogenase mimics. The interaction between the photosensitizer and HER catalyst is still largely diffusion controlled, which might not be very efficient for electronic communication considering the size of these species. Inspired by Pickett's work discussed above, Wu *et al.* recently reported another promising approach to integrate HER catalysts with QD photosensitizers, *i.e.*, interface-directed assembly.[59] They envisaged that the surface affinity of water-soluble QDs could induce the interaction of the sulfurs of a water-insoluble [FeFe]-hydrogenase mimic in aqueous media by interface-directed surface binding. Indeed, when an aqueous solution of MPA-CdSe QDs (MPA: 3-mercapto-propionic acid) was stirred with a CH_2Cl_2 solution of $Fe_2S_2(CO)_6$ vigorously for 12 h at room temperature, the color of the aqueous phase turned from green to yellowish green, suggesting the formation of $CdSe/Fe_2S_2(CO)_6$ assembly (Figure 6.3b). The assembly was confirmed by the observation of carbonyl stretching peaks in infrared spectroscopy. Because of the intimate interaction between the parent catalysts and QDs, this assembly facilitated the electron transfer from the excited CdSe QDs to the catalytic centre $Fe_2S_2(CO)_6$. Furthermore, the tight binding of sulfur to CdSe also avoided the water solubility issue of the catalysts. Therefore, this interface-directed assembly achieved a remarkable TON of 8781 based on $Fe_2S_2(CO)_6$ and a TOF of 596 h^{-1} in the first 4 h of photocatalysis.

Although the systems discussed above showed improved activity in H_2 production, the microenvironments around the active sites of artificial HER catalysts were still far different from that of natural hydrogenases, whose active centres are deeply embedded within protein matrix. In order to better mimic the reaction of hydrogenases within restricted environments, polymer-based [FeFe]-hydrogenases mimics were designed and studied by the same Wu group.[60] The utilized polymer was hydrophilic poly(acrylic acid) (PAA), whose carboxy groups not only enhanced water solubility but also provided modification sites for functionalization. For instance, [FeFe]-hydrogenase mimics could be linked to the polymer chain to render grafted polymer PAA-*g*-Fe$_2$S$_2$ as water-soluble catalysts. It was also demonstrated that the carboxyl groups of PAA were able to coordinate the cadmium ions on the surface of CdSe (or CdTe) QDs. Because of these great features of PAA, the grafted polymer PAA-*g*-Fe$_2$S$_2$ chain could likely wrap up the QDs, which significantly shortened the distance between the photosensitizer and the grafted HER catalyst, thus promoting photo-induced electron transfer from the excited QDs to the catalyst upon light irradiation. As shown in Figure 6.4, the polymer-based catalytic system (PAA-*g*-Fe$_2$S$_2$) was synthesized by stirring deprotected Fe$_2$S$_2$-NHBoc and PAA ($M_w = 1800$) in the presence of PyBOP ((benzotriazol-1-loxy)tripyrrolidinophosphonium hexafluorophosphate). Still using ascorbic acid as the proton source and the sacrificial electron donor, this polymer-based [FeFe]-hydrogenase mimic system exhibited extremely high activity for photocatalytic H_2 evolution with a TON of 27 135 per Fe$_2$S$_2$ unit and a TOF of 3.6 s^{-1} (12 960 h^{-1}).

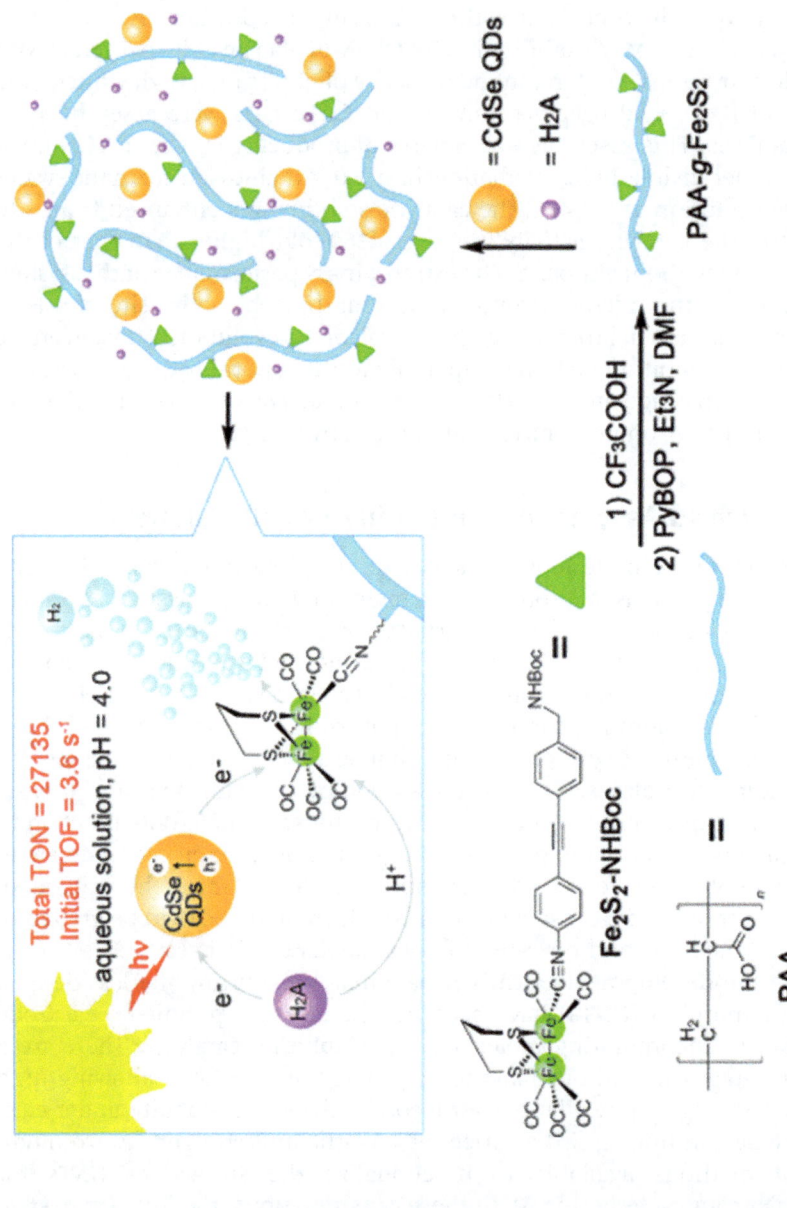

Figure 6.4   The synthetic route of PAA-*g*-Fe$_2$S$_2$ for photocatalytic H$_2$ production. Reproduced from ref. 60 with permission from John Wiley & Sons, Inc.

   Replacing poly(acrylic acid) with chitosan, a simplified but more efficient HER system was published by Wu's group recently (Figure 6.5).[61] Chitosan contains a significant amount of primary amines and hydroxyl groups. When the amines are protonated, chitosan bears a polycationic character. Due to the hydrophobic and electrostatic interactions, [FeFe]-H$_2$ase mimics are likely to be incorporated within chitosan. In addition, the negatively charged surface of MPA-stabilized CdTe QDs will also be able to interact with cationic chitosan. Therefore, in the presence of chitosan, a highly integrated system of [FeFe]-hydrogenase mimics and CdTe QDs were assembled for HER catalysis. This assembly was capable of producing H$_2$ with a TON up to 52 800 under visible light irradiation in a 1.0 g/L chitosan methanol/water (1 : 3, v/v) solution at pH 4.5. The catalytic stability was enhanced from 8 to 60 h, and the catalytic activity was over 4160-fold higher than that of the same system without chitosan. The extraordinary performance of this system indicates that the microenvironment surrounding the active site of a HER catalyst plays a crucial role in the photocatalytic H$_2$ evolution. Thus, to create an active H$_2$ evolution system comparable to the natural hydrogenases, one needs to mimic not only the structure of the active sites but also the biological local environment surrounding the active sites.

6.2.2   Hybrid Assemblies Containing Cobalt Catalysts

Despite not found in the active sites of natural hydrogenases, cobalt has long been investigated as a promising element in HER catalysis.[62–64] Cobalt azamacrocycles were among the earliest non-noble metal catalysts for H$_2$ generation. Based on coordinating ligands, molecular cobalt HER catalysts can be summarized in the following eight categories: N4-macrocyclic complexes,[65,66] hexaamino complexes,[67] porpridine complexes,[68] phthalocyanine complexes,[69] cyclopentadienyl complexes,[70] glyoxime complexes,[71,72] polypyridine complexes,[73–75] and cobalt complexes with one or two base-containing diphosphine ligands.[76] However, most of them only function in non-aqueous media in the presence of organic acids, requiring moderate to high overpotentials for H$_2$ production. In order to meet the criteria of practical applications, those molecular cobalt catalysts need to be grafted to electrode materials and operate in fully aqueous electrolytes.

   Along the development of carbon nanomaterials during the last decade, carbon nanotubes (CNTs) have been considered as a promising electrode material for incorporating molecular water splitting catalysts. There are a few advantages of carbon nanotubes: (i) large surface area allowing high catalyst loading; (ii) excellent electric conductivity and stability under electrocatalytic conditions; (iii) a variety of versatile and straightforward modification methods available to functionalize the surface of CNTs for molecular catalyst tethering. With these considerations in mind, Artero *et al.* went forward to attach a molecular cobalt diimine–dioxime catalyst to functionalized CNTs.[77] The synthetic methodology of this electrocatalytic CNT/Co system is detailed in Figure 6.6. Firstly, an azide group was tethered to the diimine-dioxime ligand, followed by metalation of cobalt in air to

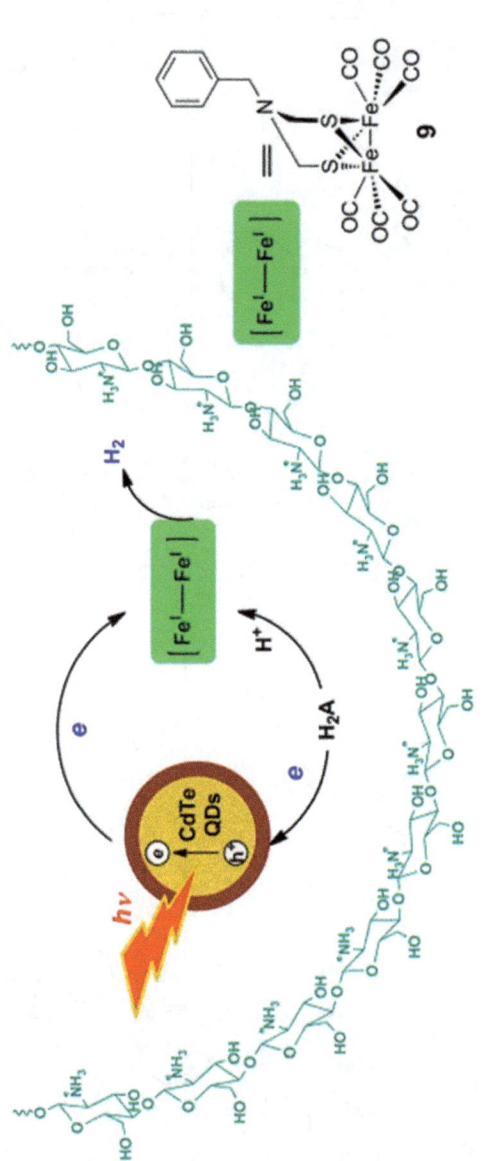

Figure 6.5 A schematic of the H$_2$ photogeneration of a chitosan-confined mimic of the diiron subsite of [FeFe]–hydrogenase in the presence of CdTe quantum dots and ascorbic acid. Reproduced from ref. 61 with permission from Nature Publishing Group.

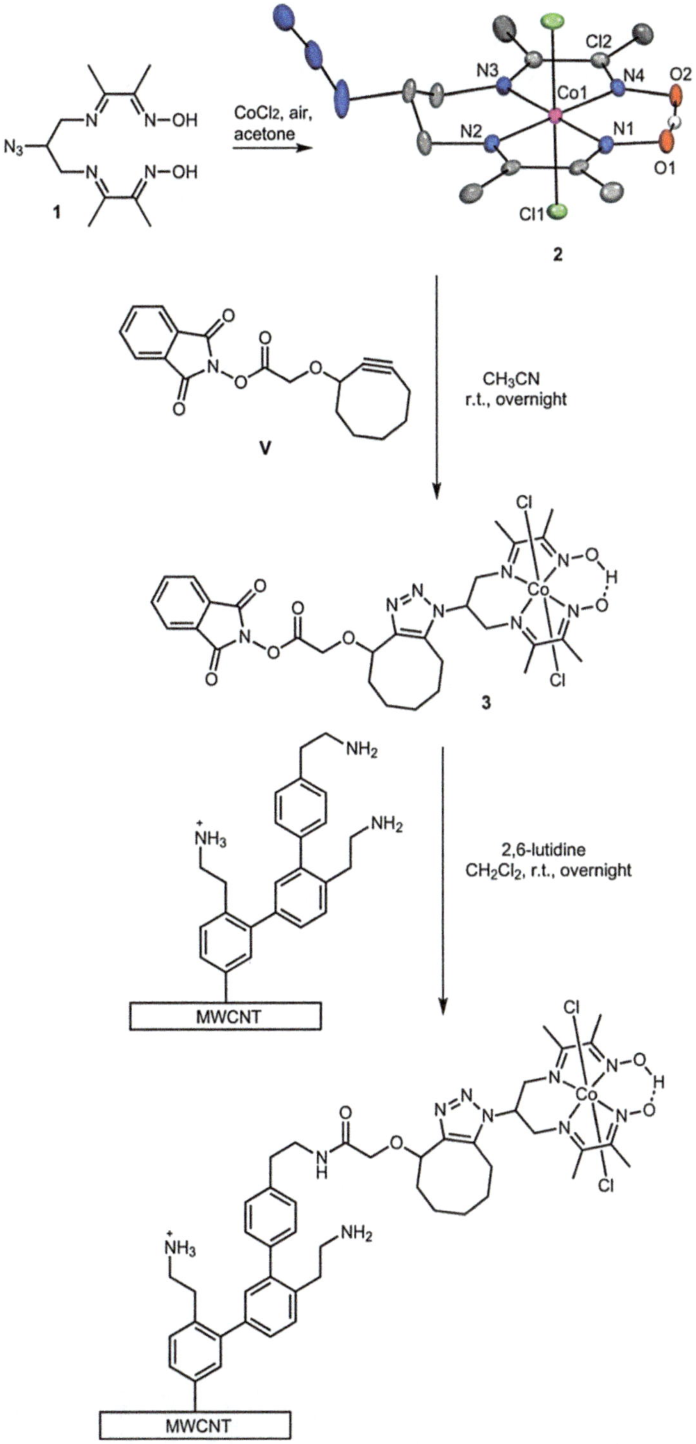

Figure 6.6   Synthetic methodology for the preparation of the CNT/Co material.
Reproduced from ref. 77 with permission from Nature Publication
Group.

form complex **2**. Later, Bertozzi's copper-free click chemistry was employed,[78] where *N*-hydroxy-phthalimide-activated ester was clicked with complex **2** to yield complex **3**. In the meantime, a multi-walled CNTs were deposited on a gas-diffusion layer electrode with the help of Nafion, and then decorated with a poly-4-(2-aminoethyl)phenylene layer through the electroreduction of the corresponding aryldiazonium salt. Finally, the cobalt complex **3** was attached covalently to the modified CNTs *via* an amide linkage formed during the reaction of the activated ester **3** with the grafted amine residue on CNTs. Cyclic voltammogram of the functionalized electrode proved the attachment of the intact cobalt complexes on CNTs. The cathodic and anodic current densities of the $Co^{III/II}$ redox couple (-1.08 V *vs.* $Fc^{+/0}$) were linearly proportional to the scan rate, which confirmed the immobilization of the cobalt complexes on the electrode surface. In addition, X-ray photoelectron spectroscopy also detected the signal of Co^{III} on the modified electrode with an estimated ratio of one cobalt atom per 310 carbons.

It has been reported that related diimine-dipyridine cobalt complexes may go through reductive processes at the metal site where transient carbon-based α-imino radical species could form and lead to the homo-coupling products of dimer complexes, which are undesirable for catalysis.[79] Therefore, molecular cobalt diimine-dioxime catalysts usually show poor turnover numbers. For example, a related $[Co(DO)(DOH)pnBr_2]$ $((DOH)_2pn = N_2,N_2$-propanediylbis(2,3-butanedione 2-imine 3-oxime) exhibited significant degradation after 50 turnovers.[80] Remarkably, those aforementioned immobilized cobalt catalysts on CNTs demonstrated much stronger robustness. Over a 7-h controlled potential electrolysis (-0.59 V *vs.* RHE) in an acetate buffer of pH 4.5, a TON of 55 000 ($\pm5\%$) was achieved with an initial TOF of 8 000 ($\pm5\%$) h^{-1}. The authors attributed the enhanced activity of the CNT/Co system to (i) the isolation of molecular cobalt complexes on the surface of CNTs, which avoided the bimolecular interaction for homo-coupling product formation, and (ii) efficient electron transfer from CNTs to the catalysts, which limited the lifetime of the reduced species for degradative side reactions.

In addition to grafting molecular cobalt catalysts to conductive materials for electrochemical H_2 production, photochemical molecular-semiconductor hybrid systems have also been reported. Moore and his co-workers reported a two-step approach to bind a molecular cobalt catalyst to the GaP surface.[81] The involved catalyst belongs to the cobaloxime class which showed promising electrocatalytic performance for H_2 production; while *p*-type GaP is a visible-light-absorbing semiconductor with an appropriate band gap ($E_g = 2.26$ eV) for solar light absorption. Figure 6.7 presents the synthetic route of the cobaloxime-modified GaP photocathode. Etching of GaP *via* buffered hydrofluoric acid removed the surface oxide film and generated hydroxyl-terminated surface sites. Subsequent UV-light irradiation induced the self-initiated photografting and photopolymerization of 4-vinylpyridine with the hydroxylated GaP surface to render

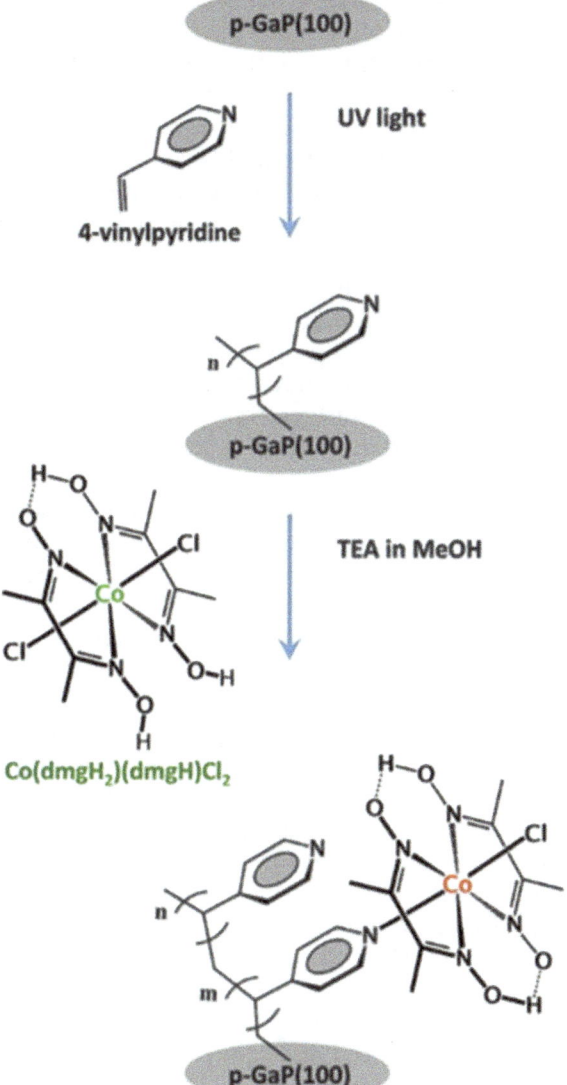

Figure 6.7   Schematic representation of the attachment method used to assemble a cobalt catalyst-tethered *p*-GaP photocathode.
Reproduced from ref. 81 with permission from the American Chemical Society.

pyridine-tethered *p*-GaP. Replacing one of the axial chloride ligands of $Co(dmgH_2)(dmgH)Cl_2$ (dmgH = dimethylglyoximate monoanion; $dmgH_2$ = dimethylglyoxime) with the surface attached pyridine on GaP led to the formation of the final molecular catalyst-semiconductor assembly. A suite of surface characterization techniques were utilized to confirm the successful construction of this hybrid photocathode. One advantage of this

polymer-based photocathode is that the multiple pyridine binding sites per chain will dramatically enhance the catalyst loading amount. In the present case, a 12 nm thick dry polymer layer had a maximum pyridyl sites of 7.9×10^{15} cm^{-2} (13 nmol cm^{-2}), significantly larger than conventional self-assembled monolayers of molecular linkers. The catalyst-coated photocathode showed enhanced photocatalytic activity for H$_2$ production than that of a non-modified counterpart in neutral water.

6.2.3 Hybrid Assemblies Containing Nickel Catalysts

Owing to the existence of nickel in the active site of one of those three hydrogenases, nickel has received much attention as a suitable metal for electrocatalytic H$_2$ generation.[82] Most efforts have been devoted to exploring nickel-based complexes mimicking the subsite of [Ni-Fe]-hydrogenases. Similar to iron-based hydrogenase mimics, the majority of those biomimetic nickel catalysts only function under pure organic conditions.[83] There are only a few reports about the catalytic activity of biomimic nickel complexes in aqueous media. In the early 1980s, Fisher *et al.* reported that Ni tetra-azamacrocycle was a competent electrocatalyst for proton reduction in water/acetonitrile mixture.[84] During the last decade, Dubois and his co-workers have conducted several systematic studies on nickel bisphosphine-based complexes for H$_2$ evolution, whose catalytic performance was enhanced when water was introduced in acidic acetonitrile.[85,86] The beauty of Dubois's catalysts mainly result from the delicate control of intra- and intermolecular proton transfer by amines in the secondary coordination sphere of nickel.

With the aim of improving the water compatibility and stability of nickel-based catalysts, Artero and his co-workers explored a method to anchor Dubois's nickel catalyst onto amino-functionalized multiwall carbon nano-tubes (MWCNTs) through a stable covalent amide linkage.[87] Activated phthalimide ester units were introduced in the para position of nitrogen-bound phenyl residues of the designed nickel catalyst. MWCNTs were deposited on an indium-doped tin oxide (ITO) electrode by the reported soluble membrane technique.[88] Covalent attachment of 4-(2-aminoethyl)phenyl groups on the surface of MWCNTs was achieved *via* electro-reduction of the corresponding protonated dianonium tetrafluoroborate between 0.0 and 0.8 V *vs.* NHE. Finally, the modified nickel catalyst was successfully anchored to the MWCNT electrode through the formation of amide linkages between the ligand and the grafted amine residues. Compared to coating on a bare ITO electrode, the presence of MWCNTs increased the catalyst loading concentration by two orders of magnitude (1.5 ± 0.5 nmol cm^{-2}). In an aqueous sulfuric acid solution, the catalyst-coated MWCNT electrode exhibited a low overpotential (20 mV) for the catalytic onset of proton reduction and a 10 h electrolysis at -300 mV *vs.* NHE resulted in a TON more than 100 000 ($\pm 30\%$). In addition, the anchored nickel catalyst was not only active for H$_2$ production, but also competent for H$_2$ oxidation. The catalytic

current of H_2 oxidation achieved at 300 mV overpotential in the same electrolyte led to a TON of 35 000 over a 10-h electrolysis. In other words, this biomimetic nickel modified electrode could reversibly catalyze the H^+/H_2 interconversion.

Following this work, Artero *et al.* reported another direct and straightforward method to decorated MWCNTs with a pyrene-functionalized nickel complex *via* non-covalent π–π stacking interaction (Figure 6.8).[89] First, MWCNTs were deposited on the surface of commercial gas diffusion layers (GDL) *via* filtration. Then pyrene-amended nickel complexes were slowly filtered through the MWCNTs/GDL electrode. This facile approach allows efficient electronic communication between MWCNTs and immobilized catalysts as well as more catalyst loading. Another advantage is CO tolerant. CO is a major impurity in H_2 fuels due to the reformation of hydrocarbon and biomass. Bulk electrolysis conducted at −300 mV *vs.* NHE in 0.5 M H_2SO_4 showed no loss of catalytic activity for 6 h with a TON of 85 000. This new nickel-based MWCNTs could also function bidirectionally for H^+/H_2

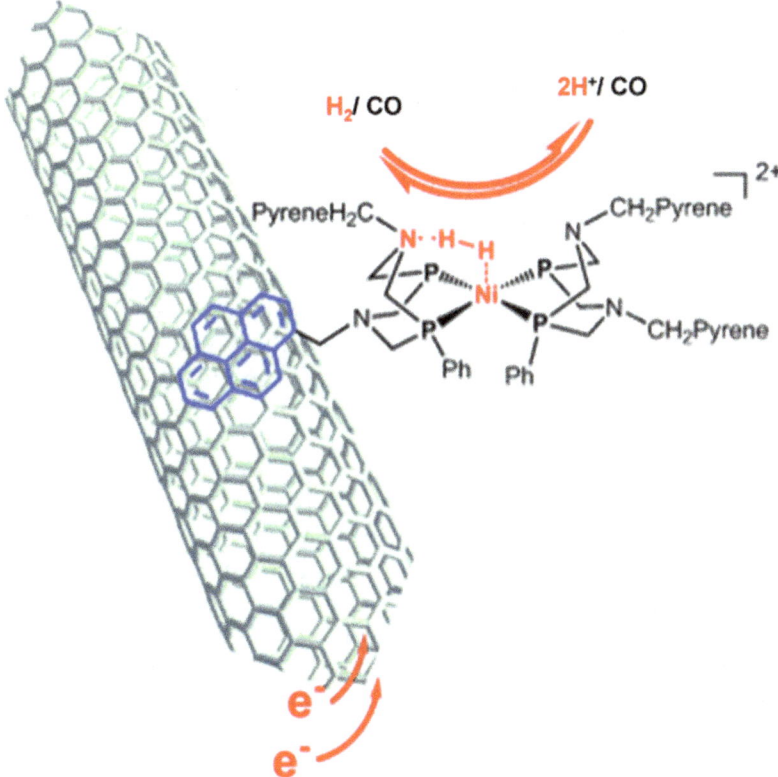

Figure 6.8   A schematic demonstration of a pyrene-functionalized nickel biphosphine catalyst attached to MWCNTs *via* π–π stacking for H_2 evolution and oxidation.
Reproduced from ref. 89 with permission from John Wiley & Sons, Inc.

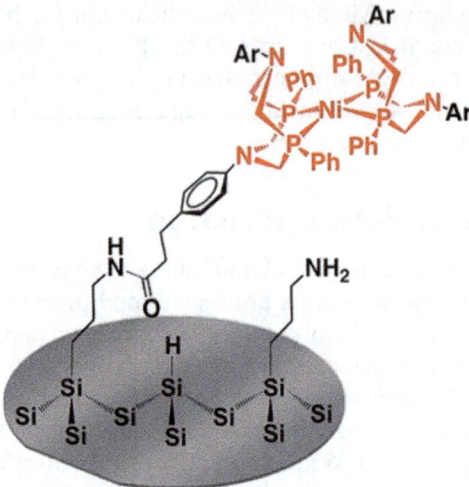

Figure 6.9   A modified nickel bisdiphosphine catalyst covalently linked to the surface of silicon.
Reproduced from ref. 90 with permission from the American Chemical Society.

interconversion at an extremely low overpotential. These remarkable results represent a promising strategy which connects noble metal-free molecular catalysts to suitable nanoscale supports. Such a strategy can be used to design efficient and durable devices for electro- and photo-induced evolution and oxidation of H_2.

The elegant bioinspired nickel bisdiphosphines can not only be grafted on CNTs for electrocatalytic H_2 evolution, but also be attached onto the surface of semiconductors serving as photocatalysts for H_2 generation. Moore and co-workers developed a facile method to connect a nickel bisdiphosphine-based catalyst to the surface of p-type silicon (Figure 6.9).[90] First *N*-allyl-2,2,2-trifluoro-acetamide was photochemically grafted onto the surface of semiconductors. The semiconductors were masked with a layer of amine groups which were able to link nickel bisdiphosphine-based catalysts *via* covalent binding. However, the catalytic performance of this photocathode has not been reported.

A superior nickel catalyst-based photocatalytic system came out from the collaboration of Eisenberg, Holland, and Krauss groups. In 2012, they reported a robust and highly active assembly consisted of [Ni]-hydrogenase mimics as catalysts and CdSe QDs as photosensitizers.[91] Dihydrolipoic acid (DHLA) was attached to the surface of CdSe QDs to form water-soluble photosensitizers. Subsequently, Ni(NO$_3$)$_2$ was added as a catalyst precursor which might interact with DHLA capped on the CdSe QDs. Under the optimized condition, this photocatalytic system yielded impressive H_2 production in pH 4.5 aqueous solution when irradiated with a 520 nm light source. A TOF of over 7 000 h^{-1} and a TON of 600 000 or more were

obtained with a quantum yield of $\sim 36\%$. The undiminished catalytic activity of this assembly could last at least 360 h. This simple but highly effective strategy did not require complicated catalyst synthesis, but used readily available nickel salts, representing a huge advantage compared to those hybrid assemblies discussed above.

6.3   Oxygen Evolution Reaction

Nature evolved a delicate biological machinery to convert solar energy, water, and CO_2 into carbohydrates *via* a process named photosynthesis. There are mainly two components involved in photosynthesis: photosystem I and photosystem II. The most chemically challenging step occurs at the O_2 evolution center of photosystem II, where water oxidation to O_2 takes place. As a four-electron/four-proton process, O_2 evolution from water bears a high activation barrier and has to go through multiple intermediate states. Recently, the crystal structure of the O_2 evolution center, which contains a Mn_4CaO_5 cluster, was obtained with a 1.9 Å resolution.[92] The cubane-type Mn_4CaO_5 cluster has been an inspiration source for many scientists aiming at developing competent O_2 evolution catalysts. Inspired, but not constrained, by nature, a large library of molecular catalysts have been reported,[93] ranging from early focus on ruthenium and iridium complexes to more recent efforts on earth-abundant transition metals, like manganese, iron, cobalt, nickel, and copper species.[94-96] Similar to the strategy employed in the advance of H_2 evolution catalysis, hybrid systems integrating well-defined molecular catalyst units and high-surface-area nanomaterial scaffolds have attracted increasing attention. The following sections highlight a few representative examples in this direction.

6.3.1   Hybrid Assemblies Containing Ruthenium Catalysts

Ruthenium complexes have been the most frequently and successfully studied catalysts for the O_2 evolution reaction. Nearly 33 years ago, Meyer *et al.* reported the first molecular ruthenium complex, blue Ru dimer (*cis,cis*-$\{[Ru(bpy)_2(H_2O)]_2(\mu\text{-}O)\}^{4+}$, as a chemically-driven O_2 evolution catalyst.[97] The detailed mechanism of the blue dimer has been elucidated by a series of chemical, electrochemical, and photochemical experiments.[98] As shown in Figure 6.10, the blue dimer features an oxidative activation by sequential $4e^-/4H^+$ loss to form $[(bpy)_2(O)Ru^VORu^V(O)(bpy)_2]^{4+}$ as a kinetic intermediate, which undergoes water attack on one of the $Ru^V=O$ sites to produce a peroxide intermediate, $[(HOO)Ru^{IV}ORu^{IV}(OH)]^{4+}$. The peroxide intermediate is in acid–base equilibrium with $[(HOO)Ru^{III}ORu^V(O)]^{3+}$, which triggers rapid oxygen evolution. As implied in this mechanism, later it was revealed that a single site of ruthenium was enough to catalyze O_2 evolution. Since the discovery of blue dimer, a variety of mononuclear ruthenium complexes have been reported as competent water oxidation catalysts, such as those reported by Thummel *et al.*[99-103] In order to enhance the stability of

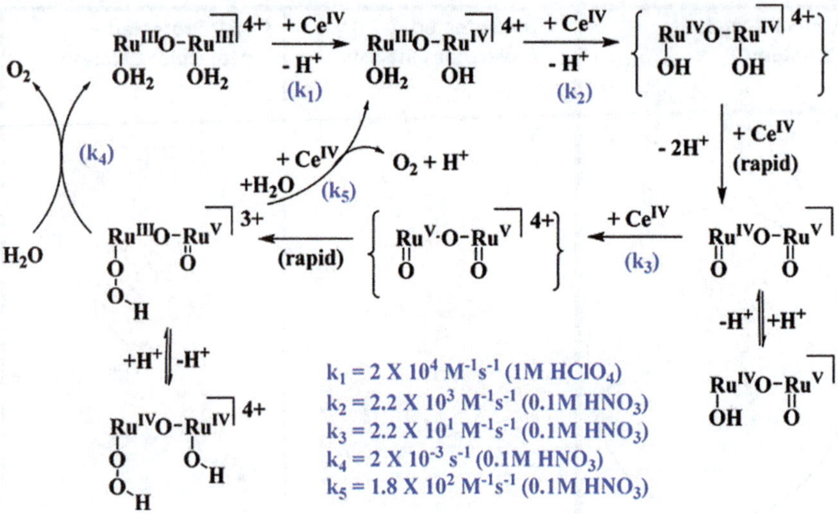

Figure 6.10 Mechanistic steps of O$_2$ evolution catalyzed by the ruthenium blue dimer along with the reaction rate of each step.
Reproduced from ref. 98 with permission from the American Chemical Society.

those mononuclear ruthenium catalysts for water oxidation, surface-binding substituents, like methylenephosphonate derivatives, have been amended on the ancillary ligands. For instance, a redox mediator-water oxidation catalyst assembly, $[\{[4,4'-(HO)_2OPCH_2]_2bpy\}_2Ru^{II}(bpm)Ru^{II}(Mebimpy)(OH_2)]^{4+}$, immobilized on ITO, was reported to exhibit sustained O$_2$ evolution catalysis under acidic condition for over a 13-h period with a TON of 28 000 and a TOF of 0.6 s^{-1}.[104] A mononuclear ruthenium catalyst, $[Ru(Mebimpy)(4,4'-((HO)_2OPCH_2)_2bpy)(OH_2)]^{2+}$, without redox mediator was also described by Meyer *et al.* to bind ITO, FTO, and nanoTiO$_2$ through phosphate groups.[105,106] It was found that the surface-bound ruthenium complex retained its chemical and physical properties, as well as catalytic capability of water oxidation.

Unfortunately, although phosphonate-derivatized catalysts and molecular assemblies provide a basis for sustained water oxidation on electrodes in acidic solution, they are unstable toward hydrolysis and loss from surfaces as pH is increased. Meyer's group recently reported an alternative protection strategy to enhance the surface binding stability by atomic layer deposition (ALD) of an overlayer of TiO$_2$.[107] As shown in Figure 6.11, the authors first bound a phosphonate-derivatized ruthenium catalysts onto the ITO electrode surface, followed by coating with a conformal nanoscale TiO$_2$ overlayer applied by layer-by-layer atomic layer deposition in a home-built flow tube reactor. ALD is a self-limiting thin-film deposition technique in which reactive vapor phase precursors are sequentially exposed to a substrate surface. With sequential precursor delivery and controlled surface reactions, ALD allows precise control over layer thickness and conformality for complex

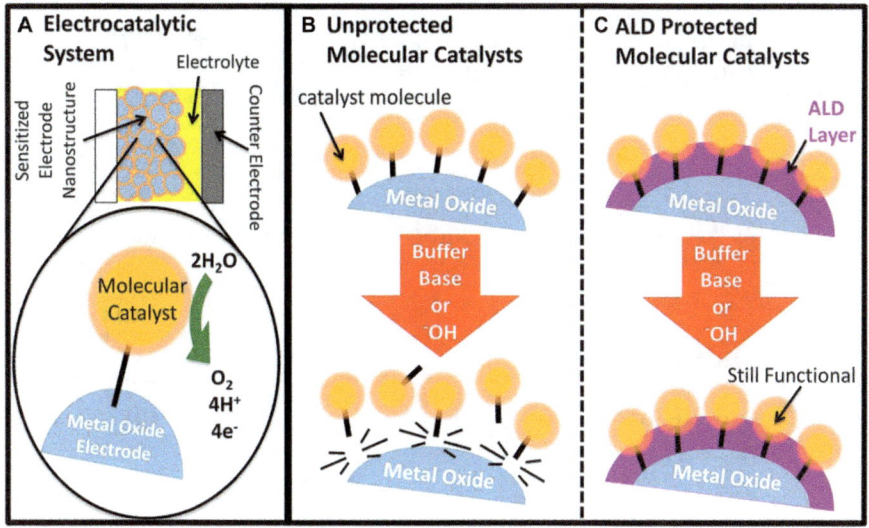

Figure 6.11   Schematic representation of the ALD strategy for a ruthenium catalyst attached to ITO protected by TiO_2. (A) Illustration of the electrochemical device architecture showing the surface-derivatized electrode and water oxidation. (B) Underivatized electrodes exposed to basic aqueous conditions showing detachment of the catalysts from the electrode surface. (C) ALD protection of surface attachment under basic aqueous conditions.
Reproduced from ref. 107 with permission from the National Academy of Sciences of the United States of America.

nanoarchitectures,[108] which represents a unique catalyst preparation/protection technique in a variety of electrochemical applications. In this case, Meyer *et al.* used $TiCl_4/H_2O$ precursors to deposit TiO_2 films with a thickness of $0.5\,\text{Å}$ per ALD cycle. Owing to the additional protection of the TiO_2 overlayer, it was possible to conduct water oxidation catalysis by this ruthenium catalyst at higher pHs. Even more remarkable is that at pH 12 the catalytic current enhancement compared with that at pH 1 is $\sim 10^6$ higher with an increase in reaction rate from $10^{-2}\,\text{s}^{-1}$ to $10^4\,\text{s}^{-1}$ at an overpotential of 1.02 V.

In contrast to the common six-coordinate intermediate of ruthenium catalysts for water oxidation, a rare seven-coordinate species was isolated in the solid state by Sun *et al.* in 2009.[109] Sun and his co-workers isolated a seven-coordinate Ru(iv) dimer complex with [HOHOH]-bridging ligand as an intermediate for catalytic water oxidation. The catalyst precursor Ru(bpa)-(pic)$_2$ (H_2bpa = 2,2'-bipyridine-6, 6'-dicarboxylic acid; pic = 4-picoline) had catalytic onset potential of ~ 0.98 V *vs.* NHE at neutral pH, which meant in principle it could be oxidized by photogenerated $[Ru(bpy)_3]^{3+}$ in a homogenous system. Hence, Sun *et al.* designed a photoelectrochemical device for visible light driven water splitting, utilizing Ru(bpa)(pic)$_2$ as the catalyst combined with dye-sensitized nanostructured TiO_2 as the anode, and Pt as

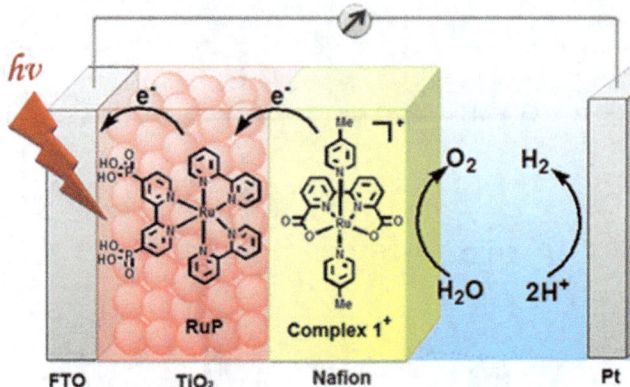

Figure 6.12 A photoelectrochemical device consisting of an anode based on ruthenium dye-sensitized TiO₂ film on FTO coated with a Nafion film incorporating a catalyst of [RuL(pic)₂]⁺, a Pt cathode, and an aqueous electrolyte, for light-driven water splitting catalysis.
Reproduced from ref. 110 with permission from The Royal Society of Chemistry.

the cathode (Figure 6.12).[110] $[Ru(bpy)_2(4,4'-(PO_3H_2)_2bpy)]^{2+}$ was employed as the photosensitizer, because of its absorbance of visible light, strong attachment to the surface of TiO₂ through the phosphate substituents, efficient charge separation, and suitable oxidation potential. Nafion membrane was included to immobilize the hydrophobic ruthenium catalyst. Nafion contains a considerable number of sulfonic groups and are electric conductive and of high chemical and thermal stability.[111–116] Upon one-electron oxidation by Ce(IV), the oxidized $[Ru(bpa)(pic)_2]^+$ could be successfully penetrated into Nafion film as confirmed by UV-vis absorbance. Since the water oxidation of Ru(bpa)(pic)₂ was highly pH dependent with a higher onset potential at low pH and the native Nafion had a pH value of 2, in order to make the catalytic onset less positive than the redox potential of the ruthenium photosensitizer, the Nafion film had to be neutralized prior to utilization. When the neutralized Nafion was used, efficient charge transfer between the catalyst and the photosensitizer was obtained as reflected by the transient short-circuit responses to the on-off cycles of illumination. Water oxidation by this photoelectrochemical device was confirmed by the formation of O₂. Assuming all the ruthenium catalysts participate in water oxidation, a TON of 16 was obtained with a corresponding TOF of 27 h⁻¹, under the illumination by a 500 W Xe lamp through a 400 nm cut off filter.

To further improve the surface reactivity of the robust ruthenium catalyst, Sun *et al.* developed a derivative of Ru(bpa)(pic)₂, which showed a remarkable electrocatalytic performance for water oxidation.[117] MWCNTs were coated on an ITO glass electrode *via* electrophoretic deposition. Such a modified electrode possessed not only excellent chemical stability, outstanding electronic and mechanical properties, but also high specific surface area. To facilitate the adsorption of the catalyst on the MWCNT sidewalls,

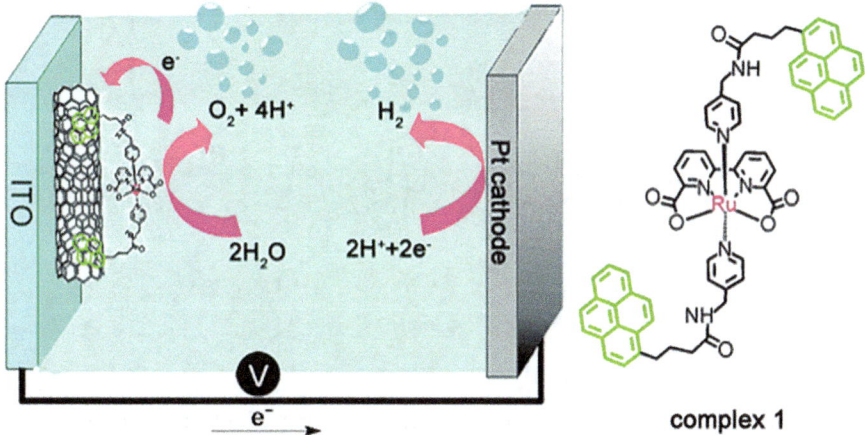

complex 1

Figure 6.13   Electrochemical cell for water splitting and structure of the molecular ruthenium catalyst.
Reproduced from ref. 117 with permission from John Wiley & Sons, Inc.

complex **1** shown in Figure 6.13 was designed and prepared by attaching pendant pyrene moieties to the axial pyridine ligands. The immobilization of the ruthenium catalyst was achieved by soaking the MWCNT-modified electrode in a stock solution of complex **1** in methanol overnight. The surface coverage of the electroactive catalysts was about $2 \pm 0.5 \, nmol \, cm^{-2}$ as estimated from the cyclic voltammograms. A 10 h electrolysis at 1.4 V *vs.* NHE in neutral aqueous solution resulted in a total TON of 110 000, an average TOF of *ca.* $0.3 \, s^{-1}$, and a Tafel slope of 160 mV/decade. When NaCl was used as the electrolyte, controlled potential electrolysis at 1.4 V did not form any chlorine gas, which indicated that the new electrodes had a high selectivity for O_2 evolution and could directly use sea water as the water source.

Besides immobilization of molecular catalysts *via* π–π stacking, carbon nanotube can also be modified to attach variable functional groups to bind water oxidation catalysts. Bonchio and his co-workers decorated MWCNTs with polyamidoamine ammonium dendrimers, followed by electrostatic scavenging at pH 5 to bind a polyanionic ruthenium-containing cluster. The cluster is a totally inorganic and highly robust tetraruthenate moiety, $Li_{10}[Ru_4(H_2O)_4(\mu\text{-}O)_4(\mu\text{-}OH)_2 \, (SiW_{10}O_{36})_2]$, featuring a multimetal oxide structure and nanoscale dimensions. Water oxidation catalysis by this ruthenium cluster in the presence of Ce(IV) had been reported with a re- markably high TOF $(450 \, h^{-1})$.[118–120] Once immobilized on the carbon nanotubes, several advantages were achieved. The carbon nanotubes pro- vided a heterogeneous support to the ruthenium catalyst for electrocatalysis of water oxidation, controlled the material morphology, increased the sur- face area of catalyst assembly, and facilitated the sequential electron transfer to the electrode and hence favored energy dispersion and relieved catalytic fatigue. Various surface characterization techniques were utilized to confirm

the integrity of the ruthenium cluster during and after the assembly process. During controlled potential electrolysis, TOF values of 36–306 h^{-1} were obtained at pH 7 and reached a peak performance of 306 h^{-1} at 0.60 V.

Graphene is another appealing material to support water oxidation catalysts. Bonchio *et al.* extended their ruthenium cluster research to functionalized graphene nano-sheets which provided a sp^2 platform to anchor tetra-ruthenium polyoxometalate catalyst {Ru$_4$(μ-O)$_4$(μ-OH)$_2$(H$_2$O)$_4$[γ-SiW$_{10}$ O$_{36}$]$_2$}$^{10-}$ (Ru$_4$POM @d-G) shown in Figure 6.14.[121] In order to preserve the sp^2 carbon network at a large extent, they decided to process pristine graphene *via* direct 3-dipolar cyclo-addition, following a two-step protocol to anchor a first generation polyaminoamide Dendron. The functionalized graphene was able to host the polyanionic Ru$_4$POM catalyst *via* a cooperative interplay of electrostatic forces and hydrogen bonds. Graphene's flat confinement zones were shaped to host the inorganic catalyst domains, modulate the electron transfer kinetics through a multilayer morphology, and set the highest specific surface area potentially available. This functionalized graphene nano-sheets resulted in the improvement of material

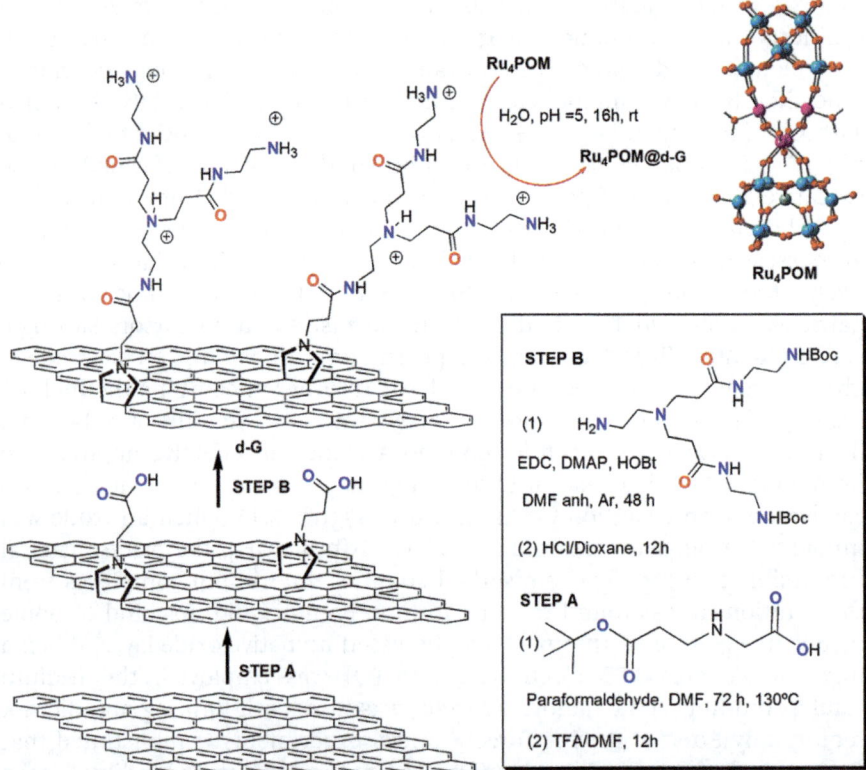

Figure 6.14 Synthesis of graphene nano-platform supporting Ru$_4$POM. Reproduced from ref. 121 with permission from The American Chemical Society.

wettability, gas-permeation, and storage properties, facilitating the overall O_2 evolution process. The assembly was characterized by a myriad of spectroscopic and electron microscopy techniques, confirming that the ruthenium clusters were preserved during the assembling procedure. The O_2 evolution activity of this system was evaluated under electrocatalytic conditions with cyclic voltammetry and chronoamperometry. The hybrid material displayed O_2 evolution at very low overpotential, 300 mV at pH 7 with negligible loss of performance after 4 h testing and enhanced the TOF by one order of magnitude with respect to the isolated catalyst and a two-fold enhancement compared to the nanotube analogues mentioned above.

6.3.2 Hybrid Assemblies Containing Iridium Catalysts

Iridium complexes is another class of O_2 evolution catalysts consisting of expensive metals which have been intensively studied. In 2007, Bernhard and co-workers reported a series of cyclometalated iridium (III) aquo complexes which were able to catalyze water oxidation efficiently.[122] Several advantages were found in using the cyclometalated iridium complexes for catalytic water oxidation, such as simple ligand design, straightforward synthesis, and easy tuning of their electronic properties through substitution on cyclometalated ligands. Later, Crabtree and Brudvig's groups developed other highly active mononuclear iridium water oxidation catalysts capped with Cp* (pentamethylcyclopentadienyl) ligands.[123] They identified several highly active homogeneous O_2 evolution catalysts that are formed by the oxidative removal of Cp*, an ancillary organic ligand, from well-studied Cp*Ir-based precursors. The compounds that formed from these precatalysts all possessed a single bidentate chelate ligand per iridium that was oxidatively stable and prevented the formation of IrO_x-based films or nanoparticles under oxidative conditions. In contrast, Cp*Ir precursors lacking a stable bidentate ligand anodically deposited amorphous IrO_x on electrode to give heterogeneous O_2 evolution catalysts. In order to combine the high efficiency of those molecular iridium catalysts with the stability of bulk metal oxides, Brudvig *et al.* recently reported a facile but effective approach to immobilize the molecular iridium catalysts on metal oxides for electrocatalytic oxygen evolution.[124] As shown in Figure 6.15, when an oxide was immersed in an aqueous solution of [Cp*Ir(pyalc)OH], the oxide material irreversibly and rapidly chemisorbed some of the iridium complexes from the solution. It was found that the iridium complex did not bind to noble metals like gold or platinum which possessed no native oxide layer. When a high-surface area ITO electrode (nanoITO) was employed, the iridium catalyst binding to the nanoITO surface was rapid, self-limiting and did not require any external driving force. Control experiments demonstrated that the removal of the Cp* ligand was necessary for the surface binding to take place. A variety of surface characterization techniques, including TEM, SEM, EDX, and XPS, and UV-vis absorbance were utilized to confirm the existence of iridium and the distinctive nature of the absorbed monolayer from that of IrO_x.

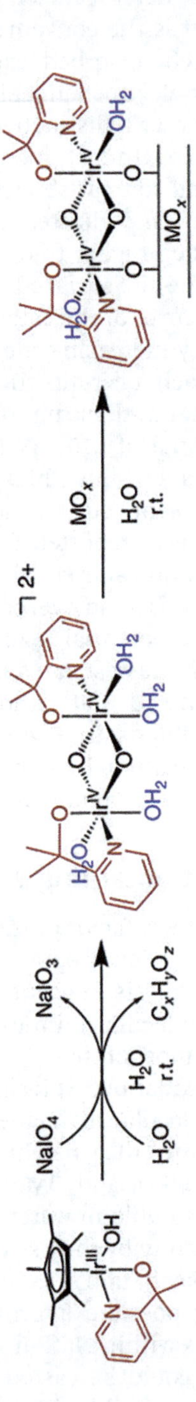

Figure 6.15 Formation of the homogenous diiridium catalyst and the proposed molecular structure of the adsorbate on metal oxides. Reproduced from ref. 124 with permission from Nature Publishing Group.

An apparent advantage of this heterogenized iridium catalyst compared to its homogeneous counterpart is the convenient construction for electrocatalytic water oxidation. The chemisorbed iridium catalyst exhibited exceptional activity when driven electrochemically. The onset of the water oxidation catalytic wave occurred at a distinctively lower potential than the $Ir^{V/IV}$ redox couple in IrO_x samples. In fact, integration of the $Ir^{IV/III}$ wave of the absorbed iridium catalyst and comparison with the total iridium loading derived from the UV-vis spectrum indicated that >90% iridium on the electrode were electroactive, much higher than those of IrO_x nanoparticles. Another advantage of using single-site surface-bound molecular catalysts for water oxidation was accurate control of the catalytic overpotential by tuning the electroactive surface area. By increasing the absorbed layer's thickness, the catalytic overpotential to reach a certain current density could be easily tuned. For instance, a typical 3-μm-thick nanoITO with absorbed iridium catalysts required an overpotential of 275 mV to attain a catalytic current density of 0.5 mA/cm^2, whereas a 18 μm thick film only needed an overpotential <160 mV to reach the same current density. These overpotentials are among the smallest values ever reported for water oxidation catalysts. Furthermore, the adsorbed iridium catalysts also exhibited excellent stability. Controlled potential electrolysis at an overpotential of 250 mV reached a TON > 10^6 while at a higher overpotential (520 mV) a TOF of 7.9 s^{-1} was measured, which was one of the fastest rates ever reported for water oxidation catalysis. It should be noted that although the proposed surface-bound structure of iridium catalysts in Figure 6.15 are consistent with available experimental data, it cannot be considered as a definitive conformation. Other catalytic iridium structures may exist as well.

6.3.3 Hybrid Assemblies Containing Manganese Catalysts

As the only transition metal in the cofactor of the oxygen evolution centre in photosystems II, manganese has been a popular target metal in the development of water oxidation catalysis. Recent years have witnessed the emergence of several delicate molecular mimics of the Mn_4CaO_5 cubane in photosystem II.[125–127] However, most of them have shown none or negligible catalytic capability for water oxidation. Spiccia, Dismukes, and Swiegers *et al.* reported a prototypical molecular manganese-oxo cube $[Mn_4O_4]^{n+}$ in a family of "cubane" complexes $[Mn_4O_4L_6]$, in which L$^-$ is a diarylphosphinate ligand $(p\text{-R-}C_6H_4)_2PO_2^-$ (R = H, alkyl, and OMe).[128,129] When R = H, the resulting manganese cluster is insoluble in water and most organic solvents. Spiccia *et al.* overcame this difficulty by the use of a two-phase water/Nafion system.[130] This allows the photooxidation system to function in the presence of water, as well as providing possible alternative binding sites on the polymer-bound sulfonate anions within Nafion. It was rationalized that the perfluorinated polymer Nafion could be casted as a membrane that comprised hydrophobic domains separated by ionisable, hydrated head groups (sulfonic acids) that formed the aqueous channels (reported diameter

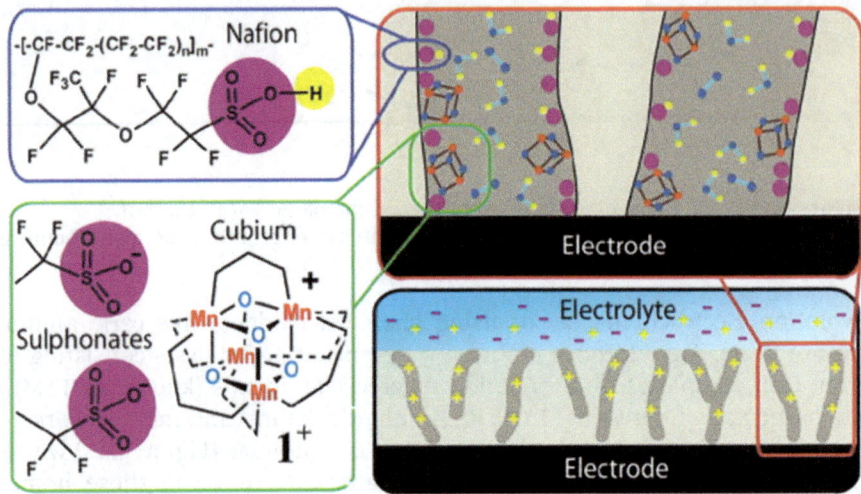

Figure 6.16 Schematic representation of the manganese cubium, $[Mn_4O_4L_6]^{1+}$ $(L = (p\text{-MeO-Ph})_2PO_2)$, suspended in a Nafion membrane. Reproduced from ref. 130 with permission from Elsevier.

ca. 20 nm) traversing the polymer domains (Figure 6.16). These channels are good proton conductors that are permeable to cations but not to anions. These properties make Nafion an excellent medium within which to immobilize hydrophobic cations, whilst maintaining direct contact with an aqueous electrolyte. Spiccia *et al.* found that the manganese cluster (Cubium) readily exchanged with protons in Nafion membranes (3–8 μm thick) and subsequent immersion in water prevented its leaching from the Nafion layer. When the Cubium/Nafion/glassy carbon electrode was polarized at 1.00 V *vs.* Ag/AgCl upon visible light irradiation, a sustainable photocurrent was obtained. During a 65 h photoelectrolysis, the membrane passed a net charge of 0.163 C in 0.1 M Na_2SO_4, which led to an approximate TON of 1 000 per Cubium.

Besides photoactive for O_2 evolution in pure water, this system also facilitated photocatalytic water oxidation from non-potable water sources like seawater with little chlorine formation on the anode. This result was due to the electrostatic repulsion and block transport of anionic Cl^- ions by Nafion as a cation exchange resin.

6.3.4 Hybrid Assemblies Containing Iron Catalysts

Being one of the most abundant terrestrial elements and widely involved in many enzymes for oxidation catalysis, iron is another popular metal for O_2 evolution catalysis. Although common iron oxides are not very active for water oxidation, iron impurities in other metal oxides have been reported to significantly enhance the catalytic performance.[131,132] On the other hand, a few molecular iron complexes were reported to show catalytic activity for O_2

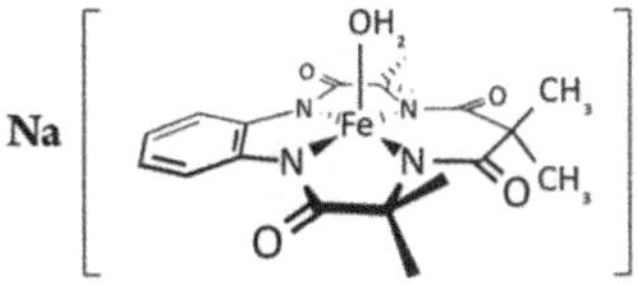

Figure 6.17   The molecular structure of the prototype Fe-TAML catalyst.
                  Reproduced from ref. 134 with permission from the American Chemical
                  Society.

evolution electrochemically or using chemical oxidants like ceric ammonium nitrate. Among those reported water oxidation catalysts consisting of iron, iron complexes of tetra-amido macrocyclic ligands (known as TAML) are particularly interesting. In 2010, Bernhard, Collins, and their co-workers reported a series of mononuclear Fe[III]-TAML complexes (Figure 6.17) which could act as molecular O_2 evolution catalysts.[133] However, these homogeneous catalysts only exhibited low turnover numbers. In order to enhance the stability of these Fe-TAML catalysts for water oxidation, Collins *et al.* furthered their study in 2014 by mixing the molecular iron catalysts within Vulcan XC-72 carbon black and depositing it on a glassy carbon electrode with Nafion.[134] The catalyst/carbon black was first sonicated for 1 h, followed by filtering and drying. Subsequently, the catalyst ink was prepared by mixing the iron/carbon black with Nafion in a methanol/isopropanol mixture. After sonication, the catalyst ink was directly deposited onto either a glassy carbon disk electrode or a Toray carbon paper for electrochemical studies. Cyclic voltammetry and real-time quantification of generated O_2 demonstrated that iron oxidation underwent a Fe[IV/III] redox process followed by a rapid O_2 evolution. Control experiments with common iron salts under the same condition did not produce measurable O_2, corroborating the catalytic role of the Fe-TAML as the active catalyst within the catalyst ink. Long-term controlled potential electrolysis resulted in a $\sim$45% Faradaic efficiency and an estimated turnover frequency of 0.081 s^{-1}, which represented a large increase compared to the experiments with chemical oxidants under homogeneous conditions. The authors found that the selectivity of O_2 evolution was limited by parasitic carbon oxidation whereas the iron catalyst itself was stable despite the oxidation of the supporting materials. Therefore, immobilizing Fe-TAML in more oxidatively robust and electrically conducting supports was anticipated to improve the overall performance for electrocatalytic O_2 production.

6.4   Conclusion

The integration of well-defined molecular catalysts on solid-state nanoscale materials for both half reactions of water splitting has resulted in many elegant hybrid systems, most of which indeed exhibit improved catalytic activities compared to their solely homogeneous counterparts. Alongside the

advancement of materials science, especially in materials of high conductivity and large surface area, it is anticipated that novel and improved hybrid molecular-nanomaterial assemblies of water splitting catalysis will be developed. Although significant progress has been achieved, much further efforts are still required to realize the real application of those hybrid systems at an industrial scale. For instance, whether the intrinsic turnover frequency of an immobilized catalysts is superior to that of a homogenous counterpart is not always assured. Another aspect deserving special attention is whether the molecular nature of the active site of a hybrid system preserves during the process of water splitting, which is usually conducted under harsh (highly acidic or basic) conditions. It is not uncommon that original active sites with well-defined structures transform to metallic species or metal oxides upon strong reducing or oxidizing potentials, respectively. In fact, the decomposition products of some molecular catalysts can also act as active water splitting catalysts.[135–137] Systematic and detailed *in situ* spectroscopic studies of hybrid molecular-nanomaterials are of pivotal importance for understanding the real active species, which is highly desirable and also recommended for future efforts in hybrid systems of electro- and photocatalytic water splitting.

References

1. V. Artero and M. Fontecave, *Chem. Soc. Rev.*, 2013, **42**, 2338–2356.
2. R. Cammack, *Nature*, 1999, **397**, 214–215.
3. M. Frey, *ChemBioChem*, 2002, **3**, 153–160.
4. E. Reisner, *Eur. J. Inorg. Chem.*, 2011, **2011**, 1005–1016.
5. J. W. Tye, J. Lee, H.-W. Wang, R. Mejia-Rodriguez, J. H. Reibenspies, M. B. Hall and M. Y. Darensbourg, *Inorg. Chem.*, 2005, **44**, 5550–5552.
6. S. Canaguier, M. Field, Y. Oudart, J. Pecaut, M. Fontecave and V. Artero, *Chem. Commun.*, 2010, **46**, 5876–5878.
7. J. Chen, A. K. Vannucci, C. A. Mebi, N. Okumura, S. C. Borowski, M. Swenson, L. T. Lockett, D. H. Evans, R. S. Glass and D. L. Lichtenberger, *Organometallics*, 2010, **29**, 5330–5340.
8. A. Begum, G. Moula and S. Sarkar, *Chem. – Eur. J.*, 2010, **16**, 12324–12327.
9. B. E. Barton and T. B. Rauchfuss, *J. Am. Chem. Soc.*, 2010, **132**, 14877–14885.
10. B. E. Barton, M. T. Olsen and T. B. Rauchfuss, *Curr. Opin. Biotechnol.*, 2010, **21**, 292–297.
11. H. I. Karunadasa, C. J. Chang and J. R. Long, *Nature*, 2010, **464**, 1329–1333.
12. H. I. Karunadasa, E. Montalvo, Y. Sun, M. Majda, J. R. Long and C. J. Chang, *Science*, 2012, **335**, 698–702.
13. P. Connolly and J. H. Espenson, *Inorg. Chem.*, 1986, **25**, 2684–2688.
14. X. Hu, B. M. Cossairt, B. S. Brunschwig, N. S. Lewis and J. C. Peters, *Chem. Commun.*, 2005, 4723–4725.

15. X. Hu, B. S. Brunschwig and J. C. Peters, *J. Am. Chem. Soc.*, 2007, **129**, 8988–8998.

16. M. Wang, L. Chen and L. Sun, *Energy Environ. Sci.*, 2012, **5**, 6763–6778.

17. P.-A. Jacques, V. Artero, J. Pécaut and M. Fontecave, *Proc. Natl. Acad. Sci. U. S. A.*, 2009, **106**, 20627–20632.

18. W. R. McNamara, Z. Han, P. J. Alperin, W. W. Brennessel, P. L. Holland and R. Eisenberg, *J. Am. Chem. Soc.*, 2011, **133**, 15368–15371.

19. W. R. McNamara, Z. Han, C.-J. Yin, W. W. Brennessel, P. L. Holland and R. Eisenberg, *Proc. Natl. Acad. Sci. U. S. A.*, 2012, **109**, 15594–15599.

20. B. D. Stubbert, J. C. Peters and H. B. Gray, *J. Am. Chem. Soc.*, 2011, **133**, 18070–18073.

21. B. J. Fisher and R. Eisenberg, *J. Am. Chem. Soc.*, 1980, **102**, 7361–7363.

22. D. E. Brown, M. N. Mahmood, A. K. Turner, S. M. Hall and P. O. Fogarty, *Int. J. Hydrogen Energy*, 1982, **7**, 405–410.

23. R. M. Kellett and T. G. Spiro, *Inorg. Chem.*, 1985, **24**, 2378–2382.

24. U. Koelle and S. Paul, *Inorg. Chem.*, 1986, **25**, 2689–2694.

25. J. P. Collin, A. Jouaiti and J. P. Sauvage, *Inorg. Chem.*, 1988, **27**, 1986–1990.

26. P. V. Bernhardt and L. A. Jones, *Inorg. Chem.*, 1999, **38**, 5086–5090.

27. U. J. Kilgore, J. A. S. Roberts, D. H. Pool, A. M. Appel, M. P. Stewart, M. R. DuBois, W. G. Dougherty, W. S. Kassel, R. M. Bullock and D. L. DuBois, *J. Am. Chem. Soc.*, 2011, **133**, 5861–5872.

28. B. D. Stubbert, J. C. Peters and H. B. Gray, *J. Am. Chem. Soc.*, 2011, **133**, 18070–18073.

29. W. M. Singh, T. Baine, S. Kudo, S. Tian, X. A. N. Ma, H. Zhou, N. J. DeYonker, T. C. Pham, J. C. Bollinger, D. L. Baker, B. Yan, C. E. Webster and X. Zhao, *Angew. Chem., Int. Ed.*, 2012, **51**, 5941–5944.

30. F. Quentel, G. Passard and F. Gloaguen, *Energy Environ. Sci.*, 2012, **5**, 7757–7761.

31. T. F. Jaramillo, K. P. Jorgensen, J. Bonde, J. H. Nielsen, S. Horch and I. Chorkendorff, *Science*, 2007, **317**, 100–102.

32. D. Merki, S. Fierro, H. Vrubel and X. Hu, *Chem. Sci.*, 2011, **2**, 1262–1267.

33. D. Merki and X. Hu, *Energy Environ. Sci.*, 2011, **4**, 3878–3888.

34. D. Merki, H. Vrubel, L. Rovelli, S. Fierro and X. Hu, *Chem. Sci.*, 2012, **3**, 2515–2525.

35. C. G. Morales-Guio and X. Hu, *Acc. Chem. Res.*, 2014, **47**, 2671–2681.

36. P. D. Tran, M. Nguyen, S. S. Pramana, A. Bhattacharjee, S. Y. Chiam, J. Fize, M. J. Field, V. Artero, L. H. Wong, J. Loo and J. Barber, *Energy Environ. Sci.*, 2012, **5**, 8912–8916.

37. D. Kong, J. J. Cha, H. Wang, H. R. Lee and Y. Cui, *Energy Environ. Sci.*, 2013, **6**, 3553–3558.

38. C. Di Giovanni, W.-A. Wang, S. Nowak, J.-M. Grenèche, H. Lecoq, L. Mouton, M. Giraud and C. Tard, *ACS Catal.*, 2014, **4**, 681–687.

39. Y. Sun, C. Liu, D. C. Grauer, J. Yano, J. R. Long, P. Yang and C. J. Chang, *J. Am. Chem. Soc.*, 2013, **135**, 17699–17702.

40. N. Jiang, L. Bogoev, M. Popova, S. Gul, J. Yano and Y. Sun, *J. Mater. Chem. A*, 2014, **2**, 19407–19414.
41. B. You, N. Jiang, M. Sheng and Y. Sun, *Chem. Commun.*, 2015, **51**, 4252–4255.
42. H. Vrubel and X. Hu, *Angew. Chem., Int. Ed.*, 2012, **51**, 12703–12706.
43. L. Liao, S. Wang, J. Xiao, X. Bian, Y. Zhang, M. D. Scanlon, X. Hu, Y. Tang, B. Liu and H. H. Girault, *Energy Environ. Sci.*, 2014, **7**, 387–392.
44. E. J. Popczun, J. R. McKone, C. G. Read, A. J. Biacchi, A. M. Wiltrout, N. S. Lewis and R. E. Schaak, *J. Am. Chem. Soc.*, 2013, **135**, 9267–9270.
45. Y. Xu, R. Wu, J. Zhang, Y. Shi and B. Zhang, *Chem. Commun.*, 2013, **49**, 6656–6658.
46. Z. Xing, Q. Liu, A. M. Asiri and X. Sun, *ACS Catal.*, 2014, **5**, 145–149.
47. Z. Xing, Q. Liu, A. M. Asiri and X. Sun, *Adv. Mater.*, 2014, **26**, 5702–5707.
48. J. M. McEnaney, J. Chance Crompton, J. F. Callejas, E. J. Popczun, C. G. Read, N. S. Lewis and R. E. Schaak, *Chem. Commun.*, 2014, **50**, 11026–11028.
49. F. H. Saadi, A. I. Carim, E. Verlage, J. C. Hemminger, N. S. Lewis and M. P. Soriaga, *J. Phys. Chem. C*, 2014, **118**, 29294–29300.
50. P. Jiang, Q. Liu, Y. Liang, J. Tian, A. M. Asiri and X. Sun, *Angew. Chem., Int. Ed.*, 2014, **53**, 12855–12859.
51. N. Jiang, B. You, M. Sheng and Y. Sun, *Angew. Chem., Int. Ed.*, 2015, DOI: 10.1002/anie.201501616.
52. Q. Lu, G. S. Hutchings, W. Yu, Y. Zhou, R. V. Forest, R. Tao, J. Rosen, B. T. Yonemoto, Z. Cao, H. Zheng, J. Q. Xiao, F. Jiao and J. G. Chen, *Nat. Commun.*, 2015, **6**, 6567.
53. M. S. Passos, M. A. Queiros, T. Le Gall, S. K. Ibrahim and C. J. Pickett, *J. Electroanal. Chem.*, 1997, **435**, 189–203.
54. C. Tard, X. Liu, S. K. Ibrahim, M. Bruschi, L. D. Gioia, S. C. Davies, X. Yang, L.-S. Wang, G. Sawers and C. J. Pickett, *Nature*, 2005, **433**, 610–613.
55. S. K. Ibrahim, X. Liu, C. Tard and C. J. Pickett, *Chem. Commun.*, 2007, 1535–1537.
56. L. Sun, B. Akermark and S. Ott, *Coord. Chem. Rev.*, 2005, **249**, 1653–1663.
57. T. Nann, S. K. Ibrahim, P.-M. Woi, S. Xu, J. Ziegler and C. J. Pickett, *Angew. Chem., Int. Ed.*, 2010, **49**, 1574–1577.
58. F. Wang, W.-G. Wang, X.-J. Wang, H.-Y. Wang, C.-H. Tung and L.-Z. Wu, *Angew. Chem., Int. Ed.*, 2011, **50**, 3193–3197.
59. C.-B. Li, Z.-J. Li, S. Yu, G.-X. Wang, F. Wang, Q.-Y. Meng, B. Chen, K. Feng, C.-H. Tung and L.-Z. Wu, *Energy Environ. Sci.*, 2013, **6**, 2597–2602.
60. F. Wang, W.-J. Liang, J.-X. Jian, C.-B. Li, B. Chen, C.-H. Tung and L.-Z. Wu, *Angew. Chem., Int. Ed.*, 2013, **52**, 8134–8138.
61. J.-X. Jian, Q. Liu, Z.-J. Li, F. Wang, X.-B. Li, C.-B. Li, B. Liu, Q.-Y. Meng, B. Chen, K. Feng, C.-H. Tung and L.-Z. Wu, *Nat. Commun.*, 2013, **4**, 2695.

62. B. J. Fisher and R. Eisenberg, *J. Am. Chem. Soc.*, 1980, **102**, 7361–7363.
63. R. M. Kellett and T. G. Spiro, *Inorg. Chem.*, 1985, **24**, 2373–2377.
64. R. Abdel-Hamid, H. M. El-Sagher, A. M. Abdel-Mawgoud and A. Nafady, *Polyhedron*, 1998, **17**, 4535–4541.
65. B. J. Fisher and R. Eisenberg, *J. Am. Chem. Soc.*, 1980, **102**, 7361–7363.
66. X. Hu, B. S. Brunschwig and J. C. Peters, *J. Am. Chem. Soc.*, 2007, **129**, 8988–8998.
67. P. V. Bernhardt and L. A. Jones, *Inorg. Chem.*, 1999, **38**, 5086–5090.
68. R. M. Kellett and T. G. Spiro, *Inorg. Chem.*, 1985, **24**, 2373–2377.
69. F. Zhao, J. Zhang, T. Abe, D. Wöhrle and M. Kaneko, *J. Mol. Catal. A: Chem.*, 1999, **145**, 245–256.
70. D. E. Brown, M. N. Mahmood, M. C. M. Man and A. K. Turner, *Electrochim. Acta*, 1984, **29**, 1551–1556.
71. S. Losse, J. G. Vos and S. Rau, *Coord. Chem. Rev.*, 2010, **254**, 2492–2504.
72. V. Artero, M. Chavarot-Kerlidou and M. Fontecave, *Angew. Chem., Int. Ed.*, 2011, **50**, 7238–7266.
73. J. P. Bigi, T. E. Hanna, W. H. Harman, A. Chang and C. J. Chang, *Chem. Commun.*, 2010, **46**, 958–960.
74. Y. Sun, J. P. Bigi, N. A. Piro, M. L. Tang, J. R. Long and C. J. Chang, *J. Am. Chem. Soc.*, 2011, **133**, 9212–9215.
75. Y. Sun, J. Sun, J. R. Long, P. Yang and C. J. Chang, *Chem. Sci.*, 2013, **4**, 118–124.
76. G. M. Jacobsen, J. Y. Yang, B. Twamley, A. D. Wilson, R. M. Bullock, M. Rakowski DuBois and D. L. DuBois, *Energy Environ. Sci.*, 2008, **1**, 167–174.
77. E. S. Andreiadis, P.-A. Jacques, P. D. Tran, A. Leyris, M. Chavarot-Kerlidou, B. Jousselme, M. Matheron, J. Pécaut, S. Palacin, M. Fontecave and V. Artero, *Nat. Chem.*, 2013, **5**, 48–53.
78. E. M. Sletten and C. R. Bertozzi, *Acc. Chem. Res.*, 2011, **44**, 666–676.
79. E. B. Hulley, P. T. Wolczanski and E. B. Lobkovsky, *J. Am. Chem. Soc.*, 2011, **133**, 18058–18061.
80. P.-A. Jacques, V. Artero, J. Pecaut and M. Fontecave, *Proc. Natl. Acad. Sci. U. S. A.*, 2009, **106**, 20627–20632.
81. A. Krawicz, J. Yang, E. Anzenberg, J. Yano, I. D. Sharp and G. F. Moore, *J. Am. Chem. Soc.*, 2013, **135**, 11861–11868.
82. R. Cammack, *Nature*, 1999, **397**, 214–215.
83. S. Canaguier, V. Artero and M. Fontecave, *Dalton Trans.*, 2008, 315–325.
84. B. J. Fisher and R. Eisenberg, *J. Am. Chem. Soc.*, 1980, **102**, 7361–7363.
85. M. Rakowski Dubois and D. L. Dubois, *Acc. Chem. Res.*, 2009, **42**, 1974–1982.
86. U. J. Kilgore, J. A. S. Roberts, D. H. Pool, A. M. Appel, M. P. Stewart, M. R. DuBois, W. G. Dougherty, W. S. Kassel, R. M. Bullock and D. L. Dubois, *J. Am. Chem. Soc.*, 2011, **133**, 5861–5872.
87. A. Le Goff, V. Artero, B. Jousselme, P. D. Tran, N. Guillet, R. Métayé, A. Fihri, S. Palacin and M. Fontecave, *Science*, 2009, **326**, 1384–1387.

88. Z. Wu, Z. Chen, X. Du, J. M. Logan, J. Sippel, M. Nikolou, K. Kamaras, J. R. Reynolds, D. B. Tanner, A. F. Hebard and A. G. Rinzler, *Science*, 2004, **305**, 1273–1276.

89. P. D. Tran, A. Le Goff, J. Heidkamp, B. Jousselme, N. Guillet, S. Palacin, H. Dau, M. Fontecave and V. Artero, *Angew. Chem., Int. Ed.*, 2011, **50**, 1371–1374.

90. G. F. Moore and I. D. Sharp, *J. Phys. Chem. Lett.*, 2013, **4**, 568–572.

91. Z. Han, F. Qiu, R. Eisenberg, P. L. Holland and T. D. Krauss, *Science*, 2012, **338**, 1321–1324.

92. Y. Umena, K. Kawakami, J.-R. Shen and N. Kamiya, *Nature*, 2011, **473**, 55–60.

93. M. D. Kärkäs, O. Verho, E. V. Johnston and B. Åkermark, *Chem. Rev.*, 2014, **114**, 11863–12001.

94. R. Brimblecombe, G. F. Swiegers, G. C. Dismukes and L. Spiccia, *Angew. Chem., Int. Ed.*, 2008, **47**, 7335–7338.

95. S. M. Barnett, K. I. Goldberg and J. M. Mayer, *Nat. Chem.*, 2012, **4**, 498–502.

96. A. Singh and L. Spiccia, *Coord. Chem. Rev.*, 2013, **257**, 2607–2622.

97. S. W. Gersten, G. J. Samuels and T. J. Meyer, *J. Am. Chem. Soc.*, 1982, **104**, 4029–4030.

98. J. J. Concepcion, J. W. Jurss, M. K. Brennaman, P. G. Hoertz, A. O. T. Patrocinio, N. Y. Murakami Iha, J. L. Templeton and T. J. Meyer, *Acc. Chem. Res.*, 2009, **42**, 1954–1965.

99. R. Zong and R. P. Thummel, *J. Am. Chem. Soc.*, 2005, **127**, 12802–12803.

100. H.-W. Tseng, R. Zong, J. T. Muckerman and R. Thummel, *Inorg. Chem.*, 2008, **47**, 11763–11773.

101. J. L. Boyer, D. E. Polyansky, D. J. Szalda, R. Zong, R. P. Thummel and E. Fujita, *Angew. Chem., Int. Ed.*, 2011, **50**, 12600–12604.

102. D. E. Polyansky, J. T. Muckerman, J. Rochford, R. Zong, R. P. Thummel and E. Fujita, *J. Am. Chem. Soc.*, 2011, **133**, 14649–14665.

103. N. Kaveevivitchai, R. Zong, H.-W. Tseng, R. Chitta and R. P. Thummel, *Inorg. Chem.*, 2012, **51**, 2930–2939.

104. J. J. Concepcion, J. W. Jurss, P. G. Hoertz and T. J. Meyer, *Angew. Chem., Int. Ed.*, 2009, **48**, 9473–9476.

105. Z. Chen, J. J. Concepcion, J. W. Jurss and T. J. Meyer, *J. Am. Chem. Soc.*, 2009, **131**, 15580–15581.

106. Z. Chen, J. J. Concepcion, J. F. Hull, P. G. Hoertz and T. J. Meyer, *Dalton Trans.*, 2010, **39**, 6950–6952.

107. A. K. Vannucci, L. Alibabaei, M. D. Losego, J. J. Concepcion, B. Kalanyan, G. N. Parsons and T. J. Meyer, *Proc. Natl. Acad. Sci. U. S. A.*, 2013, **110**, 20918–20922.

108. G. N. Parsons, S. M. George and M. Knez, *MRS Bull.*, 2011, **36**, 865–871.

109. L. Duan, A. Fischer, Y. Xu and L. Sun, *J. Am. Chem. Soc.*, 2009, **131**, 10397–10399.

110. L. Li, L. Duan, Y. Xu, M. Gorlov, A. Hagfeldt and L. Sun, *Chem. Commun.*, 2010, **46**, 7307–7309.

111. I. Gillaizeau-Gauthier, F. Odobel, M. Alebbi, R. Argazzi, E. Costa, C. A. Bignozzi, P. Qu and G. J. Meyer, *Inorg. Chem.*, 2001, **40**, 6073–6079.

112. I. Ogino, K. Nagoshi, M. Yagi and M. Kaneko, *J. Chem. Soc., Faraday Trans.*, 1996, **92**, 3431–3434.

113. M. Yagi, S. Tokita, K. Nagoshi, I. Ogino and M. Kaneko, *J. Chem. Soc., Faraday Trans.*, 1996, **92**, 2457–2461.

114. M. Yagi, K. Nagoshi and M. Kaneko, *J. Phys. Chem. B*, 1997, **101**, 5143–5146.

115. R. Brimblecombe, D. R. J. Kolling, A. M. Bond, G. C. Dismukes, G. F. Swiegers and L. Spiccia, *Inorg. Chem.*, 2009, **48**, 7269–7279.

116. R. Brimblecombe, G. C. Dismukes, G. F. Swiegers and L. Spiccia, *Dalton Trans.*, 2009, 9374–9384.

117. F. Li, B. Zhang, X. Li, Y. Jiang, L. Chen, Y. Li and L. Sun, *Angew. Chem., Int. Ed.*, 2011, **50**, 12276–12279.

118. F. M. Toma, A. Sartorel, M. Iurlo, M. Carraro, P. Parisse, C. Maccato, S. Rapino, B. R. Gonzalez, H. Amenitsch, T. Da Ros, L. Casalis, A. Goldoni, M. Marcaccio, G. Scorrano, G. Scoles, F. Paolucci, M. Prato and M. Bonchio, *Nat. Chem.*, 2010, **2**, 826–831.

119. A. Sartorel, M. Carraro, G. Scorrano, R. D. Zorzi, S. Geremia, N. D. McDaniel, S. Bernhard and M. Bonchio, *J. Am. Chem. Soc.*, 2008, **130**, 5006–5007.

120. Y. V. Geletii, B. Botar, P. Kögerler, D. A. Hillesheim, D. G. Musaev and C. L. Hill, *Angew. Chem., Int. Ed.*, 2008, **47**, 3896–3899.

121. M. Quintana, A. M. López, S. Rapino, F. M. Toma, M. Iurlo, M. Carraro, A. Sartorel, C. Maccato, X. Ke, C. Bittencourt, T. Da Ros, G. Van Tendeloo, M. Marcaccio, F. Paolucci, M. Prato and M. Bonchio, *ACS Nano*, 2012, **7**, 811–817.

122. N. D. McDaniel, F. J. Coughlin, L. L. Tinker and S. Bernhard, *J. Am. Chem. Soc.*, 2007, **130**, 210–217.

123. J. F. Hull, D. Balcells, J. D. Blakemore, C. D. Incarvito, O. Eisenstein, G. W. Brudvig and R. H. Crabtree, *J. Am. Chem. Soc.*, 2009, **131**, 8730–8731.

124. S. W. Sheehan, J. M. Thomsen, U. Hintermair, R. H. Crabtree, G. W. Brudvig and C. A. Schmuttenmaer, *Nat. Commun.*, 2015, **6**, 6469.

125. J. S. Kanady, E. Y. Tsui, M. W. Day and T. Agapie, *Science*, 2011, **333**, 733–736.

126. E. Y. Tsui, R. Tran, J. Yano and T. Agapie, *Nat. Chem.*, 2013, **5**, 293–299.

127. J. S. Kanady, P.-H. Lin, K. M. Carsch, R. J. Nielsen, M. K. Takase, W. A. Goddard and T. Agapie, *J. Am. Chem. Soc.*, 2014, **136**, 14373–14376.

128. R. Brimblecombe, G. F. Swiegers, G. C. Dismukes and L. Spiccia, *Angew. Chem., Int. Ed.*, 2008, **47**, 7335–7338.

129. R. Brimblecombe, A. M. Bond, G. C. Dismukes, G. F. Swiegers and L. Spiccia, *Phys. Chem. Chem. Phys.*, 2009, **11**, 6441–6449.

130. R. Brimblecombe, J. Chen, P. Wagner, T. Buchhorn, G. C. Dismukes, L. Spiccia and G. F. Swiegers, *J. Mol. Catal. A: Chem.*, 2011, **338**, 1–6.

131. M. S. Burke, M. G. Kast, L. Trotochaud, A. M. Smith and S. W. Boettcher, *J. Am. Chem. Soc.*, 2015, **137**, 3638–3648.
132. A. M. Smith, L. Trotochaud, M. S. Burke and S. W. Boettcher, *Chem. Commun.*, 2015, **51**, 5261–5263.
133. W. C. Ellis, N. D. McDaniel, S. Bernhard and T. J. Collins, *J. Am. Chem. Soc.*, 2010, **132**, 10990–10991.
134. E. L. Demeter, S. L. Hilburg, N. R. Washburn, T. J. Collins and J. R. Kitchin, *J. Am. Chem. Soc.*, 2014, **136**, 5603–5606.
135. Z.-J. Li, X.-B. Li, J.-J. Wang, S. Yu, C.-B. Li, C.-H. Tung and L.-Z. Wu, *Energy Environ. Sci.*, 2013, **6**, 465–469.
136. J. J. Stracke and R. G. Finke, *J. Am. Chem. Soc.*, 2011, **133**, 14872–14875.
137. A. M. Ullman, Y. Liu, M. Huynh, D. K. Bediako, H. Wang, B. L. Anderson, D. C. Powers, J. J. Breen, H. D. Abruña and D. G. Nocera, *J. Am. Chem. Soc.*, 2014, **136**, 17681–17688.

CHAPTER 7

Hierarchical Nanoheterostructures for Water Splitting

MD MONIRUDDIN,[a] SARKYT KUDAIBERGENOV[b] AND NURXAT NURAJE*[a]

[a] Department of Chemical Engineering, Texas Tech University, Lubbock, TX 79409, USA; [b] Institute of Polymer Materials and Technology, Panfilov Str.52/105, 050004, Almaty, Kazakhstan
*Email: nurxat.nuraje@ttu.edu

7.1 Introduction

The use of fossil fuels as an energy source in various sectors, such as power generation, transportation, *etc.*, releases greenhouse gases. The detrimental influence of greenhouse gases on the environment, especially global warming, and the depletion of finite sources of fossil fuels have intensified the concern to shift energy sources from fossil fuels to alternative green energy carriers.[1-3] In this context, the technology of harvesting energy directly from sunlight can be a strong contender towards fulfilling the need for clean energy with negligible environmental effect. Solar energy is a decentralized and inexhaustible natural source with a magnitude of 9.5×10^4 TW. Currently, the world utilizes energy around 1.2×10^1 TW, which is about 0.01% of the solar energy reaching the Earth's surface.[4-6] However, before any solar energy revolution, some key factors need to be addressed first – such as harnessing solar energy and converting it efficiently

RSC Green Chemistry No. 42
Green Photo-active Nanomaterials: Sustainable Energy and Environmental Remediation
Edited by Nurxat Nuraje, Ramazan Asmatulu and Guido Mul
© The Royal Society of Chemistry 2016
Published by the Royal Society of Chemistry, www.rsc.org

$$H_2O \xrightarrow[\text{Catalyst}]{\text{hγ}} H_2 + \frac{1}{2}O_2 \qquad \text{(A)}$$

$$CH_4 + 2H_2O \longrightarrow 4H_2 + CO_2 \quad \Delta G° = 131 \text{ KJ mol}^{-1} \qquad \text{(B)}$$

Scheme 7.1 Hydrogen production process by: (A) water splitting and (B) methane reforming reaction.

into chemical fuel that can be stored, transported, and used to meet energy demands in a cost-effective way.

Inspired by the natural photosynthesis process, much progress has been made in harvesting and storing solar energy in chemical bonds, especially hydrogen, which can be produced by the water photoelectrolysis process using semiconductors. Additionally, hydrogen is the most viable choice for storing solar energy in the absence of sunlight because of its high energy density. Currently, hydrogen-based energy systems are of great interest because they are environmentally benign for nature. For example, the proton exchange membrane fuel cell (PEMFC) uses hydrogen in a highly efficient way in transportation and power generation with only water as a by-product. As shown in Scheme 7.1, solar energy is converted and stored in hydrogen chemical bonds. Today, hydrogen is mostly produced from fossil fuels using the methane reforming reaction. During this process, CO_2 is one of the products generated, along with hydrogen gases. This process is not an environmentally friendly technique, and hydrogen is best produced using artificial photosynthesis.

In the early 1970s, Fujishima and Honda[7] first demonstrated photo-assisted water splitting using a photoelectrochemical (PEC) cell consisting of titania as the photoanode and platinum as the counter electrode under UV irradiation and with an external bias. This work has stimulated research involving overall water splitting using particulate photocatalysts. The maximum efficiency obtained during the overall water-splitting process is around 5.9%; however, these results still do not satisfy the requirement for practical application (10%).[8] In recent years, nanotechnology offers the opportunity to provide light energy harvesting assemblies and an innovative strategy for desired energy conversion devices. Nanomaterials, as building blocks for solar energy conversion devices, are used in the following ways:

(a) assemblies of donor–acceptor molecules and clusters that mimic photosynthesis
(b) the production of solar fuel using semiconductor-assisted photocatalysis
(c) uses of nanostructured semiconductor materials in solar cells.

Among the nanostructured materials, metal oxide-based photocatalysts are advantageous because they offer high chemical stability, have a negative band gap position for hydrogen generation, and use currently available natural resources.

7.2 Water Splitting by Photocatalysts

7.2.1 Thermodynamic and Kinetic Considerations

In water splitting reactions, pure water decomposes into hydrogen and oxygen (Scheme 7.2). At standard temperature and pressure, the Gibbs free energy change of the water decomposition reaction is positive ($\Delta G = +237.2$ kJ mol^{-1}).[8] So thermodynamically, the overall water splitting process is unfavourable. For photocatalysts to make this reaction favorable under light irradiation, the photocatalysts must absorb radiant light with photon energies.

Generally, materials having semiconductive properties can act as photocatalysts. An ideal semiconductor should possess a suitable band gap (difference between valence band and conduction band) so that it can generate electrons and holes on its surfaces by absorbing sunlight or photon energy. Another key property of semiconductors is the relative position of conduction and valence bands with respect to the water oxidation and reduction potentials, respectively. When the energy of incident light is larger than that of the band gap, electrons in the valence band become excited and shift to the conduction band, simultaneously creating holes in the valence band (Figure 7.1). The photo-generated electrons and holes are responsible for redox reactions that are similar to electrolysis.

Oxidation: H_2O (l) + 2h$^+$ $\longrightarrow$ 2H$^+$ (aq) + $\frac{1}{2}$ O_2 (g)   **(A)**

Reduction: 2H$^+$ (aq) + 2e$^-$ $\longrightarrow$ H_2 (g)   **(B)**

Overall reaction: H_2O (l) $\longrightarrow$ H_2 (g) + $\frac{1}{2}$ O_2 (g)   **(C)**

Scheme 7.2 Water splitting reaction steps.

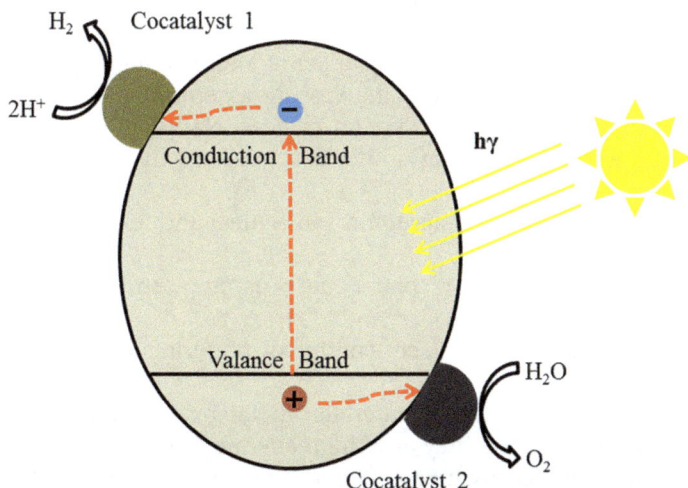

Figure 7.1 Schematic diagram of the overall water-splitting reaction on a solid photocatalyst.

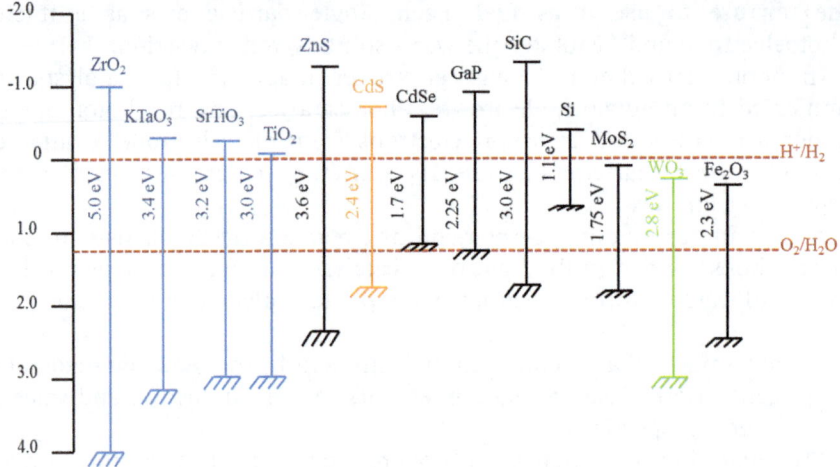

Figure 7.2 Band edge positions of some semiconductors relative to the position of redox potentials of water splitting.
(Reproduced from ref. 9 with permission from the Royal Society of Chemistry. © Royal Society of Chemistry 2009.)

In water-splitting reactions under irradiation, water molecules are reduced by the electrons to form $H_2(g)$ and oxidized by the holes to form $O_2(g)$ for overall water splitting. For the semiconductor, the bottom level of the conduction band has to be more negative than the redox potential of H^+/H_2 (0 V), while the top level of the valence band has to be more positive than the redox potential of O_2/H_2O (1.23 V). Hence, the photocatalyst must absorb radiant light with photon energies of greater than 1.23 eV, which corresponds to light of about 1100 nm $\left(\text{band gap in eV} = \dfrac{1240}{\lambda \text{ in nm}} \right)$ to split water molecules. Additionally, the band gap of a semiconductor must overlap the reduction and oxidation potentials of water, which are +0 and +1.23 V, respectively, *versus* a normal hydrogen electrode (NHE) at a zero reactant solution pH.[9] Band edge positions of few representative semiconductors in contact with aqueous electrolyte relative to the redox potential of the water-splitting reaction are given in Figure 7.2.

7.2.2 The Mechanism of Photocatalytic Water Splitting

Generally, there are two approaches to utilize photocatalysts for water splitting:

(a) particulate photocatalytic system
(b) photoelectrochemical cells.[10]

In particulate photocatalytic systems, particulate suspensions of photocatalysts or semiconductors are used, where the mixture of H_2 and O_2 gases is generated from the bulk suspensions and H_2 need to be separated from

the mixture to use it as fuel. Each single particle acts as a micro-photoelectrode and facilitates the water-splitting redox reaction.

In photoelectrochemical cells, electrodes made of photocatalysts are immersed in an aqueous electrolyte. Photocatalysts absorb photon energy under irradiation and transport electrons from one electrode to another through an external circuit. Here, H_2 and O_2 gases are generated in two separate electrodes.

A good understanding of water-splitting reaction mechanisms is important to choose or design the right photocatalyst material. The steps involved in overall water splitting reactions using photocatalytic materials are:

(1) absorption of a photon from incident light by the semiconductor, and generation of electron and hole pairs in the conduction and valence bands, respectively
(2) separation and transfer of photo-excited carriers to the surface active sites
(3) consumption of photo-excited carriers by water redox reaction on the active site (Figure 7.3).

Recombination of charges may occur along with the reaction. The semiconductor material properties including crystallinity, particle size and shape, and crystal structure play an important role to prevent recombination phenomena. Furthermore, due to the existence of defects in the bulk and surfaces of the nanomaterials, the oppositely charged electrons and holes can be recombined and not take part in the water-splitting reactions. Nano size particles enhance the photocatalytic activity of semiconductors because they have high surface areas and this results in increased reaction active

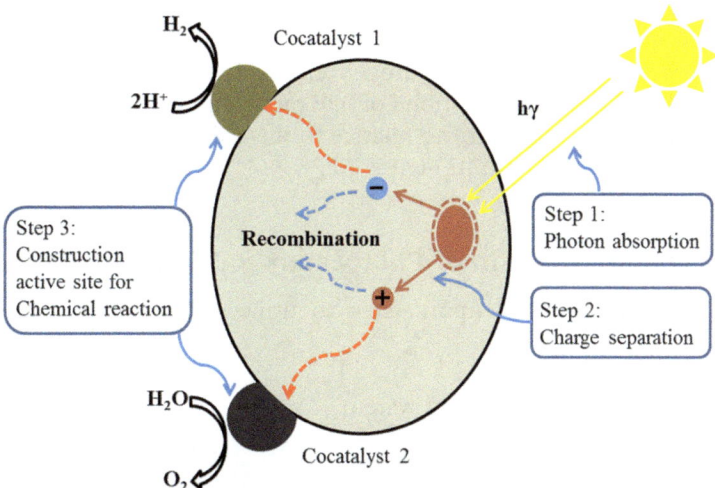

Figure 7.3 The electron transport processes during water-splitting reactions.

sites. Photocatalysts modified with suitable cocatalysts help to prevent charge separation, improve electron and hole transport processes, and facilitate the surface reaction by reducing the overpotential. Generally, noble metals such as Pt and RuO_2 are broadly used as cocatalysts.

During the half reaction of hydrogen and oxygen production, sacrificial agents are used as electron donors or acceptors to increase the efficiency of the products. Methanol, diethanol amine, triethanol amine, ethanol, *etc.* are used for hydrogen production half reactions. In the presence of UV irradiation, the semiconductor absorbs the photon energy and creates an electron/hole pair. The photo-excited electrons migrate to the surface of the semiconductor and are entrapped by the cocatalysts such as Pt (Figure 7.4A). The electron eventually becomes responsible for the hydrogen production at the interface of the cocatalyst and thin layer of water. Some dye elements may be used to absorb the visible light to enhance the absorption of sunlight because most metal oxide particles' band gaps lie in the UV range. A sacrificial agent such as methanol has the standard electrode potential for reduction of carbonic acid to methanol H_2CO_3/CH_3OH (+0.04 V). This is more negative than that for reduction of oxygen to water O_2/H_2O (+1.23 V *versus* NHE). Therefore, thermodynamically methanol is oxidized more easily than water by donating electrons to the positive hole in the valence band of the host photocatalyst to complete the water redox reaction. This reaction mechanism continues until the exhaustion of sacrificial agents.

During the water oxidation half reaction (Figure 7.4B), the photo-excited holes from the semiconductor valence band are entrapped by a cocatalyst such as RuO_2 or IrO_2. These holes are responsible for water oxidation reactions which occur at the surface of the cocatalysts. With a sacrificial agent, for example $AgNO_3$, the standard electrode potential for reduction of silver cation to metallic silver Ag^+/Ag (+0.8 V *versus* SHE) is less negative than reduction potential of proton to hydrogen H^+/H_2 (0 V *versus* NHE).

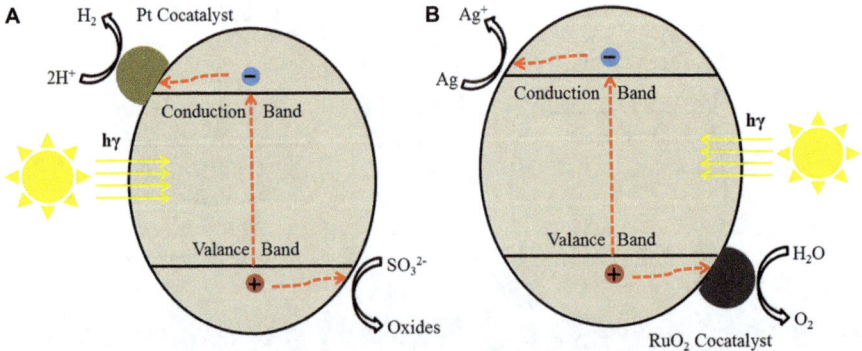

Figure 7.4 (A) Half reaction for hydrogen production using a photocatalyst. (B) Half reaction for oxygen evolution using a photocatalyst.

Therefore, the silver cation is reduced more easily than protons by accepting electrons in the conduction band of the host photocatalyst to compete with the water reduction reaction.

Based on the above understanding, now we will discuss two strategies for overall water splitting under solar irradiation. The first one is the dual-bed configuration or Z scheme and the second one is dye-sensitized photo-electrochemical cells. In the Z scheme, two semiconductors with appropriate band gap energy are working in series for water splitting (Figure 7.5). One semiconductor's (catalyst-A) conduction band position needs to be more positive than the water reduction potential, so that it can reduce water molecules to generate hydrogen gas. In contrast, the other semiconductor's (catalyst-B) valence band position needs to be more negative than the water oxidation potential to oxidize water for oxygen generation. During hydrogen generation, photo-excited electrons reduce water to hydrogen and photo-excited holes oxidize redox mediator-R (electron donor) to redox mediator-O (electron acceptor) on the surface of catalyst A. Concurrently, on the surface of catalyst B, photo-excited electrons reduce redox mediator-O to redox mediator-R and photo-excited holes oxidize water to oxygen.

In dye-sensitized photoelectrochemical cells, photoelectrodes are loaded with dye molecules that absorb visible-light photons to excite the electrons, which are then injected into the conduction band of the metal oxides. Eventually, the electrons proceed to the counter electrode through conductive wires and facilitate the water reduction into hydrogen gas. The oxidation reaction occurs on the surface of the cocatalyst, such as IrO_2 or RuO_2. After the process, dye molecules need to be regenerated, and the regeneration is done by using electrons from water oxidation reactions.[11]

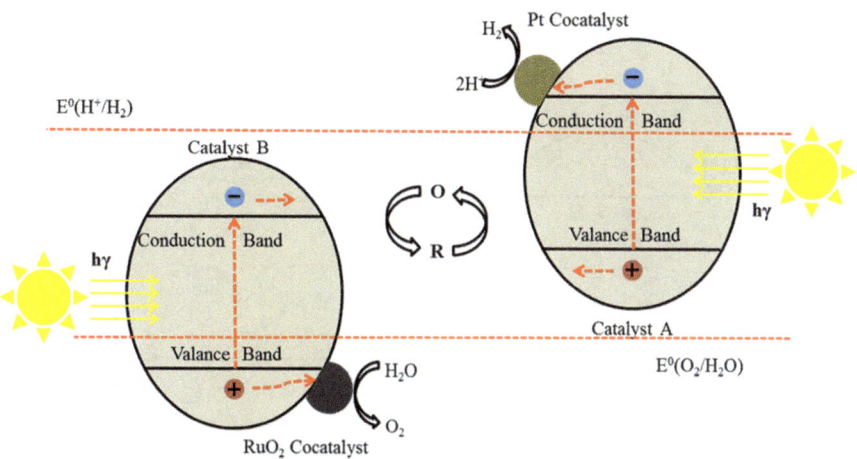

Figure 7.5   Schematic diagram of the Z-scheme for an overall water-splitting reaction on two different solid photocatalysts.

7.2.3 Standard of Measurement

There are two common ways to evaluate the photocatalytic activity of the water-splitting system. One is by direct measurement of the amounts of hydrogen, and the other one is an indirect method that utilizes the transported electrons from the semiconductor to the water within a certain time period under light irradiation. It is difficult to directly compare the results from different research groups and the photocatalytic hydrogen generation systems for the same catalyst, even if they test the same photocatalyst, owing to the different experimental setups.

Solar-to-hydrogen (STH) efficiency can be used as a practical standard to measure the performance of photocatalysts and photoelectrodes. The STH efficiency can be calculated as shown in eqn (7.1).

$$STH = \frac{\text{output energy as } H_2}{\text{energy of incident solar light}} = \frac{r_{H_2} \times \Delta G}{P_{Sun} \times S} \tag{7.1}$$

where r_{H2}, P_{Sun}, S, and ΔG are the rate of hydrogen production, energy flux of the sunlight, area of the reactor, and gain in Gibbs free energy, respectively. Considering the ASTM-G173 AM1.5 global tilt, solar irradiation has an energy flux of 1.0×10^3 W m^{-2}, and its power spectrum is well defined.

In PEC systems, the applied bias photon-to-current efficiency (ABPE) is often used for two-electrode measurements.

$$ABPE = \frac{|j| \times (V_{th} - V_{bias})}{P_{Sun}} \tag{7.2}$$

where j is the photocurrent density, and V_{th} and V_{bias} are the theoretical water-electrolysis voltage (1.23 V), and the applied voltage, respectively. The general expression of STH *via* PEC water splitting under sunlight is the applied-bias-compensated solar-to-hydrogen (AB-STH) efficiency that usually is defined as energy conversion efficiency or quantum efficiency.

$$AB\text{-}STH = \frac{\eta_F \times |j| \times (V_{th} - V_{bias})}{P_{Sun}} \tag{7.3}$$

where η_F is the faradic efficiency, the ratio of current contributing to water splitting into hydrogen and oxygen to the observed current. From eqn (7.2) and (7.3), we can see that ABPE is a special case of AB-STH when η_F is unity and V_{bias} is zero.

The efficiency also can be calculated as hypothetical half-cell solar-to-hydrogen (HCSTH) efficiency utilizing the three-electrode configuration. At the photoanode, the gain in potential by PEC water splitting could be estimated by the difference between the photoanode potential and the oxygen equilibrium potential. In a photocathode, the gain in potential is the difference between the photocathode potential and the hydrogen equilibrium potential.

$$\text{Photoanode: HC-STH} = \frac{|j| \times \left(E_{O_2/H_2O} - E_{RHE}\right)}{P_{Sun}} \qquad (7.4)$$

$$\text{Photocathode: HC-STH} = \frac{|j| \times \left(E_{RHE} - E_{H^+/H_2}\right)}{P_{Sun}} \qquad (7.5)$$

where E_{O_2/H_2O} and E_{H^+/H_2} stand for the equilibrium potentials of oxygen evolution ($+1.23$ V *versus* RHE) and hydrogen evolution (0 V *versus* RHE), respectively. If η_F of PEC water splitting is unity and water splitting is performed by counter electrodes without any over-potential, HC-STH efficiency indicates a convenient estimate of maximum ABSTH efficiency of a half-cell. However, in practical applications, the potential of a counter electrode cannot reach equilibrium with the potential of the redox reaction on the counter electrode. Thus, HCSTH efficiency should not be regarded as ABSTH efficiency.

Usually, the rate of gas (O_2 and H_2) evolution (units such as μmol h^{-1} and μmol h^{-1} g^{-1} catalyst) is applied to make the relative comparison between different photocatalysts under similar experimental conditions. Hence, the quantum yield is always used to make a direct comparison. A thermopile or Si photodiode can be utilized to determine the incident photons. The real quantity of absorbed photons is difficult to measure because of dispersion-system scattering. The real quantum yield (eqn (7.6)) is larger than the apparent quantum yield (eqn (7.7)) because the number of absorbed photons is usually smaller than that of the incident light; however, the apparent quantum yield is usually reported.

$$\text{overall quantum yield } (\%) = \frac{\text{number of reacted electrons}}{\text{number of absorbed photons}} \times 100\% \qquad (7.6)$$

$$\text{apparent quantum yield } (\%) = \frac{\text{number of reacted electrons}}{\text{number of incident photons}} \times 100\% \qquad (7.7)$$

7.3 Heterostructure Photocatalysts for Water Splitting

As was mentioned earlier, metal oxides are the most common photocatalysts for the water-splitting process due to their negative band gap position for hydrogen generation. Metal oxides are relatively easy to synthesize and convenient to introduce as photocatalysts in the particulate photocatalytic system to generate hydrogen from water. The particulate photocatalytic system is a closed system where both hydrogen and oxygen are generated due to the redox reaction of water molecules. Hydrogen then needs to be separated from oxygen to be used as a clean energy carrier. There are lot of different binary and ternary metal oxide photocatalysts that have been

investigated over the years. They have been categorized into three different metal oxide groups based on the electronic configuration:

- d^0 (Ti^{4+}, Zr^{4+}, Nb^{5+}, Ta^{5+}, W^{6+}, Mo^{6+})
- d^{10} (In^{3+}, Ga^{3+}, Ge^{4+}, Sn^{4+}, Sb^{5+})
- f^0 (Ce^{4+}).

However, most of the metal oxide photocatalysts are active in the UV region that covers only a small fraction (4%) of the incoming solar energy.[12] The visible light and infrared regions cover the rest of 46% and 50% of the incoming solar energy, respectively. Therefore it is important to understand the semiconductive properties, structure, and light sensitivity of different metal oxides to design photocatalysts that utilize a wide region of solar energy.

7.3.1 UV Active Photocatalysts

7.3.1.1 *Binary Metal Oxide Photocatalysts*

Most of the metal oxides containing d^0 electronic configured metal ions are photosensitive in the UV region. In general, the valence band and conduction band of these photocatalysts are composed of O 2p orbitals and d orbitals, respectively. Nb_2O_5, ZrO_2, Ta_2O_5, and WO_3 are some bimetallic oxides having d^0 electronic configured metal ions, which have the photocatalytic property under irradiation. Among them, TiO_2 (band gap = 3.2 eV; crystal structure = anatase) has a lower band energy gap than the other metal oxides. TiO_2 was the first bimetallic oxide to be identified as a photocatalyst for water splitting under UV irradiation. To absorb photon energy under UV irradiation, the photocataltytic materials' band gap energy must be less than the band gap energy of UV light. But, most of the d^0 metal oxides have sufficient or higher band gap energy than UV light. Therefore, the metal oxides with lower band gap energy must be modified to split water molecules. For example, colloidal TiO_2, modified with Pt and RuO_2 particles as cocatalysts, enhances the photocatalytic water splitting reaction with high quantum yield.[13] The addition of NaOH or Na_2CO_3 into the TiO_2/Pt photocatalyst system enhances the water splitting reaction.[14] Addition of Na_2CO_3 significantly accelerates the O_2 desorption from the semiconductor surfaces through the formation of peroxycarbonate intermediates. These intermediates are formed due to the reaction between surface carbonate species and positive holes in the valance band of photocatalysts.[15] Additionally, TiO_2 combines with a second metal oxide semiconductor to display increased photocatalytic activity under UV irradiation. Metal oxides like SnO_2, Ag_xO, ZrO_2 etc. can be combined individually with TiO_2 to form heterophase structures that increase the rate of photocatalytic hydrogen evolution from an aqueous solution containing electron donors.[16–18] Nb-based metal oxides, such as Nb_2O_5 (band gap = 3.4 eV) modified with Pt as a cocatalyst and in

presence of methanol as an electron donor, can efficiently decompose water into H_2.[19] Without any modification, Ta_2O_5 (band gap: 4.0 eV) can produce a small amount of hydrogen but no oxygen. It shows greater photocatalytic activity for overall water splitting when modified with NiO and RuO_2 as cocatalysts.[20] W- and Mo-based bimetallic oxides have not demonstrated satisfactory photocatalytic activity for overall water splitting.

Various metal oxides containing d^{10} electronic configured metal ions like Ga^{3+}, Zn^{2+}, Ge^{4+}, In^{3+}, Sn^{4+}, and Sb^{5+} exhibit effective photocatalytic activity for water splitting. One of the important metal oxides of this configuration is Ga_2O_3 (band gap = 4.6 eV) which, when loaded with Ni, shows decent photocatalytic performance[21] and can be effectively enhanced by the addition of Zn, Ca, Sr, Cr, Ba, and Ta ions for the overall water-splitting process. For example, Zn ion-doped Ga_2O_3 with Ni as cocatalyst provides an excellent photocatalytic activity with an apparent quantum yield of 20%.[22] Usually, f^0 block metal oxides are used as cocatalysts with other photocatalysts. However, Sr^+ doped CeO_2 can act as an active photocatalyst for water splitting when RuO_2 is added as a promotor.[23]

7.3.1.2   *Ternary Metal Oxide Photocatalysts*

We have already discussed a number of binary metal oxides with d^0, d^{10}, and f^0 metal ions that can act as photocatalysts during water splitting reactions. There are also several ternary metal oxides that possess remarkable photocatalytic behaviors. Ti-based ternary metal oxides like white titanate are worked efficiently for water splitting reactions under UV exposure. Titanates consist of titanium oxide layers and interlayers. The layered structure titanates, like $Na_2Ti_3O_7$, $K_2Ti_2O_5$, and $K_2Ti_4O_9$, have the photocatalytic activity to generate hydrogen from aqueous methanol solutions even without the use of Pt as a cocatalyst.[24] Layered structures of cesium compounds like $Cs_2Ti_nO_{2n+1}$ ($n = 2$, 5, and 6) have also shown photocatalytic properties to generate H_2 and O_2 from aqueous solutions. Among them, $Cs_2Ti_2O_5$, with five-coordinate structures consisting of TiO_5 units, is highly photo-active due to the unsaturated coordination state of the coordinate structures, which act as active sites during photocatalytic reactions.[25] Layered perovskite structures of lanthanum compounds, including La_2TiO_5, $La_2Ti_3O_9$, and $La_2Ti_2O_7$, have much higher photocatalytic activities under UV irradiation in comparison to bulk $LaTiO_3$.[26] In addition, $LaTiO_3$ doped with alkaline earth materials (Ba, Sr, and Ca) shows higher photoactivity. In particular, NiO-modified Ba-doped $LaTiO_3$ with alkaline hydroxide as an additive, increases the photoactivity for overall water-splitting reactions with high quantum yield.[27]

The synthesis technique used for of ternary metal oxides may change their photoactivities by varying the size, shape, and surface area of synthesized materials. For example, the perovskite structure of $BaTiO_3$ (3.22 eV) and $La_2Ti_2O_7$ exhibits improved photocatalytic behavior while synthesizing with a polymerized complex technique instead of traditional solid-state reaction

methods.[28] NiO-loaded $SrTiO_3$ powders can act as photocatalysts to generate H_2 and O_2 from water under UV irradiation.[29,30] Pre-treating the photocatalyst with H_2 and using concentrated NaOH aqueous solutions improve the photoactivity of NiO-loaded $SrTiO_3$ for the overall water splitting reaction.[30] Additionally, the photocatalytic performance of $SrTiO_3$ may also be enhanced by suitable metal cation doping, such as La^{3+}, Ga^{3+}, and Na^+.[31,32] Similarly, the perovskite structure of $CaTiO_3$ (band gap = 3.5 eV) loaded with Pt catalysts acts as a good photocatalytic catalyst under UV light.[33] $CaTiO_3$ doped with Zr^{4+} metal ions shows higher photoactivity with quantum yields up to 1.91% and 13.3% to generate H_2 from pure water and an aqueous ethanol solution, respectively.[34]

The layered structure of niobate (layers of niobium oxides) like $K_4Nb_6O_{17}$ (band gap = 3.4 eV) has shown excellent photoactivity for H_2 evolution from an aqueous methanol solution without any cocatalysts. K^+ ions are attached to the two different interlayers of niobium oxide sheets; one interlayer consists of K^+ ions and water molecules, whereas the other one contains only K^+ ions.[35,36] K^+ ions of niobium oxide interlayers can exchange with other cations such as Cr^{3+} and Fe^{3+} to enhance the photoactivity. In particular, the H^+-exchanged $K_4Nb_6O_{17}$ demonstrates the maximum photoactivity for H_2 evolution from an aqueous methanol solution with the quantum yield of 50%.[37] $K_4Nb_6O_{17}$ modified with NiO, Pt, Cs, and Au as cocatalysts becomes more efficient for hydrogen production.[38–40]

Tantalate metal oxide compounds, such as $LiTaO_3$ (band gap = 4.7 eV), perovskite $NaTaO_3$ (band gap = 4.0 eV), and $KTaO_3$ (band gap = 3.6 eV), show increased water-splitting activities under UV irradiation.[9,41] The photocatalytic behavior of tantalates mostly depends on alkaline cations due to the different band angles of Ta–O–Ta. The moment the band angle turns into 180°, excited electron–hole pairs are easily transported, and the band gap becomes much smaller. But the conduction band position of $NaTaO_3$ is higher than that of NiO. Hence, it is easy for electrons to migrate from the higher conduction band ($NaTaO_3$) to the lower (NiO). Although the electron transfer in the $KTaO_3$ structure is easier, the photocatalytic activity is significantly low, unexpectedly, because the conduction bands of $KTaO_3$ and NiO are mismatched. The photocatalytic activity of $NiO/NaTaO_3$ improves significantly when lanthanum ions are used as dopants.[42] La-doped $NaTaO_3$ offers smaller particle size and ordered surface structure, which are assumed to be the main reasons for its increased photocatalytic activity. Some of the W- and Mo-based heterogeneous photocatalysts are found to be active for water splitting only under UV irradiation. For example, $PbWO_4$ (band gap = 3.9 eV) incorporated with WO_4 tetrahedron has shown high photocatalytic activity for water splitting.[16] Under UV exposure, the scheelite structure of $PbMoO_4$ (band gap = 3.31 eV) is capable of producing H_2 from an aqueous methanol solution and oxygen from an aqueous silver nitrate solution.[43] The photocatalytic properties of indates with the octahedrally In^{3+} d^{10} configuration ion may possibly be employed for water decomposition.[44–47]

7.3.2 Visible-light-sensitive Photocatalysts

WO_3 (band gap = 2.8 eV) is one of the widely studied metal oxide photocatalysts for O_2 evolution under visible light, where Ag^+ or Fe^{3+} are used as sacrificial agents.[48] For example, both Bi_2WO_6 (band gap = 2.8 eV) and Bi_2MoO_6 (band gap = 2.7 eV) have aurivillius crystal structures and show photoactivity for O_2 evolution. They do not show photoactivity for H_2 evolution because they have low-conduction band positions. Another example of a visible light photocatalyst is α-Fe_2O_3 (band gap = 2.2 eV) which is cheap, but has the disadvantages of a high resistivity and a high recombination rate of photo-excited charge carriers.[49–51] As the above example suggests, the main concerns for visible-light photocatalysts are improving the photoconductivity and reducing the recombination rate of charge carriers. To address these issues, there are numerous ways to make the photocatalysts active under the visible-light spectrum for water-splitting reactions.

One of the most effective ways to develop visible-light-sensitive photocatalysts is metal or non-metal ion doping. In this technique, the doping ions help to engineer the band gap level of photocatalysts. The doping ions either act as electron donors, whose band gap levels are above the original photocatalyst's valence band, or act as electron acceptors, whose band gap levels are below the original conduction band. This makes the wide band gap metal oxides photoactive in the visible-light region. Several wide band gap metal oxides such as TiO_2, $SrTiO_3$, and $La_2Ti_2O_7$, when doped with metal ions, can act as photocatalysts in the visible light region. TiO_2, for example, when doped with Cr^{5+}, is capable of generating hydrogen and oxygen through water splitting under visible light (400–550 nm) irradiation.[52] Several transitional metal ions, such as V, Ni, Cr, Mo, Fe, Sn, Mn, Fe, Co, and Ni, are used as dopants with TiO_2 to improve the visible light absorption and photocatalytic activities. Pt^{4+}-doped and Ag^+-doped TiO_2 nanoparticles improve photocatalytic activities under both visible and UV irradiations.[53,54] This improvement could be explained by the recombination inhibition effect, since these metal ions could contribute to visible light absorption by inhibiting the recombination of electrons and holes. Non-metal ions such as C^-, N^-, or S^-, can also act as dopants because of their optical and photocatalytic properties.[55–57] Their absorption spectra are red-shifted, which further enhances their photocatalytic activities.

The addition of dopant helps to form recombination sites in the catalyst. For this, the formation of a valence band in the oxide photocatalysts is critical for the design of visible-light-driven photocatalysts.[58] The orbitals of Pb 6s in Pb^{2+}, Bi 6s in Bi^{3+}, Sn 5s in Sn^{2+}, and Ag 4d in Ag^+ are used to build a new valence band above the previous valence band consisting of O 2p orbitals.[58] The conduction band of $BiVO_4$ (2.4 eV) is composed of V 3d, like other d^0 oxide photocatalysts. The new valence band formed with Bi 6s orbitals retains the potential to split water.

Another important technique is using dye to sensitize metal oxides in the preparation of visible-light-sensitive photocatalysts for water splitting.

Typically, the sensitization reduces the wide band gap of semiconductor photocatalysts. Dye-loaded TiO_2 and $K_4Nb_6O_{17}$ are capable of producing H_2. During this process, the excited electron from the highest occupied molecular orbital (HOMO) is shifted to the lowest occupied molecular orbital (LUMO) of a dye and then transferred to the conduction band of semiconducting photocatalysts. During this process, H_2 evolution takes place on the surface of these wide band gap photocatalysts. Ruthenium(II) complex dyes are extensively used to sensitize the wide band gap semiconductors for the photocatalytic evolution of H_2. In the sensitization process, $Ru(bpy)_3^{2+}$/ $K_4Nb_6O_{17}$ thin film is used as an electrode to deliver a photocurrent that responds to visible light. Duonghong and co-workers[59] observed that Pt/RuO_2-loaded TiO_2 particles with either $Ru(bpy)_3^{2+}$ or its amphiphilic derivatives as a sensitizer can act as efficient photocatalysts for the water-splitting process under visible light. Additionally, Pt/TiO_2 with a polymer-pendant $Ru(bpy)_3^{2+}$ complex as a sensitizer has the ability to produce H_2 in the presence of a sacrificial donor, such as EDTA, in presence of visible light.[60] The photoactivity of $K_4Nb_6O_{17}$ significantly changes when $Ru(bpy)_3^{2+}$ complex is intercalated into the $K_4Nb_6O_{17}$ interlayers. The intercalation facilitates a fast and efficient electron transfer process between the $Ru(bpy)_3^{2+}$ complex and $K_4Nb_6O_{17}$, and exponential decay of the transient bleaching of the $Ru(bpy)_3^{2+}$ complex.[61] Furthermore, several transitional metal complexes – such as polypyridine complexes, alizarine, phthalocyanine, and metalloporphyrins with metal centers like Pt^{II}, Co^{II}, Zn^{II}, and Cr^{II} – have the ability to sensitize the wide band gap photocatalyst for H_2 evolution under visible light. Metal-free dyes (porphine, xanthene, melocyanine, and coumarin, *etc.*) are attractive as sensitizers for water splitting because they are less expensive than transition metal complexes such as Ru complexes. For example, Pt-loaded semiconductor catalysts sensitized by xanthene dyes, such as rose bengal, erythrosine, and eosine bluish, produce H_2 with a high quantum yield.[62]

7.3.3 Photocatalysts for the Z-scheme Reaction

Although many binary and ternary metal oxides decompose water in a single photocatalyst in both the UV and visible-light ranges, this process suffers from the following weaknesses:

(a) a backward reaction of water formation
(b) a limited range of light absorption up to the bandgap of the photocatalysts.

The Z-scheme offers a system comprising two photocatalysts and a charge-carrying redox couple. In this scheme, O_2 is generated in the oxygen-photocatalyst while the electron is passed on to the redox couple, which is consumed in the hydrogen-photocatalyst for H_2 evolution. The Z-scheme has the advantages of limiting the backward water formation to enhance

the photocatalytic efficiency and of broadening the range of light to be absorbed.

As an example of a Z-scheme system, Pt-loaded ZrO_2/TaON and Pt-loaded WO_3 photoctalysts are used for the H_2 evolution and O_2 evolution reactions, respectively.[63] The I^-/IO_3^- redox couple (I^- = donor and IO_3^- = acceptor) carries the charges between the two different photocatalysts. Together, the system exhibits its highest apparent quantum yield (AQY), 6.3% at 420 nm.[63] The side reactions of redox couple degradations with photocatalysts are significantly reduced due to Pt loading into the catalyst. ZrO_2 modifies the TaON n-type semiconductor characteristics and extends the lifetime of the photo-excited holes within the photocatalyst. The amounts of photocatalysts and redox couples are actively tested to achieve the highest AQY. In the end, a weight ratio of 10 : 20 : 3 among the hydrogen photocatalyst (0.5% Pt), oxygen photocatalyst, and redox couple can reach the optimal conversion efficiency.

Another example of a Z-scheme systems, Ru-loaded $SrTiO_3$ doped with Rh as a hydrogen photocatalyst and $BiVO_4$ as an oxygen photocatalyst with a redox couple of $[Co(bpy)_3]^{3+/2+}$. This system gives an AQY of 2.1% at 420 nm.[64] This system is also sensitive to the pH of aqueous solution because pH condition can shift the electronic property of the redox couple and eventually alter the photocatalytic activity of the catalyst. When the pH decreases and becomes more acidic, the potential of $[Co(bpy)_3]^{3+/2+}$ shifts to become more negative and enhances the H_2 evolution of $SrTiO_3$. However, high pH shrinks the energy difference between the redox couple and $BiVO_4$, thus it makes oxygen generation difficult due to the smaller energy offset. Hence, a neutral pH condition provides the most balanced oxygen and hydrogen evolution for the system. One interesting observation is that the evolution of oxygen and hydrogen continue even in the absence of a redox couple at a pH of 3.5. This happened due to the aggregations of both photocatalysts. The aggregation leads the direct physical contact between the two photocatalysts, and the oxidation state of Rh on the surface of $SrTiO_3$ is reversible and serves as an electron shuttle. As a result, this Z-scheme exhibits photocatalytic activity even without the need of redox couple, $[Co(bpy)_3]^{3+/2+}$. Isoelectric points (IPs) cause the aggregations. The IP of $SrTiO_3$ is 4 and the IP of $BiVO_4$ is 2. Hence, at a pH between 2 and 4, $SrTiO_3$ is positively charged and $BiVO_4$ is negatively charged, resulting in their aggregation. More detail is given elsewhere.[64]

7.4   Photoelectrochemical Cells for Water Splitting

7.4.1   Principle of Photoelectrochemical Cells

The photoelectrochemical cell for water splitting is a device that utilizes light energy to decompose water into hydrogen and oxygen within a cell that consists of two photoelectrodes immersed in an aqueous electrolyte. Semiconductors, the main photoactive materials that convert incident photons to electron–hole pairs, are used as photoelectrodes in the PEC cell.

A simple PEC cell device for water splitting can be assembled:

(a) using a single n-type semiconductor as a photoanode and a cathode made of metal
(b) using a single p-type semiconductor as a photocathode and an anode made of metal or
(c) connecting both n-type and p-type semiconductors in a series of photoanodes and photocathodes, respectively (Figure 7.6).

To design the PEC cell device, it is important to understand the physics of a semiconductor–liquid junction. When a semiconductor electrode is immersed in an electrolyte solution containing a redox couple (comprising an electron donor, R, and acceptor, R^-), electrons will start flowing between the semiconductor and the solution until the Fermi level of the semiconductor comes into equilibrium with the electrochemical potential of the redox couple. The charge transfer creates an interfacial electric field whose electrostatic potential balances the initial Fermi level differences between the solution and the semiconductor. In a PEC water-splitting device, the redox couples of interest are the H^+/H_2 couple and O_2/H_2O couple for a p-type semiconductor photocathode and n-type semiconductor, respectively. When a typical n-type semiconductor (Figure 7.7) is placed in contact with the solution containing the redox couple (*e.g.* O_2/H_2O), the electrons are injected from the conduction band to that couple in the solution until the Fermi level is neutralized. The electron transfer creates two regions: one is the depletion layer on the semiconductor side, and the other is the Helmholtz layer on the electrolyte side. The positive charges as minor carriers spread out over the depletion region, D, whereas the negative charges spread over the Helmholtz region. This results in an electric potential gradient or band bending in the semiconductor. Under irradiation with adequate photon energy, the minority charge (holes) moves into the negatively charged redox couple, and the photo-excited electron transfers to the counter electrode.

The phenomena of p-type semiconducting electrodes are analogous to the n-type semiconductor, where electrons as minor carriers accumulate on the semiconductor side (accumulation layer, A) and the Helmholtz layer contains the positive charge. Hence, in p-type semiconductors, electrons move into the positively charged acceptor at the interface upon illumination. In both cases, photo-excited minority carriers drive the PEC reactions on photoelectrodes. Under photoexcitation, the counter electrode electron potential is identical to the Fermi level of the photoelectrode. If the Fermi level of the photoelectrode is not situated at a desirable potential level between a photoelectrode and a counter electrode, an external voltage is applied to make up the potential deficiency and drive the redox reactions in a counter electrode. In such a case, energy conversion efficiency is determined by excluding the external power input, which is the product of the current and the applied voltage.[65] The interfacial potential energy barrier in the semiconductor can be determined by the initial energy difference between the

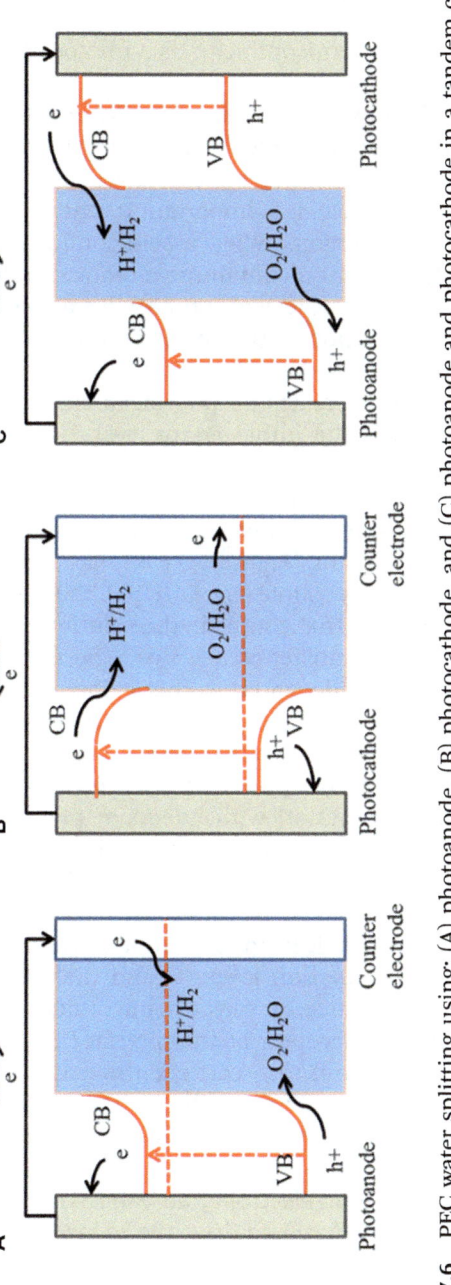

Figure 7.6   PEC water splitting using: (A) photoanode, (B) photocathode, and (C) photoanode and photocathode in a tandem configuration. (Reproduced from ref. 68 with permission from the Royal Society of Chemistry. © 2014 Royal Society of Chemistry.)

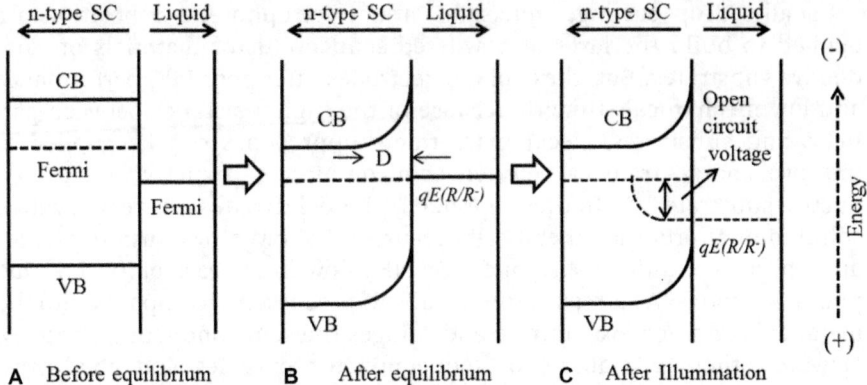

A Before equilibrium **B** After equilibrium **C** After Illumination

Figure 7.7 Semiconductor–liquid junction band energies for the n-type semiconductor, where D is the depletion layer.
(Reprinted with permission from *Chem. Rev.*, 2010, **110**, 6446–6473. © 2010 American Chemical Society.)

Fermi level of the semiconductor and the value of $-qE(R/R^-)$, where E is the Nernst potential of the redox couple and q is the unit charge. The initial energy difference and the depletion width in the semiconductor are on the order of 1 eV and on the order of hundreds of nanometres, respectively, and the electric field in the semiconductor can be up to 10^5 V cm^{-1}.[66] Generally, photo-excited charge carriers (electron–hole pairs) are very efficiently separated under this electric field due to large mobility of charge carriers (10–1000 cm^2 V^{-1} s^{-1}) in crystalline inorganic semiconductors. Their actual free energy is contingent on the kinetics of the charge carriers in the photostationary state. The free energy generated by the semiconductor is determined by the difference between the hole and electron quasi-Fermi levels under illumination, which is equal to the free-energy difference between the majority carriers and the photo-excited minority carriers. The quasi-Fermi level gives a description of the electrochemical potential of one carrier type (*i.e.*, either electrons or holes) at a time under illumination (*e.g.*, non-equilibrium) conditions by the help of Fermi–Dirac statistics which independently describe the populations of electrons and holes.[67] To generate oxygen, the valence band of the photoanode must be more positive than the oxygen evolution potential. Likewise, for a p-type semiconductor photocathode for hydrogen evolution,[68] the conduction band edge is required to be more negative than the hydrogen evolution potential.

7.4.2 Fabrication Methods for a Photoelectrode

In the photoelectrode fabrication methods for the PEC water-splitting device, it is necessary to:

(a) establish good electrical contact between the semiconductor particles/ layers and a conductive substrate and
(b) reduce grain boundaries density across the photoelectrode to minimize series resistance.

Usually, drop casting, squeezing, and electrophoretic deposition are applied to build the layer of powdered semiconductor materials on conductive substrates. But, these photoelectrodes offer poor PEC performance and low mechanical strength because of the high resistance between particles and substrates. Necking treatment approaches can be applied to improve charge transfer between semiconducting particles and a conductive substrate.[68,69] In these methods, photoelectrodes are treated with a solution of appropriate metal salt precursors followed by annealing under appropriate conditions, *e.g.*, under an NH_3 flow in the case of (oxy) nitride photoelectrodes. As a result, the metal salt precursor decomposes into the metal oxide and/or (oxy) nitride and bridges the semiconducting particles. However, thermal stability of semiconductor particles and conductive substrates is one of the major problems that can occur during the annealing process of necking treatment. For example, transparent conductive oxides, especially fluorine-doped SnO_2 and tin-doped In_2O_3, cannot be treated at high temperatures. To address this problem, the particle transfer method has been developed as an alternative to the necking treatment method.[70,71] In this method (Figure 7.8), a glass substrate covered with a semiconducting powder is sputtered with metals as a contact layer (approximately 100–300 nm). The choice of metal is important here to establish a considerable electrical contact between the semiconductor particles and the metal layer. Then, several micrometer-thick metal films are deposited continuously by the sputtering method. The resulting metal film with adequate conductivity and mechanical strength is transferred to another substrate by a resin. The primary substrate is peeled off, and followed by ultrasonication in water to remove extra or loosely bound powder. Thus, a mono-particle layer covers the metal conductor film. This method provides a good electrical conductivity between the semiconducting particles and the metal layer, and it also can be applied to a variety of powdered semiconductors.

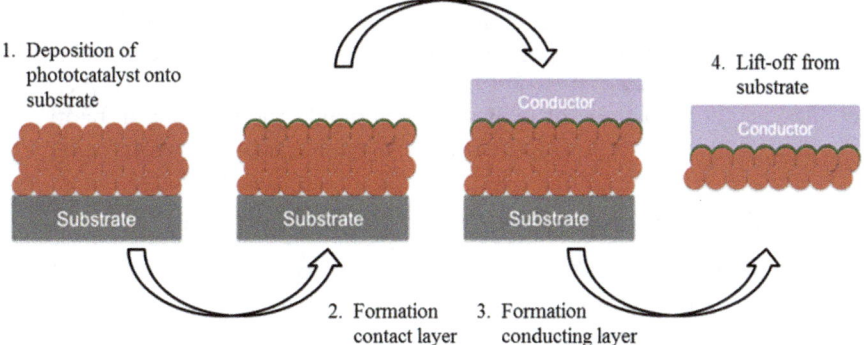

Figure 7.8   Schematic illustration of the particle transfer method.
(Reproduced from ref. 70 with permission from the Royal Society of Chemistry. © Royal Society of Chemistry 2013.)

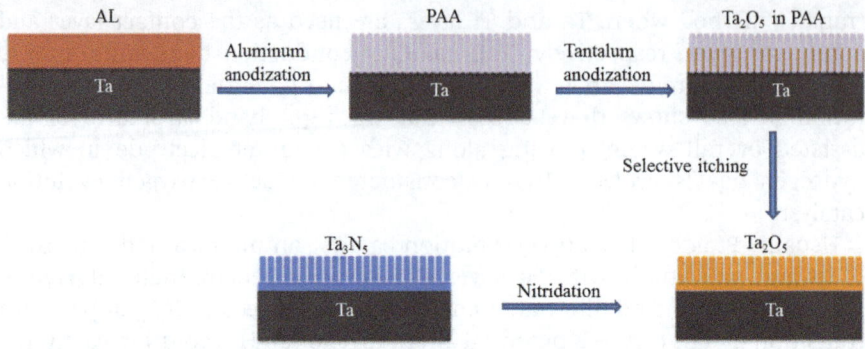

Figure 7.9 Schematic illustration of the nanorod photoelectrode fabrication process.
(Reprinted with permission from ref. 71. © 2013 Wiley-VCH Verlag GmbH & Co. KGaA, Weinheim.)

The use of nanorods vertically on conductive substrates is advantageous over other types of photoelectrodes because the photogenerated minority can reach faster at the electrolyte interface in the submicron-diameter nanorods.[71] The motion of majority carriers is faster along their long axis within the core of nanorods. For example, one-dimensional Ta_3N_5 nanorod arrays are considered as a potential candidate for solar-driven PEC water splitting.[71] Nanorod arrays of Ta_3N_5 can be produced by nitridation of Ta_2O_5 nanorod arrays that are grown *via* a through-mask anodization method. As shown in Figure 7.9, on top of the Ta substrate, a porous anodic alumina (PAA) mask is formed by anodizing an evaporated Al layer. Then, Ta_2O_5 nanorods are rooted in the nanochannels of the PAA mask by anodizing the Ta substrate through the PAA mask in an aqueous solution of boric acid (0.5 M). Ta^{5+} has low solubility in the boric acid solution and shows volume expansion during the anodization of Ta into Ta_2O_5. Thus, the Ta_2O_5 is packed into the nanochannels of the PAA and masked under a high electrical field. Subsequently, 5% phosphoric acid is applied to selectively etch the PAA mask. The resulting Ta_2O_5 nanorod arrays are nitrided into Ta_3N_5 nanorod arrays under a heated NH_3 flow.

7.4.3 Metal Oxide-based Photoelectrochemical Cells

In PEC, usually n-type semiconductors are used as photoanodes for the oxygen evolution reaction at the interfaces between the photoanodes and electrolytes. Under illumination, the photo-excited holes are generated in the n-type semiconductor and migrate to the interface to oxidize water. Proper surface modifications of photoanode assist to achieve increased photocurrent and stability. For example, one study show that $LaTiO_2N$ modified with IrO_2 can act as photoanodes in the presence of aqueous Na_2SO_4 solution.[70] In this study, $LaTiO_2N$ is synthesized by the particle

transfer method where Ta and Ti layers are used as the contact layer and conductor layer, respectively. The $LaTiO_2N$ conduction band minimum is found to be more negative than the equilibrium potential of hydrogen evolution. It also shows that $LaTiO_2N$ has the right band structure for unassisted overall water splitting along with a counter electrode in which hydrogen gas is evolved. IrO_2 is considered an active oxygen evolution catalyst.

Usually, Pt is used as an H_2 evolution catalyst on photocathodes because of its high catalytic activity. However, Pt is a very expensive material. Hence, it is necessary to find alternative materials at a low cost. MoS_2 doped with transition metals can be a potential alternative for an H_2 evolution catalyst.[72] In PEC, p-type semiconductors are also used as photocathodes to evolve hydrogen at the interface between semiconductor particles and electrolytes through reduction reactions. The possibility of oxidative degradation is lower than that for photoanodes. However, inherent p-type semiconductors are not very common. Most oxide, nitride, and sulfide semiconductors are n-type semiconductors due to anion defects. Some Cu-based chalcogenides have p-type semiconductor properties and display remarkable performance in photovoltaic applications. Layered structures of p-type and n-type semiconductors can act as an efficient hydrogen evolution catalyst. For example, when the n-type semiconductor CdS is deposited on the Pt modified $CuGaSe_2$ photoelectrode, the cathodic photocurrent is noticeably improved compared with the only Pt modified $CuGaSe_2$.[73] The photocurrent improves because CdS forms a good p-n heterojunction with $CuGaSe_2$, which results in the band formation between CdS and $CuGaSe_2$. Another similar approach can also be applied for n-type photoelectrode chemical cell systems. For example, $CaFe_2O_4$ (p-type) is used to coat the TaON (n-type) photoelectrode which form a p-n heterojunction. This heterojunction exhibits an enhanced photocurrent during oxygen evolution.[74]

Most of the currently available p-type semiconductors' valence band positions are more negative than the potential of oxygen evolution (1.23 V *versus* RHE). Hence, the photocathodic current is difficult to obtain for p-type photoelectrodes. Since the potential of the counter electrode is not positive enough, it is difficult to split water for them in a single step. In order to split water only by the energy of light, p-type photoelectrodes in combination with n-type photoelectrodes and/or photovoltaic cells are required. The combined system can split water without electrical power if the photocurrents of the photoanode and photocathode are balanced at a certain potential.

7.4.4 Dye-assisted Photoelectrochemical Cells

The dye-assisted photoelectrochemical cell is a new generation of solar cell based on the cheap material TiO_2. A simple dye-assisted solar cell for water splitting consist of a dye-sensitized wide band gap semiconductor electrode, an aqueous electrolyte comprising of redox pairs, and a Pt electrode

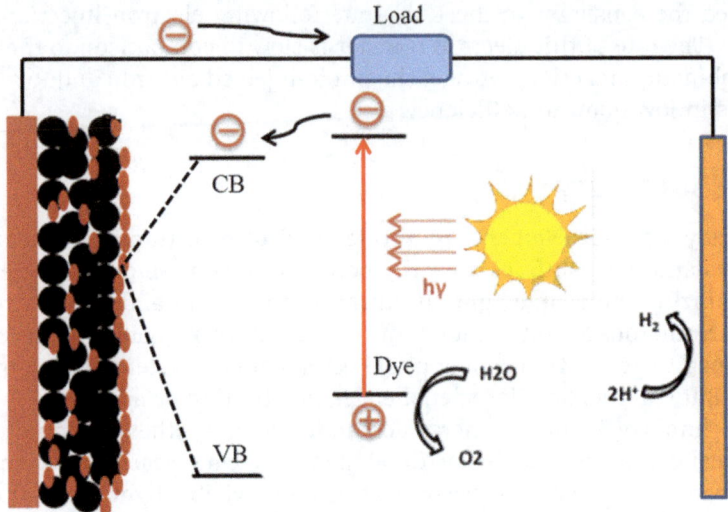

Figure 7.10 A dye-sensitized photoelectrochemical cell for overall water splitting.

(Figure 7.10). Various semiconductor nanoparticles such as TiO_2, ZnO, or SnO_2 are loaded with dye molecules that sensitize the semiconductor to capture solar energy. Under visible light exposure, the excited dyes can inject electrons to CB of semiconductors to initiate the catalytic reactions. The CB electrons can then be transferred to the Pt electrode to initiate water reduction and produce $H_2(g)$. Simultaneously, dye molecules are regenerated by using electrons which are produced from water oxidation reaction during $O_2(g)$ evolution. The dye-assisted solar cell basically mimics the photosynthesis process of plants in which the dye molecules behave like the pigments (*e.g.* chlorophyll). The main purpose of the dye is to create a wider absorption spectrum for a longer period of time to generate more energy during its lifespan. Usually, the functional complex of Ru and Zn is used as a dye to sensitize the semiconductor in the solar cell. An example is rose bengal, $Ru^{II}L_2(NCS)_2$, and black dye. For example, a dye-assisted photoelectrochemical cell studied by Youngblood *et al.*[11] is illustrated in Figure 7.10. In this study, the photoanode consist of mesoporous TiO_2 and the cathode is a Pt wire electrode which is externally connected by a wire. Electrodehydrated iridium oxide ($IrO_2 \cdot nH_2O$) nanoparticles are attached to the Ru complex ($[Ru(bpy)_3]^{2+}$) sensitizer to serve as an O_2 evolution catalyst. Under illumination, a photo-excited electron LUMO is injected into the conduction band of TiO_2 and transferred to a counter electrode (Pt) through the outer circuit to reduce water to H_2. The dye is returned to the ground state after accepting an electron from the $IrO_2 \cdot nH_2O$ nanoparticle where water oxidizes to O_2. Youngblood *et al.*[11] studied this system and found steady state quantum efficiency to be about 1%. In this systems, the dye photodegrades rapidly because of back electron transfer and decomposition of the oxidized dye molecules.[11] Here, the electron donor is Ir^{IV}, which

, reduced the sensitizer to the Ru^{III} state following electron injection to the anode. The rate of this electron transfer is slow in comparison to the charge recombination reaction between the photo-injected electron and Ru^{III} which results in low quantum efficiency.

7.5 Summary

Generally, d^{10}-type element oxynitrides and d^0-type transition metal (oxy) nitrides and (oxy) sulfides have the potential to achieve overall water splitting utilizing the visible-light spectrum up to 500 nm and 600 nm, respectively. Numerous combinations of semiconductor materials have been examined to generate H_2 using photelectrochemical water splitting over the years. But still there are lot of challenges that need to be addressed to utilize this technology for future energy demand. Several synthesis techniques and chemical additives have been developed in recent years to enhance the photocatalytic activity of semiconductors under visible-light irradiation. The major hurdles are low H_2 production rate due to quick charge recombination, quick backward reactions, and an inability to harness visible light efficiently. The problem associated with d^0-type transition metal (oxy) nitrides is defect formation during the preparation process. Various post-treatments processes can reduce the defect density and improve crystallinity. However, overall water splitting is still difficult to obtain with d^0-type transition metal (oxy) nitride photocatalysts. The reason most likely lies with the thermal instability of these (oxy) nitrides compared to the corresponding oxides. But, in some cases, the enhancement of photocatalytic activity has also been observed. Photocatalytic reaction mechanisms on surfaces and the effect of individual defect species are not well understood. In order to achieve a better understanding of this process, newly developed techniques are required to explore individual defect species quantitatively and their influence on the photocatalytic reaction qualitatively.

References

1. A. J. Nozik, *Annu. Rev. Phys. Chem.*, 1978, **29**, 189–222.
2. N. S. Lewis, *MRS Bull.*, 2007, **32**, 808–820.
3. D. G. Nocera, *ChemSusChem*, 2009, **2**, 387–390.
4. M. G. Walter, E. L. Warren, J. R. McKone, S. W. Boettcher, Q. Mi, E. A. Santori and N. S. Lewis, *Chem. Rev.*, 2010, **110**, 6446–6473.
5. N. Nuraje, R. Asmatulu and S. Kudaibergenov, *Curr. Inorg. Chem.*, 2012, **2**, 23.
6. T. R. Cook, D. K. Dogutan, S. Y. Reece, Y. Surendranath, T. S. Teets and D. G. Nocera, *Chem. Rev.*, 2010, **110**, 6474–6502.
7. A. Fujishima and K. Honda, *Nature*, 1972, **238**, 37–38.
8. R. M. Navarro Yerga, M. C. Álvarez Galván, F. del Valle, J. A. Villoria de la Mano and J. L. G. Fierro, *ChemSusChem*, 2009, **2**, 471–485.
9. A. Kudo and Y. Miseki, *Chem. Soc. Rev.*, 2009, **38**, 253–278.

10. M. Gratzel, *Nature*, 2001, **414**, 338–344.
11. W. J. Youngblood, S. H. A. Lee, K. Maeda and T. E. Mallouk, *Acc. Chem. Res.*, 2009, **42**, 1966–1973.
12. K. Maeda and K. Domen, *J. Phys. Chem. Lett.*, 2010, **1**, 2655–2661.
13. D. Duonghong, E. Borgarello and M. Graetzel, *J. Am. Chem. Soc.*, 1981, **103**, 4685–4690.
14. A. Kudo, K. Domen, K.-i. Maruya and T. Onishi, *Chem. Phys. Lett.*, 1987, **133**, 517–519.
15. H. Arakawa and K. Sayama, *Catal. Surv. Asia*, 2000, **4**, 75–80.
16. N. Saito, H. Kadowaki, H. Kobayashi, K. Ikarashi, H. Nishiyama and Y. Inoue, *Chem. Lett.*, 2004, **33**, 1452–1453.
17. Q. Yuan, Y. Liu, L.-L. Li, Z.-X. Li, C.-J. Fang, W.-T. Duan, X.-G. Li and C.-H. Yan, *Microporous Mesoporous Mater.*, 2009, **124**, 169–178.
18. J.-W. Park and M. Kang, *Int. J. Hydrogen Energy*, 2007, **32**, 4840–4846.
19. X. Chen, T. Yu, X. Fan, H. Zhang, Z. Li, J. Ye and Z. Zou, *Appl. Surf. Sci.*, 2007, **253**, 8500–8506.
20. H. Kato and A. Kudo, *Chem. Phys. Lett.*, 1998, **295**, 487–492.
21. T. Yanagida, Y. Sakata and H. Imamura, *Chem. Lett.*, 2004, **33**, 726–727.
22. Y. Sakata, Y. Matsuda, T. Yanagida, K. Hirata, H. Imamura and K. Teramura, *Catal. Lett.*, 2008, **125**, 22–26.
23. H. Kadowaki, N. Saito, H. Nishiyama and Y. Inoue, *Chem. Lett.*, 2007, **36**, 440–441.
24. M. Shibata, A. Kudo, A. Tanaka, K. Domen, K.-i. Maruya and T. Onishi, *Chem. Lett.*, 1987, **16**, 1017–1018.
25. A. Kudo and T. Kondo, *J. Mater. Chem.*, 1997, **7**, 777–780.
26. J. Kim, D. Hwang, S. Bae, Y. Kim and J. Lee, *Korean J. Chem. Eng.*, 2001, **18**, 941–947.
27. J. Kim, D. Hwang, H. Kim, S. Bae, J. Lee, W. Li and S. Oh, *Top. Catal.*, 2005, **35**, 295–303.
28. Y. Yamashita, K. Yoshida, M. Kakihana, S. Uchida and T. Sato, *Chem. Mater.*, 1998, **11**, 61–66.
29. K. Domen, S. Naito, M. Soma, T. Onishi and K. Tamaru, *J. Chem. Soc., Chem. Commun.*, 1980, **12**, 543–544.
30. K. Domen, S. Naito, T. Onishi and K. Tamaru, *Chem. Phys. Lett.*, 1982, **92**, 433–434.
31. Y. Qin, G. Wang and Y. Wang, *Catal. Commun.*, 2007, **8**, 926–930.
32. T. Takata and K. Domen, *J. Phys. Chem. C*, 2009, **113**, 19386–19388.
33. H. Mizoguchi, K. Ueda, M. Orita, S.-C. Moon, K. Kajihara, M. Hirano and H. Hosono, *Mater. Res. Bull.*, 2002, **37**, 2401–2406.
34. W. Sun, S. Zhang, C. Wang, Z. Liu and Z. Mao, *Catal. Lett.*, 2007, **119**, 148–153.
35. K. Domen, A. Kudo, M. Shibata, A. Tanaka, K.-I. Maruya and T. Onishi, *J. Chem. Soc., Chem. Commun.*, 1986, **23**, 1706–1707.
36. K. Domen, A. Kudo, A. Shinozaki, A. Tanaka, K.-i. Maruya and T. Onishi, *J. Chem. Soc., Chem. Commun.*, 1986, 356–357.

37. A. Kudo, A. Tanaka, K. Domen, K.-i. Maruya, K.-i. Aika and T. Onishi, *J. Catal.*, 1988, **111**, 67–76.
38. K. Sayama, A. Tanaka, K. Domen, K. Maruya and T. Onishi, *Catal. Lett.*, 1990, **4**, 217–222.
39. K. Sayama, A. Tanaka, K. Domen, K. Maruya and T. Onishi, *J. Phys. Chem.*, 1991, **95**, 1345–1348.
40. K. Sayama, K. Yase, H. Arakawa, K. Asakura, A. Tanaka, K. Domen and T. Onishi, *J. Photochem. Photobiol., A*, 1998, **114**, 125–135.
41. H. Kato and A. Kudo, *J. Phys. Chem. B*, 2001, **105**, 4285–4292.
42. H. Kato, K. Asakura and A. Kudo, *J. Am. Chem. Soc.*, 2003, **125**, 3082–3089.
43. A. Kudo, M. Steinberg, A. J. Bard, A. Campion, M. A. Fox, T. E. Mallouk, S. E. Webber and J. M. White, *Catal. Lett.*, 1990, **5**, 61–66.
44. J. Sato, H. Kobayashi, N. Saito, H. Nishiyama and Y. Inoue, *J. Photochem. Photobiol., A*, 2003, **158**, 139–144.
45. J. Sato, H. Kobayashi and Y. Inoue, *J. Phys. Chem. B*, 2003, **107**, 7970–7975.
46. J. Sato, N. Saito, H. Nishiyama and Y. Inoue, *J. Phys. Chem. B*, 2003, **107**, 7965–7969.
47. J. Sato, N. Saito, H. Nishiyama and Y. Inoue, *Chem. Lett.*, 2001, **30**, 868–869.
48. J. R. Darwent and A. Mills, *J. Chem. Soc., Faraday Trans. 2*, 1982, **78**, 359–367.
49. V. R. Satsangi, S. Kumari, A. P. Singh, R. Shrivastav and S. Dass, *Int. J. Hydrogen Energy*, 2008, **33**, 312–318.
50. W. B. Ingler, J. P. Baltrus and S. U. M. Khan, *J. Am. Chem. Soc.*, 2004, **126**, 10238–10239.
51. V. M. Aroutiounian, V. M. Arakelyan, G. E. Shahnazaryan, G. M. Stepanyan, J. A. Turner and O. Khaselev, *Int. J. Hydrogen Energy*, 2002, **27**, 33–38.
52. E. Borgarello, J. Kiwi, M. Graetzel, E. Pelizzetti and M. Visca, *J. Am. Chem. Soc.*, 1982, **104**, 2996–3002.
53. S. Kim, S.-J. Hwang and W. Choi, *J. Phys. Chem. B*, 2005, **109**, 24260–24267.
54. S. Rengaraj and X. Z. Li, *J. Mol. Catal. A: Chem.*, 2006, **243**, 60–67.
55. C. Burda, Y. Lou, X. Chen, A. C. S. Samia, J. Stout and J. L. Gole, *Nano Lett.*, 2003, **3**, 1049–1051.
56. Y. Choi, T. Umebayashi and M. Yoshikawa, *J. Mater. Sci.*, 2004, **39**, 1837–1839.
57. T. Ohno, T. Mitsui and M. Matsumura, *Chem. Lett.*, 2003, **32**, 364–365.
58. M. Kitano and M. Hara, *J. Mater. Chem.*, 2010, **20**, 627–641.
59. D. Duonghong, E. Borgarello and M. Graetzel, *J. Am. Chem. Soc.*, 1981, **103**, 4685–4690.
60. T. Nakahira, Y. Inoue, K. Iwasaki, H. Tanigawa, Y. Kouda, S. Iwabuchi, K. Kojima and M. Grätzel, *Makromol. Chem., Rapid Commun.*, 1988, **9**, 13–17.

61. A. Furube, T. Shiozawa, A. Ishikawa, A. Wada, K. Domen and C. Hirose, *J. Phys. Chem. B*, 2002, **106**, 3065–3072.
62. T. Shimidzu, T. Iyoda and Y. Koide, *J. Am. Chem. Soc.*, 1985, **107**, 35–41.
63. K. Maeda, M. Higashi, D. Lu, R. Abe and K. Domen, *J. Am. Chem. Soc.*, 2010, **132**, 5858–5868.
64. M. Higashi, R. Abe, T. Takata and K. Domen, *Chem. Mater.*, 2009, **21**, 1543–1549.
65. T. F. J. Z. Chen, T. G. Deutsch, A. Kleiman-Shwarsctein, A. J. Forman, N. Gaillard, R. Garland, K. Takanabe, C. Heske, M. Sunkara, E. W. McFarland, K. Domen, E. L. Miller, J. A. Turner and H. N. Dinh, *J. Mater. Res.*, 2010, **25**, 3–16.
66. N. S. Lewis, *Inorg. Chem.*, 2005, **44**, 6900–6911.
67. M. X. Tan, C. N. Kenyon, O. Krüger and N. S. Lewis, *J. Phys. Chem. B*, 1997, **101**, 2830–2839.
68. T. Hisatomi, J. Kubota and K. Domen, *Chem. Soc. Rev.*, 2014, **43**, 7520–7535.
69. M. Higashi, K. Domen and R. Abe, *Energy Environ. Sci.*, 2011, **4**, 4138–4147.
70. N. N. T. Minegishi, J. Kubota and K. Domen, *Chem. Sci.*, 2013, **4**, 1120.
71. Y. Li, T. Takata, D. Cha, K. Takanabe, T. Minegishi, J. Kubota and K. Domen, *Adv. Mater.*, 2013, **25**, 125.
72. B. Hinnemann, P. G. Moses, J. Bonde, K. P. Jørgensen, J. H. Nielsen, S. Horch, I. Chorkendorff and J. K. Nørskov, *J. Am. Chem. Soc.*, 2005, **127**, 5308–5309.
73. M. Moriya, T. Minegishi, H. Kumagai, M. Katayama, J. Kubota and K. Domen, *J. Am. Chem. Soc.*, 2013, **135**, 3733–3735.
74. E. S. Kim, N. Nishimura, G. Magesh, J. Y. Kim, J.-W. Jang, H. Jun, J. Kubota, K. Domen and J. S. Lee, *J. Am. Chem. Soc.*, 2013, **135**, 5375–5383.

CHAPTER 8

Nanophotocatalysis in Selective Transformations of Lignocellulose-derived Molecules: A Green Approach for the Synthesis of Fuels, Fine Chemicals, and Pharmaceuticals

JUAN CARLOS COLMENARES

Institute of Physical Chemistry, Polish Academy of Sciences, ul. Kasprzaka 44/52, 01-224 Warsaw, Poland
Email: jcarloscolmenares@ichf.edu.pl

8.1 Introduction

The use of solar energy to drive organic syntheses is not a novel concept. The idea was originally proposed by Ciamician as early as 1912.[1] However, the common use of the generally accepted term *photocatalysis* and significant developments in this field properly started in the 1970s after the discovery of photo-induced electrochemical photolysis of water on a TiO_2 electrode by Fujishima and Honda.[2]

RSC Green Chemistry No. 42
Green Photo-active Nanomaterials: Sustainable Energy and Environmental Remediation
Edited by Nurxat Nuraje, Ramazan Asmatulu and Guido Mul
© The Royal Society of Chemistry 2016
Published by the Royal Society of Chemistry, www.rsc.org

Photocatalysis, in which solar photons are used to drive redox reactions to produce chemicals (*e.g.*, fuels), is the central process to achieve the principles of green and sustainable chemistry.[3] Despite significant efforts to date, a practically viable photocatalyst with sufficient efficiency, stability, and low cost is yet to be demonstrated. It is often difficult to simultaneously achieve these different performance metrics with a single material component. The ideal heterogeneous photocatalysts with multiple integrated functional components could combine individual advantages to overcome the drawbacks of single-component photocatalysts. For further information, readers are kindly referred to recent overviews in the field on the development of advanced photocatalysts.[4,5]

As a fundamental and applied field of science, heterogeneous photochemistry continues to be an important component of modern chemistry in the 21st century. Research in this field has significantly evolved during the last four decades (especially titanium oxide chemistry), with enhanced knowledge on mechanisms, development of new technologies for storage and conversion of solar energy, detoxification of liquid and gaseous environments, and the photochemical production of new materials. More recently, a new research avenue related to selective transformations of biomass and residues to high-added-value products has emerged as a potentially useful alternative to conventional heterogeneously catalyzed processes.[6] This chapter is intended to provide an overview of recent work conducted along the lines of selective photochemical transformations, particularly focused on heterogeneous oxidative photocatalysis for lignocellulose-based biomass valorization. Future prospects and work in progress in this field will be also emphasized. The effective utilization of clean, safe, and abundant solar energy is envisaged to provide energy and chemicals as well as solving environmental issues in the future, and an appropriate nanostructured semiconductor photo-induced mediated biomass/waste conversion can be the key for such transformations.

8.2 Basic Principles of Heterogeneous Oxidative Photocatalysis

The fundamental principles of heterogeneous photocatalysis have been extensively reported previously.[5] Briefly, a photocatalytic transformation is initiated when a photo-excited electron is promoted from the filled valence band (VB) of a semiconductor photocatalyst (*e.g.*, TiO_2) to the empty conduction band (CB) as the absorbed photon energy, $h\nu$, equals or exceeds the band gap of the semiconductor photocatalyst. As a consequence, an electron and hole pair $(e^- - h^+)$ is formed (eqn (1)–(4), Scheme 8.1).

$HOO^\bullet$ radicals (eqn (4), Scheme 8.1) have also scavenging properties similar to oxygen, thus prolonging the photohole lifetime (eqn (5) and (6), Scheme 8.2).

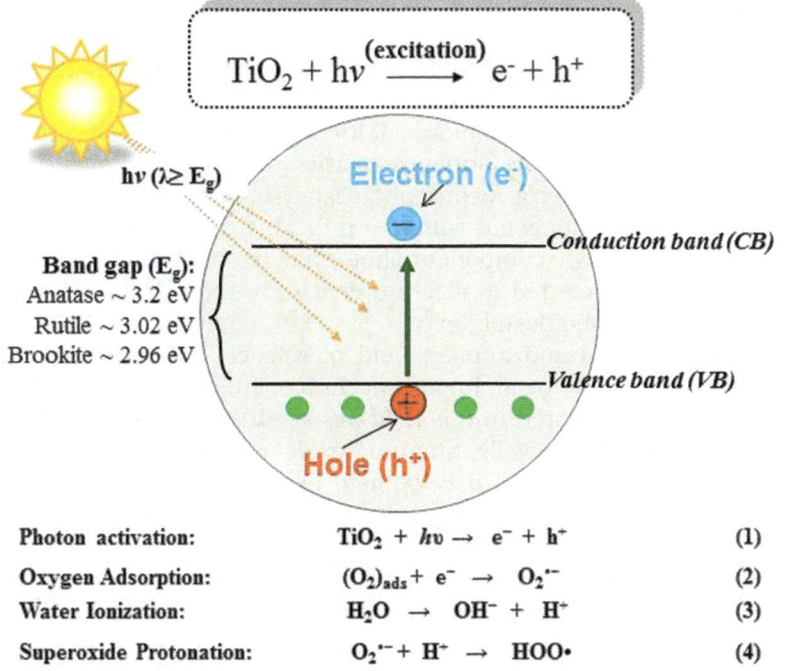

Photon activation:	$TiO_2 + h\nu \rightarrow e^- + h^+$	(1)
Oxygen Adsorption:	$(O_2)_{ads} + e^- \rightarrow O_2^{\cdot-}$	(2)
Water Ionization:	$H_2O \rightarrow OH^- + H^+$	(3)
Superoxide Protonation:	$O_2^{\cdot-} + H^+ \rightarrow HOO\cdot$	(4)

Scheme 8.1 Photo-activation of TiO_2 and primary redox reactions occurring on its surface.

$$HOO\cdot + e^- \rightarrow HO_2^- \qquad (5)$$

$$HOO^- + H^+ \rightarrow H_2O_2 \qquad (6)$$

Scheme 8.2 Extension of hole stability by HOO$^\bullet$ radicals.

Redox processes can take place at the surface of the photo-excited photocatalyst. Very fast recombination between electron and hole occurs unless oxygen (or any other electron acceptor) is available to scavenge the electrons to form superoxides ($O_2^{\bullet-}$), hydroperoxyl radicals (HOO$^\bullet$) and subsequently H_2O_2 (Schemes 8.1 and 8.2).

In contrast to conventional catalysis thermodynamics, not only spontaneous reactions ($\Delta G < 0$) but also non-spontaneous reactions ($\Delta G > 0$) can be promoted by photocatalysis. The input energy by a photo-induced step can help overcome the activation barrier in spontaneous reactions so as to facilitate photocatalysis at an increased rate or under milder conditions. Comparatively, part of the input energy is converted into chemical energy that is accumulated in the reaction products of non-spontaneous processes.

Generated holes have high potential to directly oxidize organic species or indirectly *via* the combination with $^\bullet$OH abundant in water solution (eqn (7)–(9), Scheme 8.3).[5a,b]

$$H_2O + h^+ \rightarrow OH\cdot + H^+ \tag{7}$$

$$R\text{–}H + OH\cdot \rightarrow R\cdot + H_2O \tag{8}$$

$$R\bullet + h^+ \rightarrow R^{+\cdot} \rightarrow \text{Mineralization} \tag{9}$$

Organic waste $\Rightarrow$ photocatalyst/O_2/$h\upsilon \geq E_g \Rightarrow$ Intermediate(s) $\Rightarrow CO_2 + H_2O +$ Mineral Acid (10)

Scheme 8.3 Last steps of the mineralization process.

The corresponding mineral acid of the non-metal substituent is formed as a by-product (*e.g.* in organic waste may be a different heteroatom including Cl, N, S) (eqn (10), Scheme 8.3).

8.2.1 Reactive Oxygen Species (ROS) in Photocatalysis

Several highly reactive ROS species can oxidize a large variety of organic compounds in heterogeneous oxidative photocatalysis.

8.2.1.1 •OH

Redox potential $= +2.81$ V *versus* standard hydrogen electrode, SHE

•OH radicals act as a main component during photo-mineralization reactions, particularly for substances that have weak affinity to the TiO_2 surface.[5a,b] •OH can be produced by oxidation of surface hydroxyls or adsorbed water. For many organic compounds, the primary one-electron oxidation should be initiated by free or trapped holes.

8.2.1.2 $O_2^{\bullet-}$

Redox potential $= +0.89$ V *versus* SHE/$HO_2^{\bullet}$

Superoxide anions ($O_2^{\bullet-}$), easily protonated to yield $HO_2^{\bullet}$ in acidic solution ($pK_a = 4.8$), are readily generated from molecular oxygen by capturing photo-induced electrons from the semiconductor conduction band. $O_2^{\bullet-}$ species are generally less important in initiating oxidation reactions but mainly participate in total mineralization of organic compounds *via* reaction with organoperoxy radicals and production of H_2O_2 (redox potential $= +1.78$ V *versus* SHE) by $O_2^{\bullet-}$ disproportionation.

8.2.1.3 O_3^-

O_3^- species are generated in reactions between the photo-formed hole center on the lattice oxygen (O_L^-) and molecular oxygen.

8.2.1.4 $^1O_2$

$^1O_2$ (singlet molecular oxygen) is usually formed *via* the energy transfer from the triplet state of a dye to molecular oxygen. It has been suggested that

oxidation of $O_2{}^{\bullet-}$ by holes (redox potential $= +2.53$ V *versus* SHE) at the TiO_2 surface should be a plausible mechanism to produce $^1O_2$.

Apart from the important role of ROS in photocatalytic degradation processes, the control and generation of ROS species is essential in heterogeneous photocatalysis to be potentially able to design and predict pathways in selective organic photocatalytic oxidations.

8.3   Lignocellulose as a Feedstock for Chemicals

Lignocellulose is a highly complex and rather recalcitrant feedstock comprising three major polymeric units (lignin, cellulose, and hemicellulose), representing a major fraction of above 90% of all plant biomass $(170 \times 10^9$ t/a).[7] Biomass is inexpensive and globally accessible. The available feedstocks should be sufficient to replace a significant portion of the non-renewable raw materials now required by the chemical industry.[8] Moreover, currently, a large portion of biomass originating from inedible plant material is considered to be waste and is burned to provide heat. There is an increasing need for a better utilization of these waste streams in the future. Selective biomass conversion to biofuels and useful chemicals requires significant improvements in the current chemical approaches and technologies. These novel strategies will play a key role in the implementation of the bio-based economy relying on renewable resources and have a significant environmental and societal impact.[7,9]

Lignocellulosic feedstocks (*i.e.*, forestry waste, agricultural residues, municipal paper waste, and certain food waste residues) can be converted into a variety of useful products in multi-step processes.[10,11] However, due to the large complexity of such feedstocks, lignocellulosics have to be broken down by several processes and technologies into simpler fractions which can be converted into desired products. These include major routes such as gasification, pyrolysis, and pre-treatment hydrolysis/fragmentation steps. Upon deconstruction, the obtained solid, liquid, or gaseous fractions need to be upgraded *via* various processes which yield a plethora of chemicals (Figure 8.1).[10,11]

From a sustainability viewpoint, the development of low-temperature, highly-selective (photo)catalytic routes for the direct transformation of lignocellulosics into valuable chemicals or platform molecules (*e.g.* sugars, carboxylic acids, alcohols, furans, aldehydes) is of great significance.[10] These compounds can then be subsequently converted into useful products. However, the selective conversion of lignocellulosics under mild conditions still remains a significant challenge owing to the highly recalcitrant structure of cellulose and lignin fractions. Subsequent sections of this chapter aim to illustrate the potential of photocatalytic processes for the conversion of selected lignocellulosic feedstocks into chemicals and fuels.

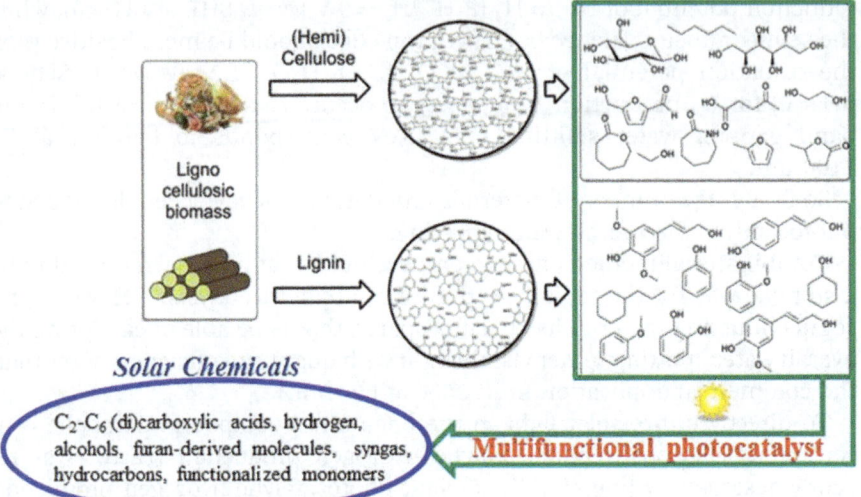

Figure 8.1 Proof of concept of the process of lignocellulose valorization through a photocatalytic route.
(Adapted with modifications from ref. 12).

8.4 Application of Nanostructured Photocatalysts in Lignocellulose Valorization

8.4.1 Photocatalytic Production of Hydrogen

Nanotechnology is offering a variety of possibilities for the modification of existing photocatalysts in the hydrogen economy and the discovery and development of novel nanofunctional materials. Lots of papers have studied the effect of different nanostructured materials on the performance of photocatalysts, since their energy conversion efficiency is principally influenced by nanoscale properties.[13]

8.4.1.1 Water Splitting by Heterogeneous Photocatalysis

In 1972 Fujishima and Honda[2] reported the splitting of water by the use of a semiconductor electrode of TiO_2 (rutile phase) connected through an electrical load to a platinum black counter-electrode. TiO_2 electrode irradiation with UV-A light caused electrons to flow from it to the platinum counter-electrode *via* the external circuit.

Hydrogen production by splitting of water using solar light offers a promising method for solar energy storage and photochemical conversion. The photocatalytic hydrogen production can occur when the semiconductor conduction band (CB) bottom edge must be more negative than the

reduction potential of H^+ to H_2 ($E\ H^+/H_2 = 0$ V *versus* SHE at pH = 0), while the semiconductor valence band (VB) top edge should be more positive than the oxidation potential of H_2O to O_2 ($E\ O_2/H_2O = 1.23$ eV *versus* SHE at pH = 0) for O_2 production from water to occur. This theoretical minimum band gap for water splitting of 1.23 eV corresponds to light of about 1100 nm.

So far a huge number of materials and derivatives have been developed to photocatalyze the overall water splitting.[13]

Combinatorial methods have been developed that have been proved to be a suitable way for the rapid selection of nanophotocatalysts.[14] However, no semiconducting material has been prepared that is be able of catalyzing the overall water splitting under visible light with quantum efficiency larger than the commercial application limit 30% at 600 nm.[15]

To directly utilize solar light in the open air, a compound parabolic concentrator (CPC)-based photocatalytic hydrogen production reactor was recently designed by Jing *et al.*[16] Efficient photocatalytic hydrogen production under direct solar light was accomplished by coupling a tubular reactor with the CPC concentrator. This demonstration drew attention for further studies in this promising direction. Nevertheless, both for material and reactor design, reduction of cost will have to be given special priority before the final utilization of semiconductor-based photocatalytic hydrogen generation.[17]

8.4.1.2 Production of Hydrogen by Photocatalytic Reforming of Lignocellulose-based Organic Waste

Biomass sources have been utilized for the sustainable production of hydrogen.[18,19] Several processes have been developed for this purpose (*e.g.*, steam gasification,[20] fast pyrolysis,[21] and supercritical conversion[22]). However, these processes require drastic reaction conditions including high temperatures and/or pressures and consequently imply high costs (Figure 8.2). In comparison to these energy intensive thermochemical processes, photocatalytic reforming may be a good approach as this process can be performed at room temperature and atmospheric pressure and be driven

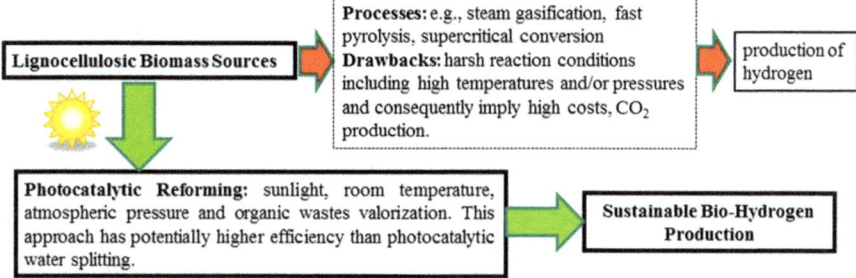

Figure 8.2 Advantages of a sustainable hydrogen production by a lignocellulose photocatalytic reforming pathway over conventional transformation routes.

by sustainable sunlight. Producing hydrogen by photocatalytic reforming of renewable organic wastes (biomass) may also be more practical and viable than that of photocatalytic splitting of water due to its potentially higher efficiency.[23] Water-splitting processes have a relatively low efficiency, as limited by the recombination reaction between photo-generated electrons and holes.[24] The thermodynamics of photochemical water splitting were investigated in detail by Bolton *et al.*[25] who concluded that it is possible to store a maximum of approx. 12% of the incident solar energy in the form of hydrogen, allowing for reasonable losses in the electron transfer steps and the catalytic reactions of water oxidation and reduction.

Lignocellulose-derived compounds can serve as sacrificial agents (electron donors) to reduce the photocatalyst recombination e^--h^+ rate. A large variety of organic compounds (most of them compounds that model the lignocellulose structure, *e.g.*: alcohols, polyols, sugars, aromatic compounds as well as organic acids) have been used as electron donors for photocatalytic hydrogen production (Table 8.1). As previously mentioned, one of the major drawbacks of semiconductor photocatalysts, especially those involving photo-induced hydrogen production from water, is their relatively low efficiency, which is mostly limited by the fast recombination reaction rate between photo-promoted electrons and holes. One way to suppress or retard this recombination rate is with the use of electron donors as sacrificial agents, the role of which is to react irreversibly with the photo-generated holes and/or oxygen thereby increasing the rate of hydrogen production. Alcohols are satisfactory hole scavengers and undergo a relatively rapid and irreversible oxidation, which results in increased quantum yields and enhanced rates of photocatalytic hydrogen production.

Pioneer studies in biomass photocatalytic reforming were conducted in 1980.[26] Kawai and Sakata reported that hydrogen could be generated from carbohydrates in the presence of $Pt/RuO_2/TiO_2$ hybrid material and under 500W Xe lamp irradiation. They also subsequently reported that hydrogen could be generated under identical conditions from other biomass sources – including cellulose, dead insects, and waste materials.[27] These studies demonstrate the feasibility of the photocatalytic production of hydrogen from biomass.

The most recent pioneering studies in biomass photocatalytic reforming for hydrogen production are presented and discussed in Table 8.1.

Taking into account the lowering cost for solar-to-H_2 energy conversion, preferential sacrificial electron donors in water-splitting systems are polluting by-products from industries and low-cost renewable biomass from animals or plants. At little or no cost, they could be exploited to accomplish both the tasks of hydrogen production and waste treatment and biomass reforming simultaneously.

Photocatalytic wet reforming of biomass is relatively in the infancy level, and at present most investigations on it are still largely on the laboratory scale. There is not yet (at the time of writing) a single report about any pilot studies on photocatalytic reforming for commercial hydrogen production.

Table 8.1 Selected pioneering research studies on photocatalytic reforming of biomass to hydrogen and photocatalytic production of high-value organic intermediates from lignocellulose-derived compounds.

Entry/ Ref.	Model biomass-based compound	Photocatalysts	Reaction conditions	Photocatalytic performances (activity/selectivity)
Hydrogen production by photocatalytic reforming				
1/28	Glucose, sucrose, starch	Various 1.0 wt.% noble metals (Pt, Pd, Au, Rh, Hg, Ru) loaded TiO_2 prepared by sol-gel and metal impregnation methods.	125 W high-pressure mercury lamp in anaerobic conditions. Pre-treatment of starch by microwave.	Hydrogen evolution rate decreases in the order: $Pt/TiO_2 > Au/TiO_2 > Pd/TiO_2 > Rh/TiO_2 > Ag/TiO_2 > Ru/TiO_2$. Conditions to reach the maximum rate of hydrogen production from glucose: $pH = 11$, pure N_2 atmosphere, 1.0 wt% Pt loading on anatase TiO_2. Microwave pre-treatment of starch (polysaccharide biomass) enhances significantly hydrogen evolution rate.
2/29	Glucose	0.2% wt Pt or NiO-doped alkali tantalates ($MTaO_3$, M = Li, Na, K) prepared by solid state reaction and metal impregnation followed by calcination (NiO) and reduction (Pt).	125 W high-pressure Hg lamp. Pure N_2 atmosphere in a closed gas-recirculation system equipped with an inner irradiation type quartz reaction cell.	NiO supported on $NaTaO_3$ showed the highest activity for H_2 production from glucose solution. For $LiTaO_3$ and $KTaO_3$ samples, Pt cocatalyst was more effective for H_2 production than NiO.
3/30	Glucose	Oxide catalysts $Bi_xY_{1-x}VO_4$ (BYV) co-doped with 1.0% wt Pt prepared by solid state reaction method and *in situ* photo-deposition of Pt.	350 W Xe lamp with a 430 nm cut-off filter, visible light.	The highest hydrogen production showed BYV with a B/Y ratio of 1:1 at $pH = 3$. After 2 or 3 hours of reaction the hydrogen production decreased, but when the gas product was replaced with N_2 atmosphere, hydrogen production recovered. It is supposed that the generated CO_2 might participate in the redox reaction, possibly forming CH_4, and in some way inhibit the hydrogen production.

4/31	Glucose, propanetriol and methanol	Pt/P25-x%R catalyst with tuned anatase-rutile structure obtained by thermal treatment of P25, 0.1% wt Pt was loaded on TiO$_2$ by *in situ* photo-deposition method.	300 W Xe lamp with a shutter window filled with water to remove infrared light illumination.	The highest hydrogen production was observed for Pt/P25-74%R and compared to P25 was enhanced approx. 3–5 times. It is suggested that the anatase-rutile phase structure cannot only enhance the charge separation and consequently the activity, but also adjust the surface acid/base property which remarkably suppresses the CO formation.
5/32	Glucose	Pt/TiO$_2$ and Pd/TiO$_2$ systems prepared by sol–gel method with application of ultrasonic treatment and calcined at high temperatures and in a redox atmosphere.	125 W Hg lamp ($\lambda_{max} = 365$ nm), under Ar atmosphere. Catalyst in suspension (glucose solution)	Pt/TiO$_2$ calcined at 850 °C gives the most effective system (it was observed a strong metal-support interaction effect, SMSI, for this system) even though oxidation/reduction treatment had led to a decrease in a surface area and transformation from anatase to rutile.
6/33	Glucose	Pt/ZnS-ZnIn$_2$S$_4$ synthesized in methanol by solvothermal technique, 0.5% wt of Pt was deposited by *in situ* photoreduction.	400 W metal halide lamp, visible light ($\lambda > 420$ nm).	ZnS-coated ZnIn$_2$S$_4$ has better activity for hydrogen evolution than pure ZnIn$_2$S$_4$. This enhancement is explained by the adsorption of glucose by ZnS on the ZnIn$_2$S$_4$ surface. Maximum hydrogen production is promoted over Pt/ZnS(17% mol)-ZnIn$_2$S$_4$.
7/34	Olive mill wastewater	Mesoporous TiO$_2$ synthesized by sol–gel technique ($S_{BET} = 59$ m^2/g and 19.5 nm of anatase crystal size).	150 W Hg lamp ($\lambda = 100$–280 nm).	The highest hydrogen production (38 mmol after 2 h of reaction) was evolved at: TiO$_2$ dosage 2g/L, pH = 3.

Table 8.1 (Continued)

Entry/Ref.	Model biomass-based compound	Photocatalysts	Reaction conditions	Photocatalytic performances (activity/selectivity)
8/35	Sugars	TiO_2 dispersed on SiO_2; TiO_2-SiO_2 and TiO_2 modified with metals (Cr^{2+}, Mn^{4+} and V^{5+}) and non-metals N and S; Au-deposited on TiO_2-SiO_2 and TiO_2. Incipient wetness impregnation, impregnation with urea or thiourea, and deposition-precipitation were used as preparation techniques.	200 W Ne lamp under N_2 atmosphere, UV-visible light.	It was observed that doping with non-metals is more effective than doping with metals in the case of TiO_2 based photocatalysts and more effective for metals in the case of TiO_2-SiO_2 based photocatalysts. However doping TiO_2 with non-metals gives the most effective catalyst, because the non-metals are less effective in forming recombination center than metals.
9/36	Glycerol	Pt/TiO_2 prepared by wet impregnation method, metal loading 0.05–5.0 wt%.	Xe-arc lamp with a water filter for the elimination of infrared radiation, UV-visible light.	The best photocatalytic performance for hydrogen production is obtained for samples containing 0.1–0.5 wt% Pt. The reaction proceeds with intermediate production of methanol and acetic acid and eventually results in complete conversion of glycerol to H_2 and CO_2. The rate of reaction is higher in neutral and basic solutions, and increases with increasing temperature from 40 to 60–80 °C.
10/37	Methanol	0.5 wt% Pt/TiO_2 P25 prepared by hydrothermal treatment of TiO_2 P25 and photochemical deposition of Pt.	UV light adjusted ($\lambda = 300$–400 nm).	Hydrogen evolution rate increases with increasing methanol concentration. The highest activity in hydrogen evolution exhibits Pt-loaded TiO_2UV100. Detected reaction products were: formaldehyde, formic acid and carbon dioxide.
11/38	Methanol, Ethanol	TiO_2 films synthesized by radio frequency magnetron sputtering (RF-MS) deposition method, thickness about 3μm.	500 W Xe arc lamp, UV light and visible light (>450 nm). 10 vol% of alcohol concentration.	Hydrogen production from methanol aqueous solution proceeds on TiO_2 thin films under visible light irradiation, while this reaction doesn't proceed on TiO_2 thin films under UV light.

12/39	Acetic acid	CuO/SnO$_2$ prepared by co-precipitation method.	300 W high pressure Hg lamp, UV light.	Photocatalytic H$_2$ production increases with the increment of CuO amount and the optimum CuO content is about 33.3 mol%.
13/40	Methanol	Different noble metal (Ag, Au, Ag–Au alloy and Pt) modified TiO$_2$ prepared by flame spray pyrolysis or by deposition of noble metal.	Iron halogenide mercury arc lamp emitting at $\lambda = 350$–450 nm.	Methanol underwent oxidation up to CO$_2$ through the formation of formaldehyde and formic acid. In this reaction carbon monoxide, methane, methyl formate, acetaldehyde and dimethyl ether were identified as by-products. Hydrogen evolved at constant rate, which significantly increased upon noble metal addition, Pt being the most effective cocatalyst, followed by gold and silver, according to their work function values.
14/41	Methanol	Au/TiO$_2$ prepared by deposition-precipitation technique on commercial TiO$_2$ Evonik P25	300 W Xe lamp with a filter to remove infrared illumination from the lamp.	Hydrogen production is remarkably increased when gold particle size is reduced from 10 nm to smaller than 3 nm. Both the rate of H$_2$ production and the CO selectivity increase with pH value up to the neutral value and decrease thereafter. Carbon monoxide by-product is probably formed *via* decomposition of the intermediate formic acid species.
15/42	Methanol	Series of ALaTa$_{x/3}$Nb$_{2-x/3}$O$_7$ (A = K, H; $x = 0$, 2, 3, 4 and 6) synthesized by solid-state method and Pt cocatalyst prepared by deposition of Pt particles.	100 W high-pressure Hg lamp (UV light).	HLaTa$_{2/3}$Nb$_{4/3}$O$_7$/Pt shows the best photocatalytic performances. The photocatalytic hydrogen evolution rate reaches 136 cm^3 g^{-1} h^{-1}, which is 45.3 times larger than that of TiO$_2$(P-25) ($ca.$ 3 cm^3 g^{-1} h^{-1}).
16/43	Oxygenated hydrocarbons, glycerol, glucose, sucrose.	(B, N)-co-doped TiO$_2$ prepared by hydrothermal synthesis and platinized by photo-deposition method.	300 W Xe lamp.	The biomass-derived fuels serve as both hydrogen sources and electron donors and they are oxidized into CO$_2$ as an ultimate product. At 5 wt% glycerol concentration is achieved the maximum hydrogen amount.

Table 8.1 (*Continued*)

Entry/ Ref.	Model biomass-based compound	Photocatalysts	Reaction conditions	Photocatalytic performances (activity/selectivity)
17/44	Glycerol and several biomass-derived compounds	0.5 wt% Pt/TiO$_2$ synthesized by impregnation of TiO$_2$ (Evonik P25).	300 or 450 W Xe arc lamp (filtration of IR radiation), solar light source.	The overall reaction is non-selective with respect to the organic substrate used and, therefore, practically all biomass-derived molecules in solution or in suspension may be used as feedstock.
18/45	Glucose	Different Me/TiO$_2$ systems (Me: Pt, Rh, Ru, Ir, Au, Ni, Cu). Ni/TiO$_2$ and Cu/TiO$_2$ were prepared by impregnation method while the other catalysts by *in situ* photodeposition method.	300 W high-pressure Hg lamp	Rh/TiO$_2$ photocatalyst is the most active for H$_2$ production and with the most extremely low CO concentration. The rate of H$_2$ production increases in the following order: Ir < Ru < Au < Ni ≈ Cu ≈ Pt < Rh. The optimum loading of supported metal is 0.3%.
19/46	Methanol	Different mesoporous-assembled SrTiO$_3$ catalysts with loaded metal cocatalyst: Au, Pt, Ag, Ni, Ce and Fe, synthesized by sol–gel technique assisted by a structure-directing surfactant.	176 W Hg lamp with the main emission λ = 254 nm and 300 W Xe lamp with λ > 400 nm. Reaction temperature under UV light source is 45 °C and under visible light is room temperature.	Au, Pt, Ag and Ni loadings have a positive effect on the photocatalytic activity enhancement, whereas the Ce and Fe loadings detrimental effect. The best loaded metal was found to be Au due to its electrochemical properties compatible with the SrTiO$_3$-based photocatalyst and its visible light harvesting enhancement. 1 wt% Au/SrTiO$_3$ exhibited the highest photocatalytic hydrogen production activity for both, UV and visible light.

20/47	Methanol	Mesoporous-assembled 0.93TiO₂–0.07ZrO₂ loaded by non-precious metals: Ag, Ni and Cu, synthesized by a sol-gel procedure with the aid of a structure-directing surfactant and photochemical deposition method.	A set of 16 Hg lamps (11 W, UV light). 50% vol of methanol aqueous solution.	The most efficient supported metal is Cu due to its suitable physical, chemical, and electrochemical properties with the $0.93TiO_2$–$0.07ZrO_2$ mixed oxide-based photocatalyst. 0.15 wt% $Cu/0.93TiO_2$–$0.07ZrO_2$ mixed oxide material exhibited the highest photocatalytic hydrogen production. The following UV light-harvesting ability is observed: Cu-loaded photocatalyst > Ni-loaded photocatalyst > Ag-loaded photocatalyst.
21/48	Methanol	ZnO/TiO_2, SnO/TiO_2, CuO/TiO_2, Al_2O_3/TiO_2 and $CuO/Al_2O_3/TiO_2$ nanocomposites synthesized by mechanical mixing.	15 W black light, UV light ($\lambda = 352$ nm).	The highest hydrogen production was observed for $CuO/Al_2O_3/TiO_2$ nanocomposites and the optimal component was 0.2 wt% $CuO/0.3$ wt% Al_2O_3/TiO_2.
22/49	Glycerol	Series of Me/TiO_2 materials (Me = Au, Pt, Pd)	400 W Xe arc lamp	The activity of Pd/TiO_2 was 1.7 times higher than that of Au/TiO_2 and was almost the same as that over Pt/TiO_2.
23/50	Ethanol, glycerol	CuO_x/TiO_2 photocatalyst obtained by water-in-oil microemulsion technique. The nominal Cu loading was 2.5% wt.	125 W medium pressure Hg lamp ($\lambda = 365$ nm).	The investigation indicates appreciable advantages of the photocatalyst preparation procedure for hydrogen evolution by photo-reforming of ethanol and glycerol water solutions with respect to conventional impregnated materials.
24/51	Glycerol	0.5% wt Pt/TiO_2 modified with heteropoly blue (HPB) prepared by photodeposition method.	250 W high-pressure Hg lamp for UV light and 300 W halogen lamp for visible light ($\lambda > 420$ nm).	The electrons formed by the excitation of HPB upon visible light irradiation would migrate to TiO_2, which would enable the hydrogen production.
25/52	Glucose, methanol, ethanol and glycerol	Highly isolated RuO_2 nanoparticles were first deposited on anatase TiO_2 by a wetness impregnation method, and then, the hydrogen evolution sites, single-site tin, were grafted onto the TiO_2 surface by the use of surface organometallic chemistry.	365 nm UV light irradiation	An optimal hydrogen evolution rate is achieved at loading 0.68 wt% of RuO_2 and grafting 0.24 wt% of Sn onto anatase TiO_2 (31 mmol per g h of hydrogen from glycerol).

Table 8.1 (*Continued*)

Entry/ Ref.	Model biomass-based compound	Photocatalysts	Reaction conditions	Photocatalytic performances (activity/selectivity)
26/53	Methanol	Cu-doped TiO$_2$ obtained by different doping methods	125 W medium pressure Hg lamp at 365 nm	From the structural and surface analysis of the catalysts the authors stated that the occurrence of highly disperse and reducible Cu^{2+} species is directly related to the photocatalytic activity for the H$_2$ production reaction. Highly active materials have been obtained from a chemical reduction method leading to 18 mmol per h g for 3 mol% copper loading. Highly disperse Cu^{2+} species would be easily reduced during the reaction forming Cu^0 sites that will act as an effective cocatalyst for a water reduction reaction. Moreover, the presulfation treatment of the TiO$_2$ support clearly induces a higher photocatalytic activity.
27/54	Aqueous cellulose suspension	0.5 wt% Pt/TiO$_2$ (P25) was prepared by a photochemical deposition procedure.	366 nm (UV-A) or under sunlight	The mechanism of the photocatalytic process relies on the TiO$_2$-mediated cellulose hydrolysis, under irradiation. The polysaccharide depolymerisation generates water-soluble species and intermediates, among them 5-hydroxymethylfurfural (HMF) was identified. These intermediates are readily oxidized following the glucose photoreforming, thus enhancing water hydrogen ion reduction to give gas-phase H$_2$ (43 μmoles after 4h, 2g/L of catalyst under UV-A)

Selective photocatalytic production of high value organic molecules

	Substrate	Catalyst	Conditions	Results
28/61	Ethanol, 1-propanol, 2-propanol, 2-butanol	TiO_2 rutile phase (3.6 and 7.7 m² g⁻¹).	Catalyst suspensions irradiated with 366 nm UV light	Selective production of ethanal, propanal and propanone with 0.101, 0.104 and 0.101 quantum yields at 20 °C, respectively.
29/62	2-Propanol, crotyl alcohol	TiO_2, Me-TiO_2 (Me = Pd, Pt, Zr, Fe, Zn, Ag, Au). Catalysts prepared by ultrasound- and microwave-assisted sol-gel procedure.	UV light ($\lambda_{max.}$ = 365 nm). Reaction in liquid phase (for crotyl alcohol) with a solid catalyst in suspension. Reaction in gas phase (for 2-Propanol) with a fix bed of the catalyst. Water used for cooling was thermostated at 10 °C for reactions with crotyl alcohol and 20 °C for 2-propanol).	For crotyl alcohol, conversions between 8% and 38% for $t = 30$ min or 32% and 95% for $t = 300$ min for the two extreme catalysts (TiO_2:Fe and TiO_2:Au, respectively). When the influence of the metal is considered, iron, palladium and zinc exhibit lower conversions than the corresponding bare-titania, whereas the presence of silver, zirconium and especially gold is beneficial to photoactivity. In the case of 2-propanol, platinum-containing solids which showed quite high selectivity values to acetone (in the 78–80% range at 22–28% conversion).
30/63	Malic acid	TiO_2-guanidine-(Ni,Co)-Fe_2O_4	Catalysts suspended in malic acid aqueous solution and illuminated under visible light (150 W Quartz Halogen Lamp, $\lambda > 400$ nm).	Selectivity close to 80% to formic acid could be achieved in less than 2 hours of reaction. Efficient separation of the photocatalyst after reaction.
31/64	Acetic, propionic, n-butanoic and n-pentanoic acids	Pt/TiO_2 (rutile). Pt was deposited by illuminating each powdered semiconductor suspended in water-ethanol solution.	30 mL of water-organic acid (6:1 v/v) mixture and catalyst in suspension irradiated with a 500 W Xe lamp, pH <2.0.	In the absence of oxygen, aliphatic carboxylic acids (especially C₄-C₅ acids) are decarboxylated to the corresponding reduced hydrocarbons: acetic (156 micromol RH/10 h), propionic (1470 micromol RH/10 h), n-butanoic (996 micromol RH/10 h), n-pentanoic (1018 micromol RH/10 h) acids.

Table 8.1 (*Continued*)

Entry/ Ref.	Model biomass-based compound	Photocatalysts	Reaction conditions	Photocatalytic performances (activity/selectivity)
32/65 and 67	Glucose	Bare-TiO$_2$ and supported titania on zeolite Y. Catalysts were synthesized by a modified ultrasound-assisted sol–gel method.	Catalysts suspended in a glucose solution (50%H$_2$O/50%CH$_3$CN solvent composition) and illuminated with 125 W mercury lamp ($\lambda_{max} = 365$ nm).	High photoselectivity for glucaric and gluconic acid production (68.1% total selectivity, after 10 min. illumination time) especially for TiO$_2$ supported on zeolite Y. Apart from photocatalyst properties, it was found that reactions conditions, especially solvent composition and short illumination times, also have considerable effect on the activity/ selectivity of tested photocatalysts.
33/68	Glucose	Fe-TiO$_2$ and Cr-TiO$_2$ supported on zeolite Y (SiO$_2$:Al$_2$O$_3$ = 80) and prepared by ultrasound-assisted wet impregnation method (rotary evaporator was coupled with ultrasonic bath)	Catalysts suspended in a glucose solution (50%H$_2$O/50%CH$_3$CN solvent composition) and illuminated with 125 W mercury lamp ($\lambda_{max} = 365$ nm).	Fe-TiO$_2$ zeolite-supported systems total selectivity for GUA + GA is 94.2% after 20 min of illumination. Cr-TiO$_2$ zeolite-supported systems total selectivity for GUA + GA is 99.7% after 10 min of illumination.
34/73	Methanol	Anatase-type TiO$_2$ particles (ST-01) having a large surface area of 300 m^2 g^{-1}	UV light (UV intensity: 1800 μW cm^{-2}). Reaction in gas phase in presence of air with a fix bed of the catalyst.	High selective (91 mol%) photooxidation of methanol to methyl formate with no catalyst deactivation. The conversion of methanol increased with elevation of the reaction temperature up to 250 °C (conversion is three times higher than at room temperature but the selectivity decreases).
35/76	Lignocellulose	Commercial TiO$_2$	Solid state mixture of titania and lignocellulose UV irradiated ($\lambda_{max} = 360$ nm).	Photocatalysis pre-treatment used for an efficient biological sacharification of lignocellulosic material. Titania did not disturb the biological reactions by the cellulase and yeast.
36/77	Kraft lignin	Ta$_2$O$_5$-IrO$_2$ thin film (prepared using thermal decomposition technique) as electrocatalyst and TiO$_2$ nanotube (prepared using electrochemical anodization) as photocatalyst	Photochemical-electrochemical process in liquid phase. The intensity of the UV light was *ca.* 20 mW cm^{-2} (main line of emission, 365 nm).	Oxidation of lignin gave vanilin and vanilic acid.

Nevertheless, since the process is frequently carried out at room temperature and both the energy and feedstock are from renewable sources, photocatalytic reforming can be considered to be especially promising for sustainable large-scale production of green biohydrogen.

8.4.2 Photocatalytic Upgrading of Lignocellulose-based Molecules: Production of High-value Chemicals

The utilization of heterogeneous photocatalysis for environment protection (both in the liquid phase and the gas phase) has been extensively investigated as photo-activated semiconductors have proven ability to unselectively mineralize various types of toxic, refractory, and non-biodegradable organic pollutants under mild conditions.[55,56] Closely related to selective synthesis, nanostructured photocatalytic systems have also been employed for the oxofunctionalization of hydrocarbons *via* selective oxidations.[57] Reported organic photosynthetic reactions include oxidations, reductions, isomerizations, substitutions, polymerizations, and condensations. These reactions can be carried out in inert solvents and/or their combinations with water. For more details on photocatalytic synthetic transformations based on different photo-active solid materials, readers are kindly referred to leading reviews in the field.[5c,58]

8.4.2.1 Photocatalytic Selective Oxidation of Cellulose-based Chemicals

Catalytic oxidations have traditionally been carried out in environmentally harmful chlorinated organic solvents at high temperatures and pressures by employing stoichiometric amounts of various inorganic oxidants as oxygen donors (*e.g.* chromate and permanganate species). Such oxidants are expensive and toxic but they also produce large amounts of hazardous waste, and therefore are needing to be replaced by safer systems.

Similarly, a photocatalytic process can bring about significant benefits in terms of milder and more environmentally sound reaction conditions and better selectivities to the desired product. Substances unstable at high temperatures may be synthesized *via* selective light-assisted processes.

Alcohol oxidation to their corresponding ketones, aldehydes, and/or carboxylic acids is one of the most important transformations in organic synthesis.[59] Semiconductor photocatalysts have not been frequently employed in selective synthetic oxidation processes as the replacement of traditional oxidation methods has been mostly covered with heterogeneously catalyzed systems.

According to thermodynamics, an alcohol molecule with singlet electronic configuration cannot directly react with an unactivated dioxygen molecule, which has a triplet electronic configuration.[60] The working mechanism in metal oxidation catalysts involves an electron transfer mediated step by the

metal which thereby induces the formation of singlet oxygen species.[60] In this context, the search for an oxidation photocatalyst capable of directly activating dioxygen under solar light is an interesting as well as challenging task.

Catalytic selective photo-oxidation of cellulose-based molecules can provide a wide range of high added-value compounds, as is shown in Table 8.1 (entries 28–36) which summarizes and discusses recent selected studies on photocatalytic selective oxidation of lignocellulose-based model compounds to valuable chemicals.

Photo-assisted catalytic dehydrogenation reactions that take place at room temperature and ambient pressure offer an interesting route for aldehyde synthesis. C_1–C_4 alcohols are easily converted in the liquid and gas phase and in the presence of oxygen into their corresponding aldehydes or ketones, which may be further transformed by non-catalytic processes into acids.[61,62]

A range of different nano-titania-based systems synthesized through sol-gel processes varying the precursor and/or the ageing conditions (magnetic stirring, ultrasound, microwave, or reflux) was recently reported for the liquid-phase selective photo-oxidation of crotyl alcohol to crotonaldehyde. The gas-phase selective photo-oxidation of 2-propanol to acetone was also chosen as a model reaction (Table 8.1, entry 29).[62] Both reactions showed relatively similar results in terms of influence of precursor and metal, despite having very different reactant/catalyst ratios and contact times. The presence of iron, palladium, or zinc in the systems was found to be detrimental for the activity. Zirconium and particularly gold improved the results as compared to pure titania (Table 8.1, entry 29).

A higher reactivity has been generally reported for primary alcohols[61] (Table 8.1, entry 28), which opens up a promising selective synthesis method for the production of hydroxycarboxylic acids. The selective dehydrogenation of various secondary alcohols was also possible. If the alcohol is unsaturated, isomerization may occur, yielding the corresponding saturated aldehyde. The photocatalytic oxidation of alcohols is highly dependent on the type of alcohol. Generally, the conversion per pass of primary alcohols is low (with a slightly higher value for secondary alcohols), but high selectivities are generally achieved (>95%).

Aliphatic carboxylic acids can be transformed to shorter chain acids (e.g. malic acid to formic acid)[63] or decarboxylated to the corresponding reduced hydrocarbons or hydrocarbon dimers in the absence of oxygen and in pure aqueous or mixed aqueous/organic solutions by means of photo-Kolbe-type processes (Table 8.1, entry 31).[64] As an example, the selective aqueous conversion of malic acid into formic acid has been conducted under visible light irradiation using a magnetically separable TiO_2-guanidine-(Ni,Co)-Fe_2O_4 nanocomposite (Table 8.1, entry 30). The photocatalyst featured a simple magnetic separation and offered the possibility to work under visible and sunlight irradiation due to titania modification with guanidine, which remarkably decreased the band gap of the metal oxide semiconductor.[63] Comparably, acetic, propionic, butanoic, and n-pentanoic acids could be decarboxylated to hydrocarbons in the absence of oxygen. For the case of

acetic acid (a common product of biological processes), the aforementioned photo-Kolbe reaction could be combined with biological waste treatments to generate combustible fuels.

The efficiency of heterogeneous nano-TiO_2 catalysts in the selective photocatalytic oxidation of glucose into high-value organic compounds has been also recently reported (Scheme 8.4A).[65] This reaction was found to be highly selective (>70%) towards two organic carboxylic acids, namely glucaric (GUA) and gluconic acids (GA). These carboxylic acids are important building blocks for pharmaceutical, food, perfume, or fuel industries.[66] Among all photocatalytic systems tested, the best product selectivity was achieved with titania synthesized by an ultrasound-modified sol–gel methodology (TiO_2(US)).[65] Solvent composition and short illumination times were proved to have a considerable effect on photocatalyst activity/selectivity. Total organic compound selectivity was found to be 39% and 71% for liquid-phase reactions using 10% water/90% acetonitrile and 50% water/50% acetonitrile, respectively. These values were obtained using the optimum nano-TiO_2(US) photocatalyst. In further studies, those results were systematically improved by homogeneously supporting nano-TiO_2(US) on a zeolite type Y (total selectivity of GUA and GA was *ca.* 68% after 10 min illumination time using a 1:1 H_2O/acetonitrile solvent composition).[67] Further photocatalyst optimization *via* development of transition metal (Fe or Cr)-containing supported nanotitania materials,[68] provided advanced systems able to achieve improved selectivities to carboxylic acids. No metal leaching (Fe, Cr, Ti) was detected after photoreaction, with Fe-TiO_2 systems being most selective (94% after 20 min of illumination under similar conditions previously reported, Table 8.1, entry 33).

Such results suggested that synthesized nano-TiO_2 materials could be in principle used in the decomposition of waste from the food industry with

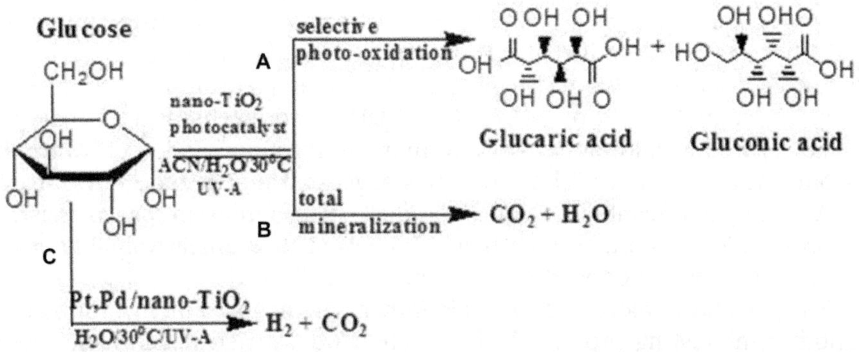

Scheme 8.4 Schematic representation of different photocatalytic glucose conversion pathways obtained by designed nanostructured titania and the selection of proper reaction conditions: (A) selective production of carboxylic acids, (B) total mineralization, and (C) hydrogen production.

the simultaneous production of high-value chemicals when residues (here glucose) act as electron donors.

The oxidation mechanism, efficiency, and selectivity of the photocatalytic oxidation of alcohols using TiO_2 Evonik P25 was reported to strongly depend on the nature of the dispersing medium. Shiraishi *et al.*[57] and Morishita *et al.*[69] demonstrated the positive influence of acetonitrile on epoxide formation in the selective photo-oxidation of alkenes. The addition of small amounts of water to CH_3CN strongly inhibited alcohol adsorption and its subsequent oxidation as evidenced by ESR-spin trapping investigations.[70] The reactivity of alcohols on the surface of photo-excited TiO_2 was also found to be affected by the nature of their hydrophobic aliphatic chain.[70] Molecules, including geraniol and citronellol, were observed to be more susceptible to water content as compared to shorter chain analogs such as *trans*-2-penten-1-ol and 1-pentanol. Optimum reaction conditions were achieved in the photocatalytic oxidation of geraniol, citronellol, *trans*-2-penten-1-ol, and 1-pentanol to the corresponding aldehydes with good selectivity (>70%).

Along these lines, a partial photo-oxidation of diols including 1,3-butanediol, 1,4-pentanediol, and vicinal diols (*e.g.* 1,2-propanediol) could also be effectively conducted using TiO_2 in dichloromethane.[71] CO_2 was not observed as a reaction product but the observed main reaction products included two hydroxy-carbonyl compounds for each 1,*n*-diol. 1,2-Propanediol mainly gave rise to hydroxyacetone (90% selectivity), with only traces of pyruvic acid. Comparatively, 1,3-butanediol was converted into 3-hydroxybutyraldehyde and 4-hydroxy-2-butanone while 1,4-pentanediol gave 4-hydroxypentanal (75% selectivity) and certain quantities of 3-acetyl-1-propanol.

Similarly, glycerol is a relevant polyol currently produced in large quantities as by-product of the biodiesel industry. The mechanism of the selective photocatalytic oxidation of glycerol was also recently reported in presence of TiO_2 Evonik P25 and Merck TiO_2 (Scheme 8.5).[72] The product distribution observed at low glycerol concentration (glyceraldehyde and dihydroxyacetone) changes after a sharp maximum giving formaldehyde and glycolaldehyde as main products for P25 Evonik (mechanism derive from a direct electron transfer). Interestingly, mainly glyceraldehyde and dihydroxyacetone were observed on Merck TiO_2 (•OH-based mechanism), a material characterized by a lower density and more uniform population of hydroxyls groups at surface sites. These findings suggested that photocatalyst structures can significantly influence product distribution in photo-assisted processes. This example is particularly relevant to engineer valuable products from a biofuel-derived by-product.

The gas-phase selective photo-oxidation of methanol to methyl formate is another interesting process.[73] The reaction was carried out in a flow-type reactor, to avoid deep oxidation of methanol, in the presence of TiO_2 particles and under UV light irradiation. A high selectivity to methyl formate was observed (91%) with no catalyst deactivation observed under the investigated conditions (Table 8.1, entry 34).

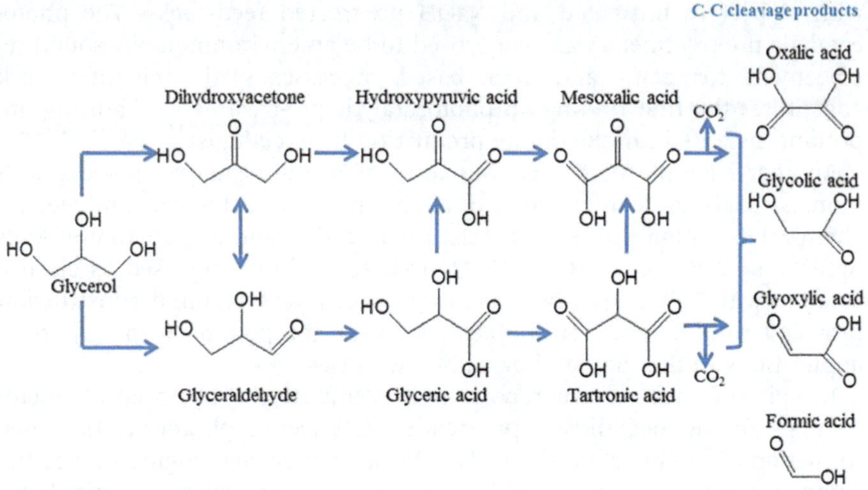

Scheme 8.5 Pathways for glycerol photocatalytic oxidation to valuable chemicals. (Adapted from ref. 6).

Detailed mechanistic studies on photocatalytic oxidation of related organic compounds (*e.g.*, methane to methanol, partial oxidation of geraniol to produce only citral, glycerol and glucose oxidation, and propene epoxidation among others) are obviously out of the scope of the present contribution, but readers are kindly referred to a recently reported comprehensive overviews on this subject.[74]

8.4.2.2 Photocatalytic Selective Oxidation of Aromatic Alcohols as Lignin Model Molecules

Thermocatalysis has occasionally been used to upgrade lignin-based intermediates after lignocellulose depolymerization. One example from literature is the selective defunctionalization by TiO_2 of monomeric phenolics from lignin pyrolysis into simple phenols.[75] This study was focused on defunctionalizing monomeric phenolics from lignin into simple phenols, but the drawback of this method is the high temperature used (550 °C) and the low yields obtained.

Photocatalytic pre-treatments of lignocellulosics have rarely been reported in the literature. Interestingly, Yasuda *et al.*[76] have recently devised a two-step process for biological bioethanol production from lignocellulose *via* feedstock pre-treatment with titania photocatalyst under UV illumination. Selected lignocellulosic feedstocks were napier grass (*Pennisetum purpureum Schumach*) and silver grass (*Miscanthus sinensis Anderss*) as summarized in Table 8.1, entry 35. The photocatalytic pre-treatment did not significantly affect the final product distribution, demonstrating that TiO_2 was not interfering with biological reactions promoted by cellulases and yeast. Most importantly, the photocatalytic pre-treatment was remarkably effective to reduce the time in enzymatic saccharification and fermentation reactions

with respect to untreated and NaOH pre-treated feedstocks. The photo-catalytic pre-treatment was also proved to be an environmentally sound replacement for acid- and alkali-based processes. This scientific work constitutes the first finding on photocatalytic pre-treatments featuring important insights into bioethanol production from cellulose.

Another example of the combination of a photocatalytic process with biomass oxidation comprised an integration of photochemical and electro-chemical oxidation processes for the modification and degradation of kraft lignin (Table 8.1, entry 36).[77] Ta_2O_5–IrO_2 thin films were used as electro-catalysts with TiO_2 nanotube arrays as photocatalysts. Lignin deconstruction provided vanillin and vanillic acid, relevant compounds with important applications in the food and perfume industries.

In spite of these recent reports that exemplify the potential of photo-catalysis for lignocellulosics pre-treatment, selective photocatalytic transformations of lignocellulosic feedstocks are rather challenging due to the complex structure of lignocellulose, particularly related to the highly recalcitrant lignin fraction.

In comparison, relevant research has been conducted on model compounds which have been selected on the basis of representing structural motifs present in lignin. Aromatic alcohols have been approached as model lignin compounds to be converted into various compounds *via* photo-catalytic processing (Table 8.2). Among selective oxidation processes of alcohols, the conversion of benzyl alcohol to benzaldehyde is particularly attractive. Benzaldehyde is in fact the second most important aromatic molecule (after vanillin) used in the cosmetics, flavor, and perfumery industries. Synthetic benzaldehyde is industrially produced *via* benzyl chloride hydrolysis derived from toluene chlorination or through toluene oxidation. In this regard, alternative protocols able to selectively produce benzaldehyde from benzyl alcohol are in demand.

While most of the research on selective production of benzaldehyde has been reported *via* heterogeneously catalyzed protocols, there are some relevant examples of such a reaction under photo-assisted conditions. Aerobic aqueous suspensions of Au supported on cerium(IV) oxide (Au/CeO_2) were reported to be able to quantitatively oxidize benzyl alcohol to benzaldehyde under green light irradiation for 36 h reaction time (Table 8.2, entries 4–7).[78] In comparison, the oxidation of benzyl alcohol was performed using differently prepared TiO_2 catalysts, reaching selectivities three- to seven-fold superior to those of commercial TiO_2 catalysts.[79] In terms of benzyl alcohol derivatives, the aqueous phase photocatalytic oxidation of 4-methoxybenzyl alcohol was achieved by using uncalcined brookite TiO_2. A maximum selectivity of *ca.* 56% towards the aldehyde (*i.e. ca.* 3 times higher than that obtained with commercial TiO_2) was reported.[79]

Heteropolyoxometalate (POM) catalysts of the type $[S_2M_{18}O_{62}]^{4+}$ (M = W, Mo) have also been developed to carry out the photo-oxidation of aromatic alcohols under sunlight and UV/Vis light in acetonitrile.[80] Mechanistic investigations of photocatalysis by these nontoxic compounds indicate that

Table 8.2 Selected representative examples on selective photocatalytic oxidation of model aromatic compounds simulating the structure of lignin.

Entry	Reactants	Products	Conversion [%]	Selectivity [%]

Immobilized TiO_2 on silica cloth. Reaction conditions: alcohol, 1.43 mmol/min; O_2/alcohol $= 22$; 2.5-L annular photoreactor (average contact time 32 s); Temperature $= 190\,°C$; Conversion after 2 h illumination (UV light) time.[91]

Entry	Reactants	Products	Conversion [%]	Selectivity [%]
1	CH₂OH structure	CHO structure	35	>95
2	OH CH–CH₃ structure	O C–CH₃ structure	97	7 (83% styrene)
3	CH₂–CH₂OH structure	CH₂–CHO structure	53	26

Oxidation over 1 wt% Au/CeO_2 under green light. Reaction conditions: benzyl alcohols 33 µmol; Au/CeO_2 50 mg; 5 mL water; 1 bar O_2; 25 °C; LED 530 nm; Conversion after 20 h illumination time.[78]

Entry	Reactants	Products	Conversion [%]	Selectivity [%]
4	CH₂OH structure	CHO structure	>99	>99
5	CH₂OH, CH₃ structure	CHO, CH₃ structure	>99	>99
6	CH₂OH, CH₃ structure	CHO, CH₃ structure	>99	>99
7	CH₂OH, H₃C structure	CHO, H₃C structure	>99	>99

Carbon quantum dots (CQDs) under NIR light irradiation. Reaction conditions: 10 mmol alcohol; H_2O_2 (10 mmol, 30 wt% in water); 8 mg of CQDs catalyst; 60 °C; NIR irradiation for 12 h.[85]

Entry	Reactants	Products	Conversion [%]	Selectivity [%]
8	CH₂OH structure	CHO structure	92	100
9	CH₂OH, H₃C structure	CHO, H₃C structure	88	>99

Table 8.2 (*Continued*)

Entry	Reactants	Products	Conversion [%]	Selectivity [%]
10			86	>99
11			90	>96
12			88	>99

$Au_2(DP_{673})/P25$ catalyst. Reaction conditions: 5 mL toluene (solvent); 10 μmol alcohol; 20 mg catalyst; 1 bar O_2; 30 °C; 4 h irradiation time (98% of light involves the visible range).[86]

Entry	Reactants	Products	Conversion [%]	Selectivity [%]
13			79	100
14			85	93
15			>99	>100
16			>99	91
17			83	99
18			>99	>100
19			81	99

Table 8.2 (*Continued*)

Entry	Reactants	Products	Conversion [%]	Selectivity [%]

Multi-Pd core/CeO$_2$ shell nanocomposite. Reaction conditions: 20 h of visible light ($\lambda > 420$ nm) irradiation time.[87]

20	benzyl alcohol (CH$_2$OH)	benzaldehyde (CHO)	28	100
21	4-methylbenzyl alcohol (H$_3$C, CH$_2$OH)	4-methylbenzaldehyde (H$_3$C, CHO)	10	100
22	4-methoxybenzyl alcohol (H$_3$CO, CH$_2$OH)	4-methoxybenzaldehyde (H$_3$CO, CHO)	12	100

Nb$_2$O$_5$ (100 mg). Reaction conditions: 16 h of visible light ($\lambda > 420$ nm) irradiation time, alcohol (1 mmol), PhCF$_3$ solvent (10 mL), oxygen pressure (1 bar).[88]

23	benzyl alcohol (CH$_2$OH)	benzaldehyde (CHO)	80	>99
24	4-methylbenzyl alcohol (H$_3$C, CH$_2$OH)	4-methylbenzaldehyde (H$_3$C, CHO)	96	97
25	4-methoxybenzyl alcohol (H$_3$CO, CH$_2$OH)	4-methoxybenzaldehyde (H$_3$CO, CHO)	99	>99

CdS (8 mg). Reaction conditions: 10 h of visible light ($\lambda > 420$ nm) irradiation time, alkyl aromatic (0.1 mmol), PhCF$_3$ solvent (1.5 mL), oxygen pressure (1 bar).[89]

26	toluene (CH$_3$)	benzaldehyde (CHO)	33	100
27	4-methyltoluene (H$_3$C, CH$_3$)	4-methylbenzaldehyde (H$_3$C, CHO)	39	100
28	4-methoxytoluene (H$_3$CO, CH$_3$)	4-methoxybenzaldehyde (H$_3$CO, CHO)	36	100

near-UV/Vis irradiation of a POM solution results in an oxygen-to-metal charge-transfer excited state that has strong oxidation ability, responsible for the oxidation of organic substrates. Photo-excited POMs were observed to be reduced by the transfer of one or two electron(s) from the organic substrate. Other types of POMs including water-tolerant materials by combining homogeneous POMs with photo-active and photo-inactive supporting

materials (*e.g.*, mesoporous molecular sieve, titania in amorphous phase) and silica-encapsulated $H_3PW_{12}O_{40}$ have also been reported in similar chemistries.

The acceleration of the aerobic photo-oxidation of similar alcohols has been also described to take place on TiO_2 or SiO_2/TiO_2 samples without any loss of selectivity *via* surface loading of Brønsted acids (reaction time 2–7 h, conversion of the starting molecules *ca.* 42–100% mol, selectivity 73–100% mol).[81] The effect of Brønsted acids was confirmed and enhanced when a small quantity of SiO_2 was incorporated into TiO_2 after the acid pre-treatment, due to the presence of more acid sites on the TiO_2 surface.

A similar selective photocatalytic oxidation of aromatic alcohols to aldehydes was systematically studied under visible-light irradiation utilizing anatase TiO_2 nanoparticles.[82] The unique features of the protocol rely on the possibility to use visible irradiation apart from UV due to the formation of surface complex species by the adsorbed aromatic alcohol. Studies using fluorinated TiO_2 pointed to a dramatic decrease of photocatalytic activity under visible irradiation which suggested a significant role played by the adsorption of the aromatic alcohol species on the surface of the solid involving the surface OH species.[82]

Benzyl alcohol and some of its derivatives (*para*-structures) could be converted into their corresponding aldehydes at high conversions and selectivities (*ca.* 99%) both under UV and visible irradiation. The only exception to this behavior was 4-hydroxybenzyl alcohol, which was oxidized to 4-hydroxy benzaldehyde (selectivity of *ca.* 23% at *ca.* 85% conversion) along with some unidentified products. The authors claimed that OH groups from the TiO_2 surface reacted with the aromatic compound to form Ti–O–Ph species, exhibiting strong absorption in the visible region by ligand to metal charge transfer. These findings confirmed the results reported by Kim *et al.*[83] which proposed a direct electron transfer from the surface complexes to the conduction band of the TiO_2 upon absorbing visible light. Importantly, results showed that visible light can induce reactions by substrate–surface complexation enabling the visible light absorption. Similar observations were recently reported by Li and co-workers[84] with nanorods of rutile titania phase synthesized by a hydrothermal reaction using rutile TiO_2 nanofibers obtained from calcination of composite electrospun nanofibers. The authors proposed a tentative mechanism where selective photocatalytic activities were proposed to be due to the visible-light absorption ability of benzyl alcohol–TiO_2 nanorod complex and the unique properties of rutile TiO_2 nanorods. Excelling properties of these materials including high surface-to-volume ratio, unidirectional 1D channels and superior survivability of electrons may contribute to more efficient electron transport, further required for benzaldehyde formation (>99% selectivity to benzaldehyde).

An innovative class of carbon nanomaterials denoted as carbon quantum dots (CQDs) with sizes <10 nm can also function as an effective near-infrared (NIR) light-driven photocatalyst for the selective oxidation of benzyl alcohol to benzaldehyde.[85] Based on the NIR light-driven photo-induced

electron transfer property and its photocatalytic activity for H_2O_2 decomposition, this metal-free catalyst could efficiently provide high yields of benzaldehyde under NIR light irradiation (Table 8.2, entries 8–12). Such metal-free photocatalytic systems also selectively convert other alcohol substrates to their corresponding aldehydes with high conversion, demonstrating a potential application of accessing traditional alcohol oxidation chemistry.

Plasmonic photocatalysis (hybrids of noble metal/semiconductor) have been reported for selective formation of organic molecules.[86] Nano-sized (<5 nm) gold nanoparticles supported on anatase-rutile interphase (TiO_2 Evonik P-25) by a deposition–precipitation technique could take advantage of plasmonic effects to achieve the photocatalytic selective production of several aromatic aldehydes from their corresponding aromatic alcohols (Table 8.2, entries 13–19). This photocatalysis can be promoted *via* plasmon activation of the Au particles by visible light followed by consecutive electron transfer in the Au/rutile/anatase interphase contact. Activated Au particles transfer their conduction electrons to rutile and then to adjacent anatase which catalyzes the oxidation of substrates by the positively charged Au particles along with reduction of oxygen by the conduction band electrons on the surface of anatase titania. Oxygen peroxide species have been found to be responsible for alcohol oxidation.

Recent progress on metal core-semiconductor shell nanohybrids has also demonstrated that these systems can be potentially utilized as photocatalysts for selective organic oxidations.[87] A (Pt)-CeO_2 nanocomposite has been prepared in an aqueous phase with tunable core–shell and yolk–shell structure *via* a facile and green template-free hydrothermal approach.[87] Authors found that the core–shell nanocomposite could serve as an efficient visible-light-driven photocatalyst for the selective oxidation of benzyl alcohol to benzaldehyde using molecular oxygen as a green oxidant. The same authors also prepared various hybrids based on other metals (*e.g.*, Pd) and carried out photocatalytic selective oxidation of benzylic alcohols over the multi-Pd core/CeO_2 shell nanocomposite (Table 8.2, entries 20–22). Electron–hole pairs are produced under visible light irradiation. Electrons are then trapped by the Pd cores and adsorbed benzyl alcohol interacts with holes to form the corresponding radical cation. Further reaction with dioxygen or superoxide radical species will lead to the formation of the corresponding aldehydes, as the plausible mechanism.

Metal oxides with large band gap energies (*e.g.* $Nb_2O_5 > 3.2$ eV) can be used in selective photo-oxidation of aromatic alcohols, as was demonstrated recently.[88] A strong interaction between the alcohol and Nb_2O_5 generates a donor level within the forbidden band of Nb_2O_5, which provides a visible-light-harvesting ability and high activity/selectivity (Table 8.2, entries 23–25).

Prof. Y-J Xu's group has reported a very simple room temperature method to prepare a cubic phase, sheet nanostructured semiconductor CdS.[89] The as-prepared CdS is able to be used as a visible-light-driven photocatalyst for the selective oxidation of saturated primary C–H bonds in alkyl aromatics with

high activity and selectivity using molecular oxygen as a benign oxidant and benzotrifluoride as the solvent under ambient conditions (Table 8.2, entries 26–28). The ease of preparation of this CdS semiconductor and its highly active, selective visible light photoactivity for the selective oxidation of quite inert primary C–H bonds in alkyl aromatics points to its promising potential in photocatalytic selective activation of C–H bonds to fine chemicals.[89]

These photocatalytic protocols account for laboratory-scale systems with relatively restricted applications. However, the possibility of scaling up photocatalytic processes to selective large-scale processes has also been attempted in recent years.[90]

With regards to photocatalytic selective oxidation processes, various aromatic alcohols could be converted in a gas-phase 2.5-liter annular photoreactor using immobilized TiO_2 catalyst and under UV light (Table 8.2, entries 1–3).[91] The system was found to be specifically suited for the selective oxidation of primary and secondary aliphatic alcohols to their corresponding carbonyl compounds. Benzylic alcohols gave higher conversions, however, with more secondary reaction products. The presence of oxygen was found to be critical for the photo-oxidation. One disadvantage of the system relates to catalyst deactivation, which is attributed to the surface accumulation of reaction products. However, the catalyst could be regenerated by calcination in air for 3 h at 450 °C, providing similar activities to those of fresh catalysts.[91]

Another similar scale-up protocol was developed in a solar pilot plant reactor using TiO_2/Cu^{II} photocatalyst.[92] A maximum yield of 53.3% for benzaldehyde could be obtained with respect to initial benzyl alcohol concentration (approx. 63% selectivity, reaction time approx. 6 hours), operating with an average temperature of 38 °C. The authors claimed that in the absence of oxygen, one of the photochemical pathways is inhibited, leading to unselective OH radical production. In fact, the formation of superoxide radical ($O_2^{\bullet-}$) or hydrogen peroxide (H_2O_2) is not possible in the absence of oxygen, whose photolysis can generate OH radicals. In this case, OH radicals should form just as result of the reaction of water with positive holes.

8.5   Outlook and Future Challenges

The idea of taking use of the solar irradiation at room temperature and atmospheric pressure *via* one-pot photocatalytic process to selectively convert lignocellulose to important chemicals and valuable products is a highly innovative approach that can bring several benefits from the energetic and environmental viewpoints. As such, photocatalytic selective processes are able to keep product selectivities to reasonable levels at increasing conversions in the systems, as opposed to conventional thermally activated heterogeneous catalysis.

Photocatalytic reactions also require milder conditions to those generally present in thermal processes, which may allow the development of short and efficient reaction sequences, minimizing side processes, and making use of sunlight as a completely renewable source of energy (waste-free energy). For

the coming decade, multi-integrated systems of simultaneous photocatalyst-mediated chemical evolution and water purification may be possible. These processes could then be efficiently coupled to scale-up facilities (*e.g.*, solar plants) able to provide continuous flow equipment and large reactors for important progress in this area.

The perfect photocatalysts for commercial applications in the upgrading of lignocellulose residues should be designed to be able to maximize conversion with a compromised selectivity to the desired product under solar light irradiation in aqueous solution (or without solvent), avoiding complete mineralization. However, even though most reported syntheses can be virtually carried out under sunlight, there are only very few reports on sunlight utilization for such transformations due to reproducibility issues in measurements.

This chapter has aimed to illustrate that novel nanotechnology and nanomaterials design are currently bringing innovative concepts, ideas, and protocols to the design and preparation of well-defined heterogeneous multifunctional photocatalysts with highly controllable and desirable properties. By combining these new synthetic routes with surface science, fundamentals of heterogeneous photocatalysis, and theoretical mechanistic studies, we should possibly develop a new generation of highly stable and selective photocatalysts for oxidative organic transformations including those of lignocellulosic derivatives.

Acknowledgements

Dr Colmenares would like to thank the Institute of Physical Chemistry of the Polish Academy of Sciences for all support.

References

1. G. Ciamician, *Science*, 1912, **36**, 385.
2. A. Fujishima and K. Honda, *Nature*, 1972, **238**, 37.
3. (a) P. T. Anastas and J. C. Warner, *Green Chemistry: Theory and Practice*, ed. P. T. Anastas and J. C. Warner, Oxford University Press, Oxford, UK, 1998; (b) P. T. Anastas and J. B. Zimmerman, *Sustainability Science and Engineering: Defining Principles*, ed. M. A. Abrahams, Elsevier, Amsterdam, The Netherlands, 2006, pp. 11–32.
4. S. Rawalekar and T. Mokari, *Adv. Energy Mater.*, 2013, **3**, 12.
5. (a) M. Fagnoni, D. Dondi, D. Ravelli and A. Albini, *Chem. Rev.*, 2007, **107**, 2725; (b) Y. Qu and X. Duan, *Chem. Soc. Rev.*, 2013, **42**, 2568; (c) H. Kisch, *Angew. Chem., Int. Ed.*, 2013, **52**, 812; (d) K. Liu, M. Cao, A. Fujishima and L. Jiang, *Chem. Rev.*, 2014, **114**(19), 10044.
6. J. C. Colmenares and R. Luque, *Chem. Soc. Rev.*, 2014, **43**, 765.
7. B. Kamm, P. R. Gruber and M. Kamm, *Biorefineries – Industrial Processes and Products*, Wiley-VCH, Weinheim, 2006, vol. 2.
8. P. N. R. Vennestøm, C. M. Osmundsen, C. H. Christensen and E. Taarning, *Angew. Chem., Int. Ed.*, 2011, **50**, 10502.

9. (a) J. H. Clark, R. Luque and A. Matharu, *Annu. Rev. Chem. Biomol. Eng.*, 2012, **3**, 183; (b) M. Besson, P. Gallezot and C. Pinel, *Chem. Rev.*, 2014, **114**, 1827; (c) N. Yan and P. J. Dyson, *Curr. Opin. Chem. Eng.*, 2013, **2**, 178.

10. M. Stoecker, *Angew. Chem., Int. Ed.*, 2008, **47**, 9200.

11. (a) C. S. K. Lin, L. A. Pfaltzgraff, L. Herrero-Davila, E. B. Mubofu, S. Abderrahim, J. H. Clark, A. Koutinas, N. Kopsahelis, K. Stamatelatou, F. Dickson, S. Thankappan, Z. Mohamed, R. Brocklesby and R. Luque, *Energy Environ. Sci.*, 2013, **6**, 426; (b) I. Delidovich, K. Leonhard and R. Palkovits, *Energy Environ. Sci.*, 2014, 7, 2803; (c) P. Gallezot, *Chem. Soc. Rev.*, 2012, **41**, 1538; (d) R. A. Sheldon, *Green Chem.*, 2014, **16**, 950.

12. P. J. Deuss, K. Barta and J. G. de Vries, *Catal. Sci. Technol.*, 2014, **4**, 1174.

13. (a) A. Kudo and Y. Miseki, *Chem. Soc. Rev.*, 2009, **38**, 253; (b) R. M. Navarro, M. C. Alvarez, F. del Valle, J. A. Villoria de la Mano and J. L. G. Fierro, *ChemSusChem*, 2009, **2**, 471; (c) M. Kitano and M. Hara, *J. Mater. Chem,.*, 2010, **20**, 627; (d) X. Chen, S. Shen, L. Guo and S. S. Mao, *Chem. Rev.*, 2010, **110**, 6503.

14. M. Seyler, K. Stoewe and W. Maier, *Appl. Catal., B*, 2007, **76**, 146.

15. K. Maeda and K. Domen, *J. Phys. Chem. C*, 2007, **111**, 7851.

16. D. Jing, H. Liu, X. Zhang, L. Zhao and L. Guo, *Energy Convers. Manage.*, 2009, **50**, 2919.

17. D. Jing, L. Guo, L. Zhao, X. Zhang, H. Liu, M. Li, S. Shen, G. Liu, X. Hu, X. Zhang, K. Zhang, L. Ma and P. Guo, *Int. J. Hydrogen Energy*, 2010, **35**, 7087.

18. M. Ni, D. Y. C. Leung, M. K. H. Leung and K. Sumathy, *Fuel Process. Technol.*, 2006, **87**, 461.

19. W. Iwasaki, *Int. J. Hydrogen Energy*, 2003, **28**, 939.

20. S. Rapagna, N. Jand and P. U. Foscolo, *Int. J. Hydrogen Energy*, 1998, **23**, 551.

21. S. G. Li, S. P. Xu, S. Q. Liu, C. Yang and Q. H. Lu, *Fuel Process. Technol.*, 2004, **85**, 1201.

22. X. H. Hao, L. J. Guo, X. Mao, X. M. Zhang and X. J. Chen, *Int. J. Hydrogen Energy*, 2003, **28**, 55.

23. Y. Ma, X. Wang, Y. Jia, X. Chen, H. Han and C. Li, *Chem. Rev.*, 2014, **114**, 9987.

24. M. Ni, M. K. H. Leung, D. Y. C. Leung and K. Sumathy, *Renewable Sustainable Energy Rev.*, 2006, **11**, 401.

25. J. R. Bolton, S. J. Strickler and J. S. Connolly, *Nature*, 1985, **316**, 495.

26. T. Kawai and T. Sakata, *Nature*, 1980, **286**, 474.

27. (a) M. Kawai, T. Kawai and K. Tamaru, *Chem. Lett.*, 1981, **8**, 1185; (b) T. Kawai and T. Sakata, *Chem. Lett.*, 1981, **1**, 81; (c) T. Sakata and T. Kawai, *Nouv. J. Chem.*, 1981, **5**, 279.

28. X. Fu, J. Long, X. Wang, D. Y. C. Leung, Z. Ding, L. Wu, Z. Zhang, Z. Li and X. Fu, *Int. J. Hydrogen Energy*, 2008, **33**, 6484.

29. X. Fu, X. Wang, D. Y. C. Leung, W. Xue, Z. Ding, H. Huang and X. Fu, *Catal. Commun.*, 2010, **12**, 184.

30. D. Jing, M. Liu, J. Shi, W. Tang and L. Guo, *Catal. Commun.*, 2010, **12**, 265.
31. Q. Xu, Y. Ma, J. Zhang, X. Wang, Z. Feng and C. Li, *J. Catal.*, 2011, **278**, 329.
32. J. C. Colmenares, A. Magdziarz, M. A. Aramendia, A. Marinas, J. M. Marinas, F. J. Urbano and J. A. Navio, *Catal. Commun.*, 2011, **16**, 1.
33. Y. Li, J. Wang, S. Peng, G. Lu and S. Li, *Int. J. Hydrogen Energy*, 2010, **35**, 7116.
34. M. I. Badawy, M. I. Ghaly and M. E. M. Ali, *Deslination*, 2011, **267**, 250.
35. M. Ilie, B. Cojocaru, V. I. Parvulescu and H. Garcia, *Int. J. Hydrogen Energy*, 2011, **36**, 15509.
36. V. M. Daskalaki and D. I. Kondarides, *Catal. Today*, 2009, **144**, 75.
37. T. A. Kandiel, R. Dillert, L. Robben and D. W. Bahnemann, *Catal. Today*, 2011, **161**, 196.
38. S. Fukumoto, M. Kitano, M. Takeuchi, M. Matsuoka and M. Anpo, *Catal. Lett.*, 2009, **127**, 39.
39. X.-J. Zheng, Y.-J. Wu, L.-F. Wei, B. Xie and M.-B. Wei, *Int. J. Hydrogen Energy*, 2010, **35**, 11709.
40. G. L. Chiarello, M. H. Aguirre and E. Selli, *J. Catal.*, 2010, **273**, 182.
41. G. Wu, T. Chen, W. Su, G. Zhou, X. Zong, Z. Lei and C. Li, *Int. J. Hydrogen Energy*, 2008, **33**, 1243.
42. Y. Li, Y. Huang, J. Wu, M. Huang and J. Lin, *J. Hazard. Mater.*, 2010, **177**, 458.
43. N. Luo, Z. Jiang, H. Shi, F. Cao, T. Xiao and P. P. Edwards, *Int. J. Hydrogen Energy*, 2009, **34**, 125.
44. D. I. Kondarides, V. M. Daskalaki, A. Patsoura and X. E. Verykios, *Catal. Lett.*, 2008, **122**, 26.
45. G. Wu, Ch. Tao, G. Zhou, X. Zong and C. Li, *Sci. China, Ser. B: Chem.*, 2008, **51**, 97.
46. T. Puangpetch, T. Sreethawong and S. Chavadej, *Int. J. Hydrogen Energy*, 2010, **35**, 6531.
47. S. Onsuratoom, T. Puangpetch and S. Chavadej, *Chem. Eng. J.*, 2011, **173**, 667.
48. T. Miwa, S. Kaneco, H. Katsumata, T. Suzuki, K. Ohta, S. C. Verma and K. Sugihara, *Int. J. Hydrogen Energy*, 2010, **35**, 6554.
49. M. Bowker, P. R. Davie and L. S. Al-Mazroai, *Catal. Lett.*, 2009, **128**, 253.
50. V. Gombac, L. Sordelli, T. Montini, J. J. Delgado, A. Adamski, G. Adami, M. Cargnello, S. Bernal and P. Fornasiero, *J. Phys. Chem. A*, 2010, **114**, 3916.
51. N. Fu and G. Lu, *Catal. Lett.*, 2009, **127**, 319.
52. Q. Gu, J. Long, L. Fan, L. Chen, L. Zhao, H. Lin and X. Wang, *J. Catal.*, 2013, **303**, 141.
53. J. M. Valero, S. Obregón and G. Colón, *ACS Catal.*, 2014, **4**, 3320.
54. A. Speltini, M. Sturini, D. Dondi, E. Annovazzi, F. Maraschi, V. Caratto, A. Profumo and A. Buttafava, *Photochem. Photobiol. Sci.*, 2014, **13**, 1410.

55. A. Di Paola, E. García-López, G. Marcì and L. Palmisano, *J. Hazard. Mater.*, 2012, **211–212**, 3.

56. M. Pelaez, N. T. Nolan, S. C. Pillai, M. K. Seery, P. Falaras, A. G. Kontos, P. S. M. Dunlop, J. W. J. Hamilton, J. A. Byrne, K. O'Shea, M. H. Entezari and D. D. Dionysiou, *Appl. Catal., B*, 2012, **125**, 331.

57. Y. Shiraishi, M. Morishita and T. Hirai, *Chem. Commun.*, 2005, 5977.

58. (a) M. D. Tzirakis, I. N. Lykakis and M. Orfanopoulos, *Chem. Soc. Rev.*, 2009, **38**, 2609; (b) C. Gambarotti, C. Punta, F. Recupero, T. Caronna and L. Palmisano, *Curr. Org. Chem.*, 2010, **14**, 1153, and references therein.

59. S. E. Davis, M. S. Ide and R. J. Davis, *Green Chem.*, 2013, **15**, 17, and references therein.

60. R. Ho, J. Liebman and J. Valentine, *Active Oxygen in Chemistry*, ed. C. Foote, J. Valentine, A. Greenberg and J. Liebman, Springer, Berlin, 1995, pp. 1–23.

61. P. R. Harwey, R. Rudham and S. Ward, *J. Chem. Soc., Faraday Trans.*, 1983, **I79**, 2975.

62. F. J. Lopez-Tenllado, A. Marinas, F. J. Urbano, J. C. Colmenares, M. C. Hidalgo, J. M. Marinas and J. M. Moreno, *Appl. Catal., B*, 2012, **128**, 150.

63. A. M. Balu, B. Baruwati, E. Serrano, J. Cot, J. Garcia-Martinez, R. S. Varma and R. Luque, *Green Chem.*, 2011, **13**, 2750.

64. T. Sakata, T. Kawai and K. Hashimoto, *J. Phys. Chem.*, 1984, **88**, 2344.

65. J. C. Colmenares, A. Magdziarz and A. Bielejewska, *Bioresour. Technol.*, 2011, **102**, 11254.

66. J. C. Colmenares, R. Luque, J. M. Campelo, F. Colmenares, Z. Karpinski and A. A. Romero, *Materials*, 2009, **2**, 2228.

67. J. C. Colmenares and A. Magdziarz, *J. Mol. Catal. A: Chem.*, 2013, **366**, 156.

68. (a) J. C. Colmenares, A. Magdziarz, K. Kurzydlowski, J. Grzonka, O. Chernyayeva and D. Lisovytskiy, *Appl. Catal., B*, 2013, **134–135**, 136; (b) J. C. Colmenares, A. Magdziarz, O. Chernyayeva, D. Lisovytskiy, K. Kurzydłowski and J. Grzonka, *ChemCatChem*, 2013, **5**(8), 2270.

69. M. Morishita, Y. Shiraishi and T. Hirai, *J. Phys. Chem. B*, 2006, **110**, 17898.

70. A. Molinari, M. Montoncello, R. Houria and A. Maldotti, *Photochem. Photobiol. Sci.*, 2009, **8**, 613.

71. A. Molinari, M. Bruni and A. Maldotti, *J. Adv. Oxid. Technol.*, 2008, **11**, 143.

72. C. Minero, A. Bedini and V. Maurino, *Appl. Catal., B*, 2012, **128**, 135.

73. H. Kominami, H. Sugahara and K. Hashimoto, *Catal. Commun.*, 2010, **11**, 426.

74. (a) V. Augugliaro, M. Bellardita, V. Loddo, G. Palmisano, L. Palmisano and S. Yurdakal, *J. Photochem. Photobiol. C*, 2012, **13**, 224; (b) R. Chong, J. Li, Y. Ma, B. Zhang, H. Han and C. Li, *J. Catal.*, 2014, **314**, 101.

75. O. D. Mante, J. A. Rodriguez and S. P. Babu, *Bioresour. Technol.*, 2013, **148**, 508.

76. M. Yasuda, A. Miura, R. Yuki, Y. Nakamura, T. Shiragami, Y. Ishii and H. Yokoi, *J. Photochem. Photobiol., A*, 2011, **220**, 195.
77. M. Tian, J. Wen, D. MacDonald, R. M. Asmussen and A. Chen, *Electrochem. Commun.*, 2010, **12**, 527.
78. A. Tanaka, K. Hashimoto and H. Kominami, *J. Am. Chem. Soc.*, 2012, **134**, 14526.
79. G. Palmisano, E. Garcia-Lopez, G. Marcì, V. Loddo, S. Yurdakal, V. Augugliaro and L. Palmisano, *Chem. Commun.*, 2010, **46**, 7074.
80. T. Ruether, A. M. Bond and W. R. Jackson, *Green Chem.*, 2003, **5**, 364.
81. Q. Wang, M. Zhang, C. Chen, W. Ma and J. Zhao, *Angew. Chem., Int. Ed.*, 2010, **49**, 7976.
82. (a) S. Higashimoto, N. Kitao, N. Yoshida, T. Sakura, M. Azuma, H. Ohue and Y. Sakata, *J. Catal.*, 2009, **266**, 279; (b) S. Higashimoto, N. Suetsugu, M. Azuma, H. Ohue and Y. Sakata, *J. Catal.*, 2010, **274**, 76.
83. S. Kim and W. Choi, *J. Phys. Chem. B*, 2005, **109**, 5143.
84. C.-J. Li, G.-R. Xu, B. Zhang and J. R. Gong, *Appl. Catal., B*, 2012, **115–116**, 201.
85. H. Li, R. Liu, S. Lian, Y. Liu, H. Huang and Z. Kang, *Nanoscale*, 2013, **5**, 3289.
86. D. Tsukamoto, Y. Shiraishi, Y. Sugano, S. Ichikawa, S. Tanaka and T. Hirai, *J. Am. Chem. Soc.*, 2012, **134**, 6309.
87. N. Zhang, S. Liu and Y.-J. Xu, *Nanoscale*, 2012, **4**, 2227.
88. S. Furukawa, T. Shishido, K. Teramura and T. Tanaka, *ChemPhysChem*, 2014, **15**, 2665.
89. Y. Zhang, N. Zhang, Z.-R. Tang and Yi-Jun Xu, *Chem. Sci.*, 2012, **3**, 2812.
90. D. Spasiano, R. Marotta, S. Malato, P. Fernandez-Ibanez and I. D. Somma, *Appl. Catal., B*, 2015, **170**, 90.
91. U. R. Pillai and E. Sahle–Demessie, *J. Catal.*, 2002, **211**, 434.
92. D. Spasiano, L. P. Prieto Rodriguez, J. Carbajo Olleros, S. Malato, R. Marotta and R. Andreozzi, *Appl. Catal., B*, 2013, **136–137**, 56.

CHAPTER 9

Photocatalytic CO_2 Conversion to Fuels by Novel Green Photocatalytic Materials

W. A. MAZA,[a] AMANDA J. MORRIS*[a] AND GUIDO MUL*[b]

[a] Department of Chemistry, Virginia Polytechnic Institute and State University, Blacksburg, VA 24060, USA; [b] Photo Catalytic Synthesis Group, MESA+ Institute for Nanotechnology, University of Twente, P.O. Box 217, 7500AE Enschede, Netherlands
*Email: ajmorris@vt.edu; G.Mul@utwente.nl

9.1 Introduction and some Simple Back of the Envelope Calculations

Worldwide production and utilization of renewable electricity, such as generated by wind and solar, is expected to grow at an increasing rate. While this is encouraging to mitigate the problems associated with combustion of fossil fuels, in particular the emission of the greenhouse gas CO_2, a novel challenge arises. The production of solar electricity is intermittent, both on a daily and seasonal basis, and in particular decentralized solar electricity exceeds (local) demand in the summer. To utilize this surplus at times when solar electricity is not effectively produced, *i.e.* typically in winter time, long-term storage is needed. In Figure 9.1, an overview of storage methods is provided, scaled on the basis of the predicted overall capacity (in Wh) and the discharge time.

RSC Green Chemistry No. 42
Green Photo-active Nanomaterials: Sustainable Energy and Environmental Remediation
Edited by Nurxat Nuraje, Ramazan Asmatulu and Guido Mul

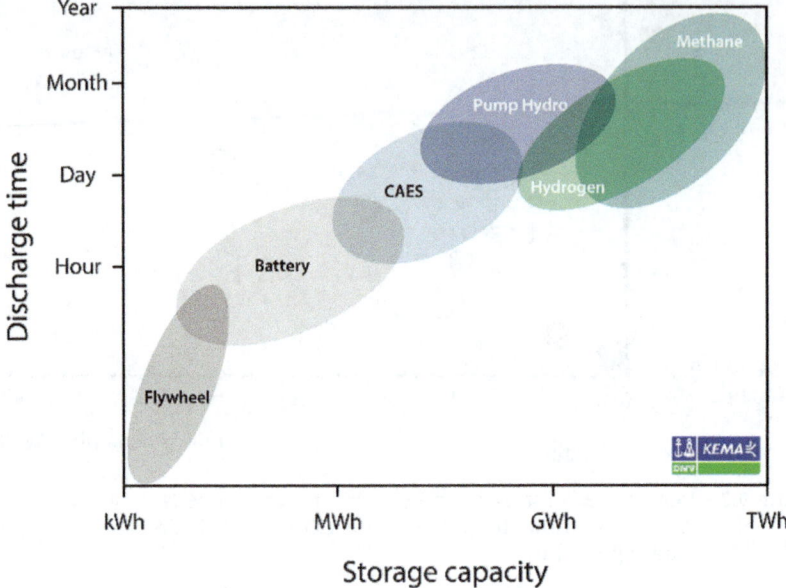

Figure 9.1 Various storage technologies shown with respect to storage capacity and discharge time (CAES = compressed air energy storage). (Source DNV KEMA, The Netherlands.)

While chemical storage is too inefficient for storage shorter than (typically) a day, storage in chemical bonds is recognized as the prime candidate for long-term (seasonal) and large-scale storage of renewable energy.[1] Of the chemicals to be produced, it is clear that hydrocarbons are the most desirable, since the energy content per weight (and per volume) is significantly larger than hydrogen (even if in compressed form). The energy content (in MJ/kg) of several hydrocarbons is compared in Figure 9.2.[2,3]

Two methodologies exist to convert solar energy to hydrocarbon bonds:

(1) the so-called indirect route, in which electricity produced *via* solar panels is used to convert CO$_2$ and water in an electrochemical reactor
(2) the direct route, in which solar energy is used in a photocatalytic reactor to convert CO$_2$ and H$_2$O (vapor) into hydrocarbons.

In this chapter we will discuss the direct route. It is obvious that for practical application of the direct route, new, preferably green, CO$_2$ sorbing materials need to be developed, which are at the same time catalytically active, to sustain the growth of renewable energy production while preventing electricity network instability. For the interested reader, Gupta *et al.* have written a review on solid materials that have been developed for capture and storage of CO$_2$.[3] Before providing an overview of the latest material discoveries, we will present a simple, back of the envelope calculation, to determine the effectivity of a catalyst (moles CO$_2$ converted per gram of

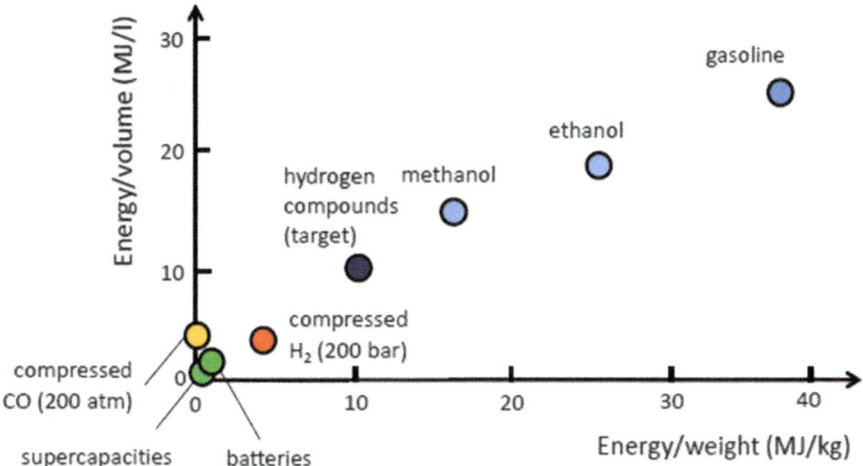

Figure 9.2   Plot of the volumetric and gravimetric energy density of several energy carriers to show the potential for CO_2-neutral hydrocarbons, based on data reported in ref. 2.

Figure 9.3   Basic concept of a device absorbing and converting CO_2 and water *via* photocatalysis. V_R indicates the volume of the reactor needed, and W the weight needed of the absorber and catalyst.

catalyst per hour), necessary to provide a household with 1 liter of methanol per day. The basic concept is illustrated in Figure 9.3.

Let's divide the day into two identical parts of 12 hours, the first being a period in the dark, and the second being the time when solar energy can be captured. We will also assume that all the CO_2 needed is contained in air at a concentration of 380 ppm, and should be accumulated in the dark. Further, we assume that sorption effectivity is 100%. Then, 31 mol of absorbed CO_2 requires 1820 m³ air (in 12 h), which results in a required air flow of 42 L/s. This is significant, but not far larger than provided by natural convection (wind). Assuming a wind flow of 10 m/s, and an inlet area of 1 m², a volumetric inlet flow of 1000 L/s can be achieved (approximately 25 times larger than required, or leading to a CO_2 capturing efficiency of 4% being sufficient). Next, we need to calculate how much sorbing material the reactor should contain. On average, typically 2.5 mole of CO_2 can be absorbed per kg of sorbent.[3] Given the 31 moles of CO_2 required for 1 kg of methanol, it can be easily calculated that 15.6 kg of sorbent needs to be contained in the converter. Assuming a typical porosity of a packed bed, this requires a reactor volume of approximately 50 L.

Table 9.1   Overview of the system parameters for a solar CO$_2$ (and H$_2$O) to methanol converter. One kg methanol has been used as basis of calculation.

Air flow CO$_2$ conc.	Absorber capacity	Weight/V_R	Solar energy harvested	Area required	Catalytic rate required
42 L s^{-1} 380 ppm	2.5 mole kg^{-1}	15.6 kg$_{Cat}$/ 50 L	1000 W m^{-2} or 2.10^{-3} Einst s^{-1} m^{-2}	15 m^2	3×10^{-4} mol kg$_{cat}$$^{-1}$ s^{-1}

Subsequently it is necessary to determine the area necessary to be exposed to the Sun. For the basis of this calculation, we have assumed a solar flux of 1000 W m^{-2} (typically used in evaluation of photoelectrochemical (PEC) cells, and equivalent to AM 1.5 radiation). Assuming an average wavelength of 400 nm, this amounts to a photon flux of 2.10^{-3} Einst s^{-1} m^{-2}. Since 31 moles of methanol in 12 hours requires a conversion rate of 5×10^{-3} mol/s, it can be determined that the exposed area should be at least 15 m^2 (6 photons need to be provided per molecule of methanol). Finally, the same rate requires a catalyst efficacy of 3×10^{-4} mol kg$_{cat}$$^{-1}$ s^{-1} (assuming 15.6 kg$_{cat}$). Summarizing (Table 9.1), the device should consist of a flat container of at least 50 L, an air inlet area in the order of 1 m^2, and a solar window of about 15 m^2. In the following, we will further evaluate if the rate of 3×10^{-4} mol kg$_{cat}$$^{-1}$ s^{-1} is something that has been achieved in the literature, and which bifunctional materials (CO$_2$ adsorption and photocatalytic conversion) are promising. We will first discuss modified TiO$_2$-based semiconductors, followed by the incorporation of Ti sites in (green) micro and mesoporous materials, and finally evaluate the performance and perspective of novel materials, such as MOFs.

9.2   TiO$_2$ and Metal-doped TiO$_2$ Composites

9.2.1   Background on TiO$_2$-based CO$_2$ Reduction

The initial steps leading to the formation of CO$_2$ reduction products over crystalline semiconductor materials include:

(1) absorption of light by the photocatalyst, resulting in an electron–hole (e$^-$–h$^+$) pair
(2) diffusion of the e$^-$ and h$^+$ to the surface of the photocatalyst
(3) separation of the e$^-$–h$^+$ pair
(4) adsorption and activation of CO$_2$ (Figure 9.4).

Titanium dioxide (TiO$_2$) nanoparticles (NPs) are attractive and widely used candidates for photocatalysis of CO$_2$ due to the abundance of titania (4th most abundant metal and the 9th most abundant element on the Earth), the low cost of titania, its relatively low toxicity (depending on the NP size) compared to other metal oxides, and its overall resistance to corrosion.[4,5] This metal oxide is a wide band gap ($E_g \sim 3.2$ eV) semiconductor whose

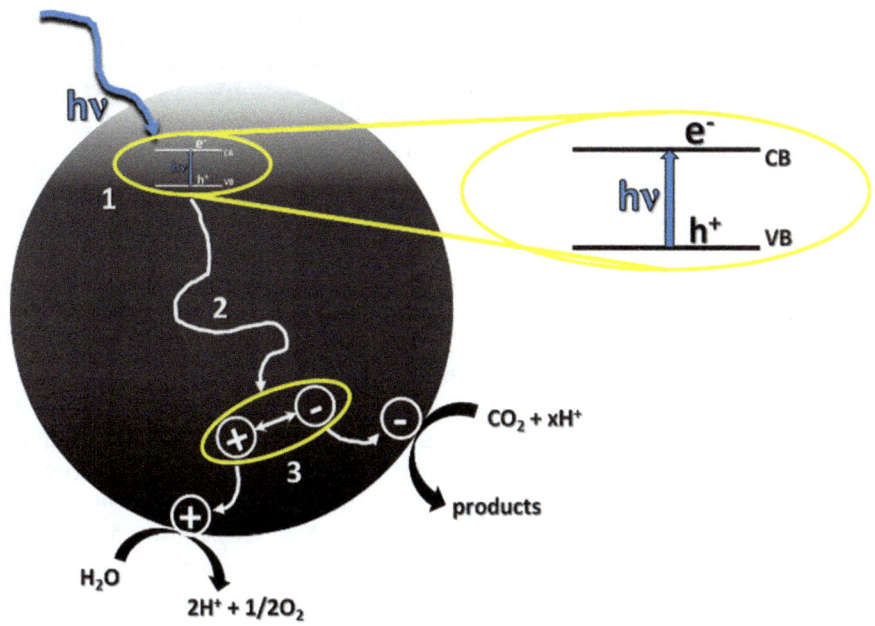

Figure 9.4   Photo-induced generation of (1) electron–hole pairs, (2) migration of the electron–hole pair to the surface of materials, and (3) separation of the electron and hole at the reduction and oxidation surface reaction sites.

valence and conduction bands lie at -7.41 eV and -4.21 eV (based on an absolute vacuum scale), respectively.[6]

There exist four polymorphs of TiO_2: anatase, brookite, rutile, and $TiO_2(B)$ (Figure 9.5). Recent evidence suggests that brookite displays more activity towards CO_2 photoreduction over anatase and rutile, whereas rutile performs the poorest.[7] Anatase TiO_2 is, by far, the most characterized and explored of the three.[7-9] The photocatalytic activity of both brookite and anatase TiO_2 is enhanced (10-fold) by introduction of oxygen vacancy defect sites in the TiO_2.[7,10] It has been proposed that the difference in photocatalytic activity between pristine and defect-containing TiO_2 is a result of increased affinity of the CO_2 for the surface of the defect-containing TiO_2 arising from increased electronegativity at the defect sites.[7,9,10] Further, the enhanced reactivity of brookite in comparison to anatase has also been attributed to interactions between the TiO_2 surface and CO_2. Brookite displays shorter interatomic distances and, therefore, may result in a higher affinity for CO_2 and facilitate stronger CO_2 binding in bridging motifs.[11]

There are several possible binding motifs of CO_2 at the TiO_2 surface. These involve:

(1) coordination at the Ti Lewis acid surface metal sites *via* one of the CO_2 oxygen atoms
(2) coordination at the Lewis basic surface oxygen sites *via* the CO_2 carbon

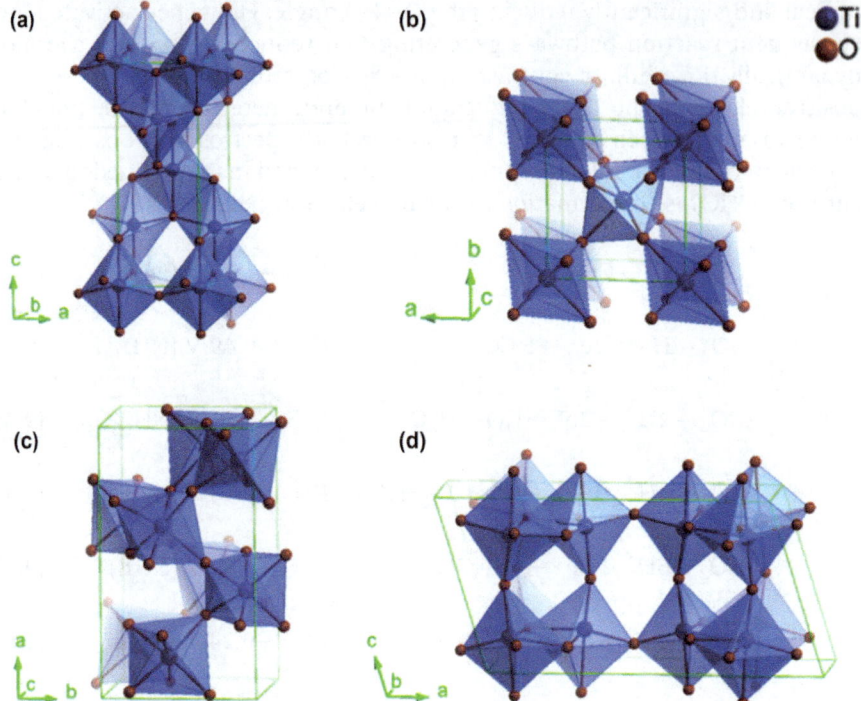

Figure 9.5 Crystal structures of the four TiO$_2$ polymorphs: (a) anatase, (b) rutile, (c) brookite, and (d) TiO$_2$(B).
Reprinted with permission from ref. 15, Copyright (2014) ACS.

(3) insertion of the CO$_2$ oxygen atoms into a TiO$_2$ oxygen vacancy
(4) bridging conformations over multiple binding sites.

Theoretical calculations suggest linear CO$_2$ interacts with the surface of oxygen-deficient reduced TiO$_2$ along a Ti^{3+}–O–Ti^{3+} surface site such that the CO$_2$ bridges the reduced Ti^{3+} metal centers.[8] Although the mode of binding of the TiO$_2$ surface by CO$_2$ has yet to be determined experimentally, what is clear from FT-IR measurements is that the bound species is CO$_2^{\delta\bullet-}$, which suggests charge transfer from the TiO$_2$ surface into the CO$_2$ upon coordination.[12] The anion radical species, CO$_2^{\delta\bullet-}$, is expected to have a bent configuration at the surface of the TiO$_2$ due to stabilization of the lowest unoccupied molecular orbital.[13] Orbital stabilization may lower the reduction potential of CO$_2$ by as much as 240 mV.[14]

The one-electron reduction of linear CO$_2$ ($E_{1/2} = -1.90$ V *versus* NHE) generating the anion radical, CO$_2^{\bullet-}$, is more negative than the conduction band of TiO$_2$ and is thermodynamically unfavorable.[6,16] Therefore, the nature of the interaction between CO$_2$ and the surface of TiO$_2$ is critical for one-electron reduction pathways. Interaction of CO$_2$ with the anionic TiO$_2$ surface must result in charge transfer into the CO$_2$, generating the anion

radical and significantly reducing the O–C–O angle. Only once activated are subsequent reaction pathways generating CO_2 reduction products thermodynamically downhill. In contrast, multi-electron reductions of CO_2 lie more positive relative to the TiO_2 conduction band and, therefore, may be possible in the absence of such binding.[16,17] However, both spectroscopic observations and theoretical calculations support that the first step in CO_2 reduction at the surface of TiO_2 is the formation of the one electron reduced $CO_2^{\bullet-}$.

$$CO_2 + e^- \rightarrow CO_2^- \qquad\qquad E^\circ = -1.90 \text{ V (NHE)} \qquad (9.1)$$

$$CO_2 + H^+ + 2e^- \rightarrow HCO_2^- \qquad\qquad E^\circ = -0.49 \text{ V (NHE)} \qquad (9.2)$$

$$CO_2 + 2H^+ + 2e^- \rightarrow CO + H_2O \qquad E^\circ = -0.53 \text{ V (NHE)} \qquad (9.3)$$

$$CO_2 + 4H^+ + 4e^- \rightarrow HCHO + H_2O \qquad E^\circ = -0.48 \text{ V (NHE)} \qquad (9.4)$$

$$CO_2 + 6H^+ + 6e^- \rightarrow CH_3OH + H_2O \qquad E^\circ = -0.38 \text{ V (NHE)} \qquad (9.5)$$

$$CO_2 + 8H^+ + 8e^- \rightarrow CH_4 + 2H_2O \qquad E^\circ = -0.24 \text{ V (NHE)} \qquad (9.6)$$

In water, the photoproducts (methanol, formic acid, methane, formaldehyde, and carbon monoxide) have been observed upon steady-state illumination of TiO_2 in the presence of CO_2.[18-28] However, the quantum yields of efficiency are known to be low due to fast electron–hole recombination relative to the rates of electron transfer into CO_2.[29] Both geminate electron–hole recombination and recombination of charges trapped at defect sites in the bulk decrease the photo-efficiency of CO_2 reduction. It is generally accepted that only electron–hole pairs that diffuse to the catalyst surface can result in reductive reactions.

Manipulation of the conditions used for the photoreaction have been shown to improve the yields at which both liquid- and gas-phase products are generated. For example, liquid- and gas-phase products were formed with higher yields at high CO_2 pressures, through a simple concentration effect.[22,30] Enhancements are also observed when solution dopants such as isopropyl alcohol are present in solvent system.[22,26] This can be attributed to the competitively easier oxidation of the solution dopants in comparison to water. Thus, these dopants act as hole scavengers, decrease recombination rates, and ultimately facilitate reduction. Considering this effect, it is important to note that many of the multiple electron reduction products of carbon dioxide (methanol, formaldehyde, and formic acid) can also act as hole scavengers. Therefore, electron donation to these products from TiO_2 and electron acceptance from these products are competitive in nature. Transient absorption and EPR studies indicate that only in the case of formic acid does the rate of electron donation effectively compete with that of

photo-oxidation.[31] Thus, methanol and formaldehyde are unlikely to serve as intermediate species in further reduction mechanisms.

9.2.2 Earth-abundant Transition Metal/Metal Oxide–TiO$_2$ Composites

Introduction of defect sites into TiO$_2$, such as oxygen deficiencies, metals, metal oxides, or non-metals, serve to affect the reactivity and improve the performance of TiO$_2$. It has been indicated above that oxygen deficiencies serve to increase the binding affinity of the TiO$_2$ surface for CO$_2$ and, thereby, lower the O–C–O bond angle, lowering the energetics of the CO$_2$ LUMO, making reduction more thermodynamically favorable. Non-metal dopants can serve to narrow the TiO$_2$ band gap by either lowering the conduction band, raising the valence band, or both.[32–36] The focus, herein, will be on transition metal and metal oxide dopants. A table is given summarizing a list of some TiO$_2$ and doped TiO$_2$ materials used for photoreduction of CO$_2$ (Table 9.2).

The preparative method used for the formation of metal-TiO$_2$ materials determines the relevant energetics, which, in turn, dictate the products observed and operative mechanisms. For example, traditional doping can be viewed as the substitution of a small number of Ti sites with the metal ions of interest. In this case, the metal atoms introduce intergap energy states that act as electron sinks/traps and reactive centers for reduction. In the case of copper, lightly doped materials result in a shift in product distribution toward CH$_3$OH as compared to undoped TiO$_2$. However, heavier levels of doping or certain preparative methods can result in pure metal oxide domains. In this case, the material is a composite of two independent semiconductors, TiO$_2$ and M$_x$O$_y$. Energy level alignment will determine the lowest energy conduction band, and it is at this material surface where reduction will occur. For clarity, the former case (light doping) will be considered M-doped TiO$_2$ and the latter (high doping) M$_x$O$_y$–TiO$_2$ composites, where x and y are determined by the oxidation state of the metal dopant.

In the latter case, coupling narrow band gap metal oxides with TiO$_2$ may potentially increase the efficacy of CO$_2$ reduction reactions on TiO$_2$ by one or more mechanisms. First, doping of narrow band gap metal oxides into TiO$_2$ extends the spectral range by which promotion of an electron into the conduction band may occur, thereby increasing photonic harvesting and energy losses due to thermalization of excitation energies smaller than the TiO$_2$ band gap. As a result, the photosensitivity and activity of the TiO$_2$/metal oxide composite photocatalyst may be extended into the visible region of the solar spectrum.

In addition, the alignment of the energy levels between TiO$_2$ and the metal oxide dopant may facilitate charge separation of the photo-generated electron–hole pairs and retard charge recombination. For efficient charge separation, energy level offsets are engineered to facilitate electron flow into material A with the more negative conduction band edge and hole transport to material B with the more positive valence band edge, Figure 9.6. As TiO$_2$ is

Table 9.2 Summary of select TiO$_2$/co-photocatalyst composites, the resulting photoproducts, and respective yields.

Photocatalyst	Illumination source	Product(s)	Yield(s)	Ref.
TiO$_2$ (P-25)	15 W UV lamp (center λ = 368 nm)	CH$_3$OH	15 mg L^{-1}	19, 50, 126
TiO$_2$ (anatase)	8 W Hg lamp (254 nm center λ)	H$_2$ CH$_4$ CH$_3$OH CO	1.4 μmol g^{-1} 0.9 μmol g^{-1} 1.1 μmol g^{-1} 1.0 μmol g^{-1}	20, 126
TiO$_2$ (cubic anatase)	300 W Xe arc lamp	CH$_4$ CH$_3$OH	4.56 μmol g^{-1} h^{-1} 1.48 μmol g^{-1} h^{-1}	20, 127, 128
Cu–TiO$_2$	Hg lamp	CH$_3$OH	<700 μmol g^{-1}	44, 127
Cu$_2$O–TiO$_2$	300 W Xe arc lamp	CH$_4$ CH$_2$CH$_2$ CH$_3$CH$_3$	21.8 μL g^{-1} 26.2 μL g^{-1} 2.7 μL g^{-1}	128
Cu$_2$O–TiO$_2$	300 W Xe arc lamp	CH$_4$ CO	28.4 ppm g^{-1} h^{-1} trace	39
Cu–TiO$_2$	10 W UV lamp	CH$_3$OH	2600 μmol g^{-1}	40–42
Cu$_2$O–TiO$_2$			1300 μmol g^{-1}	
CuO–TiO$_2$			1100 μmol g^{-1}	
Cu–TiO$_2$	8 W Hg lamp (254 nm center λ)	CH$_3$OH	119 μmol g^{-1}	42, 43
Cu–TiO$_2$	8 W Hg lamp (254 nm and 365 nm center λ)	CH$_3$OH	600 μmol g^{-1} (254 nm) 10 μmol g^{-1} (365 nm)	43, 44
Cu$_2$O–TiO$_2$	300 W Xe arc lamp	CH$_3$OH	0.16 μmol g^{-1} h^{-1}	40, 44

Material	Light source	Product	Yield/rate	References
FeTiO$_3$/TiO$_2$	500 W Xe arc lamp ($\lambda > 300$ nm and $\lambda > 400$ nm)	CH$_3$OH	0.462 μmol g^{-1} h^{-1} ($\lambda > 300$ nm), 432 μmol g^{-1} h^{-1} ($\lambda > 400$ nm)	45, 47
Fe^{3+}–TiO$_2$	15 W UV lamp (center $\lambda = 254$ nm)	CH$_3$OH		47, 129
Fe–TiO	6 W Hg lamp (365 nm center λ)	CH$_4$	0.92 mmol g^{-1} L^{-1}	49, 129
Ni–TiO$_2$	18 W cm^{-2} (center $\lambda = 365$ nm)	CH$_4$	14 μmol g^{-1} h^{-1}	46, 50
Graphene–TiO$_2$	300 W Xe arc lamp	CO, CH$_4$	8.9 μmol g^{-1} h^{-1}, 1.14 μmol g^{-1} h^{-1}	87
Graphene–TiO$_2$	300 W Xe arc lamp	CH$_4$, CH$_3$CH$_3$	10 μmol g^{-1} h^{-1}, 7.2 μmol g^{-1} h^{-1}	88
Graphene-P-25	100 W Hg lamp (center $\lambda = 365$ nm)	CH$_4$	8 μmol gm^{-2} h^{-1} (365 nm)	85
Reduced graphene oxide-P-25	60 W Halco bulb (λ range 400 nm–850 nm)		2 μmol gm^{-2} h^{-1} (365 nm)	
Reduced graphene oxide–TiO$_2$	15 W Philips bulb	CH$_4$	0.135 μmol g^{-1} h^{-1}	86
NH$_2$-MIL-125(Ti)	500 W Xe arc lamp, UV-cut on filter ($\lambda_{pass} > 420$ nm)	HCOO$^-$	8.14 μmol	115
Cu$_3$(BTC)$_2$–TiO$_2$	300 W Xe arc lamp, UV-cut off filter ($\lambda_{pass} < 400$ nm)	CH$_4$	2.64 μmol g^{-1} h^{-1}	114
UiO-66-(Zr/Ti)–NH$_2$	300 W Xe arc lamp (420 nm $< \lambda_{pass} < 800$ nm)	HCOOH	22 μmol	113

a wide band gap semiconductor with an extremely oxidizing valence band, TiO_2/metal oxide composites are typically of the form where multi-photon excitation of the composite drives migration of high-energy conduction band electrons, generated by narrow band gap absorption, into the TiO_2 and high energy holes into the metal oxide dopant. Therefore, both the valence and conduction band of the metal oxide dopant should be more negative than those of the TiO_2. Spatial separation of the charges increases the lifetime of the charge separated state. Thus, the effective concentration of charge carriers are increased, which, in turn, can lead to more efficient photocatalysis.

Green nanomaterials necessitate the use of relatively non-toxic and abundant metals. Several earth-abundant transition metals have been tested for toxicity based on the degree of urease protein inhibition and follow from least to more toxic $Mn^{II} < Fe^{II} < Co^{II} < Ni^{II} < (Zn^{II} \leq Cu^{II})$.[26] Despite its toxicity compared with other earth-abundant transition metals, more work has been dedicated to CO_2 reduction by copper-doped TiO_2 relative to other non-toxic transition metals.

9.2.2.1   Copper

Cu-promoted TiO_2 has been extensively studied for activity in photocatalytic CO_2 reduction. Before discussion of the various chemical variations in Cu–TiO_2 composites, it is important to note that this system has been instrumental to demonstrate the importance of using very pure oxide materials when evaluating photocatalytic reduction of CO_2.[37] It was clearly demonstrated by Yang and co-workers by using isotopically labelled CO_2, that if hydrocarbon contaminants are present in composites after synthesis, these contribute significantly to formation of (unlabelled) CO by reaction with H_2O.[37] It is thus recommended when analyzing data from the literature, to carefully assess the preparation method of the composites, and the steps undertaken for purification, to exclude a contribution of potential contaminants to activity and product selectivity. Besides the use of isotopically labelled $^{13}CO_2$, performing a simple experiment in which only water (vapor) is present in the feed, and verification of the absence of hydrocarbon products upon illumination, suffices to ensure product distributions are truly the result of photocatalytic CO_2 reduction.[38]

9.2.2.2   Copper Oxide Composites

The energetics of Cu–TiO_2, CuO–TiO_2, and Cu_2O–TiO_2 are suited to the mechanism of charge separation in metal oxide–TiO_2 composites, making them an attractive candidate for CO_2 photoreduction reaction. To this end, Xu, et al.[39] prepared Cu_2O–TiO_2 composite nanoparticles via a microwave-solvothermal reaction between TiO_2 and copper(II) acetylacetonate in benzyl alcohol. The absorption spectra of the Cu_2O–TiO_2 composite displayed two distinct band edges at 388 nm (3.2 eV) and 510 nm (2.1 eV) corresponding to the conduction and valence band gap for TiO_2 and Cu_2O, respectively. X-ray

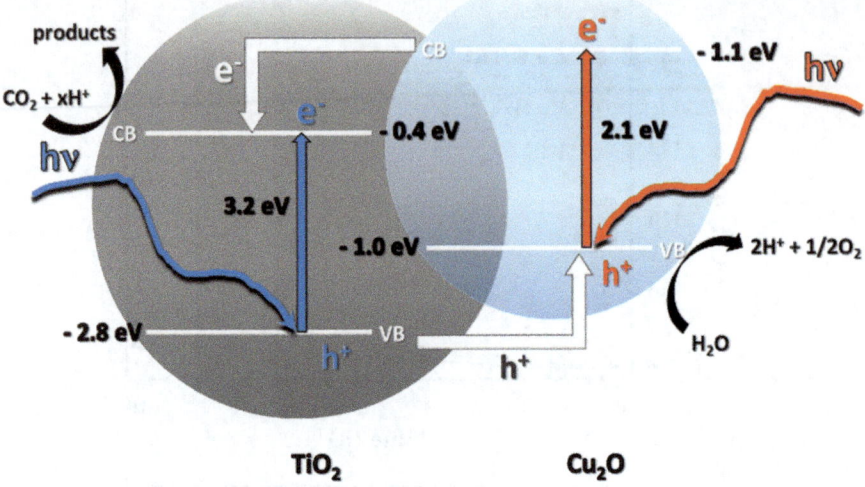

Figure 9.6 An energy diagram depicting the energy level alignment, multi-photon absorption, and resultant electron–hole gradients in a Cu$_2$O–TiO$_2$ nanoparticle composite.
Reproduced with permission from ref. 39.

photoelectron spectroscopy (XPS) data indicated the positions of the valence bands of TiO$_2$ and Cu$_2$O were 2.8 eV and 1.0 eV, respectively. From the absorption and XPS data, the conduction bands were calculated to be at −0.4 eV and −1.1 eV for TiO$_2$ and Cu$_2$O, respectively (Figure 9.6). The composite material was found to out-perform undoped TiO$_2$ in terms of photo-activity towards CO$_2$ reduction. The major reaction product for both undoped and the composite material in water was CH$_4$, with trace amounts of CO as the minor product. The charge separation mechanism and putative energy diagram is supported by these findings, as the product distribution did not vary between undoped TiO$_2$ and the composite. The Cu$_2$O/TiO$_2$ mass ratio was optimized so that a Cu$_2$O/TiO$_2$ composition with a mass ratio of 1 : 5 displayed the highest CH$_4$ yield.

Bi *et al.*[40] prepared Cu$_2$O–TiO$_2$ composites by post-synthetic modification of hollow Cu$_2$O nanospheres with TiO$_2$ *via* a deposition method. The morphology, topological features, and physical properties of the resultant composites were not significantly different from the unmodified Cu$_2$O nanospheres. It was argued that the minimal differences observed were due to low loading and amorphous nature of the deposited TiO$_2$. The atomic ratio of Ti to Cu was estimated from EDX and XPS to be between 0.15 and 0.21. Despite the low loading of TiO$_2$ onto or into Cu$_2$O, the composite displayed considerably higher activity towards photocatalytic reduction of CO$_2$ compared to undoped Cu$_2$O; the main product of the reaction was CH$_4$ (Figure 9.7). However, significant reduction of activity for both undoped Cu$_2$O and Cu$_2$O–TiO$_2$ composite nanospheres was observed over time and

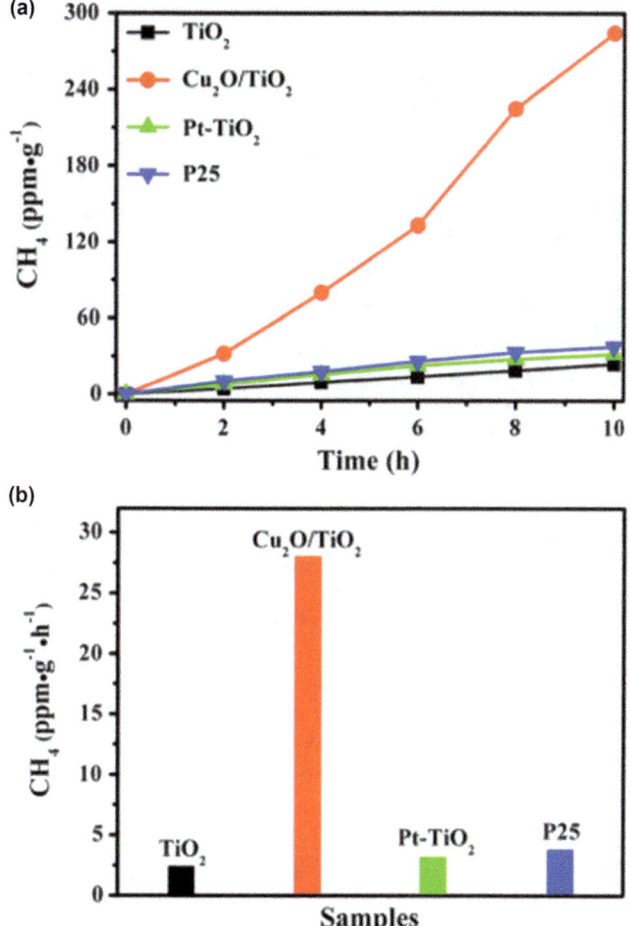

Figure 9.7   (a) CH production due to photoreduction of CO_2 over unmodified TiO_2 (black squares), Cu_2O/TiO_2 (red circles), Pt-modified TiO_2 (green triangles), and P25 (blue triangles). (b) Rates of methane production for the materials described in (a).
Reproduced with permission from ref. 39.

was attributed to photocorrosion of Cu_2O with continued water oxidation reaction, which was coupled to CO_2 reduction (Figure 9.8).

9.2.2.3   Copper Doping

The effect of the copper oxidation state in doped TiO_2 was probed by Slamet *et al.*[41,42] Cu–TiO_2, Cu_2O–TiO_2, and CuO–TiO_2 were prepared by impregnation of Degussa-P25 TiO_2 containing a mixture of $\sim 80\%$ anatase and $\sim 20\%$ rutile TiO_2. The resultant materials containing a ~ 2.2 wt% content of Cu^{m+} displayed a relatively uniform distribution of sizes between 19 and

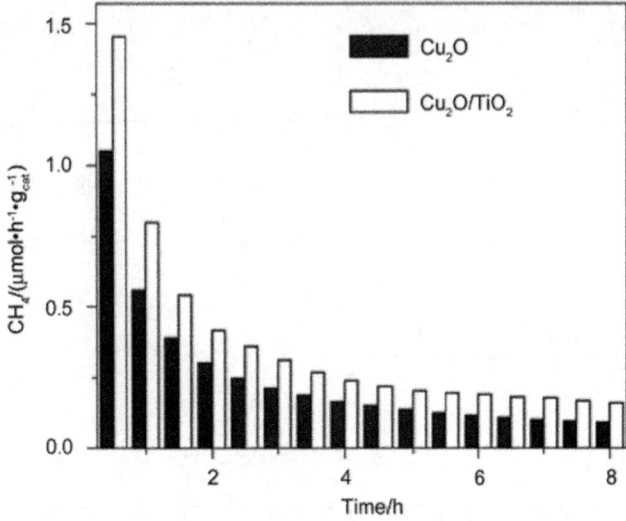

Figure 9.8 Reduction of Cu$_2$O and Cu$_2$O–TiO$_2$ composite rate of methane production from the CO$_2$ photoreduction reaction as a function of time as a result of photocorrosion.
Reproduced with permission from ref. 40.

23 nm for the anatase and 23 to 31 nm for the rutile. Interestingly, the anatase content in the composite mixture was found to decrease with addition of Cu^{m+}. Upon illumination and exposure to CO$_2$, optimum yields of the main product, CH$_3$OH, was observed by 3% CuO–TiO$_2$. The differences in photo-activity between the materials of varying copper oxidation states were explained as resulting from the effect of the copper reduction potential and its relationship to electron–hole recombination. The highly positive redox potential of Cu$^+$ ($E° = 0.52$ V) relative to Cu^{2+} ($E° = 0.34$ V) makes Cu$_2$O a comparatively effective electron sink. These electrons are more difficult to transfer into adsorbed reactants at the surface of the composite. Therefore, a significant difference in electron–hole recombination should be observed between CuO and Cu$_2$O. Based on FT-IR data, a mechanism was proposed in which the copper sites play a direct role in reduction of CO$_2$ to form CH$_3$OH (Scheme 9.1).[41]

Tseng *et al.*[43,44] have suggested that the preparative method used to synthesize Cu–TiO$_2$ could affect the activity and efficiency of CO$_2$ photoreduction. The authors proposed that by preparing the Cu-doped TiO$_2$ materials using a sol–gel they could control the dispersion of Cu throughout the TiO$_2$.[44] They found that 2 wt% Cu–TiO$_2$ composites displayed both the highest activity for CO$_2$ and the greatest dispersion of Cu ($\sim$40%). The major product, CH$_3$OH, yield was also found to increase in the presence of NaOH which was attributed to the greater solubility of CO$_2$ in the basic solution compare to neat water, and hole scavenging properties of OH$^-$ anions. By varying the copper salt used in preparation of the Cu–TiO$_2$ composites Tseng

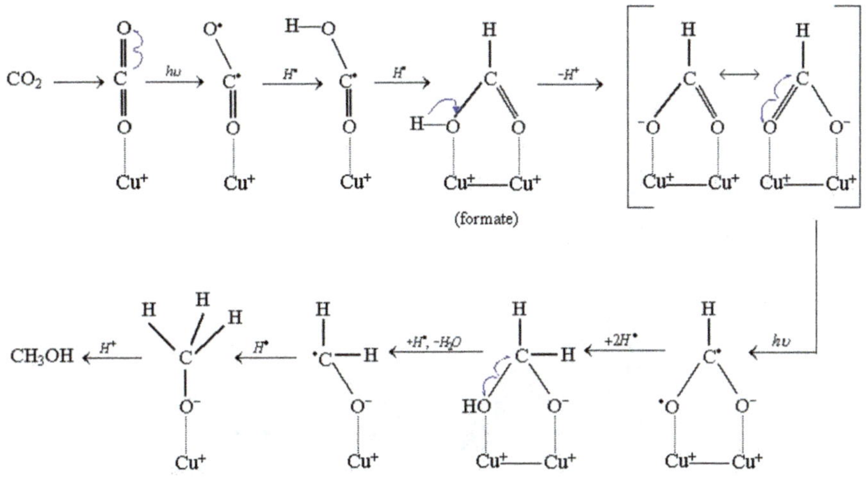

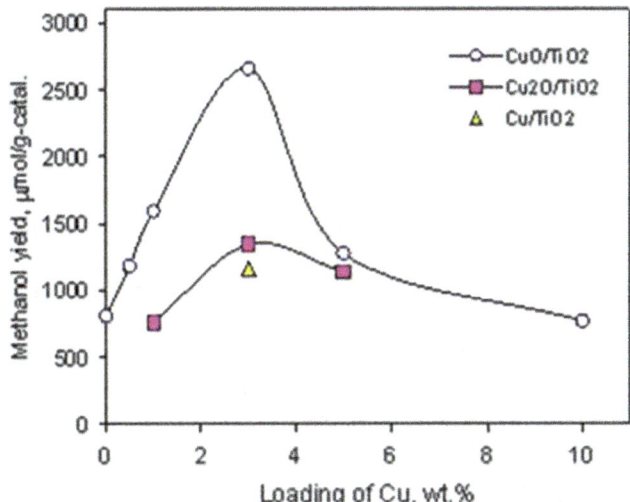

Scheme 9.1 Mechanism proposed by Slamet *et al.* for CO₂ reduction to CH₃OH at the CuO sites of a CuO–TiO₂ composite.
Reprinted with permission from ref. 42.

Figure 9.9 Methanol yield as a function of wt% loading of Cu^0, Cu^+, and Cu^{2+} into Degussa P-25 TiO_2 nanoparticles upon UV irradiation over the course of 6 hours.
Reprinted with permission from ref. 42.

et al.[43] also demonstrated that the surface charge played a large role in the efficiency of CO_2 reduction (Figure 9.9). The authors observed better yields when the composites were prepared in such a way as to yield a positive surface zeta potential (indicative of a positive surface charge). It was suggested that

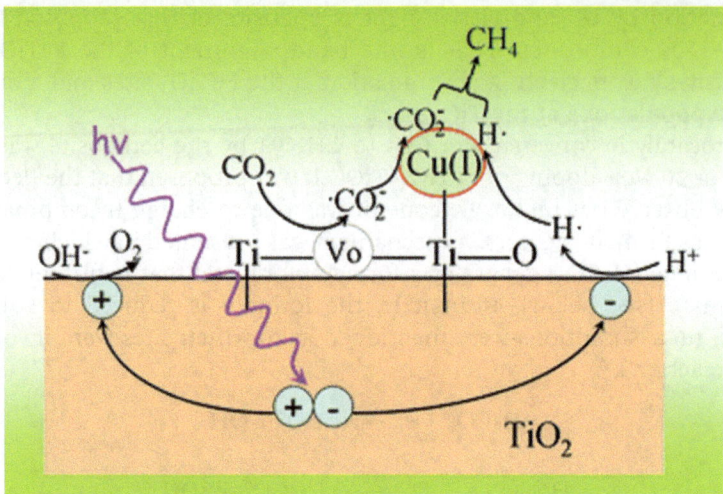

Figure 9.10   Proposed reaction mechanism for the photoreduction of CO_2 on Cu_2O–TiO_2 nanosheet composites.
Reproduced with permission from ref. 46.

the positive charge on the surface of the Cu–TiO_2 composite facilitates adsorption of CO_2 or HCO_3^-, increasing reactivity at the Cu clusters.

Highly dispersed Cu-doped TiO_2 materials consisting of dispersions of Cu_2O on TiO_2 nanosheets have been prepared by Zhu *et al.*[45] Characterization of the material suggested formation of oxygen vacancies in the TiO_2 upon incorporation of the Cu_2O cluster. These vacancy defect sites were purported to be the primary site of interaction with CO_2. Adsorption of CO_2 at the oxygen vacancy site results in formation of the $CO_2^{\bullet-}$ anion radical. The activated $CO_2^{\bullet-}$ is then thought to react with atomic hydrogen (generated from photo-generated electrons at the TiO_2 surface) at the Cu_2O sites to form CH_4. The proposed reaction mechanism is summarized in Figure 9.10.

9.2.2.4   Iron Oxide Composites

Photocatalytic reduction of CO_2 has also been carried out with iron oxide TiO_2 composite photocatalysts by Truong *et al.*[47] Interestingly, spherical 20 nm TiO_2 nanoparticles were found to grow off of $FeTiO_3$ rods 100 nm in length. The absorption band edge of the 10 wt% $FeTiO_3$–TiO_2 composite was found to be considerably red-shifted relative to undoped TiO_2 and continued to red shift well into the near IR upon increasing the Fe content of the composites where maximum absorption across the near UV to near IR was observed for 50 wt% $FeTiO_3$–TiO_2. These composites display a band gap of 2.6 eV. The valence band of the composites are nearly isoenergetic with the TiO_2 valence band (see above) lying at -2.6 eV, therefore the conduction band should lie at 0 eV, approximately 0.4 eV below the conduction band of the TiO_2.[48] It is, therefore, likely that the electrical gradient generated upon

illumination by UV and visible light is opposite of that proposed for the Cu_xO_y–TiO_2 composites. That is, the band alignment of the $FeTiO_3$–TiO_2 composites favors electron accumulation at the $FeTiO_3$ sites and a build-up of hole populations at the TiO_2 sites.

Photocatalytic conversion of CO_2 to CH_3OH by the composite was maximized at 20 wt% doping of $FeTiO_3$–TiO_2. It was proposed that the decreased activity observed at higher Fe content was due to charge recombination at the excess Fe metal centers. A mechanism was presented in which reduction of carbonate (H_2CO_3) generates a formyloxyl radical that is further reduced to formate (see below). Ultimately, the formate is reduced to CH_3OH – though no assumptions were mentioned as to which sites were involved in these reactions.

$$H_2CO_3 + e^- \rightarrow HCOO^\bullet + OH^- \tag{9.7}$$

$$HCO_3^- + H_2O + e^- \rightarrow HCOO^\bullet + 2OH^- \tag{9.8}$$

$$HCOO^\bullet + e^- \rightarrow HCOO^- \tag{9.9}$$

$$HCOO^- + H^+ \rightarrow HCOOH \tag{9.10}$$

$$HCOOH + 3H_2O + 4e^- \rightarrow CH_3OH + 4OH^- \tag{9.11}$$

For comparison, a second iron oxide TiO_2 composite, Fe_2O_3–TiO_2, was prepared *via* a solvothermal technique.[49] The Fe_2O_3–TiO_2 composite produced methane at levels lower than both $FeTiO_3$–TiO_2 and un-doped TiO_2. The maximum yield of CH_4 observed for Fe_2O_3–TiO_2 occurred at 1 mol% Fe content (6 W illumination, center wavelength 365 nm). In the Fe_2O_3–TiO_2 composite, the conduction and valence band of TiO_2 bracket the conduction and valence band of Fe_2O_3, *i.e.*, the conduction band edge of TiO_2 is more reducing and the valence band edge is more oxidizing. Therefore, upon light excitation, all charges migrate to the Fe_2O_3 material, charges are not spatially separated, and recombination persists. These results highlight the importance of energy level engineering when exploring composites of interest (Figure 9.11).

9.2.2.5 Nickel Oxide Composites

Nickel has also been used to form a composite photocatalyst for CO_2 reduction by Kwak *et al.* (Figure 9.12).[50] At Ni concentrations higher than 0.1 mol%, spectroscopic data suggested the presence of NiO domains. Upon irradiation of the composite material by near UV and visible light (center wavelength 365 nm), CO_2 was reduced to CH_4. It was found that at a Ni concentration of 0.1 mol% production of CH_4 was maximized. Product formation showed a decline at higher Ni concentrations (up to 1 mol%). The decrease in product formation at higher concentrations was argued to be

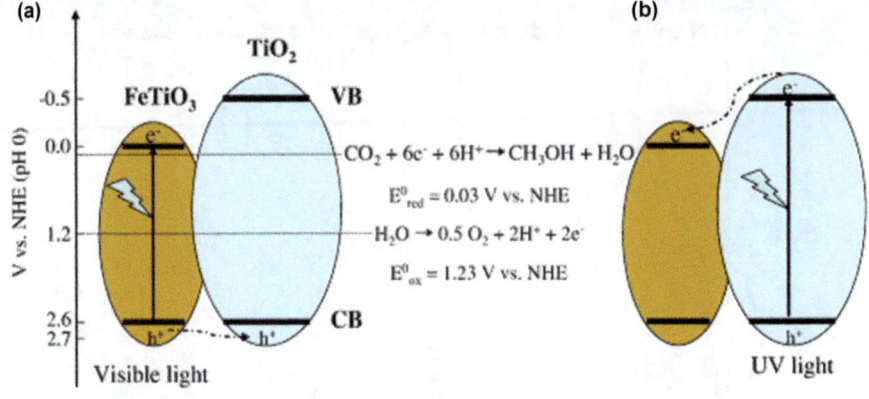

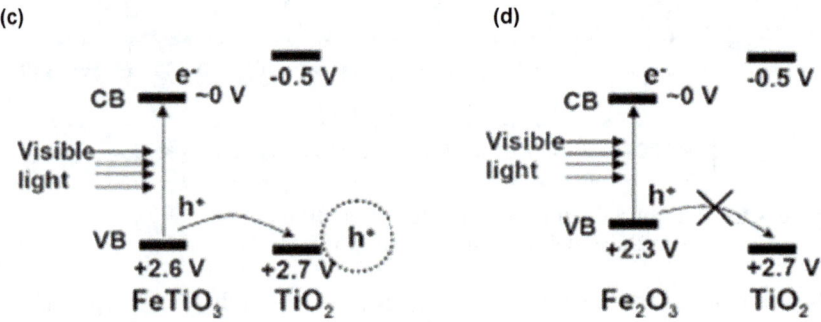

Figure 9.11 (Top) Proposed mechanism of photoreduction of CO$_2$ by a FeTiO$_3$/TiO$_2$ composite under (a) visible and (b) UV-light illumination.[47] (Bottom) Energetic differences between (c) FeTiO$_3$/TiO$_2$ and (d) Fe$_2$O$_3$/TiO$_2$ composites.
Reproduced with permission from ref. 48.

due to decreased active surface area and increased charge recombination at doped-Ni electron traps.

9.3 Supported Silica Framework Catalysts

9.3.1 Introduction

As explained in the introduction, prospective solar to fuel converters require the development of very efficient photocatalysts, capable of sorbing CO$_2$ and converting this into useful products, preferably selectively. Besides (crystalline) semiconductors previously discussed, Ti-functionalized silica-based mesoporous materials show relatively high yields in the gas-phase conversion of water (vapor) and CO$_2$, in particular on a per active site basis. Silica matrices containing isolated Ti sites were pioneered by Anpo and his co-workers in the 1990s.[27,28] In the following, we will discuss the types of

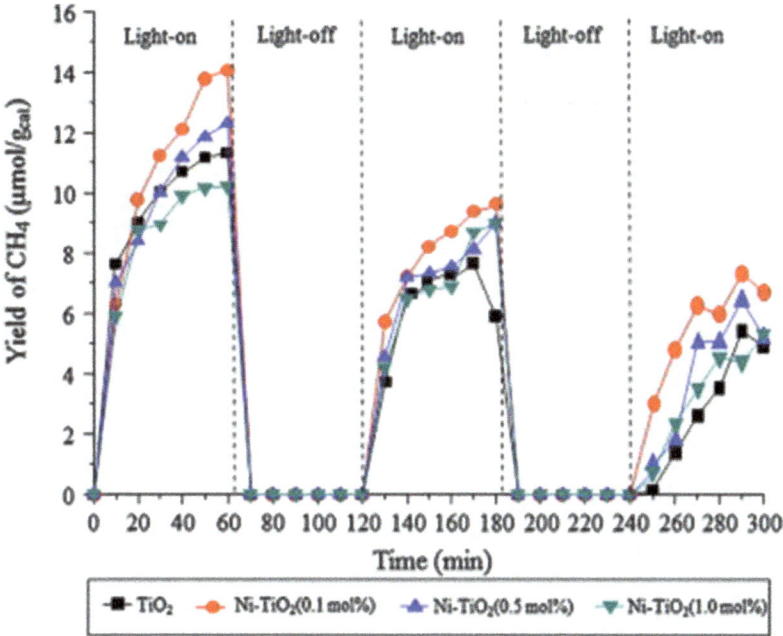

Figure 9.12 Photoproduction of CH_4 over Ni-modified TiO_2.
Reproduced with permission from ref. 50.

silica matrices that have been considered, the methods for incorporation of the active centers, and the performance achieved. We will also evaluate strategies to improve on the current standing of the intriguing catalysts on the basis of Ti sites in silica matrices by inclusion of metal to metal charge transfer complexes,[51] water oxidation functionality, using *e.g.*, Co_3O_4 microcrystals,[52,53] stimulating CO_2 sorption by basic sites,[54] and finally separation of products, to prevent back reactions of the photo-produced hydrocarbons to CO_2 and water.[28,55]

9.3.2 Synthesis of Mesoporous Materials

The isolated Ti sites in micro materials were first found more effective in photocatalytic CO_2 reduction than crystalline TiO_2 by Yamashita *et al.* in 1995, based on data for the zeolite Ti-ZSM-5.[28] Later, Anpo and co-workers showed that use of isolated Ti-centers in mesoporous silicas further increased the hydrocarbon yields by two orders of magnitude as compared to TiO_2, and the CH_3OH yield by one order of magnitude in terms of μmol produced per gram Ti per hour.[56–59] Silica-based materials that have been tested and found to enhance the activity of TiO_2 in photocatalytic CO_2 reduction are MCM-41, SBA-15, and montmorillonite. More recently novel mesoporous silicas such as KIT-6[60] and TUD-1[61] have been developed.

A common aspect of the synthesis of most mesoporous materials is that a surfactant is used in order to obtain the regular structure. Ti-SBA-15 is typically

synthesized by dissolving the surfactant, pluronic P123 (EO20PO70EO20), in diluted HCl(aq), followed by addition of TEOS (tetraethylorthosilicate) and TBOT (titanium(IV) butoxide).[62] Pure silica Si-MCM-41 is prepared according to established procedures wherein tetramethylammonium hydroxide (TMAOH) and the surfactant cetyltrimethylammonium bromide (CTAB) are simultaneously added to water. Fumed silica is added to the template solution to give a gel which is allowed to stand for 20 h at ambient temperature and then placed in Teflon-lined autoclaves and heated at 150 °C for 48 h. After cooling, the solid products are recovered by filtration, washed, and calcined for 8 h under static air conditions at 550 °C.[63] Finally, a large mesoporous silica with cubic Ia3d symmetry (designated as KIT-6) can be prepared in 0.5 M HCl aqueous solution using a 1:1 (wt%) mixture of Pluronic P123 (EO20PO70EO20, MW = 5800, Aldrich) and butanol.[64,65] Either tetraethoxysilane (TEOS) or sodium silicate is acceptable as the silica source. Hydrothermal treatment occurs at 100 °C, followed by removal of the template by washing with ethanol-HCl mixtures, and calcination at 550 °C.

From a synthetic perspective, TUD-1 can be considered as the most green nano material, since contrary to other mesoporous materials, surfactants are not needed in order to obtain a regular pore structure.[61] The key agent in the synthesis of TUD-1 is triethanol amine (TEA), which not only serves as the pore template molecule, but at the same time allows metal functionalization in a one-pot procedure. Many different metal-containing siliceous TUD-1 materials have been prepared, typically denoted as M-TUD-1. A scheme for the synthesis of TUD-1 is provided in Figure 9.13.

9.3.3   Mechanism of CO$_2$ over Isolated Ti Sites

Various mechanistic routes have been proposed in explaining the conversion of CO$_2$ over isolated centers in mesoporous materials. Anpo *et al.* proposed a mechanism for isolated excited (Ti^{+III}–O^{-I}) sites in silica scaffolds, based on EPR data,[66] involving simultaneous reduction of CO$_2$ and decomposition of H$_2$O, leading to formation of CO to C radicals, and water to H and OH radicals, respectively. Subsequently, these photo-induced C, H, and OH radicals combine to final products, such as CH$_4$ and CH$_3$OH. In an advanced IR study on Ti supported on MCM-41, CO could be demonstrated as being the primary product of the reaction. An energetic scheme was proposed as indicated in Figure 9.14.[67]

Excitation of the Ti–O ligand to metal charge transfer transition of Ti centers leads to transient Ti^{+III} and a hole on a framework oxygen (O^{-1}). Electron transfer from Ti^{+III} to CO$_2$ yields CO$_2^-$ (single electron transfer), and transfer of the hole to water generates a surface OH radical and H$^+$. CO$_2^-$ and H$^+$ combine to yield CO and a second OH radical. The OH radicals either combine to yield H$_2$O$_2$ or directly dismutate to give O$_2$ and H$_2$O. Figure 9.14 indicates the free energies associated with the formation of the stable products. Besides a role in achieving high dispersion of Ti centers, the

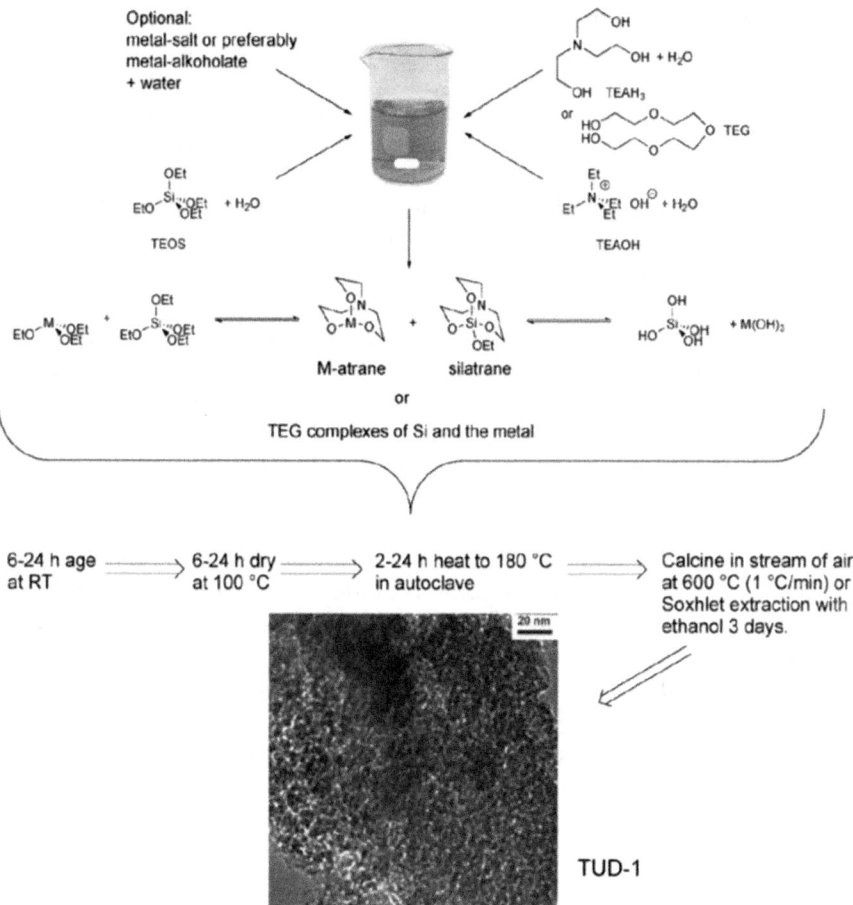

Figure 9.13 Synthesis procedure for metal promoted or all-siliceous TUD-1. The micrograph shows the porous structure of TUD-1.
Reprinted with permission from the Royal Society of Chemistry from ref. 61.

silica framework also has a perspective role in stabilizing photo-excited charges, enhancing the photocatalytic efficiency of the isolated Ti centers.

Also Yang *et al.*, based on the mechanism of Frei and co-workers, propose CO is an important intermediate in conversion of isolated Ti sites, now dispersed in SBA-15, but elaborate on the subsequent formation of hydrocarbons.[62] Based on the similar product distribution obtained as compared to CO_2, and the high reactivity of this molecule, these authors propose formaldehyde to be an important intermediate for the formation of methane and ethene/ethane. The mechanism is illustrated in Figure 9.15.

The authors propose a subsequent reaction of the peroxide proposed by Frei and co-workers (Figure 9.15) with CO and H_2O to form HCHO, in which the peroxide and water are converted to Ti–OH and one molecule of O_2.

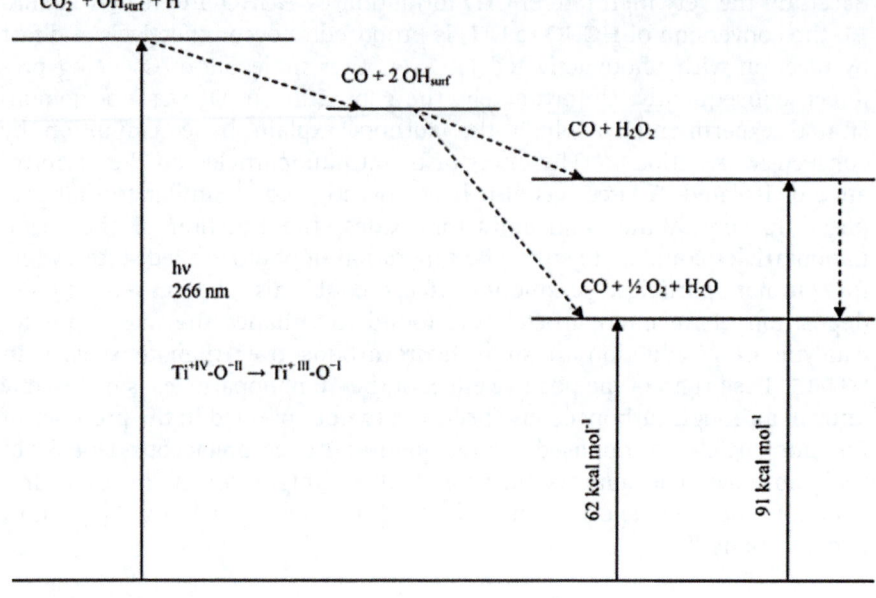

$CO_2^- + OH_{surf} + H^+$

$CO + 2\,OH_{surf}$

$CO + H_2O_2$

hv
266 nm

$CO + \tfrac{1}{2}\,O_2 + H_2O$

$Ti^{+IV}\text{-}O^{-II} \rightarrow Ti^{+\,III}\text{-}O^{-I}$

62 kcal mol⁻¹

91 kcal mol⁻¹

$CO_2 + H_2O$

Figure 9.14 Scheme for CO₂ reduction in the presence of H₂O over isolated Ti sites in the MCM-41 framework. The conversion is induced by a 266 nm laser pulse.
Reprinted with permission from ref. 67. Copyright (2004) ACS.

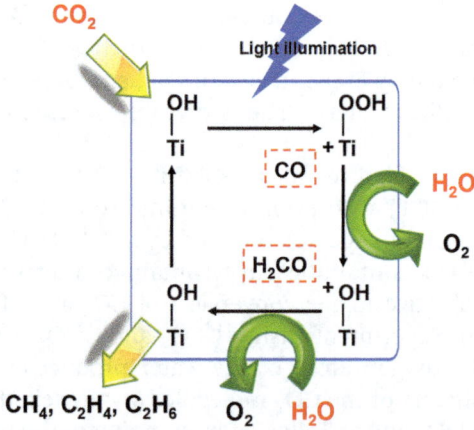

CO₂

Light illumination

OH
|
Ti

CO

OOH
|
+ Ti

H₂O

O₂

H₂CO

OH
|
Ti

+ OH
|
Ti

CH₄, C₂H₄, C₂H₆   O₂   H₂O

Figure 9.15 Scheme for CO₂ reduction in the presence of H₂O over isolated Ti sites in the SBA-15 framework. The conversion is induced by the spectrum of a mercury lamp. Formaldehyde is proposed as a reactive intermediate based on the observed high reactivity of this molecule, and the comparable product distributions in hydrocarbons obtained upon anaerobic decomposition over isolated Ti centers in SBA-15.
Reprinted with permission from ref. 62.

Based on the very high rate of CH_4 formation by HCHO-pre-treated Ti-SBA-15, the conversion of HCHO to CH_4 is proposed to occur effectively and fast by reaction with photo-activated H_2O, another molecule of O_2 being produced consequently. Unfortunately, the generation of O_2 was not demonstrated experimentally, which the authors explain by consumption by consecutive reactions.[62] The effect of metal nanoparticles on the performance of isolated Ti sites has also been investigated.[38] Similar to that proposed for crystalline semiconductor oxides, the function of the metal nanoparticles could be to stimulate separation of photo-excited states, while for Au nanoparticles, plasmonic effects could also be present. Photo-deposition of Au nanoparticles was found to enhance the rate of photo-catalytic CO_2 reduction to short hydrocarbons over titanate species in SBA-15. The origin of the positive effect of the Au nanoparticles is not exactly known, although carbonaceous species were not observed in the presence of Au nanoparticles, as opposed to large quantities of carbonaceous deposits in their absence. The authors mention that Au might contribute by hydro-genation of the species in the 'carbon pool', yielding the short hydrocarbons.[38]

9.3.4 Metal to Metal Charge Transfer Complexes

A very important disadvantage of isolated titania centers in mesoporous silica framework materials, is the low wavelength, or high energy, of the radiation that is required for photo-excitation. Typically UV absorption of isolated Ti centers requires wavelengths in the order of 275 nm.[38,62] To allow absorption of mesoporous framework catalysts, two strategies have been investigated, including the introduction of plasmonic (Au) particles,[38] and formation of metal to metal charge transfer moieties.[68–72] The first binuclear entity was investigated by Frei and co-workers, and based on a Zr–O–Cu unit in the MCM-41 silicate sieve. The structure is schematically shown in Figure 9.16.[72]

The bimetallic site features a Zr^{IV}–O–Cu^{I} to Zr^{III}–O–Cu^{II} metal-to-metal charge-transfer (MMCT) absorption extending from the UV region to about 500 nm. The Zr–O–Cu linkage was revealed by a Cu^{I}–O infrared stretch mode at 643 cm^{-1}. The Cu^{I}-containing entity obtained after synthesis was demonstrated to be effective in the conversion of CO_2 and H_2O upon laser ex-citation at 355 nm. Experiments using $^{13}CO_2$ and $C^{18}O_2$ revealed that carbon monoxide and the oxygen atom of the water product originate from CO_2. This indicates splitting of the CO_2 molecule by the excited MMCT moiety to CO and a surface OH radical, followed by trapping of the products at the Cu^{I} centers (OH is reduced to H_2O). The concurrent intensity loss of the Cu^{I}–O–Zr mode at 643 cm^{-1} (Figure 9.16) confirms that the carbon dioxide photo-reduction is accompanied by the oxidation of Cu^{I} linked to the Zr center. However, in principle the reaction is not photocatalytic, but stoichiometric. The researchers conclude that if a metal can be identified that oxidizes H_2O by a visible light-induced LMCT (ligand to metal charge transfer), thereby

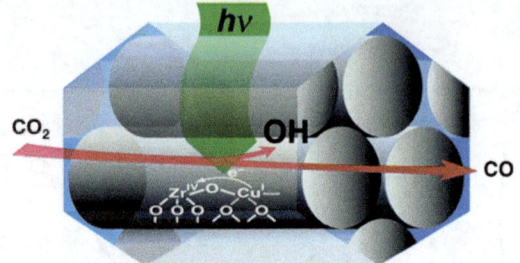

Figure 9.16 Representation of the MCM-41 silicate sieve in which the binuclear Zr–O–Cu site can be identified. The conversion of CO$_2$ to CO was demonstrated by time-resolved infrared spectroscopy.
Reprinted with permission from ref. 72. Copyright (2005) ACS.

reducing the bimetallic moiety to its original state, a Z-scheme for CO$_2$ reduction by H$_2$O under visible light can be envisioned.

Ti–O–Co^{II} is another binuclear unit that was successfully synthesized by the Frei group, again in the mesoporous framework of MCM-41. These bimetallic moieties feature an absorption extending from the UV into the visible to about 600 nm which is attributed to the Ti^{IV}–O–Co^{II} → Ti^{III}–O–Co^{III} metal-to-metal charge-transfer (MMCT) transition. Such photo-excitation would generate Co in an oxidation state that is energetically capable of driving a water oxidation catalyst, although this unit was not further used to demonstrate overall reduction of CO$_2$ (to CO) with concomitant water oxidation.[73] Instead, the authors focused on Ti–O–Mn^{II} units, since these have donor and acceptor redox potentials appropriate for driving multi-electron catalysts for water oxidation, and proton or CO$_2$ reduction. Effective synthesis was again demonstrated, and by using O$_2$ or ^{18}O$_2$ and methanol as acceptor and donor probe molecules, respectively, excitation of the Ti^{IV}–O–Mn^{II} was demonstrated to be effective in photocatalytic transformations, yielding formate and water as primary products.

9.3.5 Water Oxidation Functionality

One of the key reactions in achieving efficient CO$_2$ conversion, either in electrocatalytic reactors or photocatalytic reactors, is the oxidation of water to produce oxygen. Not only does the oxidation of water require a significant thermodynamic potential (1.23 V at reference conditions *vs.* RHE), also the so-called over-potential to overcome kinetic barriers is significant. Water oxidation catalysts have been reported to be able to reduce this over-potential to a significant degree, but the metal (oxides) usually applied are rare and expensive, such as IrO$_x$. That said, IrO$_x$ has been the metal oxide studied most intensively with respect to the mechanism of water oxidation, and the only metal oxide catalyst that performs water oxidation effectively and with high stability in acid, as well as in basic conditions.[74] The high effectivity of IrO$_x$ in oxidizing water could be demonstrated by molecularly

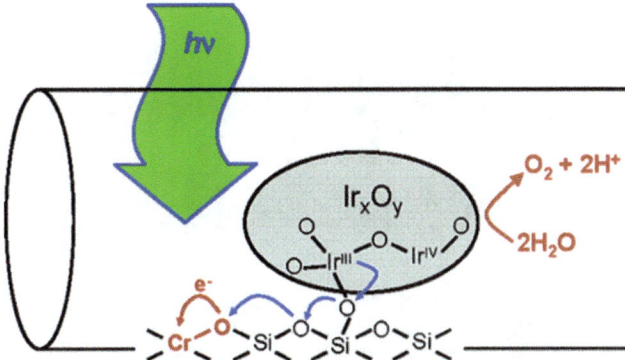

Figure 9.17 Combining photo-excitation of isolated Cr^{VI} sites with the water oxidation ability of IrO_x in a MCM-41 mesoporous silica network. Reprinted with permission from ref. 75. Copyright (2006) ACS.

coupling these oxide particles to a chromophore. To this end, again Frei and co-workers synthesized a functionalized MCM-41 with isolated Cr centers, while they achieved oxygen evolution upon photo-excitation of the chromophore.[75] The electron transfer scheme the authors envision is schematically shown in Figure 9.17.

More recently, the authors achieved anchoring of the IrO_x cluster to a $Zr–O–Co^{II}$ MMCT unit.[69] Illumination of the MMCT chromophore of the resulting $Zr–O–Co^{II}–IrO_x$ units in the SBA-15 pores loaded with a mixture of CO_2 and H_2O vapor resulted in the formation of CO and O_2 monitored by FT-IR and mass spectroscopy, respectively. Use of O-18 labeled water resulted in the formation of $^{18}O_2$ product. This is the first example of a closed photosynthetic cycle of carbon dioxide reduction by water using an all-inorganic polynuclear cluster featuring a molecularly defined light absorber.

The mechanism of water oxidation over IrO_x was recently demonstrated by rapid scan infrared spectroscopy to involve a Ir^{III}–OOH surface intermediate.[76] The formation of the peroxide intermediate is proposed to be initiated by hole-induced oxidation of Ir^{IV}=O to Ir^V=O, followed by water insertion, yielding a proton (H^+) and the $IrI^{II}OOH$ intermediate. The decomposition pathway of the $Ir^{III}OOH$ intermediate to yield oxygen still remains to be resolved.

While effective, large-scale use of IrO_x is prohibited by the scarcity and price of iridium. Recently, (oxides of) more earth-abundant elements have been found to have significant water oxidation activity, in particular when supported in, or on silica-based mesoporous materials. Co-based structures (Co_3O_4/SBA-15, Co^{II}/SBA-15, *etc.*) have been reported to be particularly effective, based on performance evaluation by the method developed by Mallouk *et al.*[77,78] This method involves a sacrificial agent ($S_2O_8^{2-}$, persulfate) and a photosensitizer ($Ru(bpy)_3$). The process is schematically illustrated in Figure 9.18.

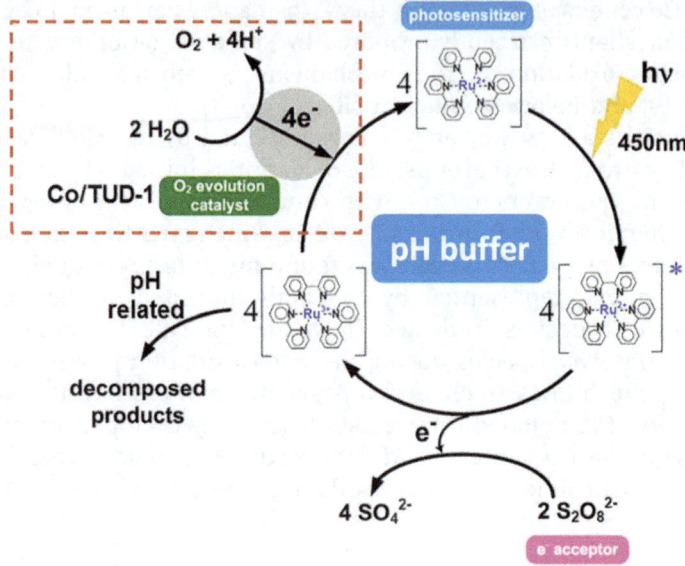

Figure 9.18   Scheme of the method typically used to evaluate efficacy in water oxidation catalysis.

The photosensitizer, $[Ru(bpy)_3]Cl_2 \cdot 6H_2O$, first absorbs the visible light at ~ 450 nm to form an excited state, which is oxidized by the sacrificial electron acceptor $S_2O_8^{2-}$. The water oxidation catalysts dispersed in the solution catalyze electron transfer from water molecules to the oxidized Ru complex, leading to the oxygen evolution from water and reduction of the Ru complex to its original state, closing the catalytic cycle. Theoretically, the Ru complex can undergo multiple cycles, until consumption of $S_2O_8^{2-}$ is completed, and a stoichiometric amount of O_2 is formed. However, due to the presence of unwanted decomposition reactions of the oxidized Ru complex into inactive components in every cycle, a considerable amount of $S_2O_8^{2-}$ is consumed, and a lower than expected conversion yield to O_2 (<100%) can usually be observed.[53] This method has been used successfully to demonstrate high performance of Co_3O_4, and also Mn_3O_4 microcrystals, in mesoporous silicas.[60,79,80]

The structure and oxidation state of the Co catalysts most favorable for water oxidation is still a matter of debate. As stated, microcrystals of Co_3O_4 were demonstrated to be very effective by Frei and co-workers,[79] while Ahn *et al.* demonstrate that the turnover frequency of isolated Co sites is actually higher than of sites present on the surface of the microcrystals.[81] The same authors recently demonstrated that the support for the isolated Co sites plays an important role in achieving high turnover frequencies. Co centers on basic supports (TiO_2 and MgO) yield the highest turnover frequencies in the order of 0.04 s^{-1}. It should be mentioned that the high performance of

isolated Co centers of presumably the $+II$ oxidation state is not in agreement with the mechanism recently proposed by Frei and co-workers for Co-catalyzed water oxidation.[74] The mechanism is schematically shown in Figure 9.19, and involves binuclear sites of Co_3O_4.

The initial state of the catalyst is represented by two Co^{III} centers terminated by two hydroxyl groups. The oxidation is initiated by oxidation (by holes) of the Co centers to Co^{IV}, with concomitant release of two protons. Water is then inserted, forming a peroxide, which after two more oxidation steps releases oxygen. This cycle was found much faster than observed for isolated centers, represented by the path indicated at the bottom of Figure 9.19. The sites indicated in yellow indicate (infrared) spectroscopically observed species during the water oxidation process, of which, as stated, the binuclear cluster appears much more effective than the isolated site. The detailed infrared study has only been performed for the microcrystalline Co centers, and future studies should reveal if similar sites are responsible for water oxidation when present on alternative supports.

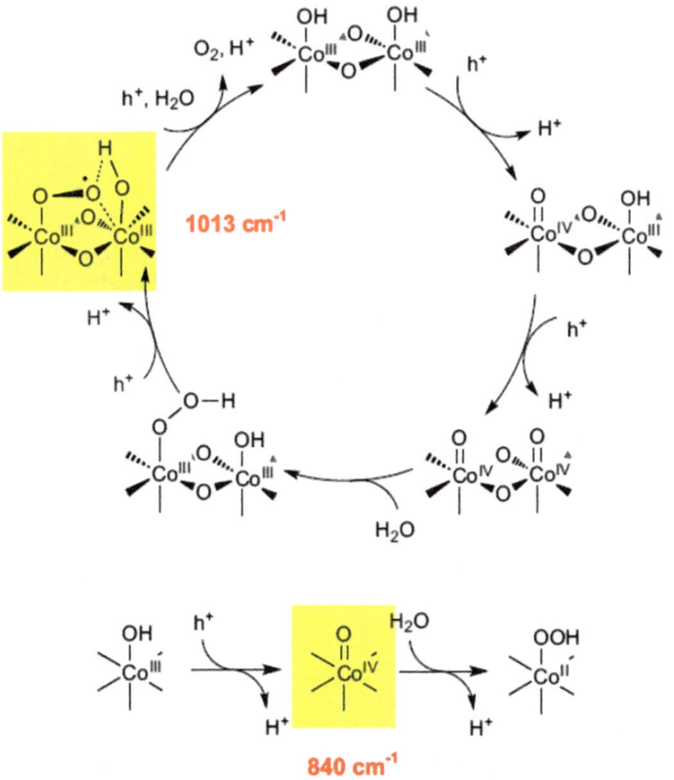

Figure 9.19 Mechanism proposed for Co_3O_4-catalyzed water oxidation. Reproduced with permission courtesy of Springer Open.[74]

9.3.6 Adsorption of CO$_2$ and (Prevention of) Back Reactions

Strunk and co-workers introduced ZnO entities in Ti-SBA-15 to stimulate CO$_2$ sorption in the siliceous framework.[82] Characterization confirmed a close proximity of ZnO$_x$ and TiO$_x$ in the subsequently grafted materials. Because of strong interactions between the Zn precursor and the SiO$_2$ surface, the order of the ZnO$_x$ and TiO$_x$ grafting steps affected the amount of Ti–O–Zn bonds formed in the materials, besides isolated species of the individual components. Irrespective of the structures in the silica pores, all TiO$_x$- and ZnO$_x$-containing samples exhibit significant improvement of the CO$_2$ adsorption capacity.

Mul and co-workers[54] investigated the potential of basic oxides in improving CO$_2$ sorption and hydrocarbon formation. In particular, ZnO was included in Ti-containing TUD-1. Contrary to what was expected, the performance of ZnO–Ti–TUD-1 was inferior to the parent Ti–TUD-1. An explanation can be found in experiments on the photocatalytic degradation of a mixture of hydrocarbons (*i.e.*, CH$_4$, C$_2$H$_4$, C$_2$H$_6$, C$_3$H$_6$, and C$_3$H$_8$) under the same illumination conditions. Ti–TUD-1 exhibits the poorest activity in hydrocarbon degradation, while ZnO–Ti–TUD-1 showed very significant degradation rates. This study clearly demonstrates the importance of evaluating hydrocarbon conversion over photocatalysts active in converting CO$_2$ to hydrocarbons (in batch reactors).

One of the species responsible for hydrocarbon oxidation efficiency could be the (hydrogen) peroxide formed in the pathway of water oxidation.[83,84]

$$CO_2 + 3H_2O \leftrightarrow CH_4 + \frac{1}{2}O_2 + H_2O_2 \qquad (9.12)$$

The question arises whether these backward reactions can be minimized. The group of Frei has proposed to create oxidation and reduction sites on opposite sides of channel walls in mesoporous silicas. Electronic 'communication' can be achieved by including electron-conducting molecules in the channel walls.[55] Although synthetically challenging, the group has succeeded creating a Co$_3$O$_4$ core/silica shell particle, with *p*-oligo(phenylenevinylene) wire molecules (three aryl units, PV3) cast into the silica shell. By various spectroscopies, and use of the Ru(bpy)$_3{}^{2+}$ complex, the hole-conducting organic wire molecules were demonstrated to offer fast, controlled charge transfer through a product-separating oxide barrier.

9.4 Graphene/Graphene Oxide–TiO$_2$ Composites

Incorporation of TiO$_2$ onto carbon nanostructural supports such as graphene and graphene oxide has been explored recently for CO$_2$ catalysis.[85–88] These composites are attractive in that they offer larger surface area for photoreactivity. The improved catalytic efficiencies over other materials is a consequence of the high conductivity observed for graphene-based materials resulting from their highly conjugated nature.[89,90] The reduction potential of

graphene (-0.8 V *versus* SHE at pH 0) lies approximately 0.24 V below the conduction band of TiO_2.[86,91–93] In this way, upon UV-light irradiation photo-generated electrons are transferred from the conduction band of TiO_2 to graphene.[89,94] Alternatively, visible light illumination of a graphene oxide–TiO_2 composite is thought to result in electron transfer from graphene oxide to the TiO_2. The e^-–h^+ separation occurs between TiO_2 and the graphene/graphene oxide supports due to the Schottky barrier formed at their interface, which effectively suppresses charge recombination.[94,95]

The photoreduction of CO_2 by graphene and graphene oxide–TiO_2 composites can result in the production of CO although other products include CH_4 and C_2H_6.[86–88] For example, the initial charge separation upon photo-excitation of a reduced graphene oxide–TiO_2 composite with visible light is thought to proceed *via* the mechanism described above from the conduction band of TiO_2 to the *reduced* graphene oxide (Figure 9.20).[86] However, little is known concerning the subsequent steps resulting in either the two-electron (CO), eight-electron (CH_4), or fourteen-electron (C_2H_6) product. It is clear from these reports that the identity of the photocatalytic product is dependent on the nature of the composite and the method of preparation. Prediction of major products for new preparative methods without information on the relationship between structural/physical properties of the materials and the mechanism of product formation is difficult. Tu *et al.*, however, suggested that the identity of the major product of the photo-catalytic reaction may be related to charge/electrical mobility through the graphene nanosheets.[87] Specifically, as electrons are more mobile and less likely to accumulate at reactive sites, less reduced products will result. This

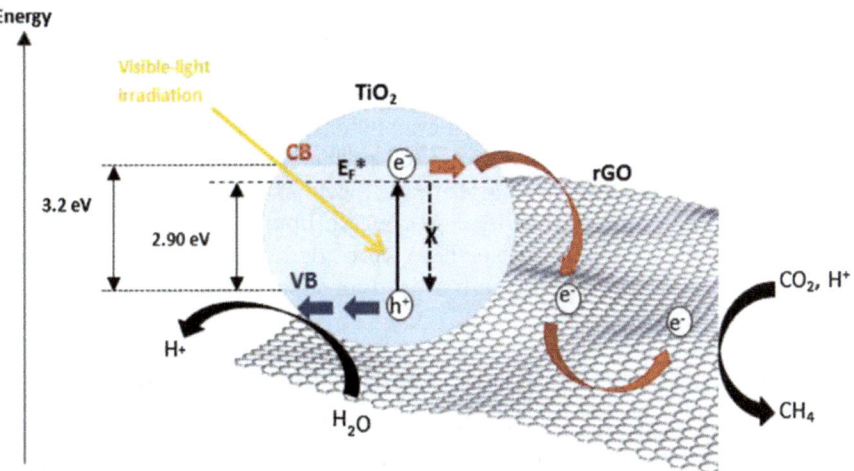

Figure 9.20   Mechanism of CO_2 reduction proposed by Tan *et al.* for reduced graphene oxide–TiO_2 composites.
Reproduced with permission courtesy of Springer Open (https://creative-commons.org/licenses/by/2.0/legalcode).[86]

is based solely on the observation that at TiO$_2$, an 8e$^-$ reduced product, CH$_4$, is preferred, whereas at TiO$_2$–graphene composites the major product is CO.

Interestingly, it has been shown that the photoreduction of CO$_2$ by graphene–TiO$_2$ composites to produce CH$_4$ is greatly improved by decreasing the number of defect sites in the graphene phase. A comparison of the photo-activity of solvent-exfoliated graphene to solvent-reduced graphene oxide shows a four-fold increase in the rate of photoreduction of CO$_2$.[85] This is probably due to better electronic coupling with TiO$_2$ resulting in improved electrical mobility and larger mean free paths for electrons to reach reaction sites. Zhang *et al.* have shown that the high mobility displayed by pristine graphene is related to the tendency of electron and hole carriers in graphene to act as massless Dirac fermions.[86,96]

9.5 Ti^{IV}-based Metal Organic Framework Catalysts

Metal organic frameworks (MOFs) are a relatively new class of three-dimensional, porous materials. These are typically comprised of metal-oxo clusters linked *via* multi-dentate organo-carboxylate linkers to form extended frameworks of varying size and porosity. The attraction for these materials is rooted in their unusually high surface areas which can be tuned by appropriately modifying the size and symmetry of the organic linker as well.[97] It is well known that the electronic properties, both ground state and excited state, are dependent on the choice of linker, metal, and oxidation state where the electronic properties of the composites do not necessarily resemble those of the individual components.[98–101]

The large surface areas afforded by MOFs have been exploited for gas storage and catalysis.[97,102–107] Extensive studies, both experimental and computational, have also been performed to understand the basis of small gas molecule–MOF interactions.[105] Indeed, a number of MOFs have been shown to display extraordinarily high affinity for CO$_2$ and, what's more, photocatalytic activity for CO$_2$.[108–112] Photocatalytic reduction of CO$_2$ has been demonstrated *via* direct interaction between CO$_2$ and the MOF (presumably vis-à-vis the metal-oxo nodes), incorporation of molecular catalysts, or by coupling reactive nanoparticles by encapsulation of the nanoparticle or as a support for the nanoparticle.[107,112–115]

Although, reports of MOFs incorporating Ti^{IV}-metal clusters as structural motifs or MOF-TiO$_2$ composites are few, three such materials have been developed and probed for the propensity to photoreduce CO$_2$.[113–115] The first is a MOF-TiO$_2$ core-shell composite in which copper(II)-trimesic acid, Cu$_3$(BTC)$_2$, microcyrstals were encapsulated within TiO$_2$ shell structures.[114] This material was shown to produce CH$_4$ as a major photoreduction product of CO$_2$ with yields up to five times that of bare TiO$_2$. A mechanism was suggested based on ultrafast spectroscopic methods involving electron transfer from the TiO$_2$ into the Cu$_3$(BTC)$_2$ resulting in reduction of CO$_2$ adsorbed to the MOF (Figure 9.21). Alternatively, Cu$_3$(BTC)$_2$ has been used as a photosensitizer for TiO$_2$ exhibiting efficient electron injection from

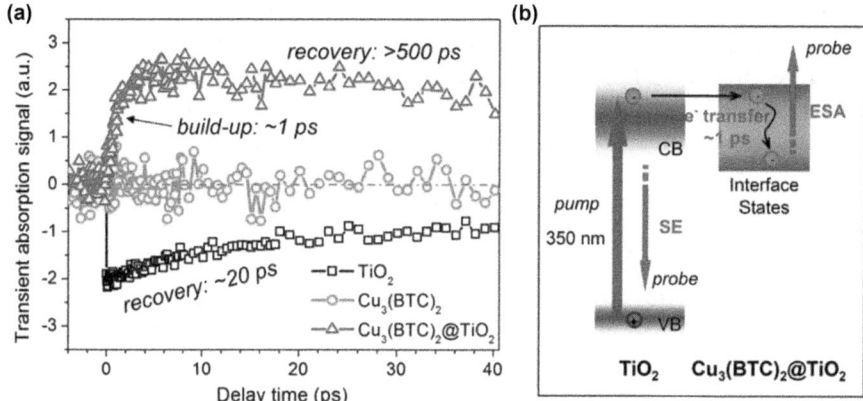

Figure 9.21   Transient absorption kinetic traces for unmodified TiO_2 (open squares, $\lambda_{obs} = 450\,nm$), $Cu_3(BTC)_2$ (open circles, $\lambda_{obs} = 600\,nm$), $Cu_3(BTC)_2$–TiO_2 composite (open triangles, $\lambda_{obs} = 600\,nm$) upon photo-excitation at 350 nm.
Reproduced with permission from ref. 114.

$Cu_3(BTC)_2$ into the TiO_2 conduction band.[114] Further studies are needed to determine the location of CO_2 reduction reactive sites, TiO_2 or $Cu_3(BTC)_2$.

Two MOF structures have been developed in which Ti^{IV}-metal sites were incorporated into the metal-oxo nodes of the MOF: NH_2–MIL-125(Ti) and UiO-66-(Zr/Ti)–NH_2.[113,115] Both are comprised of amine-functionalized terephthalic acid linkers bound to either $Ti_8O_8(OH)_4(COO^-)_6$ or $Zr_{4.3}Ti_{1.7}O_4(OH)_4(COO^-)_6$ nodes. The first MOF, N_2-MIL-125(Ti), displayed photoreactivity toward CO_2 producing $HCOO^-$ as a major product.[115] The second MOF, UiO-66-(Zr/Ti)–NH_2, also resulted in $HCOO^-$ as a major reaction product displaying a turnover number of approximately 5.[113] Both reports include photoreaction experiments conducted with isotope-labeled $^{13}CO_2$, which demonstrate the source of the $HCOO^-$ is in fact CO_2 and not breakdown of the MOF. Though little is actually known regarding the reaction mechanism, it is supposed that CO_2 reduction occurred at Ti^{III} sites of the metal-oxo clusters generated by photo-excitation of the ligands (Figure 9.21).

9.6   Conclusions and Outlook

There are a few overarching themes that arise from the presented review of relevant literature on CO_2 reduction by TiO_2 and TiO_2 composites:

(1) metal dopants act as inter-band electron traps, whereas metal oxide–TiO_2 composites and their reactivity rely on band edge alignment
(2) the development of structure–property relationships are key to understanding the observed product distributions and guiding future catalyst development

(3) photocorrosion and stability must be considered in future catalyst development.

All three of these concepts are interwoven. However, they each have direct implications for future considerations of green nanomaterials for carbon dioxide reduction.

In the cases discussed herein, when doping TiO$_2$ with a metal, the reductive chemistry was hypothesized to occur at these electron trap sites. From a cursory level analysis, one might assume that the operative mechanisms and observed photoreductive products in doped materials would mirror the chemistry at the independent metal surface. However, for copper, at no applied potential in electrochemical CO$_2$ reduction is methanol the major observed product as was reported here for Cu-doped TiO$_2$.[116] Indeed, methane, similar to bare TiO$_2$, is the major CO$_2$ electrochemical reduction product. Therefore, when proposing photoreductive mechanisms, it is important to consider those that involve a cooperative nature to or bridged intermediates between copper and titania sites,[117] as well as unique potential gradients that may be formed upon constant photo-excitation of TiO$_2$. Additionally, this is a clear example that electrocatalysis cannot always effectively model photocatalysis and comparative studies should be subject to a high level of scrutiny. However, it is clear that the addition of metal dopants can be utilized to tune product distributions away from that of undoped TiO$_2$.

In contrast, metal oxide dopants most often result in the same products observed with unmodified TiO$_2$. Considering the required energy level alignment for spatial charge separation, this result is perhaps not surprising. The highly oxidizing valence band of TiO$_2$ often precludes charge gradients that would result in electron accumulation on the metal oxide dopant. The example of FeTiO$_3$ reported here is to our knowledge the only reported example that exhibits this phenomenon. Therefore, while metal oxide dopants can effectively alter the absorption properties of the materials, specifically shifting the absorption into the visible portion of the solar spectrum, they have not effectively been used to tune reaction specificity or product distribution. Metal dopants can effect product distribution, but do not effectively change the absorption properties. A key interplay arises from these conclusions. Is it possible to modify TiO$_2$ to result in visible light absorption and selective CO$_2$ chemistry? Recent results that report a synergistic effect between band-gap engineering through nitrogen doping and product distribution by Ni^{2+} doping may provide an avenue to achieve this ultimate goal.[118]

Surface area, dopant dispersion, defect density, and composite interactions have been proposed to alter product distributions and photocatalytic efficiency. However, to properly support these hypotheses, systematic investigations coupled to detailed structural analysis in regard to each of these controlling factors is necessary for each metal oxide composite and metal dopant. While surface area and morphology have been explored,[119–121] full characterization (structure and conductivity) of nanomaterials is sometimes

overlooked in reports of CO_2 reduction. An example of the consequence is that reports of metal dopants may, in fact, be explorations into metal oxide composite materials. The structure–property relationships that result from systematic investigations are critical to guide the development of future catalyst candidates.

Lastly, stability is key. While TiO_2 is chemically stable under a wide variety of conditions, the dopants (either metal or metal oxide) have been shown to be susceptible to photocorrosion, dissolution, and/or photo-induced aggregation. Strategies to maintain structure after prolonged light exposure are needed. This can be accomplished through surface engineering (*i.e.*, deposition of thin protective layers of oxides or polymeric coatings), the immobilization of catalyst materials inside mesoporous supports, and/or the addition of photocorrosion inhibitors to the reaction solution.[17,122–125] Considering the vast scientific foundation on which to build upon, titania-based nanomaterials provide a promising platform for the development of green carbon dioxide reduction catalysts. Future investigations should consider approaches to maintain stability, elucidation of the structural requirements for efficient CO_2 photoreduction, and methods to both increase the visible light absorption and product selectivity.

References

1. E. V. Kondratenko, G. Mul, J. Baltrusaitis, G. O. Larrazabal and J. Perez-Ramirez, *Energy Environ. Sci.*, 2013, **6**, 3112–3135.
2. G. W. Crabtree and J. Sarrao, *Phys. World*, 2009, **22**, 24.
3. A. Samanta, A. Zhao, G. K. H. Shimizu, P. Sarkar and R. Gupta, *Ind. Eng. Chem. Res.*, 2012, **51**, 1438–1463.
4. O. Carp, C. L. Huisman and A. Reller, *Prog. Solid State Chem.*, 2004, **32**, 33–177.
5. S. Zhen, Q. Qian, G. Jia, J. Zhang, C. Y. Chen and Y. J. Wei, *J. Occup. Environ. Med.*, 2012, **54**, 1389–1394.
6. Y. Xu and M. A. A. Schoonen, *Am. Mineral.*, 2000, **85**, 543–556.
7. L. J. Liu, H. L. Zhao, J. M. Andino and Y. Li, *ACS Catal.*, 2012, **2**, 1817–1828.
8. V. P. Indrakanti, H. H. Schobert and J. D. Kubicki, *Energy Fuels*, 2009, **23**, 5247–5256.
9. F. De Angelis, C. Di Valentin, S. Fantacci, A. Vittadini and A. Selloni, *Chem. Rev.*, 2014, **114**, 9708–9753.
10. M. M. Rodriguez, X. H. Peng, L. J. Liu, Y. Li and J. M. Andino, *J. Phys. Chem. C*, 2012, **116**, 19755–19764.
11. W.-K. Li, X.-Q. Gong, G. Lu and A. Selloni, *J. Phys. Chem. C*, 2008, **112**, 6594–6596.
12. G. Busca and V. Lorenzelli, *Mater. Chem.*, 1982, **7**, 89–126.
13. H. J. Freund and M. W. Roberts, *Surf. Sci. Rep.*, 1996, **25**, 225–273.
14. H. Y. He, P. Zapol and L. A. Curtiss, *J. Phys. Chem. C*, 2010, **114**, 21474–21481.

15. Y. Ma, X. L. Wang, Y. S. Jia, X. B. Chen, H. X. Han and C. Li, *Chem. Rev.*, 2014, **114**, 9987–10043.

16. W. H. Koppenol and J. D. Rush, *J. Phys. Chem.*, 1987, **91**, 4429–4430.

17. S. N. Habisreutinger, L. Schmidt-Mende and J. K. Stolarczyk, *Angew. Chem., Int. Ed.*, 2013, **52**, 7372–7408.

18. T. Inoue, A. Fujishima, S. Konishi and K. Honda, *Nature*, 1979, **277**, 637–638.

19. Y. Ku, W.-H. Lee and W.-Y. Wang, *J. Mol. Catal. A: Chem.*, 2004, **212**, 191–196.

20. K. Kočí, L. Obalová, L. Matějová, D. Plachá, Z. Lacný, J. Jirkovský and O. Šolcová, *Appl. Catal., B*, 2009, **89**, 494–502.

21. S. Kaneco, H. Kurimoto, K. Ohta, T. Mizuno and A. Saji, *J. Photochem. Photobiol., A*, 1997, **109**, 59–63.

22. S. Kaneco, Y. Shimizu, K. Ohta and T. Mizuno, *J. Photochem. Photobiol., A*, 1998, **115**, 223–226.

23. S. Kaneco, H. Kurimoto, Y. Shimizu, K. Ohta and T. Mizuno, *Energy*, 1999, **24**, 21–30.

24. C.-C. Lo, C.-H. Hung, C.-S. Yuan and J.-F. Wu, *Sol. Energy Mater. Sol. Cells*, 2007, **91**, 1765–1774.

25. L. Liu and Y. Li, *Aerosol Air Qual. Res.*, 2014, **14**, 453–469.

26. G. Dey and K. Pushpa, *Res. Chem. Intermed.*, 2007, **33**, 631–644.

27. M. Anpo, H. Yamashita, Y. Ichihashi and S. Ehara, *J. Electroanal. Chem.*, 1995, **396**, 21–26.

28. H. Yamashita, A. Shiga, S.-i. Kawasaki, Y. Ichihashi, S. Ehara and M. Anpo, *Energy Convers. Manage.*, 1995, **36**, 617–620.

29. S. Leytner and J. T. Hupp, *Chem. Phys. Lett.*, 2000, **330**, 231–236.

30. T. Mizuno, K. Adachi, K. Ohta and A. Saji, *J. Photochem. Photobiol., A*, 1996, **98**, 87–90.

31. N. M. Dimitrijevic, I. A. Shkrob, D. J. Gosztola and T. Rajh, *J. Phys. Chem. C*, 2012, **116**, 878–885.

32. Y. P. Peng, Y. T. Yeh, S. I. Shah and C. P. Huang, *Appl. Catal., B*, 2012, **123**, 414–423.

33. M. Sathish, B. Viswanathan, R. P. Viswanath and C. S. Gopinath, *Chem. Mater.*, 2005, **17**, 6349–6353.

34. H. Irie, Y. Watanabe and K. Hashimoto, *Chem. Lett.*, 2003, **32**, 772–773.

35. Z. Zhao, J. Fan, J. Wang and R. Li, *Catal. Commun.*, 2012, **21**, 32–37.

36. R. Sellappan, J. F. Zhu, H. Fredriksson, R. S. Martins, M. Zach and D. Chakarov, *J. Mol. Catal. A: Chem.*, 2011, **335**, 136–144.

37. C. C. Yang, Y. H. Yu, B. van der Linden, J. C. S. Wu and G. Mul, *J. Am. Chem. Soc.*, 2010, **132**, 8398–8406.

38. B. Mei, A. Pougin and J. Strunk, *J. Catal.*, 2013, **306**, 184–189.

39. H. Xu, S. Ouyang, L. Liu, D. Wang, T. Kako and J. Ye, *Nanotechnology*, 2014, **25**, 165402.

40. F. Bi, M. F. Ehsan, W. Liu and T. He, *Chin. J. Chem.*, 2015, **33**, 112–118.

41. H. W. Slamet, E. Nasution, S. Purnama, Kosela and J. Gunlazuardi, *Catal. Commun.*, 2005, **6**, 313–319.
42. H. W. N. Slamet, E. Purnama, K. Riyani and J. Gunlazuardi, *World Appl. Sci. J.*, 2009, **6**, 112–122.
43. I. H. Tseng, W. C. Chang and J. C. S. Wu, *Appl. Catal., B*, 2002, **37**, 37–48.
44. I. H. Tseng, J. C. S. Wu and H. Y. Chou, *J. Catal.*, 2004, **221**, 432–440.
45. S. Zhu, S. Liang, Y. Tong, X. An, J. Long, X. Fu and X. Wang, *Phys. Chem. Chem. Phys.*, 2015, **17**, 9761–9770.
46. A. Nishimura, G. Mitsui, M. Hirota and E. Hu, *Int. J. Chem. Eng.*, 2010, 2010.
47. Q. D. Truong, J. Y. Liu, C. C. Chung and Y. C. Ling, *Catal. Commun.*, 2012, **19**, 85–89.
48. B. Gao, Y. J. Kim, A. K. Chakraborty and W. I. Lee, *Appl. Catal., B*, 2008, **83**, 202–207.
49. J. Y. Do, Y. Im, B. S. Kwak, J.-Y. Kim and M. Kang, *Chem. Eng. J.*, 2015, **275**, 288–297.
50. B. S. Kwak, K. Vignesh, N.-K. Park, H.-J. Ryu, J.-I. Baek and M. Kang, *Fuel*, 2015, **143**, 570–576.
51. T. Cuk, W. W. Weare and H. Frei, *J. Phys. Chem. C*, 2010, **114**, 9167–9172.
52. M. Zhang, M. de Respinis and H. Frei, *Nat. Chem.*, 2014, **6**, 362–367.
53. C.-C. Yang, T. M. Eggenhuisen, M. Wolters, A. Agiral, H. Frei, P. E. de Jongh, K. P. de Jong and G. Mul, *ChemCatChem*, 2013, **5**, 550–556.
54. M. S. Hamdy, R. Amrollahi, I. Sinev, B. Mei and G. Mul, *J. Am. Chem. Soc.*, 2014, **136**, 594–597.
55. A. Agiral, H. S. Soo and H. Frei, *Chem. Mater.*, 2013, **25**, 2264–2273.
56. H. Yamashita, Y. Fujii, Y. Ichihashi, S. G. Zhang, K. Ikeue, D. R. Park, K. Koyano, T. Tatsumi and M. Anpo, *Catal. Today*, 1998, **45**, 221–227.
57. M. Anpo, H. Yamashita, K. Ikeue, Y. Fujii, S. G. Zhang, Y. Ichihashi, D. R. Park, Y. Suzuki, K. Koyano and T. Tatsumi, *Catal. Today*, 1998, **44**, 327–332.
58. S. G. Zhang, Y. Fujii, K. Yamashita, K. Koyano, T. Tatsumi and M. Anpo, *Chem. Lett.*, 1997, 659–660.
59. M. Anpo, H. Yamashita, Y. Ichihashi, Y. Fujii and M. Honda, *J. Phys. Chem. B*, 1997, **101**, 2632–2636.
60. F. Jiao and H. Frei, *Chem. Commun.*, 2010, **46**, 2920–2922.
61. S. Telalovic, A. Ramanathan, G. Mul and U. Hanefeld, *J. Mater. Chem.*, 2010, **20**, 642–658.
62. C.-C. Yang, J. Vernimmen, V. Meynen, P. Cool and G. Mul, *J. Catal.*, 2011, **284**, 1–8.
63. A. M. Prakash, H. M. Sung-Suh and L. Kevan, *J. Phys. Chem. B*, 1998, **102**, 857–864.
64. K. Soni, K. C. Mouli, A. K. Dalai and J. Adjaye, *Catal. Lett.*, 2010, **136**, 116–125.

65. K. Soni, B. S. Rana, A. K. Sinha, A. Bhaumik, M. Nandi, M. Kumar and G. M. Dhar, *Appl. Catal., B*, 2009, **90**, 55–63.
66. M. Anpo, H. Yamashita, Y. Ichihashi and S. Ehara, *J. Electroanal. Chem.*, 1995, **396**, 21–26.
67. W. Y. Lin, H. X. Han and H. Frei, *J. Phys. Chem. B*, 2004, **108**, 18269–18273.
68. M. L. Macnaughtan, H. S. Soo and H. Frei, *J. Phys. Chem. C*, 2014, **118**, 7874–7885.
69. W. Kim, G. Yuan, B. A. McClure and H. Frei, *J. Am. Chem. Soc.*, 2014, **136**, 11034–11042.
70. X. Wu, W. W. Weare and H. Frei, *Dalton Trans.*, 2009, 10114–10121.
71. W. W. Weare, Y. Pushkar, V. K. Yachandra and H. Frei, *J. Am. Chem. Soc.*, 2008, **130**, 11355–11363.
72. W. Y. Lin and H. Frei, *J. Am. Chem. Soc.*, 2005, **127**, 1610–1611.
73. H. Han and H. Frei, *Microporous Mesoporous Mater.*, 2007, **103**, 265–272.
74. M. Zhang and H. Frei, *Catal. Lett.*, 2015, **145**, 420–435.
75. R. Nakamura and H. Frei, *J. Am. Chem. Soc.*, 2006, **128**, 10668–10669.
76. N. Sivasankar, W. W. Weare and H. Frei, *J. Am. Chem. Soc.*, 2011, **133**, 12976–12979.
77. M. Hara, J. T. Lean and T. E. Mallouk, *Chem. Mater.*, 2001, **13**, 4668–4675.
78. M. Hara, C. C. Waraksa, J. T. Lean, B. A. Lewis and T. E. Mallouk, *J. Phys. Chem. A*, 2000, **104**, 5275–5280.
79. F. Jiao and H. Frei, *Energy Environ. Sci.*, 2010, **3**, 1018–1027.
80. F. Jiao and H. Frei, *Angew. Chem., Int. Ed.*, 2009, **48**, 1841–1844.
81. H. S. Ahn, J. Yano and T. D. Tilley, *Energy Environ. Sci.*, 2013, **6**, 3080–3087.
82. B. Mei, A. Becerikli, A. Pougin, D. Heeskens, I. Sinev, W. Grunert, M. Muhler and J. Strunk, *J. Phys. Chem. C*, 2012, **116**, 14318–14327.
83. M. Anpo, *J. Phys. Chem. B*, 1997, **101**, 2632–2636.
84. O. K. Varghese, *Nano Lett.*, 2009, 9.
85. Y. T. Liang, B. K. Vijayan, K. A. Gray and M. C. Hersam, *Nano Lett.*, 2011, **11**, 2865–2870.
86. L. L. Tan, W. J. Ong, S. P. Chai and A. R. Mohamed, *Nanoscale Res. Lett.*, 2013, 8.
87. W. G. Tu, Y. Zhou, Q. Liu, Z. P. Tian, J. Gao, X. Y. Chen, H. T. Zhang, J. G. Liu and Z. G. Zou, *Adv. Funct. Mater.*, 2012, **22**, 1215–1221.
88. W. G. Tu, Y. Zhou, Q. Liu, S. C. Yan, S. S. Bao, X. Y. Wang, M. Xiao and Z. G. Zou, *Adv. Funct. Mater.*, 2013, **23**, 1743–1749.
89. G. Williams, B. Seger and P. V. Kamat, *ACS Nano*, 2008, **2**, 1487–1491.
90. H. Zhang, X. Lv, Y. Li, Y. Wang and J. Li, *ACS Nano*, 2010, **4**, 380–386.
91. X. Wang, L. Zhi and K. Müllen, *Nano Lett.*, 2008, **8**, 323–327.
92. K. Zhou, Y. Zhu, X. Yang, X. Jiang and C. Li, *New J. Chem.*, 2011, **35**, 353–359.

93. Q. Li, B. Guo, J. Yu, J. Ran, B. Zhang, H. Yan and J. R. Gong, *J. Am. Chem. Soc.*, 2011, **133**, 10878–10884.

94. Q. Xiang, J. Yu and M. Jaroniec, *Nanoscale*, 2011, **3**, 3670–3678.

95. R. Sellappan, J. Sun, A. Galeckas, N. Lindvall, A. Yurgens, A. Y. Kuznetsov and D. Chakarov, *Phys. Chem. Chem. Phys.*, 2013, **15**, 15528–15537.

96. Y. B. Zhang, Y. W. Tan, H. L. Stormer and P. Kim, *Nature*, 2005, **438**, 201–204.

97. D. Zhao, D. J. Timmons, D. Yuan and H.-C. Zhou, *Acc. Chem. Res.*, 2011, **44**, 123–133.

98. M. D. Allendorf, M. E. Foster, F. Leonard, V. Stavila, P. L. Feng, F. P. Doty, K. Leong, E. Y. Ma, S. R. Johnston and A. A. Talin, *J. Phys. Chem. Lett.*, 2015, **6**, 1182–1195.

99. Z. Hu, B. J. Deibert and J. Li, *Chem. Soc. Rev.*, 2014, **43**, 5815–5840.

100. Y. Cui, F. Zhu, B. Chen and G. Qian, *Chem. Commun.*, 2015, **51**, 7420–7431.

101. C. M. Doherty, D. Buso, A. J. Hill, S. Furukawa, S. Kitagawa and P. Falcaro, *Acc. Chem. Res.*, 2014, **47**, 396–405.

102. J. Liu, L. Chen, H. Cui, J. Zhang, L. Zhang and C.-Y. Su, *Chem. Soc. Rev.*, 2014, **43**, 6011–6061.

103. J. Gascon, A. Corma, F. Kapteijn and F. X. Llabrés i Xamena, *ACS Catal.*, 2014, **4**, 361–378.

104. U. Mueller, M. Schubert, F. Teich, H. Puetter, K. Schierle-Arndt and J. Pastre, *J. Mater. Chem.*, 2006, **16**, 626–636.

105. B. Chen, S. Xiang and G. Qian, *Acc. Chem. Res.*, 2010, **43**, 1115–1124.

106. Y. Takenaka and U. Mueller, *CSJ Curr. Rev.*, 2010, **3**, 160–164.

107. A. Aijaz and Q. Xu, *J. Phys. Chem. Lett.*, 2014, **5**, 1400–1411.

108. S. Chaemchuen, N. A. Kabir, K. Zhou and F. Verpoort, *Chem. Soc. Rev.*, 2013, **42**, 9304–9332.

109. S. Keskin, T. M. van Heest and D. S. Sholl, *ChemSusChem*, 2010, **3**, 879–891.

110. Z. Zhang, Z.-Z. Yao, S. Xiang and B. Chen, *Energy Environ. Sci.*, 2014, **7**, 2868–2899.

111. Z. Zhang, Y. Zhao, Q. Gong, Z. Li and J. Li, *Chem. Commun.*, 2013, **49**, 653–661.

112. C. C. Wang, Y. Q. Zhang, J. Li and P. Wang, *J. Mol. Struct.*, 2015, **1083**, 127–136.

113. Y. Lee, S. Kim, J. K. Kang and S. M. Cohen, *Chem. Commun.*, 2015, **51**, 5735–5738.

114. R. Li, J. Hu, M. Deng, H. Wang, X. Wang, Y. Hu, H.-L. Jiang, J. Jiang, Q. Zhang, Y. Xie and Y. Xiong, *Adv. Mater.*, 2014, **26**, 4783–4788.

115. Y. Fu, D. Sun, Y. Chen, R. Huang, Z. Ding, X. Fu and Z. Li, *Angew. Chem., Int. Ed.*, 2012, **51**, 3364–3367.

116. K. P. Kuhl, E. R. Cave, D. N. Abram and T. F. Jaramillo, *Energy Environ. Sci.*, 2012, **5**, 7050–7059.

117. I. X. Green, W. Tang, M. Neurock and J. T. Yates, *Science*, 2011, **333**, 736–739.
118. J. Fan, E.-z. Liu, L. Tian, X.-y. Hu, Q. He and T. Sun, *J. Environ. Eng.*, 2011, **137**, 171–176.
119. L. Liu, H. Zhao, J. M. Andino and Y. Li, *ACS Catal.*, 2012, **2**, 1817–1828.
120. W.-N. Wang, W.-J. An, B. Ramalingam, S. Mukherjee, D. M. Niedzwiedzki, S. Gangopadhyay and P. Biswas, *J. Am. Chem. Soc.*, 2012, **134**, 11276–11281.
121. K. L. Schulte, P. A. DeSario and K. A. Gray, *Appl. Catal., B*, 2010, **97**, 354–360.
122. L. Zhang, Q. Liu, T. Aoki and P. A. Crozier, *J. Phys. Chem. C*, 2015, **119**, 7207–7214.
123. S. Kakuta and T. Abe, *Electrochem. Solid-State Lett.*, 2009, **12**, P1–P3.
124. M. D. Hernandez-Alonso, F. Fresno, S. Suarez and J. M. Coronado, *Energy Environ. Sci.*, 2009, **2**, 1231–1257.
125. X. Chen and S. S. Mao, *Chem. Rev.*, 2007, **107**, 2891–2959.
126. Y. Ku, W. H. Lee and W. Y. Wang, *J. Mol. Catal. A: Chem.*, 2004, **212**, 191–196.
127. Q. Xu, J. Yu, J. Zhang, J. Zhang and G. Liu, *Chem. Commun.*, 2015, **51**, 7950–7953.
128. K. Adachi, K. Ohta and T. Mizuno, *Sol. Energy*, 1994, **53**, 187–190.
129. F. Jun, L. Enzhou, Z. Bo, H. Xiaoyun, C. Guoliang and Y. Junjie, *Petrochem. Technol.*, 2009, 7, 019.

CHAPTER 10

Hybrid Inorganic and Organic Assembly System for Photocatalytic Conversion of Carbon Dioxide

XIN ZHANG,[a] YU LEI[b] AND NURXAT NURAJE*[a]

[a] Department of Chemical Engineering, Texas Tech University, Lubbock, TX 79409, USA; [b] School of Engineering and Applied Sciences, Harvard University, Cambridge, MA 02138, USA
*Email: nurxat.nuraje@ttu.edu

10.1 Introduction to the Conversion of Carbon Dioxide and Industrial Applications

Carbon dioxide is a naturally occurring oxidation product. It takes part in photosynthesis and is turned into carbohydrate during daylight, and in darkness it is released through plants' respiration. Hence, carbon dioxide is ubiquitous and recycles in a steady carbon circle. However, since the Industrial Revolution, excessive anthropogenic activities, including burning of much hydrocarbon fuel, have taken place. These activities lead to CO_2 concentration in the atmosphere exceeding the normal level of 250 parts per million (ppm) (preindustrial level) and increasing to 394.5 ppm as of September 2012. It is precisely because CO_2 is the major greenhouse gas that superfluous CO_2 may cause global warming. Many unexpected environmental consequences appear in the ecological system worldwide, such as sea

RSC Green Chemistry No. 42
Green Photo-active Nanomaterials: Sustainable Energy and Environmental Remediation
Edited by Nurxat Nuraje, Ramazan Asmatulu and Guido Mul

level rising, gargantuan tsunamis, glaciers melting, animal and plant extinctions, and ocean acidification, which are partially caused by extra carbon dioxide release. Burning fossil fuels, ranging from coal, petroleum, and natural gas to biomass, is the chief culprit in emitting carbon dioxide. Destruction of natural environments, such as swamps and forests, further aggravate the CO_2 overload.

At the same time, hydrocarbon fuels are currently an important energy source on account of their stability for transportation and storage, as well as high energy density. For instance, in the United States, 84% of the total energy supply in 2011 depended heavily on hydrocarbon fuels, including coal, petroleum, and natural gas. The consequences of heavily relying on hydrocarbon fuels are:

(1) vulnerability of the security of the energy source
(2) increasing additional carbon that does not originally circulate within the atmospheric system.

Therefore, the conversion of carbon dioxide *via* the photocatalytic reduction process into energy dense portable hydrocarbon fuels, which are compatible with the current energy infrastructure, is one attractive CO_2 recycling prospect.

As is well known, the majority of popular renewable energy technologies, mainly including solar cell and wind electric power, are always located in the suburban areas, which are far away from energy consumption centers. Therefore, it will probably be expensive in transmission and distribution to deliver the energy in the form of electricity from the production site to energy consumption sites. Inefficiency in transmission and distribution results a loss of an enormous amount of energy (26% of entire generated energy) every year. What's more, both of the two renewable energy technologies are strongly weather-dependent, which causes intermittent electric energy production and thus limits their utilization in the electricity grid.

In the area of artificial photosynthesis, hydrogen production derived from water electrolysis is theoretically well researched and the related technology is relatively mature. Meanwhile, many countries have improved their hydrogen economies for many years. However, due to the evident disadvantages such as fast leakage rate and storage, the future of hydrogen production continues to face tremendous challenges.

Many technologies have been explored to turn carbon dioxide into hydrocarbon fuels, in both research and industry. One solution, called hydrogenation, is to reduce CO_2 with the aid of hydrogen, turning electricity power into chemical power with the energy supply of electricity to synthesize hydrocarbon fuels directly. One of the defects of this solution is that both the hydrogen source and the electricity source are mainly obtained from fossil fuels, which in turn generates CO_2.

Figure 10.1 shows a carbon-neutral method of manufacturing carbonaceous fuels, which means through the well-balanced cycle, the carbon

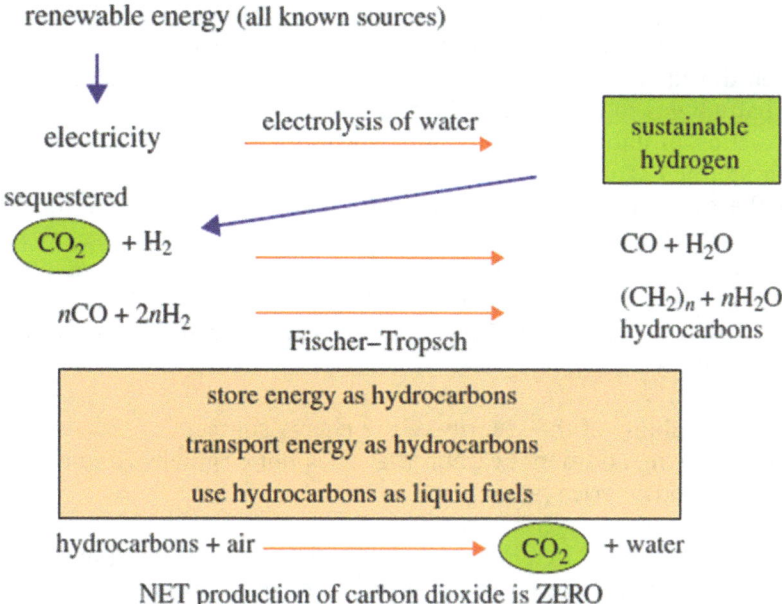

renewable energy (all known sources)

electricity   electrolysis of water

sustainable
hydrogen

sequestered

CO_2 + H_2 $CO + H_2O$

$nCO + 2nH_2$ $(CH_2)_n + nH_2O$
                                                                hydrocarbons
                    Fischer–Tropsch

store energy as hydrocarbons

transport energy as hydrocarbons

use hydrocarbons as liquid fuels

hydrocarbons + air CO_2 + water

NET production of carbon dioxide is ZERO

Figure 10.1   Energy cycle using sequestered CO_2 to yield carbonaceous fuels. Reprinted with permission from ref. 2. Copyright 2010 The Royal Society Publishing.

dioxide concentration stays almost the same. The Fischer–Tropsch approach depicted here is an essential step to synthesize the fuels. Electrolysis of water uses electricity which is generated from renewable energy. The sustainable hydrogen then joins with sequestered carbon dioxide and forms hydrocarbon fuels as a result. The products are used as liquid fuels and are finally turned into CO_2 through combustion, keeping the net production of carbon dioxide zero. Based on available renewable energy technologies, the process can occur at any time. Plus, the medium of energy storage is hydrocarbon fuel, which can be easily stored and transported. However, the feasibility of this method largely relies on chemical sequestration, which wastes an enormous amount of energy, and the multi-step process of making hydrocarbon fuels is extremely inefficient as well as time consuming.

The approach of using a photocatalyst to reduce CO_2 with the aid of free solar energy is a simple process and can produce hydrocarbon fuels which are stable, readily available, and energy dense. This unique generation technology for making fuels has already become a hot spot of research and will surely be in the spotlight of the industrial energy revolution in the future.

It has been published in several review articles as the methods of conversion to carbonaceous fuels from carbon dioxide have been widely studied.[2–9] Many of these only contain a non-photochemical process.[2,8,9] Among the other reviews referring to other photoelectrochemical processes, Roy et al.[3] reviewed inorganic semiconductor materials in photocatalytic

reduction while Morris *et al.*[10] reviewed an organic molecular approach. However, there are few publications providing a systematic summary of carbon dioxide photocatalytic reduction and the updated research progress. In this chapter, we are aiming to offer a comprehensive view and add in the latest research.

In the coming section, we plan to introduce electrochemistry, kinetics, and thermodynamics of carbon dioxide photoreduction reactions as well as show common measuring criteria for both the semiconducting particles system and the organic molecular system under UV and visible spectra. Moreover, using these theories, along with measuring criteria, we can easily judge the feasibility and efficiency of the photocatalytic reduction system. Thus, it will be convenient to select suitable materials for the photocatalytic reduction devices in the future.

10.2 Theories of Carbon Dioxide Conversion

10.2.1 Electrochemistry, Kinetics, and Thermodynamics of Photoreduction Reactions

Considering the serious greenhouse effect along with ready availability, stability, and high energy density of hydrocarbon fuels, it is of vital importance to seek an effective way to convert carbon dioxide into syngas and other carbonaceous fuels. It is reported that Fujishima and Honda[11] first discovered the photoelectrocatalytic reduction of carbon dioxide in aqueous suspensions of semiconductor powders. Ever since then, researchers have been actively looking for alternative methods of recycling carbon dioxide into readily transportable hydrocarbon fuels. With the aid of an artificial photosynthetic system, which is composed of water and carbon dioxide, together with a semiconductor photocatalyst, a photon is absorbed and energy is transferred through the redox process. Among several advantages, the most attractive one is that without generation, separation, and hydrogen storage operations, which appeared in conventional methods, this system allows direct conversion from carbon dioxide to hydrocarbon fuels.

In the case of CO_2 reduction, methanol is the main product and fuel source, along with formic acid as a by-product. Eqn 10.1 illustrates the main reaction.

$$CO_2 + 2H_2O \rightarrow CH_3OH + \frac{3}{2}O_2 \tag{10.1}$$

According to this equation, the overall process requires 713 kJ mol^{-1} energy with 6-electron transfer. Thus, the average energy needed is 119 kJ mol^{-1} per electron, which equals to the energy given out by a 900 nm light, supposing only 70% of entire solar spectrum could be utilized.

With regard to the thermodynamic properties of CO_2 reduction, the whole conversion is mainly dependent on the Gibbs free energy of both CO_2 and its reduced products. According to the Gibbs free energy of each species, shown

Table 10.1 Gibbs free energy of formation for selected chemicals with descending order (from NIST database).

Gibbs free energy of formation $\Delta G°(\text{kJ mol}^{-1})$	Selected chemicals
34.4	$C_{10}H_{22}$
17.3	C_8H_{18}
0.0	N_2
0.0	H_2
0.0	O_2
−16.6	NH_3
−23.5	C_3H_8
−32.9	C_2H_6
−50.7	CH_4
−137.2	CO
−159.2	CH_3OH
−228.4	H_2O
−394.0	CO_2

in Table 10.1, we can easily find out that all the hydrocarbon products possess higher Gibbs free energy than carbon dioxide, which results in a positive total energy for this reaction. Considering CO_2 is a stable molecule in nature, we can also arrive at the conclusion that the process of carbon dioxide reduction is an endothermic reaction, which is unspontaneous. Therefore, all carbon dioxide reductions seem to possess poor feasibility and require an input of energy in a thermodynamic sense, suggesting that a highly efficient and active photocatalyst site, which helps by decreasing activation energy and increasing reaction rate, must be fabricated *via* catalysis chemistry. Nowadays, many research institutes continue to investigate unique properties of various kinds of photocatalysts, including metal complexes, trying to make the process kinetically favorable based on experimental designs.

Solar cell devices are usually manufactured by using combinations of semiconductors including p–n and Schoctky combination of inorganic semiconductors. Due to the perfect photophysical properties of related semiconductor photocatalysts, a photon can be absorbed through illumination and light energy can be transformed into chemical energy as well. The schematic diagram of this redox reaction is illustrated in Figure 10.2. Through light illumination, the semiconductor material allows electron transition from the valence band up to the conduction band. Sufficient energy is absorbed by an electron, which helps that excited electron quickly reach the conduction band and then get to the surface of the semiconductor photocatalyst. Then the electron joins the reduction reaction, which occurs there and turns carbon dioxide, which is dissolved in the solution, into multiple hydrocarbon fuels. At the same time, water is oxidized through a redox period, giving out an electron to the vacant hole in the valence band and turning into oxygen. The whole process seems a little complicated

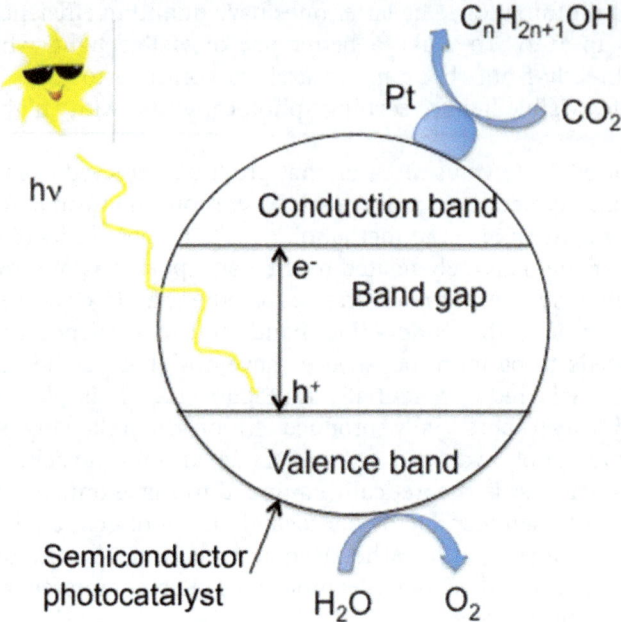

Figure 10.2 Basic diagram of photoreduction of CO_2 on Pt-coated semiconductor photocatalyst.

because of multiple-electron transfer redox coupling with light absorption. Choosing an applicable inorganic semiconducting photocatalyst or organic photocatalyst is essential for the photoreduction process. Due to various intrinsic energy band gaps as well as distinct absorption efficiencies, different wavelengths of light can be absorbed by different photocatalysts. Nowadays, researchers are managing to improve properties of semi-conductors in order to absorb as much solar energy as possible through techniques such as doping, controlling the morphology of the materials, and bundling with photosensitizers.[12–14]

Compared to previous CO_2 reduction examples, this photoreduction pro-cess mainly depends on the photon absorption part. Actually, the multiple-electron transfer reduction of CO_2 occurs simultaneously with the water oxidation process. In general, many steps through the whole reduction reaction determine the overall conversion efficiency from carbon dioxide to the corresponding photocatalytic solar fuel produced. Moreover, conversion efficiency is limited by back redox reactions and electron–hole recombin-ation, along with inefficient light absorption.

In fact, most materials selected as water-splitting photocatalysts are also capable of working as carbon dioxide reduction photocatalysts. The quan-tum efficiencies (which will be defined in Section 10.2.2) of some chosen photocatalysts with distinct absorption wavelengths have been studied well. From the results, it is obvious that photocatalysts under UV light illumin-ation show relatively high quantum efficiencies compared with those under

visible light illumination. The latter ones have quantum efficiencies ranging from 0.1% to 10%. To make a better use of visible light, which can be obtained directly from the Sun, researchers continue to focus on looking for suitable visible-light-absorbing photocatalysts with high quantum efficiency.

It is reported in previous research that products achieved from the carbon dioxide photoreduction process contain carbon monoxide, formic acid, formaldehyde, methane, and methanol.[11,15,16] The possibility of obtaining a specific solar fuel is largely related to the band positions of semiconductor materials and the relevant reaction's redox potential. In order to get certain hydrocarbon fuels, the conduction band of the semiconducting photocatalysts needs to be more negative or have a higher position than any of those solar fuels redox potentials. In Figure 10.3 it is clear that a SiC photocatalyst can potentially produce all of the following solar fuels: methane, methanol, hydrogen, formaldehyde, and formic acid. Meanwhile, a ZnO photocatalyst is theoretically equipped to make only methane.

In addition to semiconductor photocatalysts, photoreduction conversion from carbon dioxide into carbonaceous fuels can also in principle be achieved with the aid of organic molecules. For commercialization, this latter route is more convenient.

Four kinds of molecular photocatalysts are mainly investigated in the literature, and they are:

(1) metalloporphyrins or metallomacrocycles
(2) $Re(CO)_3(bpy)_x$-based compounds (bpy = 2,2'bipyrindine)
(3) metal tetra-aza macrocyclic complexes
(4) supramolecular complexes.

Many studies have been investigating the common structures of metalloporphyrins and metallomacrocycles, which include several subsidiary structures – metalloporphyrin (MP), metallocorrin (MN), metallophthalocyanine (MPc), and metallocorrole (MC). Their molecular structure can be found in Figure 10.4.

Some early molecular chemistry studies provided many important insights. Eyring et al.[17] studied the kinetics of the reduction of CO_2 to formic acid in depth. Their successful experiment in 1969 used a mercury drop electrode and obtained an almost 100% conversion efficiency in solution at pH 6.7. After that, several other research groups made attempts to reproduce this technology. Erying combined the mercury drop electrode with the lithium bicarbonate-supporting electrolyte. Other studies focused on homogeneous transitional metals that catalyze the reduction reaction. In 1975, Aresta and colleagues published a discovery of the crystal structure of CO_2 binding to a transitional metal complex through a mode of η^2-binding.[18] Meanwhile, Darensbourg et al. contributed an important finding that anionic group 6B metal hydrides enabled the formation of the metal formates with sufficient CO_2 in the surrounding system.[19]

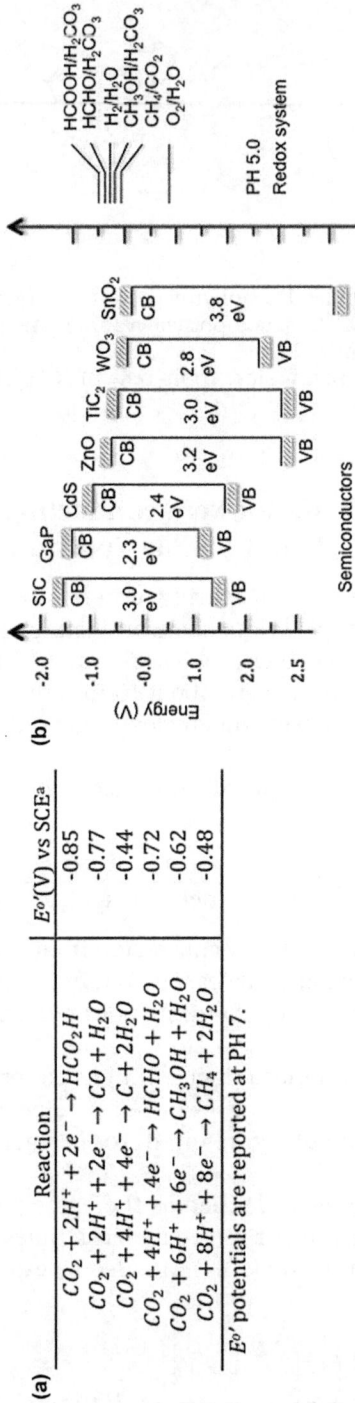

(a)

Reaction	$E^{o'}(V)$ vs SCE^a
$CO_2 + 2H^+ + 2e^- \rightarrow HCO_2H$	-0.85
$CO_2 + 2H^+ + 2e^- \rightarrow CO + H_2O$	-0.77
$CO_2 + 4H^+ + 4e^- \rightarrow C + 2H_2O$	-0.44
$CO_2 + 4H^+ + 4e^- \rightarrow HCHO + H_2O$	-0.72
$CO_2 + 6H^+ + 6e^- \rightarrow CH_3OH + H_2O$	-0.62
$CO_2 + 8H^+ + 8e^- \rightarrow CH_4 + 2H_2O$	-0.48

$E^{o'}$ potentials are reported at PH 7.

Figure 10.3 (a) Various solar fuels and their redox potentials. (b) Plot of redox potentials of solar fuels and band positions of semiconductor materials.
Reprinted with permission from ref. 3 and 10. Copyright 2009 and 2010 American Chemical Society.

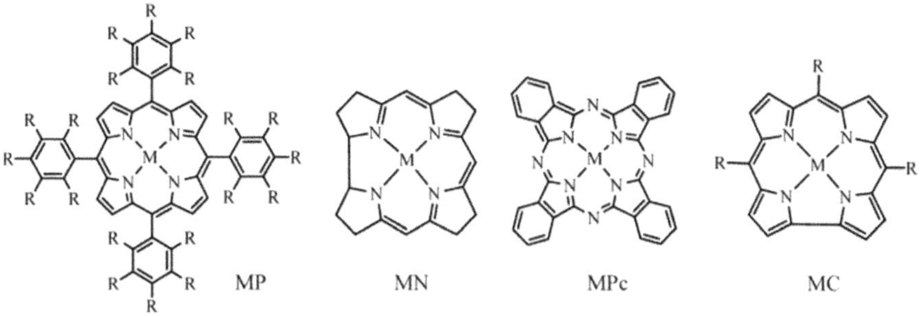

Figure 10.4   Four classes of metal porphyrin derivatives: metalloporphyrin (MP), metallocorrin (MN), metallophthalocyanine (MPc), and metallocorrole (from left to right).
Reprinted with permission from ref. 10. Copyright 2009 American Chemical Society.

10.2.2   Measuring Criteria for Semiconducting Particles System under UV and Visible Spectrum

There are large quantities of measuring criteria for the properties of photosensitivity devices based on the semiconducting particles system, including the two most popular ones, which are applied frequently in both related research and industry. We are about to introduce details of both of these two criteria, optional conversion efficiency and quantum efficiency, in the following paragraphs.

The definition equation of optical conversion efficiency is shown in eqn (10.2).

$$\text{optical conversion efficiency } (\eta) = \frac{\text{output power}}{\text{incident light power}} \times 100\% \qquad (10.2)$$

This equation indicates total energy conversion from light energy to stored chemical energy. In a particular study carried out by Nozik,[7] this equation is rewritten with the numerator as chemical power minus parasitic power (output power) and the denominator as light intensity on the semiconductor–electrolyte interface (incident light power). 'Parasitic power' refers to the generated fuels coming from non-photoenergy, consisting of external bias, which is regularly referred to in many studies as the overpotential.

Meanwhile, chemical power is defined as the enthalpy of combustion of generated fuels (normalized to per molecule basis times the rate of photogenerated fuels). In this case, eqn (10.2) can also be expressed as shown in eqn (10.3).

$$\eta = \frac{i\left[\xi\left(\dfrac{\Delta H}{n'}\right) - V_{\text{B}}\right]}{h\nu} \times 100\% \qquad (10.3)$$

In this formula, '*i*' stands for the current at potential 'V_B'; 'ξ' stands for faradic efficiency; 'ΔH' stands for the enthalpy of fuel combustion; '*n*' stands for the number of electrons required to oxidize the fuel back to carbon dioxide and water with complete combustion; 'V_B' stands for the external bias applied to the system; '*h*' stands for the Planck constant; and '*v*' stands for frequency. This is another formula widely used in publications, mainly for demonstrating the properties of the semiconducting particles system.

Eqn 10.4 is the definition of quantum efficiency.

$$\text{quantum efficiency}\,(\Phi) = \frac{\text{converted electrons}}{\text{incident photons}} \times 100\% \qquad (10.4)$$

Quantum efficiency is a measure of a device's electrical sensitivity to light. To make it clear, we select the process from carbon dioxide to hydrocarbons as an example. For the product methanol, the quantum efficiency equals to six (number of electrons transferred from CO_2 to methanol) multiplied by the mole number of CO_2 divided by the whole number of incident photons. As a result, quantum efficiency varies with different products due to their distinct electron-transfer values.

10.2.3 Measuring Criteria for Molecular Photocatalyst System under UV and Visible Spectrum

There are several measurements related to the molecular photocatalytic system. Here, we select two methods, which are in common use. The first one is called catalytic selectivity, which is defined as the molar ratio of the products obtained from the reduction of CO_2 to H_2. Since the formation of hydrogen is thermodynamically favorable, it is an inevitable product obtained during the proton reduction process. Therefore, the value of catalytic selectivity (CS) offers a quantitative measurement of the capability of the molecular photocatalyst to convert CO_2 into the reduced hydrocarbon products rather than hydrogen efficiently. One of the goals in the research is to increase the catalytic selectivity value.

$$\text{catalytic selectivity}\,(\text{CS}) = \frac{CO_2 \text{ reduced products}}{H_2} \qquad (10.5)$$

The second one is called turnover number (TON), which is defined as the number of reduction cycles that takes place per catalyst over the lifetime of the whole reaction. TON is usually obtained by calculating the amount of CO_2 reduced products per catalyst. Although theoretically infinite, TON is approximately 1 000 000 in industrial productions owing to catalyst materials degradation. In research results to be discussed in later sections, TON varies from 1 to 500. The reason why TON is relatively low in most publications is

that illumination time on the photocatalysts is shorter than the lifetime of that material due to limited experimental periods.

$$\text{turnover number} = \frac{CO_2 \text{ reduced products}}{\text{catalyst}} \qquad (10.6)$$

10.3 Inorganic Semiconducting Materials for Photocatalytic Reduction of Carbon Dioxide

A variety of semiconducting materials have been investigated and utilized as novel photocatalysts. These systems exploit either UV light illumination or visible light illumination. According to the light source they use, these materials can be subdivided into two groups as follows:

(1) photocatalytic materials for UV irradiation
(2) photocatalytic materials for visible light.

In general, there are two main methods of carrying out carbon dioxide reduction. The related approaches are electrochemical reduction and photocatalytic reduction, respectively. The former approach has a longer history than the latter. For instance, Royer's research group, which was among the earliest pioneers in electrochemical reduction in 1870, verified the feasibility of turning carbon dioxide to formic acid *via* injecting carbon dioxide into aqueous sodium bicarbonate solution charged with electricity.[20] Udupa *et al.* confirmed that bicarbonate ions in aqueous media are of vital importance to the carbon dioxide conversion process, and a decreasing current density passed on the surface of amalgamated copper electrodes could cause the increased faradaic yield.[21] Later on, other researchers carried out further studies on various reaction parameters together with diverse reaction conditions, ranging from metal electrode species, electrolyte compositions, pH value, and pressure to temperature.[2] Paik *et al.* suggested that the mechanism of the carbon dioxide reduction process depended heavily on the pH value.[22] Ito *et al.* tried various metal electrodes, such as Sn, Pb, and Zn, and discovered specific temperature and pressure conditions to optimize the yields of several carbonaceous products, formic acid included.[23,24] The theory of carbon dioxide reduction has been studied thoroughly. However, one of the problems is that in order to get electricity for the process, large quantities of fossil fuels have to be burned, which lifts energy consumption and decreases the efficiency of the cell system. Thus, we should pay more attention to the photocatalytic reduction of carbon dioxide since it is more efficient and feasible compared with the electrochemical reduction of carbon dioxide.

In 1962, Getoff *et al.*[25] demonstrated the first photocatalytic mechanism. In this research, aqueous ferrous salts were used as photocatalysts, onto which UV light was shone directly. The main products detected in this

experiment contain formic acid and formaldehyde. Motivated by the success achieved by Getoff, Halmann *et al.* published an article based on the photoelectrochemical method of reducing carbon dioxide in a similar way in 1978.[26] P-type gallium phosphide (p-GaP with a band gap of 2.25 eV) is one of the inorganic semiconducting materials first used in a carbon dioxide reduction process. It is a type of electron-deficient photocatalyst in contrast to non-doped materials. A Zn-doped single crystal semiconductor acted as photo-electrode, while a carbon rod acted as counter-electrode. Soon afterward, p-type silicon (p-Si) was also tested using the same experimental facilities. Both p-GaP and p-Si were tested under illumination and in darkness.[20] A cathodic bias of -1.0 V (SCE) was applied to electrochemical cells. Under dark circumstances, the current density arrived at a value of 0.1 mA cm^{-2}. Under illumination of UV light, the current density went up to 6 mA cm^{-2} at the start of the experiment. The current density then was observed to drop within the first 24 hours and eventually reached a plateau at 1 mA cm^{-2} after that period of decline. Two distinct UV light wavelengths were selected for comparison. Under the illumination of a 365 nm UV light wavelength, the maximum optical conversions of methanol and formaldehyde were 3.6% and 5.6%, respectively, with the aid of a bias -0.8 V voltage. At near UV wavelength (~ 315 nm) and blue region wavelength (~ 510 nm) solar irradiance,[27] the maximum optical conversions of methanol and formaldehyde were 0.61% and 0.97%, respectively. Eqn 10.7 and 10.8 show the mechanism explained by the author, which seems similar to the carbon dioxide electrochemical reduction. The most obvious distinction between the two mechanisms is that the latter one includes a termination at the formic acid formation step, while the former one can further reduce formic acid into methanol and formaldehyde, which displays more convenient features for application as hydrocarbon fuels.

$$CO_2 + e^- \rightarrow {}^{\bullet}CO_2^- \overset{H+}{\Leftrightarrow} {}^{\bullet}HCO_2 \qquad (10.7)$$

$$HCO_2 + e^- \rightarrow HCO_2^- \overset{H+}{\Leftrightarrow} HCO_2H \qquad (10.8)$$

In 1972, Fujishima's research group[28] successfully confirmed the possibility of converting solar energy to chemical energy through water decomposition into hydrogen and oxygen by the use of specific light-sensitive semiconductors with photocatalytic properties. Based on the mechanism of this photocatalytic reaction, ammonia and hydrazine were also yielded as by-products. Not satisfied with the existing achievement, Inoue *et al.* (members of Fujishima's group) followed up to investigate multiple photocatalysts species including TiO_2, SiC, ZnO, CdS, GaP, and WO_3.[11] Inoue aimed to generate carbonaceous fuels besides hydrogen and oxygen with a continuous input of CO_2. During the experiment, methane, methyl alcohol, formic acid, and formaldehyde are obtained as products from saturated carbon dioxide solution with a mercury arc lamp and Xe

lamp illuminating all the time. Inspired by the experimental results, the researchers presented a mechanism of this reduction reactions (depicted in eqn (10.9)–(10.13)) and pointed out the connection between redox potentials and semiconductor materials' band positions:

$$H_2O + 2p_{val}^+ \rightarrow \frac{1}{2}O_2 + 2H^+ \tag{10.9}$$

$$CO_2(aq) + 2H^+ + 2e_{cond}^- \rightarrow HCOOH \tag{10.10}$$

$$HCOOH + 2H^+ + 2e_{cond}^- \rightarrow HCHO + H_2O \tag{10.11}$$

$$HCHO + 2H^+ + 2e_{cond}^- \rightarrow CH_3OH \tag{10.12}$$

$$CH_3OH + 2H^+ + 2e_{cond}^- \rightarrow CH_4 + H_2O \tag{10.13}$$

In the equations above, H^+ represents a hole in the valence band and e^- represents an electron in the conduction band. The results obtained showed that using SiC photocatalysts could lead to a relatively high conversion into methyl alcohol and formaldehyde. Comparing the relative position of SiC conduction together with chemical potential levels of formaldehyde and methanol, one can easily explain this phenomenon. Inoue *et al.* had studied various theoretical band positions of different photocatalytic materials. In contrast with other photocatalysts' conduction potentials, the level of the SiC conduction band is more negative. In other words, it possesses a higher position than that of the formaldehyde redox potential. Furthermore, we can observe high yields of products such as methyl alcohol, formic acid, and formaldehyde in the experiment. In theory, because the WO_3 conduction band is much lower than all hydrocarbon fuels' redox potentials, nothing would be produced as a result. The absence of carbonaceous products with WO_3 as photocatalysts in the experiment again proved the credibility of the theory.

Monnier *et al.* followed Inoue's steps and particularly focused on the consequences of using TiO_2 particles with 1% Ru coating (wt.%) as photo-electrodes. In these series of experiments, researchers were able to detect methanol and methane as the main reducing products. In addition, a platinum amalgam electrode may generate hydrogen as a by-product.[29,30]

Taniguchi *et al.* (1982) investigated lithium carbonate electrolytes with saturated CO_2 utilizing p-GaP semiconducting material as photocatalyst products.[15] Similar to what other researchers had found previously, they were confronted with a challenge on how to improve the productive rates because of the existence of hydrogen, which competed with hydrocarbon products. Later they came up with the idea of adding various ethers into the electrolyte. This method came into effect and increased the current

efficiency together with gaining three major carbonaceous fuels – methanol, formaldehyde, and formic acid. For example, with an injection of 15-crown-5 ether, the current efficiencies of those products increased from 1% to 15% for formic acid, from 0.5% to 4% for formaldehyde, and from 6.5% to 44% for methanol.

Furthermore, Aurian-Blajeni, Halmann, *et al.* made some adjustments to their previous study on GaP and developed two kinds of photocatalysts, namely single-crystal Zn-doped p-GaP and p-GaAs. These photocatalysts were fixed on a GaIn substrate as an electrode.[31] In the process of characterizing the *I-V* curve, they found that p-GaAs possessed better properties with a faradaic yield of 39.3% in 0.1 M Li_2CO_3 media, while p-GaP only arrived at a faradaic yield of 12.6%. During the experiment, they found out that counter-electrodes also influenced the faradaic yield as well as the product generation rate. Bright platinum was discovered to be the best anode, producing 71 μmol formic acid per reaction and reaching a faradaic efficiency of 26.8%. Compared with the commonly used photocatalyst materials such as carbon rod and glassy carbon anode, bright platinum displayed superiority in the aspect of producing formic acid and arriving at over ten times the faradaic efficiency. Blajeni also reported that p-GaAs could increase the selectivity of formic acid compared with p-GaP, while for p-GaP photocatalyst, a variety of electrolytes had been tested to optimize this kind of photocatalytic material. From this came the result that 0.5 M Na_2CO_3 performed better, with 80.3% yield over the second one, not to mention $HClO_4$ and $NaHCO_3$ electrolytes.

Several researchers investigated other gallium composite electrodes and took a close examination of them. Canfield *et al.* chose p-GaAs and p-InP as a pair of electrodes for a full photoelectrochemical system to reduce carbon dioxide into methanol and formaldehyde.[32] In this closed system, CO_2 was injected into the Na_2SO_4 electrolyte until saturation. A trivial amount of methane was perceived under an electrode potential of -1.2 V. The total conversion efficiency with p-GaAs and p-InP paired electrodes reached a maximum of 0.8%.

It is known that finding the photocatalytic selectivity of diverse products is one of the most difficult challenges to overcome for researchers because all the products' redox potentials are similar to each other. Namely, the redox potentials of formic acid, formaldehyde, methane, acetaldehyde, methanol, ethene, carbon monoxide, and ethanol are extremely close in value. A large number of experimental approaches have been thought of so as to overcome kinetic and thermodynamic obstacles and favorably enhance the selectivity of a particular product from the reaction. Furthermore, a mild reaction environment will attract more attention. Thampi and Gratzel set a good example for younger generations of researchers since they managed to design a photocatalytic reaction under mild conditions and with an improved selectivity of the desired product.[33] Although producing methane from carbon dioxide reduction was feasible by comparing the relative band position of photocatalysts with the redox potential of methane, the

eight-electron transfer taking place during the reduction from CO_2 to CH_4 was kinetically unfavorable because of the conduction band stayed higher than the redox potentials of all hydrocarbon products. This most likely causes the formation of many hydrocarbon by-products.

Meanwhile, Cook *et al.* introduced a novel approach of reducing carbon dioxide in 1988, which helped the CO_2 photoelectrochemical reduction process to methane, ethane, and ethene to be optimized. The system made use of a suspended mixture of p-SiC and Cu particles as photocatalytic materials and added a mercury UV lamp for illumination.[34] The two particles were inserted into a CO_2-saturated K_2CO_3 electrolyte separately. The purpose of involving copper particles in the electrolyte was to increase methane's yield since the reduction from CO_2 to CH_4 occurs on the surface of copper particles. The whole process was limited by the rate of electron transfer to the saturated CO_2 on the surface of the copper particles. A variety of pH values were also investigated to optimize the yield of desired products. The result showed CH_4 yield was optimized and arrived at 14.9 $\mu L\,g^{-1}\,h^{-1}$ at a pH value of 5. In the same solution, C_2H_4 reached 3.9 $\mu L\,g^{-1}\,h^{-1}$, and C_2H_6 reached 2.3 $\mu L\,g^{-1}\,h^{-1}$. It was demonstrated that a low pH value always resulted in the formation of SiO_2 on the surface of SiC. At the same time, hydrogen was produced instead of hydrocarbon fuels.

In addition to several popular choices of photocatalysts including SiC, TiO_2, $SrTiO_3$, $K_4Nb_6O_{17}$, $Na_2Ti_6O_{13}$ as well as $BaTi_4O_9$, researchers had also investigated other photocatalytic materials. Sayama *et al.* provided a unique method of carbon dioxide reduction to carbon monoxide using ZrO_2 as photocatalyst, while H_2 and O_2 were also generated as by-products[35] in $NaHCO_3$ electrolyte. The photocatalyst was coated with 1% (wt.%) copper and RuO_2. Using different characterization facilities including flame ionization detector (FID), gas chromatography (GC), and thermal conductivity detector (TCD), the output gases and aqueous hydrocarbons were analyzed and identified. Finally, methanol and formic acid were observed to be the major products. Through comparing the results of ZrO_2 photocatalysts with copper-coated ZrO_2 photocatalysts, researchers found that the ZrO_2 photocatalysts performed better in the production rate of H_2, O_2, and CO. Moreover, the rates of production of H_2, O_2 together with CO were faster in Na_2CO_3, $NaHCO_3$, and $KHCO_3$ electrolyte solutions than those of pure water. The optimal result in $NaHCO_3$ solution was producing 309 $\mu mol\,g^{-1}\,h^{-1}$ of H_2, 167 $\mu mol\,g^{-1}\,h^{-1}$ of O_2, and 3 $\mu mol\,g^{-1}\,h^{-1}$ of CO.

Copper-loaded catalysts except those mentioned above have been under investigation for many years. Adachi *et al.* selected 0.5% copper loaded (wt.%) commercial anatase TiO_2 as photocatalysts, which led to the production of many more types of carbonaceous materials other than formaldehyde, formic acid, and methanol. This had never been reported before, which meant it had great research value.[36] Those particles were placed under a Xe lamp for illumination for 48 hours. Changing illumination time together with the weight of loaded copper, the optimal production yield could be found. The output materials contained

$0.46\ \mu L\,g^{-1}\,h^{-1}$ methane, $0.056\ \mu L\,g^{-1}\,h^{-1}$ ethane, and $0.32\ \mu L\,g^{-1}\,h^{-1}$ of ethene. Compared with previous publications, Adachi proposed a slightly different reaction mechanism to explain the production of additional carbonaceous fuels, which is depicted in eqn (10.14) to (10.17).

$$TiO_2 + hv \rightarrow TiO_2^*(e_{cond}^- + h_{val}^+) \tag{10.14}$$

$$H_2O \xrightarrow[TiO_2,Cu]{hv} 2H^+ + \frac{1}{2}O_2 \tag{10.15}$$

$$CO_2 \xrightarrow[TiO_2]{hv} CO \xrightarrow[TiO_2]{hv} {}^\bullet C \xrightarrow[TiO_2,Cu]{2H^++2e^-} {}^\bullet CH_2 \xrightarrow[TiO_2,Cu]{2H^++2e^-} CH_4 \tag{10.16}$$

$${}^\bullet CH_2 \xrightarrow[Cu]{{}^\bullet CH_2} CH_4 \tag{10.17}$$

Similarly, Ishitani *et al.* also published an article on TiO_2 with copper particle co-photocatalysts.[27] At the same time, Ishitani replaced Cu-loaded TiO_2 with other metal-loaded TiO_2, seeking for the best photocatalyst to optimize the production rates of methane and acetic acid. Finally, Pd-loaded TiO_2 stood out as the highest methane-yielding photocatalyst, which produced 10 times as much methane as that produced by Cu-loaded TiO_2 after an illumination of 5 hours.

Anpo *et al.* changed the structure of traditional photocatalyst Pt-TiO_2 to an innovated macrostructure.[37] Instead of a dispersed particle system that appeared in many previous studies, Anpo inserted the photocatalysts into zeolite cavity spaces prepared through ion exchange or *via* the impregnating method. Since the efficiency of the photocatalytic reduction reaction relied heavily on the structure of photocatalysts, this porous complex system's structure possessed several advantages, such as large pore size, unique internal surface topology, and superior ion-exchange capacities, which in turn helped lifting the efficiency of the whole reaction. This experiment was carried out at $55\,^\circ C$ under $280\,nm$ wavelength UV light illumination. Through detecting the output gases, they found that the major products were methane and methanol, and the trivial by-products were carbon monoxide, ethane, and ethene. The author treated the TiO_2 photocatalyst in three different ways. The first method was loaded with Pt *via* an ion-exchange method, the second method was loaded with Pt *via* impregnation method, and the third method was treated as a blank control group. Through investigation, the first treatment turned out to be the most efficient one because it had the highest methane production rate. They also aimed at utilizing the zeolite Pt-TiO_2 system to enhance the selectivity of methanol in the future.

Two major approaches had been used frequently for the study of carbon dioxide photoreduction reactions before 2002. The first approach was to use heterogeneous photoelectrochemistry involving photocatalyst particles

dispersed in liquid solutions. The most commonly used materials were SiC, CdS, and TiO_2. According to the mechanism, CO_2 absorption and reduction took place on the liquid–solid interface, thus the photocatalyst species and their morphologies were major parameters modified to optimize the photoreaction. According to the second approach, photocatalysts, like TiO_2 and so on, were encapsulated into microporous silicates.[37,38] This approach aimed at modifying the pores' macrostructure. Through increasing the complexity of the macrostructure system, the photocatalyst's packing density would increase, which in turn improved the photoreduction conversion efficiency. At the same time, the photoreduction selectivity and reaction rate increased significantly compared with the particle system.

Ikeue et al.[39] designed two kinds of Ti-containing porous silica with a similar approach to utilize microporous silicates. At $50\,^\circ C$, these photocatalysts, which possess hexagonal and cubic pore structures, were incorporated into thin films in water to reduce carbon dioxide. After an 8-hour period of illumination under UV light, methane and methanol were detected among the output gases as the main products formed. The Brunauer–Emmett–Teller (BET) surface area of that silica thin film was calculated to be $900\ m^2\,g^{-1}$. Ikeue illustrated that the catalytic activity of porous silica photocatalysts increased by two-fold when compared with previous research results. The production rate of methane was $7\ \mu mol\,g^{-1}\,h^{-1}$, while that of methanol was only $2\ \mu mol\,g^{-1}\,h^{-1}$. The quantum yield of that photocatalytic reduction reaction was 0.28%. Also, carbon nanotubes have stayed popular since their creation and drew more and more attention from many research areas. Researchers managed to incorporate them into photocatalytic materials for carbon dioxide reduction and also figured out the mechanism behind the reaction. Xia et al. managed to use a multi-wall carbon nanotube (MWCNT)-supported TiO_2 system to photo-reduce carbon dioxide.[40] This composite photocatalyst could be prepared by both sol–gel methods and hydrothermal methods. In the former method, anatase TiO_2 nanoparticles were coated on the MWCNT; while in the latter method, rutile TiO_2 nanorods were precipitated on the surface of the MWCNT. The main product gained with the sol–gel method was ethanol, while the hydrothermal method selectively formed formic acid. After anatase, TiO_2 was successfully attached onto the MWCNT, the most optimal BET surface area of photocatalyst prepared by the sol–gel method was $168\ m^2\,g^{-1}$. After illuminating for 5 hours, the products produced in the experiment were $58.7\ \mu mol\,g^{-1}$ methane, $93.35\ \mu mol\,g^{-1}$ formic acid, as well as $149.36\ \mu mol\,g^{-1}$ ethanol. The author proposed a slightly different mechanism (shown in eqn (10.18)–(10.24)) and pointed out that each product was independently produced from CO_2.

$$TiO_2 + h\nu \rightarrow e^- + h^+ \tag{10.18}$$

$$h^+ + e^- \rightarrow H^\bullet \tag{10.19}$$

$$2H_2O + 4h^+ \rightarrow O_2 + 4H^+ \qquad (10.20)$$

$$CO_2 + e^- \rightarrow {}^\bullet CO_2{}^- \qquad (10.21)$$

methane formation: $\;{}^\bullet CO_2{}^- + 8H^\bullet + h^+ \rightarrow CH_4 + 2H_2O \qquad (10.22)$

formic acid formation: $\;{}^\bullet CO_2{}^- + 2H^\bullet + h^+ \rightarrow HCOOH \qquad (10.23)$

ethanol formation: $\;2{}^\bullet CO_2{}^- + 12H^\bullet + 2h^+ \rightarrow C_2H_5OH + 3H_2O \quad (10.24)$

Recently, Barton *et al.* went back to study p-type GaP photocatalytic materials, which first appeared in early research.[41] Previous studies always indicated good yield as well as high selectivity. However, it was of vital importance to apply a higher overpotential to the system as well. What's more, although several extensive studies included reduction from carbon dioxide to carbon monoxide and formic acid, the process from carbon dioxide to methanol was rarely carried out successfully. Without extra energy input, Barton obtained almost 100% faradaic efficiency for carbon dioxide reduction. Using the chromo tropic acid method to detect the products, little formaldehyde and formic acid could be found.

10.4 General Research Guidance for Molecular Catalysts of Photocatalytic Reduction of Carbon Dioxide

In addition to the semiconducting catalysts, many studies with molecular complexes demonstrate the possibilities of the photoreduction of CO_2 to solar fuels. Grodkowski and co-workers[1] used the photochemical method to reduce several iron porphyrins in both aqueous and organic solution from Fe^{3+} porphyrins (Fe^{III}-P) to Fe^{2+} porphyrins (Fe^{II}-P) to Fe^+ porphyrins (Fe^I-P). Several iron-center porphyrins as electrocatalysts were identified as the most efficient catalysts at that time for the reduction process of CO_2 to CO. The research found that the large and rigid structure of porphyrin was the reason for improvement, and the structure destabilized the intermediate complex and improved the reduction rate. The reduction mechanism is generally divided into two steps:

(1) from Fe^{III}-P to Fe^{II}-P and
(2) from Fe^{II}-P to Fe^I-P.

The overall conversion efficiency relies on the individual efficiency, which is often measured by the turnover number (TON). The starting photocatalytic molecule was chosen to be $ClFe^{III}$-tetraphenylporphyrin (TPP) dissolved in

N,N-dimethylformamide (DMF) containing 5% tetrahydrofuran (THF), and the reduction was filtered a wavelength cutoff at 325 nm. The mechanism was proposed as follows. The first step's quantum efficiency was 0.016 in the aqueous solution and increased to 0.05 in the organic solution (5% TEA). In the second step, the quantum efficiency was as low as 0.01. This was due to Fe^I-TPP existing in the process for a very short amount of time, which was caused by the rapid conversion to $(CO)Fe^{II}$-TPP from porphyrins. This phenomenon indicated the accumulation of CO from the reduction of CO_2. At a concentration of 1×10^{-5} mol L^{-1} of Fe^{III}-TPP, the turnover number of CO per porphyrin reached a maximum value of 70 because of the destruction of the porphyrin macrocycle. In addition, the research found that adding 0.5–1% water could effectively improve the rate of production of CO from CO_2.

$$Et_3N(Cl)Fe^{3+}P \xrightarrow{h\nu} Et_3N^{\bullet+} + Cl^- + Fe^{2+}P \tag{10.25}$$

$$Et_3N + Et_3N^{\bullet+} \xrightarrow{h\nu} Et_3NH^+ + Et_2NCHCH_3 \tag{10.26}$$

$$Et_2NCHCH_3 + ClFe^{3+} \rightarrow Et_2N^+{=}CHCH_3 + Fe^{2+}P + Cl^- \tag{10.27}$$

$$Et_3NFe^{2+}P \xrightarrow{h\nu} Et_3N^{\bullet+} + (Fe^-P)^- \tag{10.28}$$

Moving away from the porphyrin catalysts, Grodkowski worked to discover a different molecular catalyst in 2000.[42] In this research, he found cobalt corrin photocatalysts could be effective in reducing CO_2. Corrins are more efficient catalysts because of their higher stability due to a more saturated structure compared to porphyrins. A photoreduction experiment was carried out in acetonitrile/methanol solutions, while using *p*-terphenyl as a photosensitizer and triethylamine as a reductive quencher. Photolysis was performed under illumination with a wavelength greater than 300 nm in the solution. Carbon monoxide, formic acid, and hydrogen were detected as a result. The reaction mechanism was discovered to be two steps, shown in the following:

(1) reduction of alkyl halides occurred with the formation of Co^+ complexes, and
(2) reduction of CO_2 happened after Co^+ was further reduced.

Among various types of corrins, cobinamide was shown to produce the highest amount of carbon monoxide at a rate of 0.48 mmol $L^{-1} h^{-1}$ and H_2 at 0.30 mmol $L^{-1} h^{-1}$. Furthermore, cyanocobalamin was the most effective catalyst to produce a large amount of H_2 at a rate of 0.46 mmol $L^{-1} h^{-1}$ as well as 0.35 mmol $L^{-1} h^{-1}$ of CO_2. Similarly to previous research findings, the reaction mechanism was proposed to explain how Co^{3+} reduced to Co^{2+} and

eventually to Co^+; the following steps of the reaction mechanism were proposed to explain how H_2 and CO were generated from CO_2 in this experiment.

$$Co^0B_{12}^{2-} + CO_2 \rightarrow (CO_2CoB_{12})^{2-} \tag{10.29}$$

$$(CO_2CoB_{12})^{2-} + H^+ \rightarrow Co^{II}B_{12} + CO + OH^- \tag{10.30}$$

$$Co^0B_{12}^{2-} + H^+ \rightarrow (HCoB_{12})^- \tag{10.31}$$

Although metalloporphyrins and other macrocycle molecules are excellent light absorbers with large extinction coefficients in the visible light region, they have the tendency to get degraded after light absorption. Typically they can only stay active for picoseconds. Moreover, because of the side reaction of hydrogenation of the macrocycle ring, the catalytic reaction prefers the reduction of protons rather than the reduction of CO_2. The typical TON reported in the literature ranges from 40 to 300.

Lehn *et al.* in 1981 explored the possibility of using a complex $Ru(2,2'\text{-bipyrindine})_3^{2+}$ in concert with $CoCl_2$ to reduce CO_2 under visible light.[43] The electrochemical reduction under a photo-induced system took place in a solution of acetonitrile, triethylamine, and water solution, and the reaction simultaneously generated CO and H_2 after reducing CO_2 and water. $CoCl_2$ was an efficient photocatalyst especially to produce carbon monoxide selectively. Both $Ru(2,2'\text{bipyrindine})_3^{2+}$ as a photosensitizer and triethanolamine as an electron donor facilitated the visible light reduction of CO_2. In preparation for the experiment, 0.43 mM $Ru(bipy)_3Cl_2$ was mixed with 1.47 mM $CoCl_2$ in 30 mL of acetonitrile, and this solution was continuously illuminated for 22 hours by a cutoff lens of 400 nm in visible light range. As a result, tripropylamine as an effective donor provided the highest yields of 2.26 mL CO and 2.66 mL H_2. Triethanolamine demonstrated an excellent selectivity toward CO production (2.93 mL) with only 0.12 mL of H_2 produced. The author pointed out that the mechanism of CO formation was under investigation. Despite this unclear mechanism, the evidence was clear that CO formation was not due to the reduction of CO_2 by H_2.

Hawecker *et al.*[44] followed up the previous work and continued with the same photocatalyst and photosensitizer $Ru(2,2'\text{-bipyrindine})_3^{2+}$ with Co^{2+}. Compared to previous results, their modifications of the experiment allowed them to achieve a significant improvement onto the production of syngas with the addition of triethanolamine and dimethylformamide (DMF) under 2-hour illumination, as the electrochemical system was able to produce 2 mL of CO and 4.5 mL of H_2. Meanwhile, when the irradiation was prolonged for 15 hours, the system could produce 8 mL of CO and 19 mL of H_2. Hence, the experiment proved the production of syngas grew linearly with time of illumination. This research again confirmed the high selectivity in the reduction of CO_2 by using Co^{2+} with rhenium complex. In addition, the authors experimented with a rhenium molecular complex (Re-) as both photocatalyst and photosensitizer as a homogenous catalyst. Rhenium

showed higher conversion efficiency than the ruthenium complex. No CO was detected when rhenium was absent, and H_2 formation by rhenium was independent of the CO_2 input. The Re-complex has the highest selectivity and quantum efficiency compared to all other molecular photocatalysts. However, the turnover number is low due to the molecular level degradation. In addition, the biggest disadvantage of Re-complex is that it only absorbs light at wavelengths shorter than 400 nm (UV region).

Craig *et al.* chose to work on various kinds of Ni^{2+} complexes [Ni(14-aneN$_4$)]$^{2+}$ or [Ni(12-aneN$_4$)]$^{2+}$ (14-aneN$_4$ = 1,4,8,11-tetraazacyclotetradecane, 12-aneN$_4$ = 1,4,7,10-tetraazacyclododecane), which were sensitized by Ru(bpy)$_3$$^{2+}$ as photocatalysts.[45] The products of the photo-induced reduction were formate and CO, and the products were verified by the labeled C14-label carbon. After Sauvage's discovery of electroreduction by a Ni complex[46] previously, Craig demonstrated a bigger potential of photochemical reduction by the Ni^{2+} complex and further extended his work with additional Ni complexes coupled with photosensitizers. The reduction reaction was operated under 440 nm monochromatic irradiation with ascorbate serving as a pH buffer (pH = 5) and sacrificial reductive quencher. A limited amount of H_2 production occurred with the presence of Ni complex. The quantum yield of CO produced from the reaction was 4.5×10^{-4}, and that of formate was 1.9×10^{-4}.

Different catalyst complexes were proposed by Fujita[47] with a Ni catalyst, including:

(1) RRSS-NiHTIM(ClO$_4$)$_2$(ClO$_4$)$_2$(HTIM1/42,3,9,10-tetramethyl-1,4,8,11-tetra-azacyclotetradecane)
(2) RSSR-NiHTIM(ClO$_4$)$_2$(ClO$_4$)$_2$ (HTIM = 2,3,9,10-tetramethyl-1,4,8,11-tetra-azacyclotetradecane)
(3) NiDMC(ClO$_4$)$_2$ (DMC = C-*meso*-5,12-dimethyl-1,4,8,11-tetraazacyclotetra-decane).

They were shown in the research to be effective and selective in catalytic activities. The structures of those complexes are shown in Figure 10.5. Among them, cyclam, which was extensively studied in many studies, was inferior to the RRSS-NiHTIM and NiDMC photocatalysts chosen in Fujita's experiment. The latter two showed larger catalytic currents and more positive potentials measure by cyclic voltammetry. After 4 hours of continuous illumination, the amount of CO collected were 2.1×10^{-4} mol by RRSS-NiHTIM, 2.0×10^{-4} mol by NiDMC, and 1.5×10^{-4} mol by Ni(cyclam), while only marginal amounts of H_2 and formate were found by each Ni complex.

Various complex structures were used for the cobalt (Co)-based photo-catalysts. This group of complexes is shown in Figure 10.6 in eight complex structures.[10] The average catalyst selectivity number was about 20 toward CO_2 reduction,[48] while L1 showed a poorer selectivity value of 0.5–10. On the other hand, Co-(cyclam)(Cl$_2$)$^+$ coupling with phenazine photosensitizer and a mixture of TEA/CH$_3$OH/CH$_3$CN was able to reduce CO_2 to [HCO$_2$]$^-$ with an

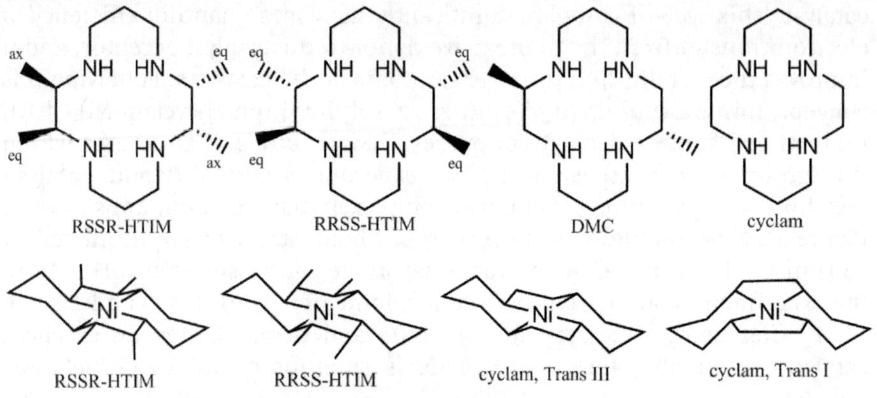

Figure 10.5 Various macrocycle structures.
Reprinted with permission from ref. 47. Copyright 1994 American Chemical Society.

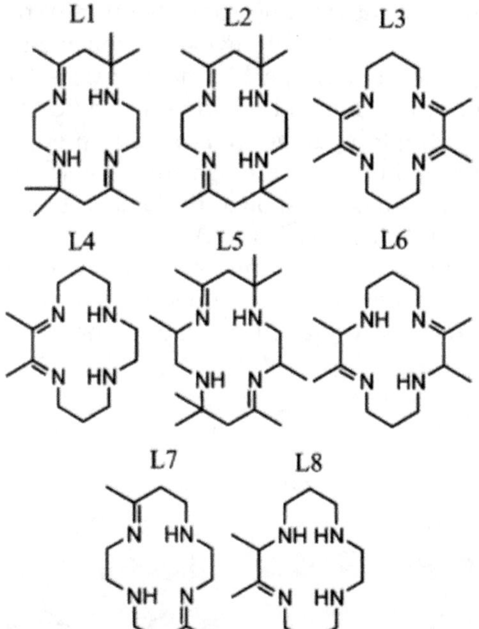

Figure 10.6 Various macrocycle structures.
Reprinted with permission from ref. 10. Copyright 2009 American Chemical Society.

extremely high selectivity value of 140.[49] The quantum yield of the reaction was 0.07 and improved to 0.13 when p-terphenyl was used.[48]

On the other hand, a supramolecular complex, constantly researched, is a large molecule containing a sensitizer that is covalent attached to the

catalyst. This type of complex significantly increases quantum efficiency of electron transfer from the photoactive donor to the catalyst acceptor, and it improves the stability and turnover numbers of the catalysts. Following this concept, Kimura *et al.* tried to synthesize $Ru(phen)_2(phen\text{-}cyclam\text{-}Ni)(ClO_4)_4$ bifunctional supramolecular complexes.[50] Many kinds of structures within this group were synthesized in the experiment with different catalyst-attaching groups. Among them, two complexes that stood out are shown in Figure 10.7 as complex-a and complex-b. Photocatalysts were irradiated by 350 nm UV light in a CO_2 saturated ascorbate buffer solution (pH = 4). In the experiment, complex-b produced a volume of 3.51 µL of CO and 1.39 µL of H_2 after being irradiated for 44 hours and demonstrated an excellent catalytic selectivity value of 2.5. This is superior complex-a, which was only 0.65.

Rhenium(I) bipyridine tricarbonyl complexes have been studied and also yield a high quantum efficiency of CO. However, the shortcomings of these complexes are that turnover number is low and absorption only exhibits in the UV region. In a 2007 publication, Sato *et al.*[51] followed up on their previous research. Both ruthenium(II) and rhenium(I) complexes combined with 1,3-bis(4′-methyl-[2,2′]bipyridinyl-4-yl)-propan-2-ol(bpyC3bpy), also called $[Ru\text{-}ReCl]^{2+}$ were illuminated under visible light. The turnover number turned out to be 160 and the quantum efficiency was 0.12, which are lower than the complexes associated with Re^+ previously produced with UV light. The author improved the structure by substituting peripheral ligands on the Re^+ site, as shown in Figure 10.8. Hence, two new supramolecules, $[Ru\text{-}ReP(OEt)_3]^{3+}$ and $[Ru\text{-}Repy]^{3+}$, were synthesized. The photo-induced reduction was illuminated under a wavelength of light greater than 500 nm. The turnover number of CO was improved to be 232. Improved from previous result of 0.12, the quantum efficiency increased to 0.21.

(a)   (b)

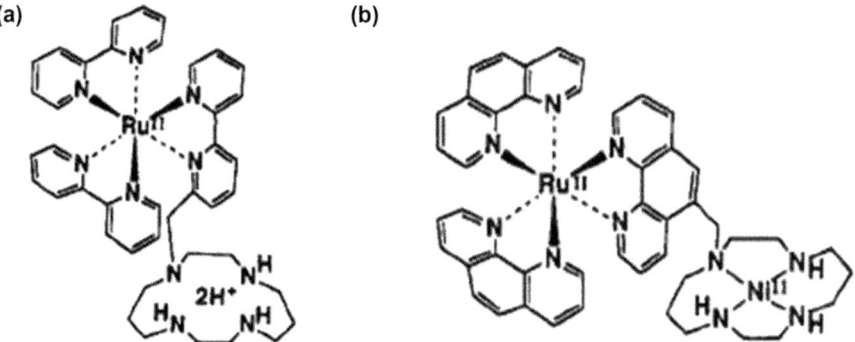

Figure 10.7   Two supramolecular complex structures.
Reprinted with permission from ref. 50. Copyright 1992 American Chemical Society.

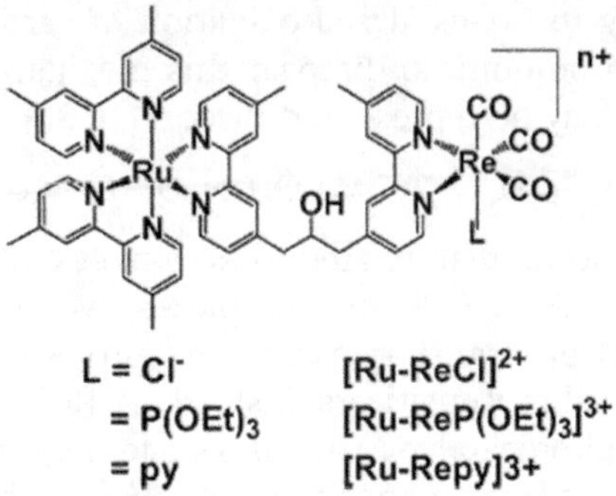

L = Cl⁻   [Ru-ReCl]²⁺
= P(OEt)₃   [Ru-ReP(OEt)₃]³⁺
= py   [Ru-Repy]3⁺

Figure 10.8   Supramolecular complex structures.
Reprinted with permission from ref. 51. Copyright 2007 The Royal Society of Chemistry and Owner Societies.

10.5 Research Cases

10.5.1 Example 1: Reduction of CO_2 to CO

Carbon dioxide fixation and reduction play an important role in producing organic fuels. Among these reduction processes, carbon monoxide is the most important intermediate product. Here, efficient photochemical reduction from carbon dioxide into carbon monoxide through visible light irradiation of systems containing $Ru(bipy)_3^{2+}-CO_2^{+}$ combinations/ $Re(bipy)(CO)_3X$ as homogeneous catalysts will be discussed.[44]

10.5.1.1 Theory

Based on the photoelectrochemical conclusions from previous experiments, we can obtain precious insights into the mechanism of the carbon dioxide photoreduction process. By adding CO_2 into $Ru(bipy)^{2+}$, which was produced using electrochemical method in acetonitrile solution, we could only gain a small yield of CO. However, after adding $Co(bipy)_3^{2+}$ into $Ru(bipy)_3^{+}$ in DMF, Co^I species formed, which was indicated by electronic absorption spectroscopy. Thus carbon dioxide dissolved subsequently in that mixture would probably afford CO with 55% yield in respect to Ru^I. Therefore, we may arrive at the photoreduction mechanism related to these results as follows.

- Firstly, reductive quenching of the excited species $*Ru(bipy)_3^{2+}$ *via* the tertiary amine led to photoproduction of $Ru(bipy)_3^{+}$.
- Secondly, with the aid of $Ru(bipy)_3^{+}$, Co^{II} ions were reduced to Co^I ions.
- Thirdly, after the Co^I species formed, newly dissolved CO_2 turned into CO.

The last process in the mechanism stood for a CO_2 activation reaction,[52] which might include similar steps but in the opposite sequence of those

described according to Co^I-catalyzed oxidation of carbon monoxide.[53] Further study needs to be done to discover this mechanism. The overall CO_2 reduction reaction may be represented by eqn (10.33).

$$CO_2 + 2H^+ + 2e^- \rightleftarrows CO + H_2O \quad E^{0'} = 0.52V \quad (10.33)$$

Referring to the mechanism related to Re(bipy)(CO)$_3$X (L = 2,2'-bipyridine or 1,10-phenanthroline; X = Cl, Br) complexes, we should highlight their photophysical, electron transfer, and redox properties. Thus, they were treated as suitable photosensitizers instead of Ru(bipy)$_3$$^{2+}$. Actually this process involved a photocatalytic rhenium cycle. Re(bipy)(CO)$_3$X complexes absorbed visible light with a maximum of 385 nm and 392 nm (ε 3100, 2680 dm^3 mol^{-1} cm^{-1}), separately, producing an excited state *Re^I. Such rhenium complexes appeared very similar to the corresponding (phen) complexes,[54] which could undergo either reductive or oxidative electron transfer quenching, largely relying on the redox potential of the quencher. Supposing that the redox potential of *Re^I(bipy)(CO)$_3$X/[Re^0(bipy)(CO)$_3$X]$^-$ complexes is in analogy with that of the (phen) group [*ca.* +1.25 V *versus* normal hydrogen electrode (NHE)],[54] reductive quenching by (HOCH$_2$CH$_2$)$_3$N (E^0*ca.* +0.8 V) would probably take place. These had already been confirmed by related quenching experiments. The [Re^0(bipy)(CO)$_3$X]$^-$ series formed later had a Re^I/Re^0 redox potential close to that of the corresponding (phen) series (−1.05 V *versus* NHE).[54] Therefore, there would be no doubt about their ability to reduce CO_2 into CO.

10.5.1.2   *Experimental Setup*

Regarding the Ru(bipy)$_3$$^{2+}$/Co^{II} complexes, extensive modifications had been made varying the solvent's nature, changing tertiary amine donor as well as altering the photosensitizer so as to make an improvement in the photo-generation efficiency together with selectivity of carbon monoxide. Thus, the experiment was carried out under 400 nm visible light irradiation (2 h, 250 W, halogen lamp) with a solution consisting of 10 mg Ru(bipy)$_3$$^{2+}$, 8 mg CoCl$_2 \cdot$6H$_2$O, 17 mg 2,9-dimethyl-1,10-phenanthroline served as ligand for the cobalt ions and a mixture of 5 mL (HOCH$_2$CH$_2$)$_3$N along with 25 mL dimethylformamide (DMF), which had 160 mL carbon dioxide dissolved in it. As a result, 2 mL CO and 4.5 mL H$_2$ were observed as main products. However, after experimental conditions were changed to irradiating with a 1000 W Xe lamp for 15 hours, the corresponding products turned to be 8 mL CO and 19 mL H$_2$. In order to verify the carbon source of the products, isotope-labeling experiments were carried out to trace the original carbon in carbon dioxide. At last, it was found out that ^{13}CO (90% enriched) came from 90% enriched ^{13}CO$_2$. This conclusion was also confirmed by gas chromatography–mass spectrometry analysis.

While referring to Re(bipy)(CO)$_3$X complexes, a 17 hour visible light irradiation was offered by 1000 W Xe lamp through a mixture of 13.3 mg Re(bipy)(CO)$_3$Br, 13.7 mg Co(bipy)$_3$$^{2+}$ and a solution of 10 mL

Table 10.2 Generation of CO by photoreduction of CO_2 *via* visible light irradiation of solutions containing $Re(L)(CO)_3X$ and CO_2 in $(HOCH_2CH_2)_3N$-DMF.[a] Reprinted with permission from ref. 44. Copyright 1969 The Royal Society Publishing.

Expt.	Complex	Additive[b]	Irradiation time/h	Vol. of CO produced/ml	Turnover number[c]
1	$Re(bipy)(CO)_3Cl$	0	1	6.5	11
2	"	0	2	9.7	16
3	"	0	4	16.8	27
4	"	Net_4Cl	2	14.5	23
5	"	Net_4Cl	4	30.0	48
6	"	Net_4ClO_4	2	6.4	10
7	"	Net_4Cl^f	2	14.0	22
8	$Re(bipy)(CO)_3Br$	0	2	7.6	14
9	"	0	4	11.4	20
10	"	NBu_4Br	2	12.0	21
11	"	NBu_4Br	4	16.0	28
12	$Re(Br\text{-}phen)(CO)_3Br$	0	2	2.7	5
13	"	NBu_4Br	2	3.7	8
14[d]	$Re(bipy)(CO)_3Br$	0	3.5	0.08	—
15[e]	$Re(bipy)(CO)_3Br$	0	6	0.04	—

[a]$Re(bipy)(CO)_3$ $Cl \times 8.710^{-4}M$; $Re(bipy)(CO)_3Br$ $7.9 \times 10^{-4}M$; $Re(Br\text{-}phen)(CO)_3Br$ $6.6 \times 10^{-4}M$; Br-phen = 5-bromo-1,10-phenanthroline. 30 ml of solution containing $Re(L)(CO)_3X$ and 160 ml CO_2 (99.8% purity) dissolved in dimethylformamide-$(HOCH_2CH_2)_3N$ (5 : 1) were irradiated with a 250 W halogen lamp (slide projector) fitted with a 400 nm cut-off filter (Schott GG 420).
[b]Net_4Cl 2×10^{-2} M; NBu_4Br 10^{-2} M.
[c]Obtained by dividing the number of moles of CO produced by the number of moles of $ReL(CO)_3X$.
[d]Experiment carried out without CO_2; formal pH of the solution adjusted to 9.5; 1.1 ml of H_2 generated.
[e]Same conditions as in experiment 14 but adjusted to 'pH' 8.5; 1.3 ml of H_2 generated.
[f]And 25 equiv. of bipy.

$(HOCH_2CH_2)_3N$ and 20 mL DMF containing 160 mL dissolved carbon dioxide. The products observed were 0.3 mL H_2 together with 11.3 mL CO, indicating a better selectivity of carbon monoxide compared with the experimental results shown above. Meanwhile, using similar irradiation conditions to the solution without $Co(bipy)_3^{2+}$, only carbon monoxide could be discovered in the products. Furthermore, a series of experiments were performed for the purpose of investigating the features of this reduction reaction and the related results are displayed in Table 10.2.

10.5.1.3 Experimental Results

(1) Among the $Re(L)(CO)_3X/(HOCH_2CH_2)_3N$ combination system, rhenium complex acted as both photosensitizer and homogeneous catalyst. It also represented a novel carbon dioxide activation reaction since CO could not be detected in the absence of the complex.

(2) When using 90.5% enriched $^{13}CO_2$, only 88% enriched ^{13}CO was obtained as shown by GC-MS analysis as well as ^{13}C NMR spectroscopy.

(3) During this reduction process, high efficiency led to high turnover numbers (illustrated in Table 10.2), and large quantities of reduced products were detected, which even outweighed the most active $Ru(bipy)_3^{2+}/Co^{II}$ complexes.

(4) This process optimized CO/H_2 selectivity, generating trace amount of H_2, though hydrogen could be formed without carbon dioxide input (shown in Table 10.2, experiments 14 and 15). However, hydrogen may be produced simultaneously *via* cobalt hydrides with the aid of $Co(bipy)_3^{2+}$.

(5) Due to the instability of ligand complexes, the activity of the whole reaction system would decrease slowly with time. This was also indicated by kinetic experiments. However, after excessive addition of bipyridine or perchlorate into the $Re(bipy)(CO)_3X$ complexes, no significant improvement was observed (compare experiment 2 with 6, and experiment 4 with 7 in Table 10.2). Meanwhile, with the addition of bromide or chloride anions into the corresponding $Re(bipy)(CO)_3X$ complexes, appreciably more CO was produced in this experiment in contrast to previous experiments over a long experimental period, predicting a remarkable increase in stability (compare experiment 2 with 4, experiment 3 with 5, experiment 8 with 10, and experiment 9 with 11 in Table 10.2).

(6) Carbon dioxide activation and reduction processes involved coordination of CO_2 to metal. Losing the halide ligand offered a free coordination site, which was confirmed by the effective improvement obtained after adding anions into the system. Actually, under those conditions shown in Table 10.2, anions exchange, such as from Cl^- to Br^-, took place in the rhenium complexes readily. Meanwhile, a single nitrogen coordination adsorbed to a sesquibipy ligand occurring in other cases[55] may also be discovered, though this would rarely happen with (phen) as ligand (experiments 12 and 13 in Table 10.2).

(7) With the aid of rhenium hydride, hydrogen formed in the absence of carbon dioxide (experiments 14 and 15 in Table 10.2), After CO_2 was inserted into Re-H bonds, formate complexes were formed subsequently, yielding a mixture containing carbonyl complexes under irradiation. However, when the irradiation was executed without excessive addition of anions (similar to experiments 1–3 in Table 10.2), the formate complex $Re(bipy)(CO)_3(O_2CH)$ was isolated and then characterized. Later, its role in carbon monoxide formation was investigated.

Inspired by the results displayed above, we could easily jump to several conclusions.

- First, these experimental methods represented novel processes for artificial photosynthesis together with light energy storage.
- Second, they enabled novel CO_2 reduction reactions *via* homogeneous catalysts, which was also useful in the electroreduction of CO_2.

- Third, single component process achieved by Re(bipy)(CO)$_3$X complexes outweighed the combined Ru(bipy)$_3{}^{2+}$–CO$_2{}^+$ system in the aspects of simplicity, efficiency and carbon monoxide selectivity.
- Fourth, metal carbonyl complexes could be used as photosensitizers and photocatalysts for other photoreduction reactions as well, particularly for the reduction from CO$_2$ to CO.

10.5.2 Example 2

As mentioned in the previous section, Kimura[50] and co-workers started to synthesize and charaterize a novel type of photoreductive catalyst. The significance of this publication provides a model for future researchers who are constantly looking for better catalytic candidates.

Kimura successfully synthesized Ru(phen)$_2$(phen-cyclam-Ni)(ClO$_4$)$_4$ bifunctional supramolecular complexes (phen = 1,10-phenanthrolin), which has a hybrid structure including Ru(phen)$_3{}^{2+}$ serving as a photosensitizer and Ni(cyclam)$^{2+}$ serving as a reduction site. Incorporating photosensitizer in the system, it could better transfer and donate electrons to the reduction center. The reduction site is mainly responsible for the reduction of CO$_2$. Previous studies of using a homogeneous combination catalyst with a photosensitizer received only modest successes. Hawecker reported the homogeneous complex of CO$_2{}^+$ and Ru (bpy)$_3{}^{2+}$, but the duration of sustainable irradiation was limited by the de-complex reaction. Kimura and co-workers looked into many possibility of complexes, and they found Ru(phen)$_3{}^{2+}$ and Ni(cyclam)$^{2+}$ provided better performance.

10.5.2.1 Experiment Preparations and Measurements

For the synthesis of Ru(phen)$_2$(phen-cyclam-Ni)(ClO$_4$)$_4$ 2H$_2$O, the preparation of 1-(1,10-phenanthrolin-5-ylmethyl)-1,4,8,11-tetraazacyclotetradecane was made first. 5-Methyl-1,10-phenanthroline in 2 mmol (7385 mg), NBS in 2 mmol (356 mg), and isobutyronitrile (20 mg) were purchased and dissolved in 20 mL of CCl$_4$ solution. The mixture was refluxed in a flash for 3 hours, and then the solids were collected after filtration and evaporation. Washed with water and CH$_2$Cl$_2$ solution several times, the residue stayed in the organic layer. After drying with anhydrous MgSO$_4$, 5-(bromomethyl)-1,10-phenanthroline was collected. This product was dissolved back to DMF and 3 mmol of cyclam solution in volume of 30 mL was added to the 20 mL DMF solution at 90 °C drop by drop. The mixture was heated and refluxed at 90 °C for 5 hours. The unreacted solids were filtered, and the rest of the products were purified by column chromatography. 1-(1,10-phenanthrolin-5-ylmethyl)-1,4,8,11-tetraazacyclotetradecane came out as an oil form.

With the aid of amberlite IRA 400 ion-exchange treatment, a solution of Ru(phen)$_2$Cl$_2$ in 25 mL total volume was added into 50 mL of 1-(1,10-phenanthrolin-5-ylmethyl)-1,4,8,11-tetraazacyclotetradecane in C$_2$H$_5$OH under

Ar gas filled environment. The mixture was refluxed for 5 hours. After filtration and evaporation, a red residue formed and was poured into water, which was maintained at pH 3 with $HClO_4$. An orange color residue formed when excessive $HClO_4$ was used, and acetone was used to re-dissolve these formed powders. After partially removing some solvents, red needle Ru(phen)$_2$(phen-cyclam-H$_2$)(ClO$_4$)$_4$3H$_2$O was obtained.

Ru(phen)$_2$(phen-cyclam-H$_2$)(ClO$_4$)$_4$ 3H$_2$O was dissolved in 10 mL of CH_3OH and added to 1 mmol NiCl$_2$ 6H$_2$O. The reaction mixture was refluxed and heated for 30 minutes. After filtration, 10 equivalent in a mole of NaClO$_4$ was added to the filtrate to obtain orange precipitates, which were Ru(phen)$_2$(phen-cyclam-Ni)(ClO$_4$)$_4$ 2H$_2$O.

The researchers used a Yanako-P1100 polarographic analyzer to perform cyclic voltammetric measurements. The tested solution was bubbled with pure Ar gas and kept at 1 mM. The three-electrode system was utilized for the testing: working electrode was a Pt disk, counter electrode used a Pt wire, and reference electrode was an Ag/AgCl electrode. The scan rate of the cyclic voltammograms was 200 mV s^{-1}.

A Shimadzu FR-5000 spectrophotometer was employed to measure emission. A Hitachi U-3200 double beam spectrophotometer was used to record UV-vis information. The quartz cell of this spectrophotometer had a 10 mm path length. The lifetime emission measurement was done at room temperature by using a Horiba NAES-550 equipped with time-correlated multiphoton counting method. A pulse Xe lamp was used as the light source, and Ar gas was used in all testing environments.

A Ushio xenon arc lamp UXL500D-0 in 500 W with a Toshiba UV-35 cut-off lens filtered the light sources at 350 nm in a photolysis cell. A Shimadzu GC-8A gas chromatography packed with 13X-S molecular sieve column was used to detect CO_2 and H_2. A Shimadzu GC-4CMPF with the same column provided a capability to detect CO.

10.5.2.2 Experimental Results

Since 1-(1,10-phenanthrolin-5-ylmethyl)-1,4,8,11-tetraazacyclotetradecane is an important intermediate product, the optimization of the synthetic procedure is crucial. The authors could create 24% yield in making this intermediate. The process from 1-(1,10-phenanthrolin-5-ylmethyl)-1,4,8,11-tetraazacyclotetradecane to the final product improved to 55% yield.

The emission maxima of the goal product Ru(phen)$_2$(phen-cyclam-Ni)(ClO$_4$)$_4$ is 580 nm. When excitation reaches 450 nm, the emission encounters a red shift from previous 580 nm to 575 nm in the conditions of vacuum and 25 °C. The authors discovered the relative emission intensity of this product is only 5% of Ru(phen)$_3$$^{2+}$. Adding external Ni(cyclam)$^{2+}$ complex could not completely quench the emission. The product complex showed short emission lifetime of a few nanoseconds where Ru(phen)$_3$$^{2+}$ shows a long emission lifetime of 1200 ns. In this complex, there is no coordinative distortion around Ru center. Hence, the quenching of the state

$Ru(phen)_3^{2+}$ in the complex is because of the energy transfer to inter-molecular electron transfer at $Ni(cyclam)^{2+}$.

The cyclic voltammogram of the product shows two features of irreversible reduction: one is quasi-reversible reduction wave, and the other is quasi-reversible oxidation wave from the Ru complex. In addition, a desorption peak was also observed. All these redox features of voltammogram are similar to $Ru(phen)_3^{2+}$. Ni^{II} to Ni^{I} was seen at a potential of -1.56 V (*versus* Ag/AgCl) in the reversible reduction curve, and this is more positive than $Ni(cyclam)^{2+}$ at -1.74 V. Similarly, Ni^{III} to Ni^{II} was seen at a potential of 1.26 V, which is 0.53 V more positive than $Ni(cyclam)^{2+}$. These observations of shifting are consistent with the expectation of stabilizing a lower oxidation state of Ni by n-alkyl substitution. However, these do not truly reflect the electrochemical interaction between $Ni(cyclam)^{2+}$ and $Ru(phen)_3^{2+}$.

The photoreduction of CO_2 took place in ascorbate buffer solution to maintain a pH 4 condition. The temperature was 25 °C and reference systems are $Ni(cyclam)^{2+}$ and $Ru(phen)_3^{2+}$. The experiment tested one-hour irradiation onto the designed catalyst, and 0.09 µL of CO was produced as a result. In comparison, the reference system took 4 hour irradiation and produced as much as the designed photocatalyst. The durability of the photocatalyst is more stable than the reference system. The selectivity of reduction of CO_2 *versus* H_2O shows this photocatalyst is better. The $[CO]/[H_2]$ from the reference system after a 44-hour irradiation is 0.57 when the photocatalyst provided a record high 2.5. Hence, the photocatalyst is more effective.

The current design of photocatalyst might not be the best since the existence of subunit $Ni(cyclam)^{2+}$ is too bulky and reduces the production ratio of $[CO]/[H_2]$. Hence, the authors are considering a new hybrid system by linking $Ru(phen)_3^{2+}$ directly to $Ni(cyclam)^{2+}$. This experimental case is introduced to teach the basic design concept and way of conducting experiments. Therefore, the practices of authors used in this publication are an example for further research.

10.6 Conclusions

Since there is a shortage of conventional fuel sources, renewable energy represents new hopes. It is considered to be a promising source of future fuel supply because of being sustainable, clean, and unlimited in reserve. Nowadays, solar and wind power are the best technology candidates and they are leading in all renewable energy productions. However, more efforts are needed to find a steady storage solution for the produced energy. Compared to the current storage technologies including pump hydro, compressed air, and lithium-ion battery, the carbonacenous fuels do have advantages of high energy density, facility of transport, and remote generation capability. Therefore, the photo-induced electrochemical reduction process will ideally be the best solution. A zero-sum in carbon accumulation can be built based on the carbonacenous fuels. The natural world can absorb as much as

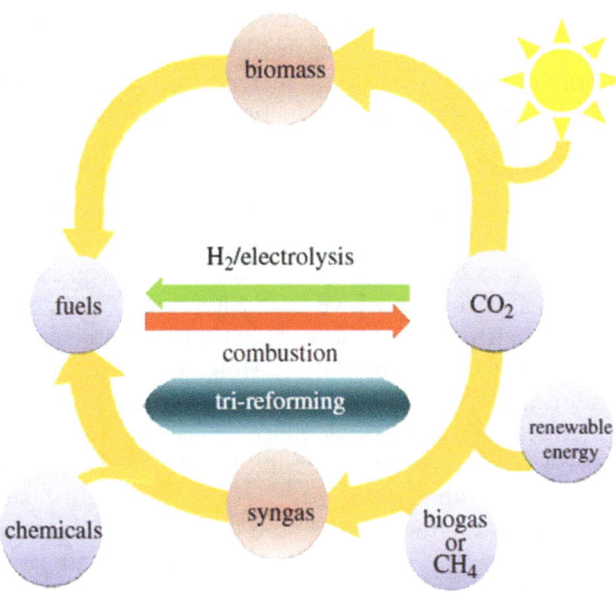

Figure 10.9 A tri-reforming schematic cycle from CO_2 to fuels.
Reprinted with permission from ref. 2. Copyright 2010 The Royal
Society Publishing.

carbon and turn it into biomass. We humans certainly have a capability to
facilitate a faster process as well since our activities right now emit too much
carbon, which cannot be digested by natural systems on the Earth. Alternatively from natural process, we can turn CO_2 into biogas, methane, syngas,
or other chemicals, which could be consumed as fuels (Figure 10.9). The
photo-induced reduction process is the easiest route to achieve compared to
the steps going through biomass or syngas.

In this chapter, we have attempted to provide a holistic review of the
current scientific studies in the areas of inorganic particle systems and
organic molecular systems, which includes the exploitation of UV and visible
light spectrum. The myriad of research mentioned in this review are
focusing on modifying morphologies of particles, microstructure, cocatalysts, and systemic structure to improve the photocatalytic capability of
the catalytic materials. Applying various characteristic tools and measurements, researchers are able to identify the types and amount of the
carbonaceous fuels produced as well as turnover number and efficiency for
benchmarking purposes. Compared to many studies, these designed systems are superior in terms of quantum efficiencies, turnover number,
catalyst selectivity, and optical conversion efficiency.

The future of research in photocatalytic reduction continues to be a
vital branch of energy research. A larger barrier resides in engineering
systems from lab scale to a commercial stage. In conclusion, in order to
transit renewable energy into primary energy for the future, the transition

from laboratory research to commercial application and the utilization of this commercial application have to be successful. This is very likely to happen with the photoelectrochemical reduction of CO_2.

Acknowledgements

The authors acknowledge the generous help of Jacopo Samson and David Gray for editing and providing numerous suggestions about writing this review.

References

1. J. Grodkowski, D. Behar, P. Neta and P. Hambright, *J. Phys. Chem. A*, 1997, **101**, 248–254.
2. Z. Jiang, T. Xiao, V. L. Kuznetsov and P. P. Edwards, *Philos. Trans. R. Soc., A*, 2010, **368**, 3343–3364.
3. S. C. Roy, O. K. Varghese, M. Paulose and C. A. Grimes, *ACS Nano*, 2010, **4**, 1259–1278.
4. J. R. Bolton, *Science*, 1978, **202**, 705–711.
5. A. J. Bard, *J. Photochem.*, 1979, **10**, 59–75.
6. J. Manassen, D. Cahen, G. Hodes and A. Sofer, *Nature*, 1976, **263**, 97–100.
7. A. J. Nozik, *Annu. Rev. Phys. Chem.*, 1978, **29**, 189–222.
8. G. A. Olah, A. Goeppert and G. K. S. Prakash, *J. Org. Chem.*, 2008, **74**, 487–498.
9. E. E. Benson, C. P. Kubiak, A. J. Sathrum and J. M. Smieja, *Chem. Soc. Rev.*, 2009, **38**, 89–99.
10. A. J. Morris, G. J. Meyer and E. Fujita, *Acc. Chem. Res.*, 2009, **42**, 1983–1994.
11. T. Inoue, A. Fujishima, S. Konishi and K. Honda, *Nature*, 1979, **277**, 637–638.
12. J. Wang, S. Yin, M. Komatsu and T. Sato, *J. Eur. Ceram. Sci.*, 2005, **25**(13), 3207–3212.
13. M. Uddin, F. Cesano, S. Bertarione, F. Bonino, S. Bordiga, D. Scarano and A. Zecchina, *J. Photochem. Photobio., A*, 2008, **196**(2-3), 165–173.
14. P. Wang, S. Zkeeruddin, J. Moser, M. Nazeeruddin, T. Sekiguchi and M. Gratzel, *Nature*, 2003, **2**, 402–407.
15. Y. Taniguchi, H. Yoneyama and H. Tamura, *Bull. Chem. Soc. Jpn.*, 1982, **55**, 2034–2039.
16. C. Wang, R. L. Thompson, J. Baltrus and C. Matranga, *J. Phys. Chem. Lett.*, 2009, **1**, 48–53.
17. W. Paik, T. N. Andersen and H. Eyring, *Electrochim. Acta*, 1969, **14**, 1217.
18. M. Aresta, C. F. Nobile, V. G. Albano, E. Forni and M. Manassero, *J. Chem. Soc., Chem. Commun.*, 1975, **636**, 6.
19. D. J. Darensbourg, A. Rokicki and M. Y. Darensbourg, *J. Am. Chem. Soc.*, 1981, **103**, 3223.
20. M. E. C. Royer, *r hebd Séanc Acad Sci, Paris*, 1870, **70**, 731–732.

21. K. S. Udupa, G. S. Subramanian and H. V. K. Udupa, *Electrochim. Acta*, 1971, **16**, 1593–1598.
22. W. Paik, T. N. Andersen and H. Eyring, *Electrochim. Acta*, 1969, **14**, 1217–1232.
23. K. Ito, S. Ikeda, T. Iida and H. Niwa, *DenkiKagaku.*, 1981, **49**, 106.
24. K. Ito, S. Ikeda, T. Iida and A. Nomura, *Denki Kagaku.*, 1982, **50**, 463.
25. N. Getoff, *Z. Naturforsch.*, 1962, **17b**, 87–90.
26. M. Halmann, *Nature*, 1978, **275**, 115–116.
27. O. Ishitani, *J. Photochem. Photobiol., A*, 1993, **72**, 269–271.
28. A. Fujishima and K. Honda, *Nature*, 1972, **238**, 37–38.
29. A. Monnier, J. Augustynski and C. Stalder, *J. Electroanal. Chem. Interfacial Electrochem.*, 1980, **112**, 383–385.
30. A. Monnier, J. Augustynski and C. Stalder, *Boulder Meeting on Photochemical Solar Energy Conversion*, 1980. p. 423.
31. B. Aurian-Blajeni, M. Halmann and J. Manassen, *Sol. Energy Mater.*, 1983, **8**, 425–440.
32. D. Canfield and K. W. Frese, *J. Electrochem. Soc.*, 1983, **130**, 1772–1773.
33. K. R. Thampi, J. Kiwi and M. Gratzel, *Nature*, 1987, **327**, 506–508.
34. R. L. Cook, R. C. MacDuff and A. F. Sammells, *J. Electrochem. Soc.*, 1988, **135**, 3069–3070.
35. K. Sayama and H. Arakawa, *J. Phys. Chem.*, 1993, **97**, 531–533.
36. K. Adachi, K. Ohta and T. Mizuno, *Sol. Energy*, 1994, **53**, 187–190.
37. M. Anpo, H. Yamashita, Y. Ichihashi, Y. Fujii and M. Honda, *J Phys Chem B.*, 1997, **101**, 2632–2636.
38. N. Ulagappan and H. Frei, *J. Phys. Chem. A.*, 2000, **104**, 7834–7839.
39. K. Ikeue, S. Nozaki, M. Ogawa and M. Anpo, *Catal. Lett.*, 2002, **80**, 111–114.
40. X.-H. Xia, Z.-J. Jia, Y. Yu, Y. Liang, Z. Wang and L.-L. Ma, *Carbon*, 2007, **45**, 717–721.
41. E. E. Barton, D. M. Rampulla and A. B. Bocarsly, *J. Am. Chem. Soc.*, 2008, **130**, 6342–6344.
42. J. Grodkowski and P. Neta, *J. Phys. Chem. A.*, 2000, **104**, 1848–1853.
43. J.-M. Lehn and R. Ziessel, *Proc. Natl. Acad. Sci.*, 1982, **79**, 701–704.
44. J. Hawecker, J.-M. Lehn and R. Ziessel, *J. Chem. Soc., Chem. Commun.*, 1983, 536–538.
45. C. A. Craig, L. O. Spreer, J. W. Otvos and M. Calvin, *J. Phys. Chem.*, 1990, **94**, 7957–7960.
46. M. Beley, J.-P. Collin, R. Ruppert and J.-P. Sauvage, *J. Chem. Soc., Chem. Commun.*, 1984, 1315–1316.
47. E. Fujita, J. Haff, R. Sanzenbacher and H. Elias, *Inorg. Chem.*, 1994, **33**, 4627–4628.
48. S. Matsuoka, K. Yamamoto, T. Ogata, M. Kusaba, N. Nakashima, E. Fujita *et al.*, *J. Am. Chem. Soc.*, 1993, **115**, 601–609.
49. T. Ogata, Y. Yamamoto, Y. Wada, K. Murakoshi, M. Kusaba, N. Nakashima *et al.*, *J. Phys. Chem.*, 1995, **99**, 11916–11922.

50. E. Kimura, X. Bu, M. Shionoya, S. Wada and S. Maruyama, *Inorg. Chem.*, 1992, **31**, 4542–4546.

51. S. Sato, K. Koike, H. Inoue and O. Ishitani, *Photochem. Photobiol. Sci.*, 2007, **6**, 454–461.

52. M. E. Vol'pin and I. S. Kolomnikov, *Pure Appl. Chem.*, 1975, **33**, 567; R. Eisenberg and D. E. Hendriksen, *Adv. Catal.*, 1979, **28**, 70; J. A. Ibers, *Chem. Soc. Rev.*, 1982, **11**, 57.

53. J. E. Bercaw, L.-Y. Goh and J. Halpern, *J. Am. Chem. Soc.*, 1972, **94**, 6534.

54. J. C. Luong, L. Nadjo and M. S. Wrighton, *J. Am. Chem. Soc.*, 1978, **100**, 5790.

55. R. C. Angelici and J. R. Graham, *J. Am. Chem. Soc.*, 1965, **87**, 5586; M. S. Wrighton and D. L. Morse, *J. Organomet. Chem.*, 1975, **97**, 405.

CHAPTER 11

Biological Systems for Carbon Dioxide Reductions and Biofuel Production

E. ASMATULU

Department of Mechanical Engineering, Wichita State University, 1845 Fairmount, Wichita, KS 67260, USA
Email: exasmatulu@wichita.edu

11.1 Introduction

11.1.1 General Background

Global warming is real, and its effects have been seen worldwide through temperature increases; drastic changes in the weather, such as extreme dry, hot, or rainy seasons; stronger hurricanes and tornedos; and faster-melting ice at the poles. Eventually all of these may cause the mass extinction of plants and animals and eventually affect human life. It is expected that this effect will continue and become even worse in the future. The accelerated rate of global warming is an outcome of greenhouse gas emissions, including 72% CO_2, 18% methane, and 9% NOx gases. The major sources of CO_2 gas include the burning of fossil fuels (e.g., coal, oil, and natural gas) as well as calcination and other industrial manufacturing processes. Algae (singular alga) are composed of eukaryotic organisms, which have the ability to capture CO_2 and convert it into useful hydrocarbons (biomass) for future fuel production.[1] In fact, algae recycle CO_2 gas for their survival and in turn protect the environment against some of its contamination.

RSC Green Chemistry No. 42
Green Photo-active Nanomaterials: Sustainable Energy and Environmental Remediation
Edited by Nurxat Nuraje, Ramazan Asmatulu and Guido Mul
© The Royal Society of Chemistry 2016
Published by the Royal Society of Chemistry, www.rsc.org

Photosynthesis is the process of converting energy from the Sun into chemical energy *via* microorganisms and plants. In this process, the raw materials are CO_2, some elements (*e.g.*, N, S, P, Fe, *etc.*) and water, while the energy source is sunlight. The end products of the photosynthesis process include pure oxygen and energy-rich hydrocarbons. Figure 11.1 shows the photosynthesis and biomass production from the algal cells.[2] Algae are the major microorganisms and recognized as biomass producers. In particular, biodiesel and other biofuels can be produced by utilizing algae biomass. During the photosynthesis process, algae absorb CO_2 and sunlight (photons) in water along with other nutrients, and produce biomass and oxygen. Lipids and other oils (mainly vegetable oils) deposited in the body of algae during the photosynthesis can be converted into biodiesel, other biofuels, and pellets or directly used for other purposes. This process has been known for a long time but has gained considerable attention recently because of global warming and environmental pollution.[3]

Nanotechnology is the development of materials, components, devices, and/or systems at the near-atomic or nanometer levels. One of the

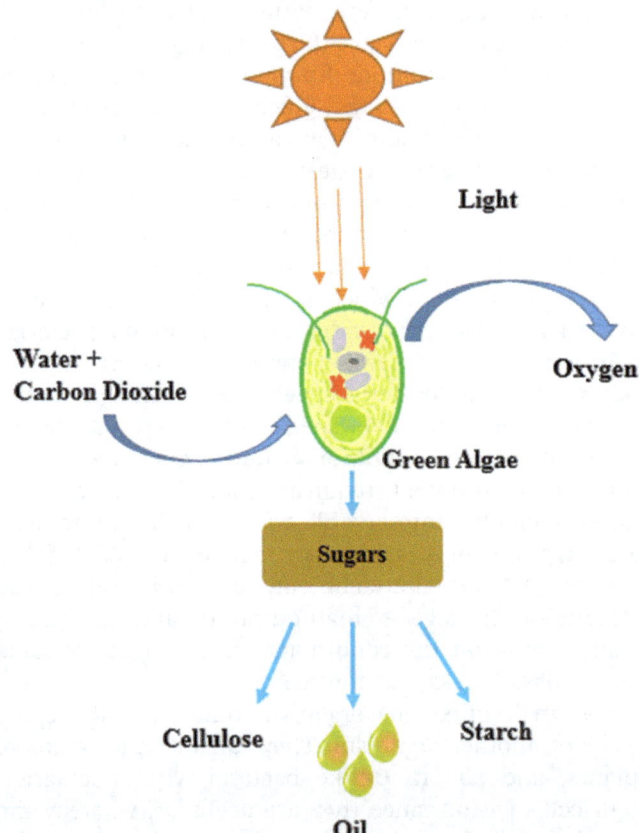

Figure 11.1 Photosynthesis and biomass production from algal cells.[2]

dimensions of nanotechnology is between 1 and 100 nm. This technology mainly involves fabricating, measuring, modeling, imaging, and manipulating matter at nanoscale. Nanotechnology consists of highly multi-disciplinary fields, including chemistry, biology, physics, engineering, and some other disciplines. For about two decades, significant progresses has been made in designing, analyzing, and fabricating nanoscale materials and devices, and this will continue for a few more decades.[4,5]

Recent studies have indicated that nanotechnology materials and processes could be applied to algae and other microorganism growth processes to potentially improve biological biomass production. This technology can significantly enhance biofuel production and biofuel conversion rates. It can also improve enzyme immobilization, lipid accumulation and extraction, enzyme loading capacity, nanoscale catalysis activities, storage capacities, separation and purification rates of liquid from other liquids and solids, and bioreactor design and applications.[6]

11.1.2 Microorganisms and Growth Conditions

Microorganisms (*e.g.*, algae, bacteria, virus, mold, and fungi) are living entities, having been in existence since the beginning of life, and have survived extreme environmental conditions for millions of years. Microorganisms usually deposit fat, lipids/oil, starch, glucose, and other hydrocarbons and organic substances in their bodies that can be extracted and converted into useful products. Bacteria are a single-cell form of life, and each individual cell is unique. Bacteria often grow into different colonies; however, each bacteria cell has its own independent life. Bacteria reproduce by a process known as cell division. A mature bacteria cell reproduces by dividing into two 'daughter cells,' each of which is identical to the other, as well as identical to the parent bacteria. Under ideal conditions, bacteria reproduce every 20 to 30 minutes into a new generation of bacteria.[7]

Thousands of different bacteria species exist throughout the world. It is estimated that more than 3000 species of bacteria are living in totally different environments and conditions. However, some of them are found only in a specific environment, requiring specialized types of food. These bacteria can be separated into aerobic types, which require oxygen to live, and anaerobic types, which can live without oxygen. Bacterial growth has several states: stage 1, initial attachment; stage 2, irreversible attachment; stage 3, maturation I; stage 4, maturation II; and stage 5, dispersion/spreading. Depending on the conditions, these stages can take seconds, minutes, hours, days, weeks, or months.[8]

A virus is a small infectious organism that can only replicate inside the living cells of another organism. They can infect all kinds of animals, bacteria, plants, and so on. Unlike bacteria, viral populations do not grow through cell division since they are acellular, whereby they use the machinery and metabolism of a host cell to produce multiple copies of themselves, and are assembled inside those cells. To date, approximately

5000 viruses have been scientifically described in detail. They mainly consist of different parts, some of which includes the genetic material made from either DNA or RNA (long molecules that carry genetic information), a protein coat that protects these genes, and an envelope of lipids that surrounds the protein coat when they are outside a cell. The shapes of viruses range from simple helical and icosahedral forms to more complex structures.[9]

A group of scientists has recently announced that they can successfully modify a virus to split water molecules, which can be an efficient and non-energy-intensive method of producing hydrogen gas (H_2). These scientists genetically modified a commonly known, harmless bacterial virus in order to assemble the components for separating water molecules into H_2 and oxygen gas (O_2) molecules, yielding a fourfold boost in production efficiency. This novel process mimics plants that use the power of sunlight to make chemical fuel for their growth. In this case, the scientists engineered the virus as a kind of biological scaffold to split a water molecule.[10,11]

Fungi store more carbon in their body and grow very fast compared to many other microorganisms. This ability can prevent and mitigate the effects of climate change by storing more carbon gas in the soil/biomass. From fungal solids, many different solid and liquid products can be produced. In addition to fungi, mold can also grow very fast and decompose organic substances into different forms, which can also be used as a renewable energy source. However, few studies have been conducted on fungi and mold related to energy conversion systems.[12]

Algae comprise several different species (~ 2800) of relatively simple living organisms all over the world, which capture light energy through photosynthesis and convert inorganic/organic substances into simple sugars using photon energy. Algae can be considered as the early stages of simple plants, and indeed some are closely related to the higher, more complex, plants as well. Some algae also appear to represent different protist groups (large and diverse groups of eukaryotic microorganisms), alongside other organisms that are traditionally considered more animal-like (*e.g.*, protozoa). Therefore, algae do not represent a single evolutionary direction but rather a level of organization that may have developed several times in the early history of microorganism life on the Earth's crust.[13]

Some microorganisms usually require the following conditions for their growth: pH: 5–9 (lower pH may be seen); presence of organic substances (waste water, city waste, leaves); temperature between 4 °C and 40 °C; sulfur-, iron-, copper-, zinc-, cobalt-, and manganese-rich conditions; and the presence of carbon and CO_2, nitrogen, phosphorus, oxygen, hydrogen, and sunlight (more sunlight, less UV rays). As an example, the growth conditions of microorganisms found in Yellowstone National Park, which is an extremely hot, and mineral- and ion-rich environment, are totally different to the growth conditions of similar species that live in coastal areas.[14]

A number of different algae with various growth conditions may be considered for energy mitigation and CO_2 reduction purposes (Figure 11.2): green algae (chlorophyll a and b, plants, and freshwater), red algae (benthic

Green Alga   Brown Alga   Red Alga

Figure 11.2   Various types of algae for CO_2 reduction and future biomass production.[15,16]

and macro-level), brown algae (benthic, macro, kelp, and marine), diatoms (single-celled with a silica cell wall), blue-green algae (exhibiting vertical migration and require freshwater conditions), and dinoflagellates (toxic, O_2 siphoners, and the cause of red tides).[15,16] *Botryococcus braunii*, a green algae with a pyramid-shaped planktonic structure, is one of the most important algae in biotechnology. These algae colonies are held together by a lipid biofilm and are found usually in temperate or tropical lakes and creeks. They will bloom in the presence of dissolved inorganic phosphorus and other nutrients in the growth condition. *B. braunii* has great potential for algae farming because of the hydrocarbons it produces, which can be chemically converted into fuels. It has been estimated that up to 86% of the dry weight of this alga can be composed of long-chain hydrocarbons, and some of its useful hydrocarbon oils can be found outside of the cell. *B. braunii* can convert 61% of its biomass into oil, which drops to only 31% without stress. It grows best between 22 °C and 25 °C, and is a great choice for biofuel production.[17]

Recently, nanotechnology-associated studies have been conducted on microorganisms to increase their efficiency of growth and rate of fuel conversion. Nanotechnology probes tap into algae and bacteria cells to extract electrical energy. It was postulated that *Chlamydomonas reinhardtii*, a single-cell alga, might be an ideal cell for the energy harvesting. Results demonstrated the feasibility of collecting high-energy electrons in steps of the photosynthetic electron transport chain prior to the downstream process. However, cells usually die after a period of time because of leaks in the membrane where the nanoscale electrodes penetrate into the body of the cell.[18]

11.2   Algae Processing

11.2.1   Algae Farming Locations and Bioreactors

Natural ponds and algae farming fields are designed to produce various forms of algae for the specific purpose of creating different biomasses.

Nitrogen, phosphorus, CO_2, some organic and inorganic substances, and water are supplied to the medium to grow algae in warm weather. A number of different venues can be considered for algae farming: near cities (water treatment plants, city sewers, and landfills, in order to clean up waste), agriculture sites (many farms and vineyards, agriculture by-products, food processing plant waste, and wood waste), lakes and swamps (salt lakes and other lakes, natural and artificial swamps, small creeks and springs, ponds, nitrogen- and phosphorus-rich wetlands, and other wetlands), and power plants and geothermal locations (hot and warm waste water during the winter/cold season).[19] Figure 11.3 shows the algae farming for CO_2 reduction and biofuel production.[20]

A bioreactor is a closed device or series of greenhouse pipes that support algae growth under sunlight. It is used to cultivate algae by consuming and converting CO_2, N, and P into biomass. Algae bioreactors are mainly used to produce fuels such as biodiesel and bioethanol, generate animal feed, and reduce environmental pollutants in power plants (CO_2 and NOx). CO_2 is dispersed into the reactor fluid to make it accessible to algae under the sunlight. The bioreactor is mainly built using a transparent material, such as Plexiglas or glass. The algae are photoautotroph organisms that perform oxygenetic photosynthesis. The equation for photosynthesis is:[21]

$$6CO_2 + 6H_2O \rightarrow C_6H_{12}O_6 + 6O_2 \quad \Delta H^0 = +2870 \text{ kJ mol}^{-1} \quad (11.1)$$

Bioreactor installations for the cultivation of algae provide fast growth and extraction of a high oil content. The basic advantages of using a bioreactor for cultivating algae growth are given below:

- bioreactor pipes can be placed horizontally or vertically in a small area, so large agricultural fields are not needed
- the bioreactor considerably reduces labor and maintenance costs and eliminates system complexities

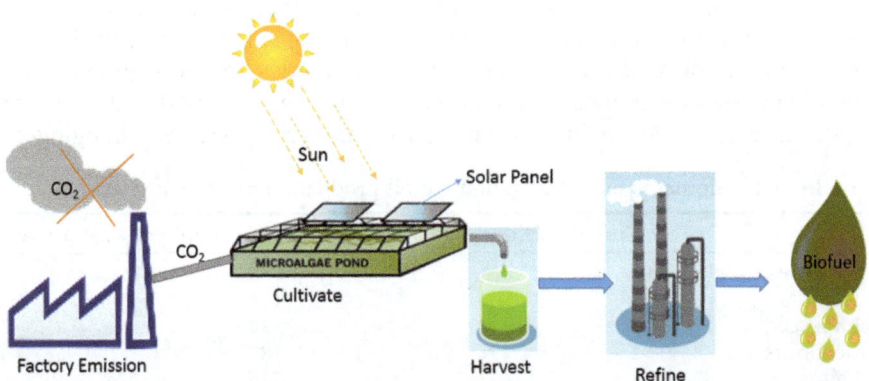

Figure 11.3 Algae farming for CO_2 reduction and biofuel production.[20]

- sunlight and other light sources in a reactor can be adjusted to increase reactor efficiency
- the system can be operated for a long time during algae production
- some bioreactors have convenient self-cleaning systems.
- the bioreactor is a closed, controlled, automated system, promoting continuous action for the hygiene of cultures
- the size of a bioreactor can be adjusted depending on the conditions needed for production.

It is shown that algae can produce approximately 300 times more oil per acre than some conventional crops such as rapeseed, palms, soybeans, jute, or jatropha. Algae can grow 20 to 30 times faster than food crops. Since algae have a harvesting cycle of 2–10 days during appropriate seasons and can accommodate several harvestings in a very short time frame. Algae can also be grown on land that is not suitable for other established crops (*e.g.*, arid land, land with excessively saline soil, and drought-stricken land). Table 11.1 provides the biomass yields and gallons of oil production per acre per year for various conventional crops as well as algae.[22]

When the algae are fully grown, flocculation and centrifugal separation techniques are applied. Flocculation is a method of using chemicals to combine microscale algae from the medium and form larger clusters. When the clusters become larger, screening or centrifugal filtration processes can be applied. The main disadvantage of the flocculation process is the chemicals that are added, because they can be difficult to remove later on. Centrifugation is a filtration method to separate algae from the medium *via* centrifugal forces. This process can be used for small- and large-scale liquid–solid separations.

11.2.2 Algae Fuels

Algae-based fuels are considered to be alternative sources of fuel in comparison to fossil fuels (*e.g.*, coal, oil, and natural gas). Like fossil fuels, harvested algae release CO_2 when burnt, but unlike fossil fuels, the CO_2 is removed from the atmosphere by the growing algae (thus contributing less to the greenhouse effect). Several private companies and government agencies have been focusing on algae-based fuels to reduce capital and operating costs and make algae fuel production commercially viable. Changing oil

Table 11.1 Biomass yields and gallons of oil production per acre per year.

Crop	Crop yields per year	Gallons of oil per acre per year
Soybeans	1	48
Peanuts	1	113
Rapeseed	1	124
Coconuts	1	287
Palms	1	635
Algae	20–30	15 000

prices, competing demands between foods and biofuel sources, and the world food crisis have all had a major interest in algaculture (farming algae) to manufacture vegetable oil, biodiesel, bioethanol, biomethanol, bio-gasoline, biobutanol, and jet fuel. The oily part of the algae biomass can be extracted using hexane or other solvents, and then converted into biodiesel through a process similar to that used for vegetable oil, or converted in a refinery into 'drop-in' replacements for petroleum-based fuels. A number of different factors affect CO_2-capturing rates:[23]

- selection of appropriate algae strain (over 2800 different strains of algae)
- injection of CO_2 into the bioreactor to maximize growth
- utilization of available sunlight during the day time
- glucose injection to the bioreactor during dark periods
- filtration of UV rays and unwanted lights to minimize damage to the algae
- continuous removal of by-product oxygen to prevent oxygen poisoning for algae
- circulation of water during growth
- monitoring the algae culture throughout production and harvesting
- maintaining the temperature and other food sources for growth
- stressing the algae prior to harvesting by limiting nutrients, temperature, or nanoparticles.

Among its many characteristics, algal fuel can have a minimal impact on fresh water resources, it is produced using ocean water as well as wastewater, and it is biodegradable and relatively harmless to the environment if accidents or spills occur. During the process of algae oil extraction, hexane removes oil from the microorganisms by evaporation. After the oil-extraction process, the pressed cake of oil can be used as a solid fuel source (it burns well), animal food, or dietary supplement (to meet future food demand for many countries). Algae oil can also be converted into biofuels using three different methods:[24]

(i) base-catalyzed transesterification of algae oil with alcohol
(ii) acid-catalyzed esterification of algae oil with methanol
(iii) conversion of the oil into fatty acids and then catalyzation of the fatty acids to alkyl esters to form algae biofuel.

As a result, algae fuels can solve many challenging world problems:

- they can displace fossil fuels: every ton of algal biofuels can replace one ton of fossil fuel-based petroleum
- they will reduce CO_2 and other emissions, and increase oxygen, in turn slowing the carbon footprint and eventually sequestering carbon in the long term

- they can provide sustainable fuel generation and economic development in many regions
- they will limit the use of freshwater, croplands, and other places
- their production facilities can be closer to industrial zones for larger manufacturing.

11.3 Parameters Affecting Algae Growth

11.3.1 Effects of Bubble Size and Mixing

Gordon and Seckbach have stated that the growth of an algal culture is a critical factor for biofuel efficiency. Illumination and proper bubble mixing are the two major components in this process.[6] The microbubble generation system could decrease the diameter of the CO_2 gas in the bioreactor and improve the biomass rate. The fluidic oscillator approach in the microbubble system could decrease friction in the pipes and cause a steady flow for greater algal growth. These authors investigated the feasibility of growing microalgae on steel plant exhaust gases using standard approaches with an airlift loop bioreactor design. Test results showed that microbubble steady growth of biomass was 100% higher for survival rates. Also, removing O_2 gas generated during growth of the algae causes high growth rates and a high density of biomass production.[25]

Tesar explored the generation of sub-millimeter air bubbles in the liquid media for algae growth. The proposed technique might offer many advantages for new algae growing stations. Sometimes reducing the bubble size to smaller diameters could make the process unfeasible. An inexpensive way of producing microbubbles could be achieved using pulsating air flow at different frequencies into a nonmoving porous media with lower pore sizes. The smaller CO_2 bubbles in bioreactors would improve the absorption rates of the bubbles for algal growth rates.[26]

11.3.2 Impacts of Nanomaterials and Processes on Lipid Accumulations

The presence of nanoscale materials in the microalgae growth medium could have a major impact on the accumulation of liquid, which may be due to stress on the microalgae. Stress in growth environments, such as low temperature ($<20\,°C$), nutrient depletion (nitrogen), and excessive metal ion concentrations (*e.g.*, Fe) could trigger lipid accumulation in the body of the algae. Nanoparticle additions of silica and iron oxide to the growth medium may result in a strong sheer between the cell and nanoparticles, which may affect nutrient absorption by the cell wall. This process influences cells to rapidly uptake available nutrients for the accumulation of lipids in the body. Zhang *et al.* investigated the effects of nanoparticles, such as silica, silica/iron oxide, and gold on *Escherichia coli* (*E. coli*) and found that these nanoparticles had no negative impact on its growth and activity.[27] This study

indicates that the type of microorganism could have a different response to nanomaterials. It was also stated that the lipid content in the microalgae could directly affect the algae oil production and biodiesel manufacturing, so more studies need to be conducted in this field to explore the major effects.[27]

Some new techniques were developed to potentially break cell walls and extract liquid from the algal body. Nanodispersion (Nano Dispersions Technology, Inc.) is a process used to mill particles into nanoscale sizes. If that method is combined with a solvent extraction process, then it may turn two process into one to increase the functionality of the new process. Electroporation might have a positive effect on lipid extraction from algae when using Nile red staining.[28] A group of scientists from the Hawaii Natural Energy Institute investigated a co-solvent system using an ionic liquid to co-extract bio-oil and protein from the body of algae. The extracted algae oil could automatically separate into the upper layer of the liquid. Recently, many other techniques, including steam explosion, a combination of microwave and ultrasonication, and an electromagnetic field have been explored by various researchers; however, no substantial improvements were observed on lipid accumulation and oil extraction.[28]

Newly developed flow cytometric procedures could improve the selection of microalgae in the process, the lipid content of the biomass, as well as other necessary characteristics and algae growth. This method consists of different steps, involving the incubation of various algae populations (or other microorganisms) in the growth medium, which could require a competitive growth and desired selection for the improved lipid production. In this way, the efficiency of biomass production and lipid generation could be enhanced thousands of folds higher compared to the other conventional methods.[29]

In order to enhance the lipid content of microalgae, a commercially available method was developed by modifying a metabolic pathway of the model marine diatom (*Thalassiosira pseudonana*). Basically, this modification accelerates lipid accumulation of the diatom by reducing lipid catabolism without any further decrease in the growth process. This modification allows for increased triacylglycerol content and increased viability of diatoms in the transformation process of cells. This method does not allow any genetic structural changes on the diatoms at the molecular level, but rather only involves 'native sequence modifications,' which can be considered as a non-genetically modified organism.[30] It was stated that theoretically this kind of modification could be possible with a native host substance through natural recombination and crossover events. During this process, some additional genes could be modified *via* a co-transfection process for improved biomass rates.[30]

Gerwick and Hildebrand investigated the biologically active constituents of different marine organisms (*e.g.*, marine algae and cyanobacteria), which are especially rich in cytotoxic natural products and can interfere with tubulin or actin polymerization process. Additionally, they focused on

diatom unicellular algae, which play essential roles in carbon fixation and other related concerns. The authors state that the diatom cell wall structure has silica mineral with micron and nanoscale features. These microalgae are attractive microorganisms for algal-based biofuel production because they can accumulate large amounts of neutral lipids in sea water in which many microorganisms cannot survive.[29,30]

11.3.3 Nanotechnology and Illumination for Bioreactor Design

Recently, nanotechnology associated studies have been conducted on the algal biomass growth during artificial illumination in closed photo-bioreactors (PBRs) but not in open PBR systems. This new generation of nanosystems could also treat the algal biomass and catalytically upgrade the crude oil of biomass. The nanoscale materials in this system can be applied to both open-pond and PBR systems.[6] It is assumed that more scientific and technological studies of nanotechnology on the photosensitivity of nano-materials, surface plasmon effects, algae growth, lipid accumulation, extraction, separation, flirtation, and biofuel production will appear in the near future.[31] Note that 'plasmon' refers to the behavior of free electrons in the structure of metals under the influence of electromagnetic radiation.[32,33]

One of the primary concerns of algal culture is the illumination of sun-light, which only substantially affects algae growth rates for 50% of the day. To be able to reach sustainable light sources for a longer period of time, the installation and maintenance of a continuous light source in the PBR are strongly desired. When the algal biomass concentration increases in the bioreactor, then uniform illumination of the culture will be a challenging issue since the self-shading as well as biofilm formation on the reactor surface will reduce light penetration.[34] Also, light sources cannot be closely attached to the PBR, because they may generate considerable heat and rays. The uniform and sustainable illumination of the PBR seriously limits the light-conversion efficiency of conventional PBRs.[6]

Using wastewater and sewer water as a source of nutrients for microalgae growth has received much attention worldwide over the past few years. However, this process requires larger open ponds, and wastewater treat-ments and investments. Pozza *et al.* developed a photobioreactor with an internal illumination system in its pipes using LED light and Plexiglas EndLighten rods.[35] The larger-scale test results obtained with *Scenedesmus* sp. and artificial wastewater indicated that the grow rate and nutrient removal with external illumination were comparable to previous results obtained on a smaller scale. This novel approach could enhance the supply of light in a more efficient and economical way, and improve the scalability of the PBR for continuous algae biomass production.[35]

Neeb also investigated the LED-based bioreactor systems to partially provide light for algae growth rates. In this method, the light source was partially submerged in the aqueous liquid with a cooling liquid for the LEDs

of the lighting system. This process will likely eliminate the additional heat of the light on the growth medium and reduce the harmful effects of rays on cell membranes. Blue LEDs, which emit light in the wavelengths of 400 to 500 nm, have a protection film on the top surface to protect the algae from excessive heat, rays, and cooling liquid. An appropriate cooling liquid has a good transparency, refractive index, and low viscosity to flow easily in the systems. In this cooling liquid, high refractive TiO_2 nanoparticles with a refractive index of about 1.8 could be dispersed in the cooling liquid to increase the refractive index of the suspension to about 1.7. This would greatly reduce some concerns for algae growth.[3]

A number of scientists from the University of Western Australia have developed an optical nanofilter system to improve the formation and yield of algae photopigments by changing the wavelengths of light absorbed by the microalgae. They used dispersed colloidal gold and silver nanoparticles in flasks to improve microalgae growth and the formation of photopigments by eliminating unwanted wavelengths of light. Light is a very important element for algae growth, so excessive light or even certain wavelengths of light can damage the algae as well as its growth and lipid accumulation. This technique involves backscattering of the wavelengths, which greatly improves algae growth. In this system, nanoparticle dispersion does not directly come in contact with the algae; thus, contamination and toxicity issues are not major concerns. This technique could increase the yield of microalgae biomass for commercial algae productions, as well as biofuels, medical antioxidants, cosmetic and pharmaceutical agents, and so forth.[36]

As stated earlier, algae production requires both sufficient sources of light with an optimum wavelength. Shorter wavelengths can be detrimental to many microalgae populations. A number of different methods have been proposed to reduce the light intensity in the growth medium. Some authors have suggested using backscattered light for promoting phototrophic algal growth. This approach can have the major advantage of localized surface plasmon resonance (LSPR) of the metallic nanoparticles (Au and Ag). Photons of the lights and surface plasmons can be utilized to increase light absorption and scattering capabilities at some specific wavelengths. In the presence of Ag nanoparticles, a strong backscattering of blue light from the suspension accelerates the photosynthetic activity of microalgae in the PBR system and improves the biomass growth rates.[37]

In the presence of Ag nanoparticles, the amount of scattered light to the microalgae bioreactor could be precisely controlled by changing the size and concentration of plasmonic nanoparticles. The frequency of scattered light could also be tuned by changing their shape and size. For instance, spherical plasmon nanoparticles can greatly scatter the more harmful blue light and enhance microalgae growth rates. Also, the shape of plasmon nanoparticles can significantly change the backscattering effects of light and help the microalgae grow faster.[38] Microalgae produce triglycerides that mainly contain fatty acids and glycerin, the former of which could be converted into biodiesel while the latter could be used as a by-product for different

industrial purposes. Tuning the light intensity and eliminating the undesired light could improve productivity.[37]

It was recently reported that metabolic activities of many photoactive microalgae were not uniform under the electromagnetic spectrum. For example, *Chlamydomonas reinhardtii*, a green microalgae, displays two different absorption peaks in the blue and red regions of the electromagnetic spectrum. The reasons for these two different peaks may be attributed to the fact that the algal photoactive pigments could work in these specific ranges of the spectrum, whereas the other may cause photoinhibition at wavelengths of 520 and 680 nm. This will allow scientists to enhance light intensity of the wavelengths favorable to microalgae growth. This can also eliminate some of the unwanted microalgae species which are not photoactive in those wavelengths and where lipid accumulation is fairly low.[38]

Torkamani *et al.* also reported that the photoactivity of green microalgae was nonmonotonic across the electromagnetic spectrum. Their test results conducted on *C. reinhardtii* (green alga) and *Cyanothece* 51142 (green-blue alga) revealed that the wavelengths and specific backscattering of the Ag nanoparticles in the blue region could be caused mainly by localized surface plasmon resonance effects. This can also promote more than 30% of microalgae growth. In addition, the wavelengths and backscattered light could be controlled by various other parameters, such as geometric features, concentration of nanoparticles and surface oxidation, tarnishing, and contamination.[38]

11.4 Nanocatalysts for Algal Fuel Productions

11.4.1 Nanocatalysts in Algae Growth, Lipid Accumulation, and Extraction

Algae fuels are of great interest worldwide because of their enormous energy potential to meet future energy demands. Algae can grow 20 to 30 times faster than food-based crops and can be cultivated anywhere in the world as long as light and a growth medium are provided. Some microalgae can produce up to 60% of their biomass in the form of oil, and their remaining biomass is mostly carbohydrates. The oil and hydrocarbons can be converted into biodiesel, alcohol, biogas, solid briquettes/pellets, as well as other products commercially produced using other natural and synthetic sources.[39]

The University of Adelaide in Australia aims at developing a nanostructured nanofiber biocatalyst assembly for the bioconversion of CO_2 into methanol. This assembly is an enzyme-on-nanofiber device to maintain various enzymatic reactions for sequential dehydrogenations of CO_2 with *in situ* regeneration. It is a promising development in the field of bionanotechnology in order to understand fundamental concepts and mechanisms of the biotransformation. The outcome of this research could offer many advantages to developing new research directions in biocatalyst systems for

the conversion of greenhouse gas into green energy and low-carbon economy in the region.[40]

Researchers at Iowa State University investigated how silica nanoparticles could be used to extract and sequester fuel-related, high-value compounds from lipids of the algae biomass. Most of the algae oil was converted to biodiesel using different catalysts.

According to Lin, the new technology was helpful during several important steps of the algae-to-biofuels supply chain process for efficient oil extraction. In this study, the solid catalyst provides a cost-effective conversion method for extracting algae oil.[41] Lin *et al.* studied acid–base mesoporous calcium oxide–silica nanoparticles to catalyze the esterification of fatty acids to methyl esters. This catalyst was highly effective for the conversion of soybean oil and poultry fat biomass feedstocks to biodiesel methyl esters. The catalyst, called the multiple cloning site (MCS), facilitates the conversion of bio-renewable feedstocks into methyl esters. The catalyst can be easily recycled and reused in the same process many times, so it is fairly economic and environmentally friendlier. Some other fine-size catalysts (Portland cement, calcium silicate hydrates, and other mixed-oxide materials) are also efficient for the esterification of soybean oil. Some of these catalytic materials consist of functional groups (*e.g.*, Lewis acidic and Lewis basic) for the synthesis of biodiesel from different free fatty acid (FFA)-containing oil feedstocks (*e.g.*, animal fats and restaurant waste oils). These functional groups can easily esterify FFAs and trans esterify other oils with short-chain alcohols (methanol and ethanol) to procedure alkyl esters (biodiesel). Some of these catalysts can be easily recycled and reused in the same or other processes.[42]

QuantumSphere, Inc. developed a breakthrough process using nanocatalysts to convert wet algae from the Salton Sea into renewable fuels. The company plans to establish an algae biogasification process to utilize nanometals as catalysts for turning similar biomass materials into methane, hydrogen, or other synthetic gases. The biogases can be used for transportation, household, industrial heating, and other purposes. QuantumSphere, Inc. claims that this process could convert any biomass, such as leaves, algae, vegetable waste, or corn stalks, into biofuels.[39]

In order to develop an inexpensive biodiesel production strategy, heterotrophic microalgae were considered because they are capable of accumulating a high lipid content (up to 57% w/w) in their body and can use complex structured carbons as nutrients to yield an equivalent quantity of oil. It was reported that various nanomaterials (CaO and MgO) could stimulate microalgae metabolism to enhance the cultivation, as well as lipid accumulation and extraction rates, of microalgae. In addition, the presence of nanomaterials could significantly improve the efficiency of lipid extraction without harming the microalgae.[43]

Zhang *et al.* reviewed the lipid accumulation of heterotrophic microalgae with a focus on the application of nanotechnology in lipid accumulation, extraction, and transesterification.[43] Nanotechnology products have several other potential applications in the algae process because of their unique

abilities and properties. The solvent extraction process can weaken or break cell walls of the microalgae and enhance lipid diffusion to the outer environment in order to dissolve the lipid. Various nanomaterials can be utilized for the immobilization of algae because of their extremely high surface area, and can be recovered easily after the process by filtration, centrifugation, or heat treatment.

A similar research study revealed that modified nanoscale silica spheres accomplished extraction from live microalgae and then were sent back to participate in the lipid accumulation process. Some nanomaterials (*e.g.*, CaO, Fe_2O_3, Al_2O_3, SiO_2, and MgO) are heterogeneous catalysts and carriers, and can have relatively high conversion rates (>99%) with a very small quantity (<1% of oil). However, bulk-scale catalytic materials have some limitations for the same applications. As a result, it is postulated that catalytic nanomaterials could have a high performance in transesterification and other algae processes.[43]

Yang *et al.* reported that different types of nanomaterials have been synthesized and used in the extraction of microalgae biomass. Using nanoscale materials was a favorable approach for lipid extraction since they might eliminate the use of toxic materials and organic solvents and eliminate demand for the solvent–lipid separation step during the conversion extraction process. Moreover, using spherical nanomaterials to extract lipids from living microalgae will not have many harmful effects on the microalgae, and could be continuously used for lipid accumulation for a long period of time.[44]

The transesterification process of an algae oil extracted from river algae was investigated using nanoscale calcium oxide catalysts. These nanoscale catalysts were manufactured by the calcination of egg shell powder in a sonochemical reactor. The transesterification tests were performed using extracted oil from algal biomass using hexane as a solvent. The effects of oil-to-methanol ratio, temperature, and catalyst concentrations were evaluated, and the optimum values were determined. Test results indicated that algal oil was a good source of oil, and nanoscale CaO catalysts were highly effective catalysts for their transesterification.[45] The following conclusions were the result of this study:

- the egg shell plays as an important role in the transesterification of algae oil in order to produce a fatty acid methyl ester (or biodiesel)
- the optimum temperature was 55 °C for the transesterification of biodiesel
- the optimum methanol to oil ratio was 9 : 1 for good conversion of fatty acids
- when the amount of catalyst was increased from 0.5 to 1.25 wt%, the transesterification was facilitated; however, after the addition of 2 wt% catalyst, the transesterification was still the same
- the collected algae samples had a total of 58.2% oil in the dry base
- using the CaO catalyst of the egg shell, 96.3% of oil was converted into biodiesel.[45]

A native microalga, called *Chlorella vulgaris* ESP-31, grown in an outdoor hollow photobioreactor with CO_2 aeration had about 63.2% oil content. This oil was converted into a biofuel using enzymatic transesterification after the immobilization process with nanocomposites (Fe_3O_4–SiO_2). The extracted algal oil and direct disrupted alga biomass products were both tested using the transesterification process. Test results showed that the second product had a higher biodiesel conversion rate (97.3 wt% oil) than the first product (72.1 wt% oil).[46]

11.4.2 Other Process Parameters Affecting Lipid Extractions

Lipid extraction is one of the major steps in biofuel production and a very costly operation. The most commonly used lipid extraction process is solvent extraction. Organic solvents, such as hexane, methanol, chloroform, or their combination, are employed to dissolve the oil from the dry or wet biomass of microalgae. Solvent type, agitation, pressure, and process temperature considerably affect the extraction yields of the lipids. It was reported that lipid extraction yields using chloroform and methanol were about 20% w/w but were considerably low using hexane in the extraction. The effects of the irradiation and ultrasonication during the extraction process improved the lipid yield, but these methods necessitate additional energy input and increase the overall extraction and biofuel production costs.[44]

Transesterification, a major step in biodiesel production, uses different oils derived from animals, plants, or oleaginous microorganisms to react with alcohol (mainly methanol) for synthesizing fatty acid methyl esters (FAMEs) (biodiesel). The reaction that occurs can take place under either extreme conditions of high temperature and pressure, or a mild condition in the presence of catalysts. Currently, biodiesel is mainly synthesized through the catalytic method because of the simplicity of the process and the cost. Four types of catalysts – acid, base, enzyme, and heterogeneous – have been studied in the synthesis of biodiesel. Also, three major ways of reducing enzymatic biodiesel production costs include the following:

(i) reducing lipid production cost through development of an inexpensive and efficient method
(ii) enhancing lipid production efficiency
(iii) reusing lipids, solvents, catalysts, and other recyclable materials in the same process.[27]

CO_2 content, organic matter, contamination, temperature, and other sources (N, P, and Fe) can drastically affect the algae biomass and lipid content. Although algae-based biofuels provide many advantages over fossil fuels with respect to their renewability, lower levels of CO_2 emissions, biodegradability, and safety, their cost is still fairly high (about $7.50 per gallon). This production cost includes cultivation, harvesting, extraction, and conversion to biodiesel. Thus, at current conditions, algal biofuels are

not economically viable. In order to reduce the overall cost and make algae-based biofuel economically viable, many other methods have been proposed over the past two decades, including illumination, nanoparticle additions, optimization of the food source and other nutrients, maximization of CO_2 concentrations, elimination of undesired light, removal of O_2 after the photosynthesis process, the recycling and reusing of catalysts and other materials, adjustment of the bioreactor temperature, the use of wastewater and sewer water, lipid extraction, the use of other materials, and biodiesel conversion.[47–49] Thus, more scientific and technological studies on algae-based biofuels will be needed, in order to match the price of fossil fuels (about \$2.50 per gallon).[46]

11.5 Conclusions

Reducing greenhouse gas emissions is a major challenging issue in the world and has been a primary concern for more than three decades. Microorganisms in general and microalgae in particular can address some concerns about global warming by converting CO_2 into biomass and biofuel. This conversion can be accelerated through nanostructured systems associated with biological systems. Even though algae biofuels have many advantages over fossil fuels, the biofuel production cost is relatively high ($\sim$\$7.50 per gallon) and needs to be reduced substantially. Some of the approaches for reducing the cost of biofuel production include using illumination, adding nanoparticle additions, optimizing the food source and other nutrients, maximizing CO_2 concentrations, eliminating undesired light, using wastewater and sewer water, employing advanced lipid extraction, and utilizing biodiesel conversion. However, microalgae-based biofuel is considerably expensive. In this chapter, we have summarized the most recent developments in the fields of CO_2 emissions reduction, microalgae growth, and applications of nano and biological systems together to address some of the concerns of greenhouse gasses and global warming issues. This information may be useful for future scientific and technological developments in the field.[46]

References

1. CORDIS, (The Community Research and Development Information Service). *Algae to capture CO₂*, 2013, available online at http://phys.org/news/2013-03-algae-capture-co2.html#jCp.
2. Sustainable Green Technologies Inc., *Why Algae for Green Oil and Biofuels Production*, 2015, available online at http://sgth2.com/algae_faq.
3. T. W. Neeb *Algae Reactor*, U.S. Pat. Application 14/031,061, 2014, available online at http://www.google.com/patents/US20140030801.
4. R. Asmatulu and H. Misak, *J. Nano Educ.*, 2011, **3**, 13–23.
5. R. Asmatulu, E. Asmatulu and B. Zhang, *Int. J. Mech. Eng. Educ.*, 2012, **40**, 1–10.

6. R. Gordon and J. Seckbach, *The Science of Algal Fuels: Phycology, Geology, Biophotonics, Genomics and Nanotechnology, Cellular Origin, Life in Extreme Habitats and Astrobiology*, Springer Science & Business Media, London, 2012, vol. 25, p. 539.

7. Cyberchem, *An explanatory guide on the function of bacterial digestants and how they work to digest waste*, 2015, available online at http://cyberchem.co.za/what-is-bio-augmentation/.

8. N. S. Jakubovics and R. J. Palmer, *Oral Microbial Ecology: Current Research and New Perspectives*, Caister Academic Press, New Britain, 2013, p. 254.

9. Z. R. Rather, *Successive Botany: for B. Sc Part 1*, Pentium Publisher, India, 2009.

10. Y. S. Nam, A. P. Magyar, D. Lee, J. W. Kim, D. S. Yun, H. Park, T. S. Pollom Jr, D. A. Weitz and A. M. Belcher, *Nat. Nanotechnol.*, 2010, **5**, 340–344.

11. Y. Lie, R. Asmatulu and N. Nuraje, *ScienceJet*, 2015, **4**, 169–173.

12. Intuitive Environmental Solutions, LLC, *What is mold?* 2008, available online at http://www.intuitiveenvsol.com/What_Is_Mold_.html.

13. H. Curtis and N. S. Barnes, *Invitation to Biology*, Macmillan, USA, 1994, p. 862.

14. H. Heasler and C. Jaworowski, *Life in Extreme Heat—National Park Service*, Yellowstone Resources and Issues Handbook, 2014.

15. Sistematica, *Protoctistas*, 2015, available online at http://sistematica.wikispaces.com/.

16. DoralBio5, *Red, Brown & Green Algae*, 2015, available online at http://doralbio5.wikispaces.com/Red%2C + Brown + %26 + Green + Algae.

17. F. E. Round, *The Ecology of Algae*, Cambridge University Press, Great Britain, 1984, p. 664.

18. M. Berger, *Nanotechnology probe taps into algae cell and saps electrical energy*, Nanowerk, 2010, available online at http://www.nanowerk.com/spotlight/spotid = 16087.php.

19. M. A. Borowitzka and N. R. Moheimani, *Algae for Biofuels and Energy*, Springer Science & Business Media, 2012, p. 299.

20. Innoventures Canada (I-CAN), *Algae system grows environmental benefits, bottom line*, 2015, available online at http://www.i-can.ca/eng/initiatives/algae.html.

21. G. Toole, S. Toole and S. M. Toole, *Essential A2 Biology for OCR*, Nelson Thornes, United Kingdom, 2004, p. 176.

22. Skimaire High Tech, *Algae waste water treatment to green energy ideas*, 2010, available online at http://skimairecanada.blogspot.com/2010/12/algae-waste-water-treatment-to-green.html.

23. BioScience, *Microalgae: The potential for carbon capture*, 2010, available online at http://bioscience.oxfordjournals.org/content/60/9/722.full.

24. V. Babu, A. Thapliyal and G. K. Patel, *Biofuels Production*, John Wiley & Sons, Salem, MA, 2013, p. 392.

25. W. B. Zimmerman, B. N. Hewakandamhy, V. Tesař, H. C. H. Bandulasena and O. A. Omotowa, *Food Bioprod. Process.*, 2009, **87**, 215–227.
26. V. Tesař, Microbubble generation by fluidics. Part 1: Development of the oscillator, *Colloquium Fluid Dynamics, 2012, Institute of Thermomechanics AS CR*, v.v.i., Prague, October 24–26, 2012.
27. X. L. Zhang, S. Yana, R. D. Tyagia and R. Y. Surampalli, *Renewable Sustainable Energy Rev.*, 2013, **26**, 216–223.
28. P. Chen, M. Min, Y. Chen, L. Wang, Y. Li, Q. Chen, C. Wang, Y. Wan, X. Wang, Y. Cheng, S. Deng, K. Hennessy, X. Lin, Y. Liu, Y. Wang, B. Martinez and R. Ruan, *Int. J. Agric. Biol. Eng.*, 2009, **2**(41), open access at http://www.ijabe.org, DOI: 10.3965/j.issn.1934-6344.2009.04.001-030.
29. William H. Gerwick and Mark Hildebrand, *Efficient Method for Selecting Microalgae Using Flow Cytometric Sorting*, patent pending, UCSD Ref. No. 2012-363, 2012, available online at http://techtransfer. universityofcalifornia.edu/NCD/23592.html.
30. William H. Gerwick and Mark Hildebrand, *Methods for Increasing Microalgae Biofuel Production and Native Sequence Modifications*, patent pending, UCSD Ref. No. 2013-251, 2013, available online at http:// techtransfer.universityofcalifornia.edu/NCD/23592.html.
31. L. J. Sherry, S. Hui Chang, G. C. Schatz, R. P. Van Duyne, B. J. Wiley and Y. Xia, *Nano Lett.*, 2005, **5**(No. 10), 2034–2038, available online at http:// sites.northwestern.edu/vanduyne/files/2012/10/2005_Sherry_3.pdf.
32. N. Nuraje, R. Asmatulu and S. Kudaibergenov, *Curr. Inorg. Chem.*, 2012, **2**, 124–146.
33. N. Nuraje, S. Kudaibergenov and R. Asmatulu, Solar energy storage with nanomaterials, in *Production of Fuels Using Nanomaterials*, ed. R. Luque and A. M. Balu, Taylor and Francis, USA, 2013, ch. 4, pp. 95–117.
34. C. Y. Chen, K. L. Yeh, R. Aisyah, D. J. Lee and J. S. Chang, *Biosens. Technol.*, 2011, **102**, 71–81.
35. C. Pozza, S. Schmuck and T. Mietzel, A novel photobioreactor with internal illumination using Plexiglas rods to spread the light and LED as a source of light for wastewater treatment using microalgae 2015, available online at http://www.qub.ac.uk/researchcentres/ATWARM/ DisseminationTemplates/Individualprojectsdisseminationoutreach outputs/24Carlo/Filetoupload,368080,en.pdf.
36. Phys.org, Enhancing microalgae growth to boost green energy production, 2013, available online at http://phys.org/news/2013-11-microalgae-growth-boost-green-energy.html#jCp.
37. A. DuChene, SU research team uses nanobiotechnology-manipulated light particles to accelerate algae growth, 2010, available online at http:// www.syr.edu/news/articles/2010/algae-biofuel-08-10.html.
38. S. Torkamani, S. N. Wani, Y. J. Tang and R. Sureshkumar, *Appl. Phys. Lett.*, 2010, **97**, 043703.
39. AZO Nano, Turning algae into biofuels using nanotechnology, 2009, available online at http://www.azonano.com/news.aspx?newsID = 10091.

40. The University of Adelaide, A nanofiber biocatalyst assembly for bioconversion of CO_2 into methanol: Turning greenhouse gas into a renewable energy, Bionanotechnology Laboratory Water Energy and Materials (BioNanoTech), 2014, available online at http://www.adelaide.edu.au/bio-nano-tech/projects/#1.1.

41. V. S. Y. Lin, Iowa State researchers part of $78 million national effort to develop advanced biofuels, 2010, available online at http://www.news.iastate.edu/news/2010/jan/biofuels#sthash.Bs7×6GYt.dpuf.

42. V. S. Y. Lin, J. A. Nieweg, J. G. Verkade, C. R. V. Reddy and C. Kern, Porous silica and metal oxide composite-based catalysts for conversion of fatty acids and oils to biodiesel, 2006, available online at https://www.google.com/patents/US7790651.

43. X. L. Zhang, S. Yana, R. D. Tyagia and R. Y. Surampalli, *Renewable Sustainable Energy Rev.*, 2013, **26**, 216–223.

44. S. T. Yang, H. El-Ensashy and N. Thongchul, *Bioprocessing Technologies in Biorefinery for Sustainable Production of Fuels, Chemicals, and Polymers*, John Wiley & Sons, 2013, p. 488.

45. S. Siva and C. Marimuthu, *Int. J. ChemTech Res.*, 2015, 7(No.4), 2112–2116.

46. D. T. Tran, K. L. Yeh, C. L. Chen and J. S. Chang, *Bioresour. Technol.*, 2012, **108**, 119–127.

47. R. Asmatulu, *Nanotechnology Safety*, Elsevier, Amsterdam, The Nederland, August, 2013.

48. A. Howel, Calcium oxide nanocrystals in creating algal biofuels, Nanotechnology in City Environments (NICE) Database by the Center for Nanotechnology in Society, Arizona State University, 2012, available online at http://nice.asu.edu.

49. E. Jurak, S. Jurak and R. Asmatulu Integrating the growing, harvesting, and processing of algae-based oil production systems for K-12 student projects, ASEE Midwest Section Conference, Fort Smith, AR, September 24–26, 2014, 8 pages.

CHAPTER 12

Organic Reactions using Green Photo-active Nanomaterials

HANYING BAI

Center for Craniofacial Regeneration, School of Dental Medicine, Columbia University, 622 W168th Street, Vanderbilt Building 12-210, New York, NY 10032, USA
Email: hb2375@cumc.columbia.edu

12.1 Introduction

Nanomaterials have been widely used in green chemistry. Green chemistry is chemical processes that reduce or eliminate hazardous substances.[1] The design of future new green chemical process follows minimum hazard as the criteria. Furthermore, green chemistry refers to resource reduction, high efficiency and other practices to reduce or avoid the creation of waste and pollution.[1-3] Organic pollutants are one main component of pollution in the water and in the air. The study of degradation and monitoring of organic pollutants has become a significant direction for green chemistry and environmental remediation.[2] Microorganisms, especially hazardous species, are one important component of pollutants. Antibacteria and antivirus processes are essential for pollution control and environmental remediation.

Photocatalytic degradation of pollutants is a promising method because light is cheap, abundant, clean, and sustainable. Furthermore, light is capable of inducing highly selective reactions.[4-6] Photocatalysis is particularly good at the degradation of persistent organic pollutants.[7]

RSC Green Chemistry No. 42
Green Photo-active Nanomaterials: Sustainable Energy and Environmental Remediation
Edited by Nurxat Nuraje, Ramazan Asmatulu and Guido Mul
© The Royal Society of Chemistry 2016
Published by the Royal Society of Chemistry, www.rsc.org

Photocatalysis works on light-induced redox reactions,[6] interferes with reactive oxygen species (ROS), and requires high-performance photoactive materials as a backbone. The effectiveness of photocatalysis on organic pollutant lies on the photophysics of the catalyst with its supplemental materials.

Nanomaterials have the advantages of a large contact surface area and high reactivity, which make them high-performance catalysts, adsorbents, and sensors in green chemical process.[2,8,9] Moreover, nanomaterials also have enhanced redox and photocatalytic capabilities compared with the same kind of material in larger sizes.[2,9]

The basic fabrication methods to make nanomaterials are:

(1) physical or chemical vapor deposition or vacuum evaporation
(2) sol–gel chemical synthesis method
(3) inter-liquid-phase synthesis method
(4) template coating or seed growing
(5) gas-phase synthesis techniques such as plasma synthesis
(6) microwave techniques
(7) delamination of layered materials.[6,11–14]

The photoactive nanomaterials for photocatalysis discussed in this chapter are of two types:

- semiconductors, such as TiO_2, ZnO, and In_2O_3
- photosensitizers, such as fullerenes (C_{60}) and carbon nanotubes (CNT).

In this chapter, we will discuss these five kinds of nanomaterials in photocatalysis of degradation of organic pollutants, as sensors of organic pollutants, and for their antimicrobiological ability. Hybridized nano-composites with other nanomaterials usually have higher properties than the pure materials, and their application in green chemistry is also introduced in this chapter.

12.2 TiO_2 Nanostructured and Hybrid Materials

Titanium dioxide (TiO_2) is the most commonly used material; it has high reactivity and chemical and biological stability.[15] It is a wide band gap semiconductor (3.18 eV for anatase and 3.03 eV for rutile).[7,16,17] Both anatase and rutile crystalline structures have a tetragonal structure, including six and twelve atoms per unit cell with an axial ratio of perpendicular axis/horizontal axis of 0.64 and 2.51, respectively. These two crystalline structures are widely applied to photocatalysis due to their easy synthesis.[18] Nano-size TiO_2 was widely used for photocatalytic degradation of organic pollutants and attracted a great deal of attention for environmental remediation.[2] The first report of nano-size TiO_2 was degrading polychlorinated biphenyls under UV irradiation in 1976.[19]

12.2.1 Photocatalytic Degradation of Organic Pollutants

Up to now, over 3000 kinds of organic refractory pollutants can be degraded using photocatalysis of nano-size TiO_2.[3,20–22] Examples of common organic pollutants are nitrobenzene (NB),[23] cyclohexane,[24] and sodium dodecylbenzene sulfonate;[25] explosive organics such as RDX (hexahydro-1,3,5-trinitro-1,3,5-triazine);[26] organophosphorus pesticides such as monocrotophos, parathion, dichorovos, and methamidophos;[27] and various dyes.[28]

Photocatalysis is a type of light-induced redox reaction and is enhanced by the catalyst. The basic mechanism of photocatalysis is summarized as follows. Charge carriers (an electron–hole pair) would generate on the catalyst surface when the catalyst was irradiated by the light with energy higher than its band gap. The charge carriers would react with the organic molecules adsorbed on the surface of the catalyst and the main reaction was oxidation. The most common oxidative species is the hydroxyl radical ($^{\bullet}OH$). These hydroxyl radicals are produced from nearby water molecules that are oxidized by the positively charged holes, and the radicals are strong oxidizing reactants. Figure 12.1 demonstrates the mechanism of photocatalysis on semiconductor nanomaterials, such as TiO_2, and the generation of oxidative species.

For small organic pollutant molecules, such as nitrobenzene (NB), their degradation process was photo-catalyzed oxidation. The basic reaction is shown in eqn 12.1, showing nitric elimination:

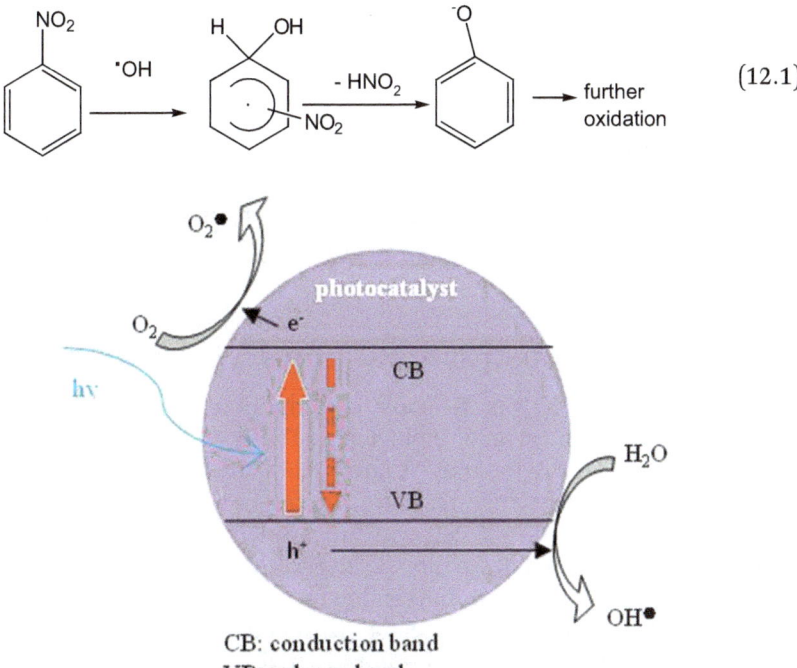

CB: conduction band
VB: valence band

Figure 12.1 Schematic illustration of the photocatalysis process on a semiconductor nanomaterial.

For persistent pollutants, *i.e.*, PFOA, the photo-catalyzed degradation procedure was that firstly, a photo-generated electron from TiO_2 initiated the decarboxylation of PFOA, as shown in eqn 12.2:

$$C_7F_{15}COOH + e^- + H^+ \rightarrow C_7F_{15}{}^{\bullet} + HCOOH \qquad (12.2)$$

Then, the C_7F_{15} radical would undergo further hydrolysis (eqn 12.3), HF elimination (eqn 12.4), and further hydrolysis (eqn 12.5) to form the PFHpA with one less CF_2 group.[10,29,30] The radical will keep being hydrolyzed and oxidized until it completely decomposes to CO_2 and F^-.

$$C_7F_{15}{}^{\bullet} + H_2O \rightarrow C_7F_{15}OH + H^{\bullet} \qquad (12.3)$$

$$C_7F_{15}OH \rightarrow C_6F_{13}COF + H^+ + F^- \qquad (12.4)$$

$$C_6F_{13}COF + H_2O \rightarrow C_6F_{13}COOH + H^+ + F^- \qquad (12.5)$$

The rapid generation of large amounts of oxidative species is essential for the performance of semiconductor catalysts. In order to enhance the photocatalysis performance of TiO_2, certain metals, non-metals, or their oxides were combined with TiO_2 nanomaterials, making doped nanomaterials.[2,3,31] The modified TiO_2 nanomaterials had lower band gaps, which benefit the electron transfer from VB to CB and facilitated the formation of oxidative species.[2,32] Nano-Ag particle-doped TiO_2 was used for the degradation of direct azo dyes.[33] FeZn–TiO_2 and SiC–TiO_2 composite nanoparticles were used for the degradation of 2-propanol.[34,35] Transition metals such as Cr^{3+}- and Ni^{2+}-doped nano-TiO_2 were used for gas-phase benzene degradation and the rate was apparently improved.[36] V-Ti nanomaterials were used to degrade chlorobenzene.[37] Zr^{4+}-doped TiO_2 and Cu-doped TiO_2 nanoparticles were used for the degradation of 4-chlorophenol.[38,39] Nano Mo-doped TiO_2 was used for the degradation of dichloroacetate (DCA).[40] The TiO_2 nanomaterials could also combine with supporting materials such as polymers to enhance the photocatalysis performance. TiO_2 hybridized with nanocomposite polymers for elevated photocatalysis has been reported: poly-1-naphthylamine (PNA)/TiO_2 for photodegradation of methylene blue (MB),[41] polyaniline (PANI)/TiO_2 for phenol,[42] and poly-3-hexylthiophene/ TiO_2 for methylene orange (MO).[43] Carbon nanotubes (CNT) were also a good support material with TiO_2 and the relevant research will be discussed in Section 12.5.

12.2.2   Sensors

TiO_2 nanomaterials have not only been used as photocatalysts for the degradation of organic pollutants, but have also been used as a component of nanosensors for specific airborne organic pollutants.[44–46] For example, polyaniline-TiO_2 nanocomposite ultra thin films were used for trimethylamine detection.[47] The TiO_2 composite nanosensor could also be used as a

biosensor due to its non-toxic nature, high surface area, high uniformity, and biocompatiblity.[2] Titanium can form coordination bonds with the amine and carboxyl groups of enzymes and it maintains the enzyme's bio-catalytic activity.[48–52]

12.2.3 Antimicrobiological Ability

TiO_2 nanomaterials were also used as antibacterials and antivirals.[53] The antibacterial activity is related to ROS (reactive oxygen species) production, especially hydroxyl free radicals and peroxide formed under UV irradiation *via* oxidative and reductive pathways.[54] The concentration of TiO_2 required to kill bacteria or viruses varies between 100 and 1000 ppm depending on the size of the nanomaterials and the intensity and wavelength of the light applied.[53] The antivirus reports included poliovirus 1,[55] the hepatitis B virus,[56] the Herpes simplex virus,[57] and MS2 bacteriophage.[58] The anti-bacterial reports included *E. coli*, *P. aeruginosa*, *S. aureus*, *Enterobacter faecium*, and *C. albicans*.[59,60]

12.2.4 Other Uses

TiO_2 nanomaterials are high-performance materials for solar cells. In the past 40 years, large amount of literature report its usage in solar cells.[6,15] However, this is not the main focus of this chapter.

12.3 ZnO Nanostructured and Hybrid Materials

Despite not having as many reports as TiO_2, zinc oxide (ZnO) nanomaterials are also high-performance materials for photocatalytic degradation of organic compounds. ZnO is a semiconductor and has a band gap of 3.3 eV.

12.3.1 Photocatalytic Degradation of Organic Pollutants

Based on similar mechanism to TiO_2, ZnO nanomaterials can be used for degrading many kinds of organic pollutants, such as toluene,[61] salicylic acid,[7] chlorinated phenols,[62] 2,4,6-trichlorophenol,[63,64] acid red 18,[50] rhodamine dyes,[65] direct blue 53,[66] ethyl violet,[67] and MB.[68] The preparation of ZnO with a flower-shape morphology resulted in a larger surface area of the catalyst and therefore enhanced photocatalytic activity.[69]

Similar to TiO_2, ZnO nanomaterials were also combined with metal/non-metal components to improve its reactivity.[70,71] There are reports about the following: lanthanum with ZnO nanoparticles for degradation of 2,4,6-trichlorophenol,[72] ZnO/SnO_2 nanocomposites used for degradation of MO,[73] and ZnO nanomaterials immobilized on Al foil for degradation of phenol.[74] Since ZnO has a larger fraction of absorption of the solar spectrum than the TiO_2, these two nanomaterials can combine together for photocatalysis.[75] Carbon nanotubes (CNT) are also a good support material with ZnO and the relevant research will be discussed in Section 12.5.

12.3.2 Sensors

ZnO nanomaterials can be applied as nanosensors for airborne volatiles, as can TiO_2. For example, there are reports about the detection of NH_3 and CO as charge donors and the detection of NO_2 and dioxin as charge acceptors on the surface of ZnO nanocomposites.[44–46]

12.3.3 Antibacterial Ability

ZnO nanoparticles showed strong antibacterial activities on a broad spectrum of bacteria.[76] ZnO nanoparticles could inhibit bacterial growth through the penetration of the cell envelope and the disorganization of the cell membrane.[77]

12.4 In$_2$O$_3$ Nanomaterials

In_2O_3 is a wide band gap semiconductor, with a direct band gap of 3.6 eV and an indirect band gap of 2.8 eV.[78] Nanoporous In_2O_3 was also used for photocatalytic degradation of perfluorooctanoic acid (PFOA).[30,79]

In_2O_3 as both single and multiple nanowires has specific selectivity to certain volatile compounds, and so could be made a nanosensor to detect some organic pollutants. For example, the In_2O_3 nanowire-fabricated sensor showed high selectivity to ethanol at 370 °C and the response time was 10 s to detected dilute ethanol at 100 ppm.[80]

12.5 Carbon Nanotube (CNT) and Hybridized Nanocomposites

Carbon nanomaterials have several forms – the most common forms are single-walled carbon nanotubes (SWCNTs) and multi-walled carbon nanotubes (MWCNTs).[81,82] CNTs can be considered as cylindrical hollow microcrystals of graphite; CNTs are graphitic carbon needles and have an outer diameter ranging from 4 nm to 30 nm and a length of up to 1 μm.[83] MWCNTs are made of concentric cylinders with spacing between the adjacent layers of about 3.4 Å, as shown as Figure 12.2.[84,85] CNTs are considered to be good support material for the enhancement of catalysts. A possible mechanism, shown in Figure 12.3, describes the enhancement in photocatalytic activity between the surface of a CNT and semiconductor nanomaterials.[86] The role of the CNT was as a highly efficient electron transfer, high electron storage, and high contact adsorbent of pollutants because of its specific surface. CNTs could also stabilize charge separation from electron–hole pairs by trapping electrons transferred from semiconductor catalyst, thereby hindering charge recombination.[86] The strong interaction between the CNTs and the catalyst could result in a close contact to form a barrier junction, which induced an effective route to reduce electron–hole recombination by improving the injection of electrons into the

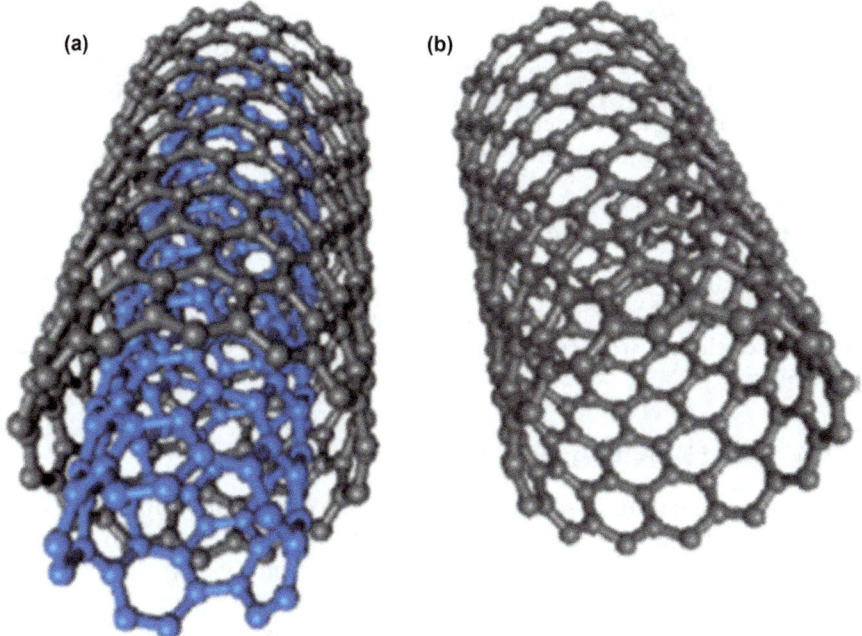

Figure 12.2   Structure of (a) a MWCNT and (b) a SWCNT84.

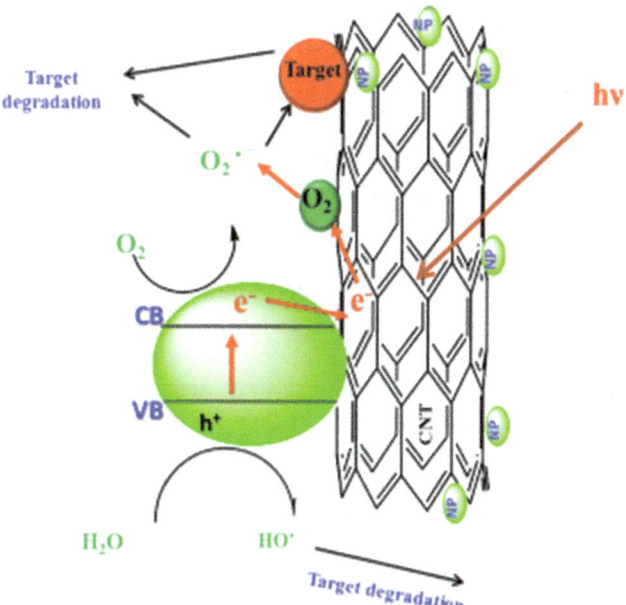

Figure 12.3   Schematic diagram of the proposed mechanism of photodegradation over CNT/semiconductor composite.[86]

CNTs. Therefore, in Figure 12.3, CNTs acted as a good photo-generated electron acceptor to promote the interfacial electron-transfer process, while the semiconductor was an electron donor under irradiation.

12.5.1 Photocatalytic Degradation of Organic Pollutants

The application of CNTs in conjunction with TiO_2 nanomaterials has gained attention for the last ten years, and a significant amount of research has been reported on photocatalysis.[86–90] There are reports about photocatalysis of MWCNT/TiO_2 composites and the degradation of acetone under irradiation of UV light, and of MWCNT/TiO_2 under visible light on the decolorization of dyes such as MB, MO, azo dyes,[91] and other dyes[92] in aqueous solutions.[93–96] Moreover, MWCNT/TiO_2 composites have been investigated in the photodegradation of phenol and photocatalytic oxidation of methanol under irradiation of visible light.[97–100] The catalysts exhibited enhanced photocatalytic activity for degradation of toluene in the gas phase under both visible and simulated solar light irradiation.[101] The ratio of CNT and TiO_2 has to be optimized in order to provide a large surface area for support and to stabilize charge separation by trapping electrons transferred from TiO_2, thereby hindering charge recombination with minimum photon scattering.[102] The composite provides a high surface area which is beneficial for photocatalytic activity as it provides a high concentration of target organic substances around sites activated by UV radiation.

CNT/ZnO nanocomposites display relatively higher photocatalytic activity than ZnO nanoparticles for the degradation of some dyes, such as rhodamine B, azo-dyes, methylene blue, and methylene orange.[103,104] The MWCNT/ZnO nanocomposites exhibit excellent photocatalytic activity toward other pollutants, such as acetaldehyde and cyanide in model solutions.[105,106] CNTs act as a photo-generated electron acceptor and slow the recombination of photo-induced electrons and holes. Adsorption and photocatalytic activity tests indicate that the CNTs serve as both an adsorbent and a visible light photocatalyst.[86]

The combination of CNTs with quantum dots have been applied to photocatalytic degradation of organic dyes. CdS-CNTs was tried on the degradation of MO under visible light irradiation.[107] The utilization of carbon nanotubes to enhance the photocatalytic activity of tungsten trioxide has also been investigated.[106,108,109] Photocatalytic activities are greatly improved when a CNT/WO_3 nanocomposite has been used for the degradation of pollutants such as rhodamine B under an ultraviolet lamp or under sunlight.

12.5.2 Antibacterial Ability

CNT/TiO_2 nanocomposites could also have potential as antibacterials. There has been a report about photoinactivation of *E.coli* through such nanocomposite under visible light irradiation.[110] The hybridized CNT with Cu nanoparticles has antimicrobial activity because of the increased contact

area. The hybridized nanocomposite could remove *E.coli* and *S. aureus* from water by photocatalysis.[111,112]

12.6   Fullerene (C$_{60}$) and Hybridized Nanocomposites

Since the discovery of fullerenes in 1985,[113] their variety of interesting properties, particularly their unique electronic properties, have gained a lot of attention.[114,115] It has been reported that fullerenes contain an extensively conjugated three-dimensional π system.[116] C$_{60}$, the most common fullerene, is described as a closed-shell configuration consisting of 30 bonding molecular orbitals with 60 π electrons, which is suitable for efficient electron transfer reductions because of the minimal changes of structure and salvation associated with electron transfer.[117] It has been suggested that intramolecular photo processes (electron or energy transfer) can occur between the peripheral C$_{60}$ subunits and the central core. Electrons in fullerenes, the photosensitizers, reach excited singlet (^{1}C$_{60}$*) and triplet (^{3}C$_{60}$*) states while staying within the same molecular orbitals (Figure 12.4). Due to its longer lifetime, the triplet state (^{3}C$_{60}$*) is the primary facilitator of energy or electron transfer to oxygen, leading to the formation of ^{1}O$_2$ or O$_2^{-\bullet}$ respectively. Some results have demonstrated that fullerene can efficiently cause a rapid photo-induced charge separation and a relatively slow charge recombination.[117,118] The fullerenes could be applied to promote efficient photocatalysis through combining with other semiconductor nanomaterials, for the degradation of organic pollutants.

12.6.1   Photocatalytic Degradation of Organic Pollutants

Because TiO$_2$ and its nanomaterials are the most commonly used photocatalytic semiconductors, there is some literature about the C$_{60}$/TiO$_2$ nanocomposites as used in the degradation of Procion red dye[119] and methylene blue[120,121] under UV irritation. The crystalline form of the TiO$_2$ nanocomposite was anatase. The fullerenes were water-soluble polyhydroxy fullerenes C$_{60}$, used in order to improve the photocatalysis in aqueous solution

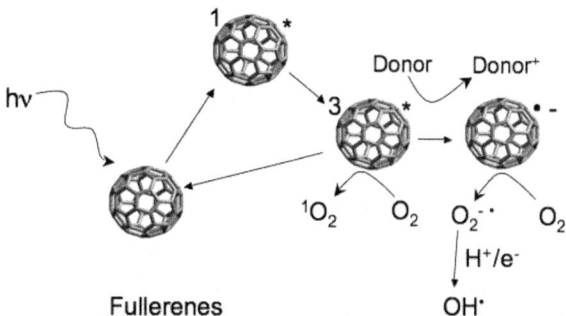

Figure 12.4   Mechanism of ROS production by fullerenes.[133]

since fullerenes are extremely hydrophobic. C_{60} combined with ZnO nano-particles were used for the degradation of organic dyes under UV-visible light; examples were MB, MO, and rohodamine B.[122–125] The C_{60}/ZrO_2 nanocomposites were also tested by the same research group for photocatalytic degradation of MB, MO, and rohodamine B under UV-visible light.[126] The immobilized C_{60} on a γ-alumina substrate was used for photocatalytic oxidation of alkenes under an oxygen atmosphere.[127]

12.6.2 Antimicrobiological Ability

The fullerene C_{60}, its derivatives, and its hybridized nanocomposites have been widely used as antibacterials. An aqueous suspension of C_{60} fullerene and C_{60} encapsulated in poly(vinyl-pyrrolidone), cyclodextrins, or poly(ethylene glycol) were tested for antimicrobial activity in aqueous solution.[128–130] The fullerene derivatives with pyrrolidine groups were cations and were applied for antibacterial usage.[131] The C_{60}/TiO_2 or C_{60}/ZnO nanocomposites were tested for their antibacterial activity.[53,132]

12.7 Conclusions

This chapter discuss two kinds of nanomaterials as applied for photocatalytic degradation of organic pollutants, sensors of organic pollutants, and antimicrobial ability. The two nanomaterials are semiconductors, the most common being TiO_2, and photosensitizers, with CNT and fullerenes being discussed. Relevant organic pollutants are airborne volatiles, dyes, medicines, explosives, and pesticides. The nanomaterials have enhance photoactivity and surface area. Hybridization with other kinds of nanomaterials lead to higher photoactivity and therefore higher selectivity for certain organic compounds and higher antibacterial ability. The hybridized nanocomposites were also used as nanosensors for specific organic compounds or as biosensors.

References

1. R. S. Varma, *Pure Appl. Chem.*, 2013, **85**, 1611–1710.
2. M. M. Khin, A. S. Nair, V. J. Babu, R. Murugan and S. Ramakrishna, *Energy Environ. Sci.*, 2012, 5, 8075–8109.
3. G. Shan, S. Yan, R. D. Tyagi, R. Y. Surampalli and T. C. Zhang, *Pract. Period. Hazard., Toxic, Radioact. Waste Manage.*, 2009, 110–119.
4. L. L. Chng, N. Erathodiyil and J. Y. Ying, *Acc. Chem. Res.*, 2012, **46**, 1825–1837.
5. C. S. Turchi and D. F. Ollis, *J. Catal.*, 1990, **122**, 178–192.
6. M. R. Hoffmann, S. T. Martin, W. Choi and D. W. Bahnemann, *Chem. Rev.*, 1995, **95**, 69–96.
7. L. Zhang, X. Wu, T. Zheng and M. Ju, *Nanotechnol. Precis. Eng.*, 2013, **11**, 511–517.

8. A. S. Nair and T. Pradeep, *Appl. Nanosci.*, 2004, 59–63.
9. L. Zhang and M. Fang, *Nano Today*, 2010, **5**, 128–142.
10. Y. Wang and P. Zhang, *J. Hazard. Mater.*, 2011, **192**, 1869–1875.
11. T. C. Zhang and R. Y. Surampalli, *Nanotechnologies for Water Environment Applications*, ASCE Publisher, 2009.
12. H. Bai, K. Xu, Y. Xu and H. Matsui, *Angew. Chem., Int. Ed.*, 2007, **46**, 3319.
13. H. Bai, F. Xu, L. Anjia and H. Matsui, *Soft Matter*, 2009, **5**, 966.
14. N. Nuraje, H. Bai and K. Su, *Prog. Polym. Sci.*, 2013, **38**, 302–343.
15. Y. Lan, Y. Lu and Z. Ren, *Nano Energy*, 2013, **2**, 1031–1045.
16. C. W. Wu, C. W. Lu and d Y. P. Lee, *J.Mater.Chem.*, 2011, **21**, 8540–8542.
17. T. Xian, H. Yang and J. F. Dai, *Nanotechnol. Precis. Eng.*, 2013, **11**, 111–117.
18. S. Nevim, H. Arzu, K. Gulin and Z. Cinar, *J. Photochem. Photobiol., A*, 2001, **139**, 225–232.
19. J. H Carey, J. Lawrence and H. M. Tosine, *Bull. Environ. Contam. Toxicol.*, 1976, **16**, 697–701.
20. P. D. Cozzoli, R. Comparelli, E. Fanizza, M. L. Curri and A. Agostiano, *Mater. Sci. Eng., C*, 2003, **23**, 701–713.
21. L. Zhang, *Langmuir*, 2003, **19**, 10372–10380.
22. K. Nagaveni, G. Sivalingam, M. S. Hegde and G. Madras, *Environ. Sci. Technol.*, 2004, **38**, 1600–1604.
23. Y. Yang, J. Ma, Q. Qin and X. Zhai, *J. Mol. Catal. A: Chem.*, 2007, **267**, 41–48.
24. X. Li, G. Chen, Y. Po-Lock and C. Kutal, *J. Chem. Technol. Biotechnol.*, 2003, **78**, 1246–1251.
25. J. Saien, R. R. Ardjmand and H. Iloukhani, *Phys. Chem. Liq.*, 2003, **41**, 519–531.
26. Z. Liu, Y. He, F. Li and Y. Liu, *Environ. Sci. Pollut. Res.*, 2006, 13.
27. L. Zhang, F. Yan, Y. Wang, X. Guo and P. Zhang, *Inorg. Mater.*, 2006, **42**, 1379–1387.
28. G. Sivalingam, K. Nagaveni, M. S. Hegde and G. Madras, *Appl. Catal., B*, 2003, **45**, 23–38.
29. H. Hori, E. Hayakawa, H. Einaga, S. Kutsuna, K. Koike, T. Ibusuki, H. Kiatagawa and R. Arakawa, *Environ. Sci. Technol.*, 2004, **38**, 6118–6124.
30. X. Li, P. Zhang, L. Jin, T. Shao, Z. Li and J. Cao, *Environ. Sci. Technol.*, 2012, **46**, 5528–5534.
31. J. Winkler, *Macromol. Symp.*, 2002, **187**, 317–323.
32. H. M. Sung-Suh, J. R. Choi, H. J. Hah, S. M. Khoo and Y. C. Bae, *J. Photochem. Photobiol., A*, 2004, **163**, 37–44.
33. N. Sobana, M. Muruganadham and M. Swaminathan, *J. Mol. Catal. A: Chem.*, 2006, **258**, 124–132.
34. H. Park, *Curr. Appl. Phys.*, 2007, 7, 118–123.
35. H. Yamashita, *Catal. Today*, 2007, **120**, 163–167.
36. K. Lee, N. H. Lee, S. H. Shin, H. G. Lee and S. J. Kim, *Mater. Sci. Eng., B*, 2006, **129**, 109–115.

37. J. L. Graham, C. B. Almquist, S. Kumar and S. Sidhu, *Catal. Today*, 2003, **88**, 73–82.
38. N. Venkatachalam, M. Palanichamy, B. Arabindoo and V. Murugesan, *J. Mol. Catal. A: Chem.*, 2007, **266**, 158–165.
39. H. S. Park, D. H. Kim, S. J. Kim and K. S. Lee, *J. Alloys Compd.*, 2006, **415**, 51–55.
40. M. S. Jeon, W. S. Yoon, H. Joo, T. K. Lee and H. Lee, *Appl. Surf. Sci.*, 2000, **165**, 209–216.
41. S. Ameen, M. Song, D. G. Kim, Y.-B. Im, Y.-S. Kim and H. S. Shin, *Theor. Appl. Chem. Eng.*, 2011, **17**, 998–1001.
42. K. Tennakone and I. R. M. Kottegoda, *J. Photochem. Photobiol., A*, 1996, **93**, 79–81.
43. D. S. Wang, J. Zhang, Q. Luo, R. Guo, X. Y. Li, Y. Duan and J. An, *J. Hazard. Mater.*, 2009, **169**, 546–550.
44. Z. Zhou, Y. Li, L. Liu, Y. Chen, S. B. Zhang and Z. Chen, *J. Phys. Chem. C*, 2008, **112**, 13926–13931.
45. L. Wang, W. Ma, L. Xu, W. Chen, Y. Zhu, C. Xu and N. A. Kotov, *Mater. Sci. Eng.*, 2010, **R70**, 265–274.
46. W. An, X. J. Wu and X. C. Zeng, *J. Phys. Chem. C*, 2008, **112**, 5747–5755.
47. J. B. Zheng, G. Li, X. F. Ma, Y. M Wang, G. Wu and Y. N. Cheng, *Sens. Actuators, B*, 2008, **133**, 374–380.
48. S. Mathur, A. Erdem, C. Cavelius, S. Barth and J. Altmayer, *Sens. Actuators, B*, 2009, **136**, 432–437.
49. E. Topoglidis, A. E. G. Cass, G. Gilardi, S. Sadeghi, N. Beaumont and J. R. Durrant, *Anal. Chem.*, 1998, **70**, 5111–5113.
50. A. F. E. Hezinger, J. Temar and A. Gopferich, *Eur. J. Pharm. Biopharm.*, 2008, **68**, 138–152.
51. C. Y. Jiang, S. Markutsya, Y. Pikus and V. Tsukruk, *Nat. Mater.*, 2004, **3**, 721–728.
52. G. A. Sotiriou, T. Sannomiya, A. Teleki, F. Krumeich, J. Voros and S. E. Pratsinis, *Adv. Funct. Mater.*, 2010, **20**, 4250–4257.
53. C. Wei, W. Lin, Z. Zainal, N. Williams, K. Zhu and A. P. Kruzic, *Environ. Sci. Technol.*, 1994, **28**, 934–938.
54. Y. Kikuchi, K. Sunada, T. Iyoda, K. Hashimoto and A. Fujishima, *J. Photochem. Photobiol., A*, 1997, **106**, 51–56.
55. R. J. Watts, S. Kong, M. P. Orr, G. C. Miller and B. E. Henery, *Water Res.*, 1995, **29**, 95–100.
56. L. Zan, W. J. Fa, T. Y. Peng and Z. K. Gong, *J. Photochem. Photobiol., B*, 2007, **86**, 165–169.
57. P. Hajkova, P. Spatenka, J. Horsky, I. Horska and A. Kolouch, *Plasma Processes Polym.*, 2007, **4**, S397–S395.
58. M. Cho, H. Chung, W. Choi and J. Yoon, *Appl. Environ. Microbiol.*, 2005, **71**, 270–275.
59. K. P. Kuhn, I. F. Chaberny, K Masholder, M. Stickler, V. W. Benz, H. G. Sonntag and L. Erdinger, *Chemosphere*, 2003, **53**, 71–77.

60. M. K. Seery, R. George, P. Floris and S. C. Pillai, *J. Photochem. Photobiol., A*, 2007, **189**, 258–263.

61. S. S. Shinde, P. S. Shinde and C. H. Bhosale, *J. Photochem. Photobiol., B*, 2011, **104**, 425–433.

62. P. V. Kamat, R. Huehn and R. Nicolaescu, *J. Phys. Chem. B*, 2002, **106**, 788–794.

63. W. J. Huang, G. C. Fang and C. C. Wang, *Colloids Surf., A*, 2005, **260**, 45–51.

64. N. Sobana and M. Swaminathan, *Sep. Purif. Technol.*, 2007, **56**, 101–107.

65. D. Yu, R. Cai and Z. Liu, *Spectrochim. Acta, Part A*, 2004, **60**, 1617–1624.

66. N. Sobana and M. Swaminathan, *Sol. Energy Mater. Sol. Cells*, 2007, **91**, 727–734.

67. C. C. Chen, *J. Mol. Catal. A: Chem.*, 2007, **264**, 82–92.

68. Y. J. Jang, C. Simer and T. Ohm, *Mater. Res. Bull.*, 2006, **41**, 67–77.

69. S. S. Shinde, P. S. Shinde and C. H. Bhosale, *J. Photochem. Photobiol., B*, 2011, **104**, 425–433.

70. H. Zhao and R. K. Y. Li, *Polymer*, 2006, **47**, 3207–3217.

71. M. A. Behnajady, N. Modirshahla, N. Daneshvar and M. Rabbani, *J. Hazard. Mater.*, 2007, **140**, 257–263.

72. S. Anandan, *Catal. Commun.*, 2007, **8**, 1377–1382.

73. C. Wang, *Appl. Catal., B*, 2002, **39**, 269–279.

74. F. Peng, H. Wang, H. Yu and S. Chen, *Mater. Res. Bull.*, 2006, **41**, 2123–2129.

75. W. Wu, Y. W. Cai, J. F. Chen, S. L. Shen, A. Martin and L. X. Wen, *J. Mater. Sci.*, 2006, **41**, 5845–5850.

76. L. K. Adams, D. Y. Lyon and P. J. J. Alvarez, *Water Res.*, 2006, **40**, 3527–3532.

77. R. Brayner, R. Ferrari-Illiou, N. Brivois, S. Djediat, M. F. Benedetti and F. Fivet, *Nano Lett.*, 2006, **6**, 866–870.

78. H. X. Bai, L. X. Zhang and Y. C. Zhang, *Mater. Lett.*, 2009, **63**, 823–825.

79. Z Li, P Zhang and T Shao, *Appl. Catal., B*, 2012, **125**, 350–357.

80. X. F. Chu, C. H. Wang, D. L. Jiang and C. M. Zheng, *Chem. Phys. Lett.*, 2004, **399**, 461–464.

81. Y. H. Li, J. Ding, Z. K. Luan, Z. C. Di, Y. F. Zhu, C. L. Xu, D. H. Wu and B. Q. Wei, *Carbon*, 2003, **41**, 2787–2792.

82. S. Iijima, *Nature*, 1991, **354**, 56–58.

83. Z. Yue and J. Economy, *J. Nanopart. Res.*, 2005, 7, 477–487.

84. Y. L. Zhao and J. F. Stoddart, *Acc. Chem. Res.*, 2009, **42**, 1161–1171.

85. T. W. Ebbesen, *J. Phys. Chem. Solids*, 1996, **57**, 951–955.

86. T. A. Saleh, in *Syntheses and Applications of Carbon Nanutubes and Their Composites*, ed. S. Suzuki, InTech, 2013, ch. 21, pp. 479–493.

87. A. Jitianu, T. Cacciaguerra, R. Benoit, S. Delpeux, F. Beguin and S. Bonnamy, *Carbon*, 2004, **42**, 1147–1151.

88. Q. Huang and L. Gao, *J. Mater. Chem.*, 2003, **13**, 1517–1519.

89. K. Woan, G. Pyrgiotakis and W. Sigmund, *Adv. Mater.*, 2009, **21**, 2233–2239.

90. W. Feng, Y. Feng, Z. Wu, A. Fujii, M. Ozaki and K. Yoshino, *J. Phys.: Condens. Matter*, 2005, **17**, 4361–4368.
91. Y. Yu, J. C. Yu, C.-Y. Chan, Y.-K. Che, J.-C. Zhao and L. Ding, *Appl. Catal., B*, 2005, **61**, 1–11.
92. C.-Y. Kuo, *J. Hazard. Mater.*, 2009, **163**, 239–244.
93. Y. Cong, X. Li, Y. Qin, Z. Dong, G. Yuan, Z. Cui and X. Lai, *Appl. Catal., B*, 2011, **107**, 128–134.
94. B. Gao, G. Z. Chen and P. G. Li, *Appl. Catal.*, 2009, **B89**, 503–509.
95. G. Hu, X. Meng, X. Feng, Y. Ding, S. Zhang and M. Yang, *J. Mater. Sci.*, 2007, **42**, 7162–7170.
96. T. A. Saleh and V. K. Gupta, *J. Colloid Interface Sci.*, 2012, **371**, 101–106.
97. W. D. Wang, P. Serp, P. Kalck and J. L. Faria, *J. Mol. Catal. A: Chem.*, 2005, **235**, 194–199.
98. Y. Yao, G. Li, S. Ciston, R. M. Lueptow and K. A. Gray, *Environ. Sci. Technol.*, 2008, **42**, 4952–4957.
99. C. Dechakiatkrai, J. Chen, C. Lynam, S. Phanichphant and G. G. Wallace, *J. Electrochem. Soc.*, 2007, **154**, A407–411.
100. G. An, W. Ma, Z. Sun, Z. Liu, B. Han, S. Miao, Z. Miao and K. Ding, *Carbon*, 2007, **45**, 1795–1801.
101. Z. Wu, D. Fan, Z. Weirong, W. Haiqiang, L. Yue and G. Baohong, *Nanotechnology*, 2009, **20**, 235701.
102. Y. Li, L. Leiyong, L. Chenwan, W. Chen and M. Zeng, *Appl. Catal., A*, 2012, **427**, 1–7.
103. K. Dai, T. Peng, D. Ke and B. Wei, *Nanotechnology*, 2012, **20**, 125603.
104. L. P. Zhu, G. Liao, W. Huang, L. Ma, Y. Yang, Y. Yu and S. Fu, *Mater. Sci. Eng. B*, 2009, **163**, 194–198.
105. T. A. Saleh, M. A. Gondal and Q. A. Drmosh, *Nanotechnology*, 2010, **21**, 8.
106. T. A. Saleh and V. K. Gupta, *J. Colloid Interface Sci.*, 2011, **362**, 337–344.
107. A. Ye, W. Fan, Q. Zhang, W. Deng and Y. Wan, *Catal. Sci. Technol.*, 2012, **2**, 969–978.
108. B. Pietruszka, F. D. Gregorio, N. Keller and V. Keller, *Catal. Today*, 2005, **102–103**, 94–100.
109. S. Wang, S. Xiaoliang, S. Gangqin, D. Xinglong, Y. Hua and W. Tianguo, *J. Phys. Chem. Solids*, 2008, **69**, 2396–2400.
110. O. Akhavan, M. Abdolahad, Y. Abdi and S. Mohajerzadeh, *Carbon*, 2009, **47**, 3280–3287.
111. L. Brunet, D. Y. Lyon, K. Zodrow, J. C. Rouch, B. Caussat, P. Serp, J. C. Remigy, M. R. Wiesner and P. J. J. Alvarez, *Environ. Eng. Sci.*, 2008, **25**, 565–576.
112. R. A. Khaydarov, R. R. Khaydarov and O. Gapurova, *Water Res.*, 2010, **44**, 1927–1933.
113. H. W. Kroto, J. R. Heath, S. C. O'Brien, R. F. Curl and R. E. C. Smalley, *Nature*, 1985, **318**, 162–163.
114. H. Fu, T. Xu, S. Zhu and Y. Zhu, *Environ. Sci. Technol.*, 2008, **42**, 8064–8069.

115. R. C. Haddon, *Science*, 1993, **261**, 1545–1550.
116. F. Hidalgo and C. Noguez, *Phys. Status Solidi B*, 2010, **247**, 1889–1897.
117. T. Hasobe, S. Hattori, P. V. Kamat and S. Fukuzumi, *Tetrahedron*, 2006, **62**, 1937–1946.
118. S. Zhu, T. Xu, H. Fu, J. Zhao and Y. Zhu, *Environ. Sci. Technol.*, 2007, **41**, 6234–6239.
119. V. Krishna, N. Noguchi, B. Koopman and B. Moudgil, *J. Colloid Interface Sci.*, 2006, **304**, 166–171.
120. Z. Meng, L. Zhu, J. Choi, C. Park and W.-c. Oh, *Cuihua Xuebao*, 2011, **32**, 1457–1464.
121. W.-C. Oh, A.-R. Jung and W.-B. Ko, *J. Ind. Eng. Chem.*, 2007, **13**, 1208–1214.
122. S. K. Hong, G. Y. Yu, C.-s. Lim and W. B. Ko, *Elastomers Compos.*, 2010, **45**, 206–211.
123. S. K. Hong, J. H. Lee and W. B. Ko, *J. Nanosci. Nanotechnol.*, 2011, **11**, 6049–6056.
124. B. H. Cho, Y. J. Oh, S. M. Mun and W. B. Ko, *J. Nanosci. Nanotechnol.*, 2012, **12**, 5907–5913.
125. B. H. Cho, K. B. Lee, K.-i. Miyazawa and W. B. Ko, *Asian J. Chem.*, 2013, **25**, 8027–8030.
126. B. H. Cho and W. B. Ko, *Asian J. Chem.*, 2013, **25**, 4577–4582.
127. M. D. Tzirakis, J. Vakros, L. Loukatzikou, V. Amargianitakis, M. Orfanopoulos, C. Kordulis and A. Lycourghiotis, *J. Mol. Catal. A: Chem.*, 2010, **316**, 65–74.
128. D. Y. Lyon, J. D. Fortner, C. M. Sayes, V. L. Colvin and J. B. Hughes, *Environ. Toxicol. Chem.*, 2005, **4**, 2757–2762.
129. Y. Kai, *Fullerenes, Nanotubes, Carbon Nanostruct.*, 2003, **11**, 79–87.
130. C. M. Sayes, J. Fortner, W. Guo, D. Lyon, A. Boyd, K. Ausman, Y. Tao, B. Sitharaman, L. Wilson, J. Hughes, J. West and V. Colvin, *Nano Lett.*, 2004, **4**, 1881–1887.
131. T. Mashino, D. Nishikawa, K. Takahashi, N. Usui, T. Yamori, M. Seki, T. Endo and M. Mochizuki, *Bioorg. Med. Chem. Lett.*, 2003, **13**, 4395–4397.
132. B. Rehn, F. Seiler, S. Rehn, J. Bruch and M. Maier, *Toxicol. Appl. Pharmacol.*, 2003, **189**, 84–95.
133. L. Brunet, D. Y. Lyon, E. M. Hotze, P. J. J. Alvarez and M. Wiesner, *Environ. Sci. Technol.*, 2009, **43**, 4355–4360.

CHAPTER 13

Hierarchical Nanoheterostructures: Layered Double Hydroxide-based Photocatalysts

LUHONG ZHANG, ZHIGANG XIONG AND GEORGE ZHAO*

School of Chemical Engineering, University of Queensland, Australia
*Email: george.zhao@uq.edu.au

13.1 Introduction

Layered double hydroxides (LDHs), also known as anionic or hydrotalcite-like clays, constitute a class of layered materials consisting of positively charged brucite-like layers and the interlayer exchangeable anions. The LDH can be represented by the general formula $[M^{2+}_{1-x}M^{3+}_x(OH)_2]^{q+}$ $[A^{n-}_{q/n} \cdot mH_2O]$, where M^{2+} and M^{3+} are divalent and trivalent metal cations such as Mg^{2+}, Ca^{2+}, Mn^{2+}, Fe^{2+}, Ni^{2+}, Cu^{2+} or Zn^{2+} and Al^{3+}, Mn^{3+}, Fe^{3+}, Co^{3+} or Ni^{3+}, respectively.[1] A^{n-} is an interlayer exchangeable anion such as CO_3^{2-}, SO_4^{2-} and NO_3^-, Cl^- and OH^-, S^{2-} etc.

The first natural LDH mineral $[Mg_6Al_2(OH)_6]CO_3 \cdot 4H_2O$ was discovered in Sweden around 1842, but the layered structure of LDHs weren't ascertained until 1969. With the swift progress of science and technology, the research on LDHs is developed rapidly after the 1990s. The structure characteristic, properties and applications of LDHs have been blossomed hugely in recent decades.[2]

RSC Green Chemistry No. 42
Green Photo-active Nanomaterials: Sustainable Energy and Environmental Remediation
Edited by Nurxat Nuraje, Ramazan Asmatulu and Guido Mul
© The Royal Society of Chemistry 2016
Published by the Royal Society of Chemistry, www.rsc.org

An advantage of LDHs is that they can be prepared in large quantities in a reliable and reproducible manner by precipitation of aqueous solutions of the corresponding metal salts by increasing the solution pH. Moreover, the atomic ratio between the divalent and trivalent metal ions of the LDHs can be varied in a wide range without altering the layered structures. By selectively changing the composition of divalent and trivalent metal ions, all the LDHs make up a rare class of clay-like materials with adjustable two-dimensional platelet structures, which could be further exfoliated into functional unilamellar nanosheets.[3] In addition, the charge-balancing anions located in the interlayers can be easily replaced *via* ion-exchange reactions. By virtue of this property, the incorporation of a wide variety of anions including simple inorganic anions (*e.g.* CO_3^{2-}, NO_3^-),[4] organic anions (*e.g.* benzoate, succinate)[5] and complex biomolecules (*e.g.* DNA)[6] into the interlayer has been realized. Such an anion-exchange property also endows LDHs the remarkable shape-selective properties that can be used for the separation of isometric drug molecules.[4–6]

The positively charged layers in LDHs are stacked together with anionic ions in the interlayers. The charged layers can be exfoliated out through swelling in delaminating agents such as formamide and butanol.[7–9] The thickness of an exfoliated LDH nanosheet is around 0.5–1 nm, which is in the range of molecular dimensions, while the lateral dimension is ranged from hundreds of nanometres to micrometres.[7] The quantum size effect[10] and the increased active sites on the surface of LDHs will remarkably broaden their potential applications.[9] The basic structure and crystal characters of LDHs are shown in Figure 13.1.[11]

Due to the large interlayer spaces and the significant number of exchangeable anions, LDHs have high surface area and high anion-exchange

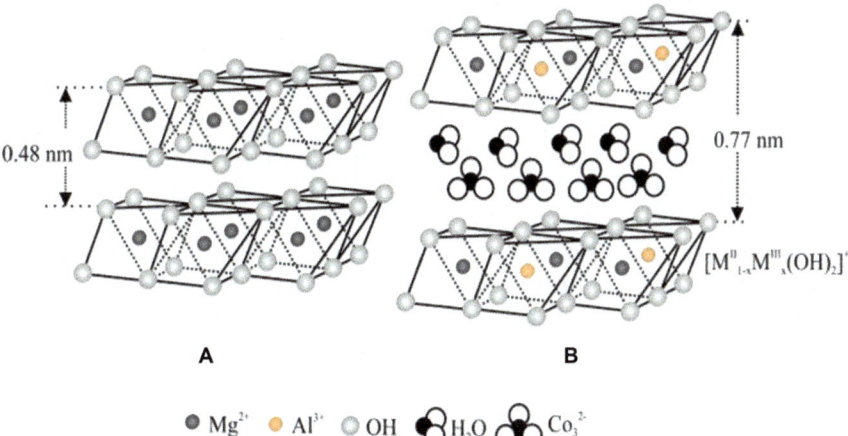

0.48 nm

0.77 nm

$[M^{II}_{1-x}M^{III}_x(OH)_2]^-$

A   **B**

● Mg^{2+} ○ Al^{3+} ○ OH   ⬤ H_2O ✿ CO_3^{2-}

Figure 13.1   Schematic representation comparing the crystal structure of brucite (A) and LDH (B).
Reproduced from ref. 11 with kind permission from Springer Science and Business Media.

capacity, and have the potential to be good ion exchangers and adsorbents. They (and their derivatives such as surfactant modified LDH and calcined LDH) have, indeed, been reported as effective adsorbents for the removal of contaminants[12] such as azo dyes, triphosphate and arsenate from aqueous solutions.[13–16] The versatility of LDHs in eliminating different contaminants enables them to be viable alternatives for environmental remediation.[14,17] Recently, LDHs and their derivatives working as photocatalytic materials have received increasing attention and have demonstrated excellent performance in photocatalytic water treatment or in photoeletrochemical processes.[18] Note that there are many reviews on LDH and derivatives in recent years.[3,7,9,19]

This chapter aims to give an overview on the most significant procedures developed in this extremely investigated field of research. It will highlight the synthetic potential, always taking into account the previously mentioned synthetic challenges. The potential applications of LDH and LDH-based heterostructures in photocatalytic water treatment, CO_2 conversion and water splitting, *etc.* will also be presented while discussing the various LDH materials.

13.2   Structure of LDH

LDH is a class of anionic clay consisting of various anions and metal hydroxide layers. The hydroxide layer is similar to the brucite layer $Mg(OH)_2$ in which each Mg^{2+} ion is surrounded by six OH^- ions in an octahedral arrangement. The partial replacement of the divalent cations in brucite by trivalent ions leads to positively charged LDH layers, which are balanced by the intercalation of interlayer anions such as CO_3^{2-}, NO_3^-, Cl^- *etc.* The identities of the di- and trivalent cations (M^{2+} and M^{3+} respectively) and the interlayer anion (A^{n-}) together with the mole ratio between M^{2+} and M^{3+} may be varied over a wide range, giving rise to a large class of isostructural materials.[20] There are a number of combinations of divalent and trivalent cations for LDH production, and the layered structures can be formed as long as the radii in octahedral coordination of M^{2+} and M^{3+} are close to those of Mg^{2+} and Al^{3+}.[21] The ionic radii of some cations incorporated in LDH materials are shown in Table 13.1.

The structure of LDH can be characterized by many techniques. Among them, X-ray diffraction (XRD) is the most widely used technique to characterize the interlayer spacing and the thickness of LDH materials. The diffraction peaks of LDH in XRD patterns normally locate in 2θ range of $5 \sim 60°$, ascribing to the characteristic diffraction peaks of the rhombohedral phase (Figure 13.2).[22,23] The peaks will shift depending on the LDH compositions. Using the (003) diffraction peak, the interlayer basal spacing of LDH can be calculated according to Bragg's law equation:

$$n\lambda = 2d \sin \theta$$

where n is the order number of diffraction, λ is the incident wavelength of radiation ray, θ is the angle diffraction occurs and d is the distance of the spacing.

Table 13.1   Ionic radii of some cations with coordinate number of 6. Reproduced from ref. 21, Copyright (2011), with permission from Elsevier.

M^{2+}	Radius/nm	M^{3+}	Radius/nm
Fe	0.061	Al	0.054
Co	0.065	Co	0.055
Ni	0.069	Fe	0.055
Mg	0.072	Mn	0.058
Cu	0.073	Ga	0.062
Zn	0.074	Rh	0.067
Mn	0.083	Ru	0.068
Pd	0.086	Cr	0.069
Cd	0.095	V	0.074
Ca	0.100	Y	0.090
Ti^{4+}	0.061	La	0.013
Sn^{4+}	0.069	Zr^{4+}	0.072

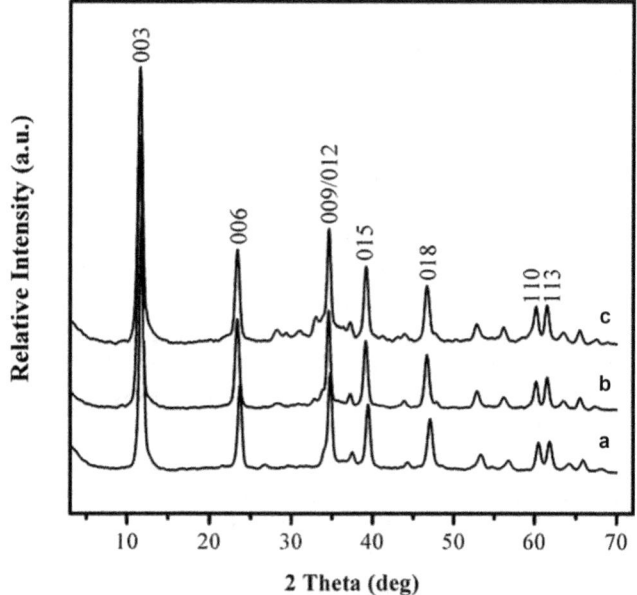

Figure 13.2   XRD patterns of (a) Zn_2Al-LDH, (b) Zn_3Al-LDH and (c) Zn_4Al-LDH. Adapted with permission from ref. 26. Copyright 2012 John Wiley and Sons.

The cell parameters of LDH can also be calculated using the following equations:[24]

$$a = 2 \times d_{(110)} \tag{13.1}$$

$$c = 3 \times d_{(003)} \text{ (or } 6 \times d_{(006)}\text{)} \tag{13.2}$$

The crystallite size of the LDH materials can be determined using the Scherrer equation in its simplified version (lattice distortions are neglected), where L, k, λ and β represent the mean size of the ordered (crystalline) domains, a dimensionless shape factor, X-ray wavelength and the line broadening at half the maximum intensity (FWHM), respectively.[25]

$$L = \frac{\lambda k}{\beta \cos \Theta} \tag{13.3}$$

LDH is also characterized by Fourier transform infrared spectroscopy (FTIR) technique to identify the anions, especially organic anions located in the interlayers. Typically, LDH materials have a strong and broad band centred around 3400 cm^{-1} attributing to the OH$^-$ stretching vibrations of the hydroxyl groups in the layers and interlayer water molecules (Figure 13.3),

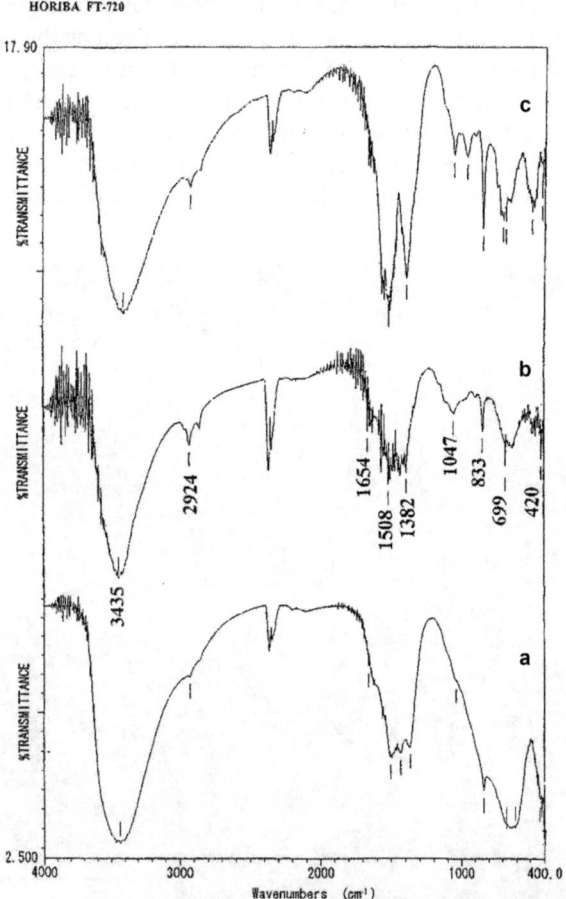

Figure 13.3 IR Spectra of (a) ZnAl-LDH, (b) ZnAlSn-LDH and (c) ZnSn-LDH. Reprinted with permission from ref. 27 with kind permission from Springer Science and Business Media.

and the water deformation band located at around $1635 \, cm^{-1}$. The vibration at around $1350 \, cm^{-1}$ indicates the presence of inorganic CO_3^{2-} or NO_3^- in the interlayers.[27]

It is known that a crystallized LDH normally has hexagonal flakes composed of octahedral units of metal hydroxide layers. By utilizing ammonia-releasing hydrolysis agents such as urea[28,29] or hexamethylenetetramine (HMT) together with hydrothermal treatment, highly crystallized CoAl-LDH with well-defined hexagonal shape (Figure 13.4) could be readily synthesized, with the lateral diameter ranging from 10 μm to 80 μm. The thickness of the flake is about several hundred nanometres due to the stacking of LDH flakes.

The nature of anions including the types and orientation of anions[13] in the LDH interlayers could affect the anion-exchange behaviours[30] and the adsorption kinetics of LDH.[31,32] It is revealed that stacking products can be derived from different interlayer contents or different orientations of the same anionic species compensating for host layer charge in LDH materials.[33] Moreover, the stacking is not only influenced by the anionic species but also influenced by the M^{2+}/M^{3+} ratio, the metallic composition ratio in starting brucite influences the staging phenomena of LDH as well. Sasaki *et al.*[34] discussed the correlation between staging product and metallic composition ratio. They discovered that for transition metals LDH, when a stoichiometric amount of iodine was used, $Co^{2+}_{2/3}Fe^{2+}_{1/3}(OH)_2$ was

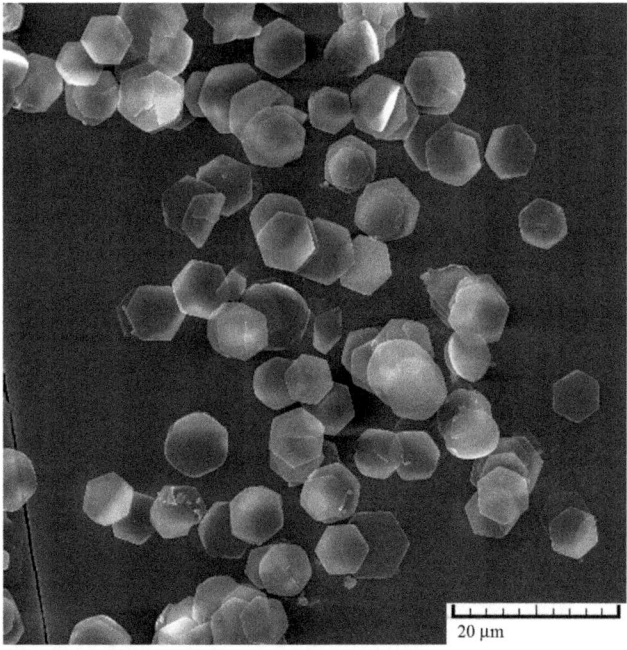

Figure 13.4   SEM micrograph of the as-prepared CoAl-CO$_3$-LDH.
Reprinted with permission from ref. 29. Copyright (2010) American Chemical Society.

completely transformed into a pure LDH phase with a different formula: $Co^{2+}_{1-x}Fe^{2+}_x(OH)_2 + 0.5xI_2 \rightarrow Co^{2+}_{1-x}Fe^{3+}_x(OH)_2I_x$.

The layer structure order and the interlayer species order are changed with the hydrothermal treatment time and temperatures.[35] Gray and co-workers[36] fabricated Mg_2Al-Cl-LDH with a fast co-precipitation process followed by controlled hydrothermal treatment, and analysed in detail the influence of hydrothermal treatment temperature and duration time on the particle size. The suspension of freshly precipitated Mg_2Al-Cl-LDH consisted of a bimodal particle size distribution with diameters of 320 nm and 2300 nm, respectively. After ageing at 50 °C overnight, the LDH aggregates decreased in size to 220–955 nm. However, the particle size of LDHs fabricated by hydrothermal treatment at 100 °C for 16 h after co-precipitation had a narrow distribution with a diameter of 114 nm (Figure 13.5). It is also noted that 2 h treatment at 100 °C was not long enough to de-aggregate all aggregates into individual LDH crystallites, while the 144 h treatment produced larger LDH crystallites. The results indicate that hydrothermal treatment is effective in de-aggregating LDHs to form nanosheets.

13.3 Synthetic Methods for LDH Materials

13.3.1 Co-precipitation Method

The conventional method for synthesizing LDH materials is the direct co-precipitation method. At a constant pH, M^{2+} and M^{3+} in their mixed ions saline solution is precipitated to form double hydroxide sediments by adding equal stoichiometric amounts of aqueous alkaline solution containing NaOH and Na_2CO_3 (or KOH and K_2CO_3).[37] Mixed metal hydroxides hydrate is dominant in the synthesis products for the simple co-precipitation step; with further ageing treatment, LDH can be gradually formed and is mainly in an aggregation state with the sizes of 1–10 μm, which contains hundreds and thousands of sheet-like LDHs nanocrystallites in each aggregate.[38] In addition, hydrothermal treatment is the most common process for the formation of hexagonal-shaped LDH plates. In this process, all reagents are put into a sealed autoclave and are hydrothermally treated at 100 to 200 °C for a certain period of time. The advantage of this method is that the reaction conditions are controllable and always generate highly crystallized LDH products.[34]

13.3.2 Other Methods

With pure Al or Zn plates as a template, Dutta *et al.*[39] reported a promising phase transition method to fabricate highly crystallized ZnAl-LDH. In this method, a certain proportion of Zn and Al ions with hexamethylenetetramine (HMTA) as buffer were prepared and adjusted to pH 7.5 by adding ammonia into the solution. The resulting solution served as the bulk solution and was then transferred to a hydrothermal bottle with a clean Al or Zn

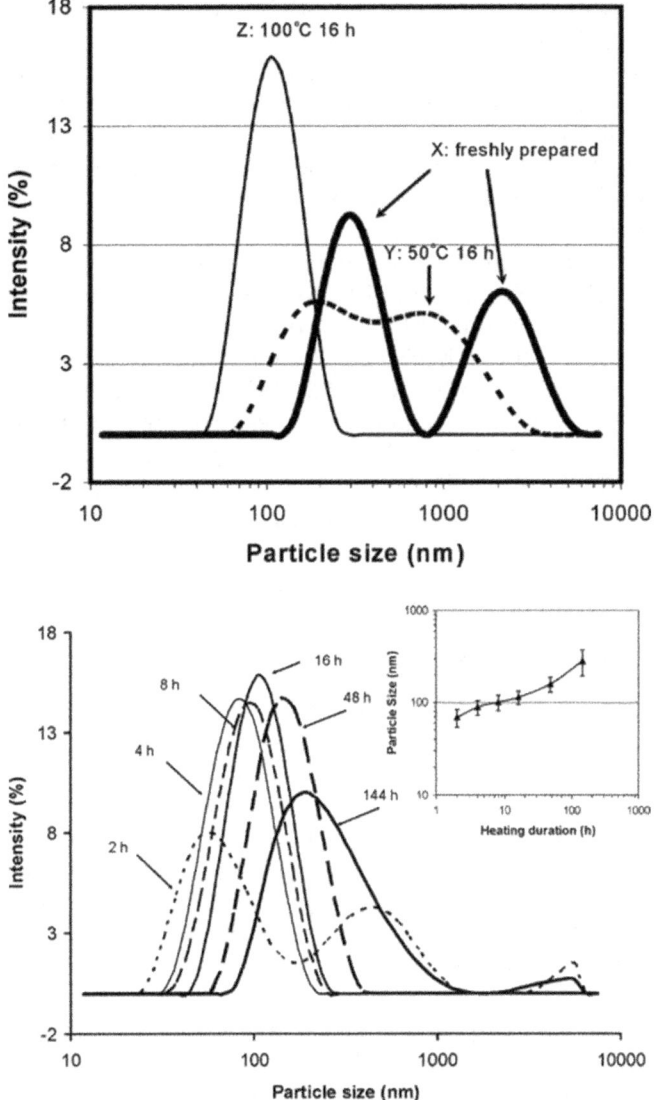

Figure 13.5 Particle size distribution of Mg$_2$Al-Cl-LDH samples collected with photon correlation spectroscopy (PCS). Top: (X) Co-precipitated and stirred for 10 min at room temperature, with two peaks at 320 and 2300 nm; (Y) sample X aged at 50 °C for 16 h, with two broad peaks at 220 and 955 nm; (Z) 100 °C hydrothermal treatment for 16 h, with one sharp peak at 114 nm. Bottom: Dispersion of Mg$_2$Al-Cl-LDH aggregates with heating duration during the hydrothermal treatment at 100 °C, where the distribution curves were obtained with PCS.

Reproduced with permission from ref. 36. Copyright (2006) American Chemical Society.

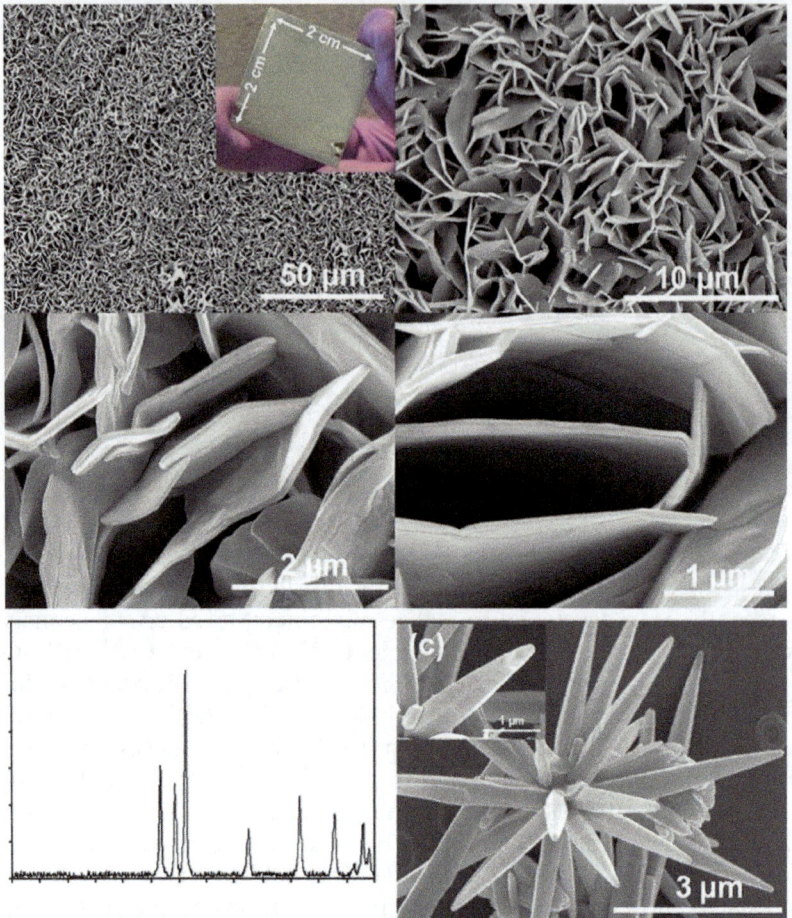

Figure 13.6 SEM images of plate LDH sample. Inset is the photographic image of 2 cm×2 cm Al plate fabricated with ZnAl-LDH assembles.
Reproduced from ref. 39, Copyright (2012), with permission from Elsevier.

plate suspended horizontally in the solution by a Teflon tread. After hydrothermal treatment, ZnAl-LDH gradually grew with its *ab* plane perpendicular to the Al or Zn substrate, *i.e.* on substrate plate vertically (c-axis parallel to the substrate) (Figure 13.6), and a proposal tentative mechanism for the formation of ZnAl-LDH platelets on Al plate is also shown in Scheme 13.1. The crystallized ZnAl-LDH has been demonstrated to be an effective photocatalyst for the decomposition of Congo red in aqueous medium.[39–42]

Sasaki and co-workers[7,32] reported the formation of LDH with transition metal ions served as the host layer composition instead of the limited M^{2+}–Al^{3+} pattern. The CoFe-LDHs were synthesized *via* an oxidative intercalation

$$C_6H_{12}N_4 + 6H_2O \rightarrow 4NH_3 + 6HCHO \qquad (2.5)$$

$$NH_3 + H_2O \rightarrow NH_4^+ + OH^- \qquad (2.6)$$

$$2Al + 2OH^- + 6H_2O \rightarrow 2Al(OH)_4^- + 3H_2 \qquad (2.7)$$

$$Al(OH)_4^- \rightleftarrows Al(OH)_3 + OH^- \qquad (2.8)$$

$$Zn^{2+} + 2OH^- \rightarrow Zn(OH)_2 + 2OH^- \rightarrow Zn(OH)_4^{2-} \qquad (2.9)$$

$$Zn^{2+} + 4NH_3 \rightarrow Zn(NH_3)_4^{2+} \qquad (2.10)$$

$$Zn(OH)_2 + NH_3 \rightarrow Zn(NH_3)_4^{2+} + 2OH^- \qquad (2.11)$$

$$Al(OH)_3/Al(OH)_4^- + Zn(OH)_4^{2-}/Zn(NH_3)_4^{2+} + OH^- + NO_3^- + H_2O \rightarrow ZnAl\text{-}LDH \quad (2.12)$$

Scheme 13.1 Proposed mechanism for self-limiting growth of LDH on Al plate. Reproduced from ref. 39, Copyright (2012), with permission from Elsevier.

reaction using an excess amount of iodine as the oxidizing agent. This synthesis method, called the topochemical oxidative intercalation method, has recently been developed to realize micrometre-sized crystals consisting of various transition metal ions such as Co^{2+}-Fe^{3+}, Co^{2+}-Co^{3+}, Ni^{2+}-Co^{3+}. In this method, mono- or bimetallic brucite-like hydroxides including $Co^{2+}(OH)_2$, $Co^{2+}_{2/3}Fe^{2+}_{1/3}(OH)_2$, and $Co^{2+}_xNi^{2+}_{1-x}(OH)_2$ were firstly precipitate via the hexamethylenetetramine (HMT) hydrolysis, which were transformed to Co^{2+}-Fe^{3+}, Co^{2+}-Co^{3+}, $Ni^{2+}(Co^{2+})$-Co^{3+} LDHs through partial oxidation of divalent transition metal ions (Fe^{2+}, Co^{2+}) to a trivalent state by employing halogens (iodine, bromine) in organic solvents, such as chloroform or acetonitrile. At the same time, halogen anions (I^-, Br^-) were intercalated between the layers. The original morphology and size of brucite-like crystals were well maintained during the process, namely a topotactic conversion. The halide-intercalating transition metal LDHs, after anion exchange into other anionic forms (e.g. CO_3^{2-}, NO_3^-, ClO_4^-), further exfoliated in formamide to produce LDH nanosheets.

As LDH could restore its original structure after a two-step annealing and reconstruction procedure, such a property is often used as an alternative method for the preparation of LDHs with different anions in the interlayers.[43] In addition, there are several reports on LDH fabrication which combine the co-precipitation process with the microwave hydrothermal treatment.[33]

A series of ZnAl-LDH hexagonal particles were synthesized using the microwave irradiation method. With the assistance of microwave irradiation, secondary growth led to the formation of ZnO nanorods on a ZnAl-LDH substrate, shown in Figure 13.7, and then to ZnO nanotube/ZnAl-LDH heterostructures with further ageing. Two-dimensional ZnO film/ZnAl-LDH sandwich-like heterostructures were also formed during secondary growth in a citrate anion-containing solution.[44]

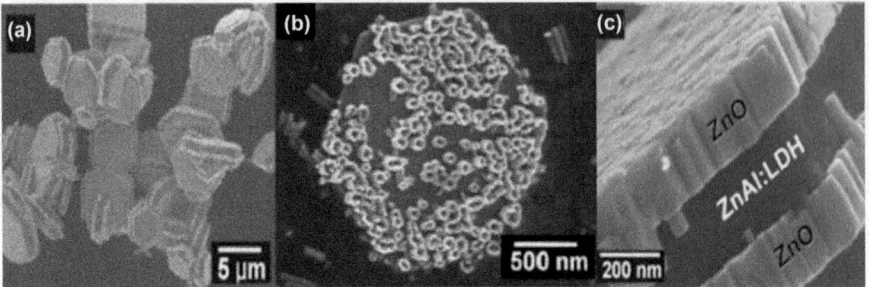

Figure 13.7   (a) ZnO nanorod/ZnAl-LDH; (b) ZnO nanotube/ZnAl-LDH; (c) ZnO film/ZnAl-LDH.
Reproduced from ref. 44.

13.4   Modification of LDH

13.4.1   Modification of LDH Structures

A diverse combination of M^{2+}-M^{3+} cations in LDH host sheets has been extensively pursued, and can be routinely attained through a convenient co-precipitation of corresponding di- and tri-valent metal salts under alkaline conditions, which produces a large family of M^{2+}-M^{3+} LDHs. However, the research interest on LDH materials has been traditionally driven and dominated by M^{2+}-Al^{3+} category, particularly Mg^{2+}-Al^{3+},[2] partly due to the fact that hydrotalcite $Mg_6Al_2(OH)_{16}(CO_3) \cdot 4H_2O$ is a widely known anionic clay found in nature. Moreover, the amphoteric feature of Al^{3+} plays a very favourable role in promoting the precipitation and crystallization of M^{2+}-Al^{3+} LDH. Besides Al^{3+}, other trivalent transition metal ions including Fe^{3+}, Cr^{3+}, Mn^{3+}, Ga^{3+}, V^{3+} and Ni^{3+},[45] and some tetravalent metal ions (such as Ti^{4+},[46] and Zr^{4+} ions[47]) have also been used to prepare LDH materials.

LDHs are well-known catalyst supports (*e.g.* Mg-Al hydrotalcite),[48] they have also been used as catalysts for different applications. By modifying the nature and mole ratio of cations in the LDH platelets, the properties of LDHs have been optimized. Silva *et al.*[49] prepared a novel series of ZnM-LDHs (M = Cr, Ti, Ce) at different Zn/M atomic ratios and tested them for visible light photocatalytic oxygen generation. They found that the most active material was ZnCr-LDH and the efficiency of these chromium LDHs for oxygen generation increases asymptotically with the Cr content. In addition to LDHs with binary metal elements, up to 30% of Al^{3+} in Al-containing LDHs could be isomorphously substituted by other tervalent or tetravalent ions to form a new ternary LDH without changing the layered structures; a lot of multi-element LDHs, such as NiMgAl-LDH, ZnMgAl-LDH and MgCrAl-LDH, have been successfully prepared.[12,24,50,51] Valente and co-workers[17] prepared MgZnAl-LDH with varying amounts of Zn, which showed that introducing a small amount of Zn in MgAl-LDH substantially modifies the bandgap energy and the adsorption capacities of these materials, thus

enhancing the photocatalytic degradation of 2,4-dichlorophenoxiacetic acid and phenol. The degradation rate for 4-chlorophenol was even faster in the presence of MgZnAl-LDH than that in the presence of Degussa P25. Other ternary LDHs such as ZnAlFe-LDH and rare earth elements doped LDHs such as Eu-doped ZnAl-LDH have also been prepared with excellent photocatalytic activities,[52] or photoluminescent properties.[53,54]

Changing the nature of the interlayer anions is another method to modify the properties of LDHs. Recently, Satpathy and co-workers[55] fabricated molybdate/tungstate-intercalated LDH using nitrate-intercalated ZnY-LDH as precursors through an ion-exchange process. The composite material demonstrated high reactivity towards the degradation of rhodamine 6G (RhG) upon visible light irradiation with a good stability. Electron paramagnetic resonance (EPR) measurements revealed that the absorption in the visible region was attributed to the metal-to-metal charge-transfer (MMCT) excitation of the oxo-bridged bimetallic linkage of Zn–O–Y in ZnY-LDH. The intercalation of MoO_4^{2-}/WO_4^{2-} ions enhanced the harvesting of visible light, lowered the bandgap of ZnY-LDH and increased the surface area by broadening the interlayer space where the photodegradation reaction occurred.[55]

Apart from changing the LDH compositions, the modification of LDH morphology provides an alternative method for extending the potential application of LDH materials. Shao et al.[56] demonstrated that using sodium dodecyl sulfonate (SDS) surfactant as the template, the MgFe-LDHs were fabricated in microsphere structures instead of being layer-by-layer structures with a tuneable interior. By controlling the concentration of SDS, the structure of the LDHs could be tailored from hierarchical flower-like solid spheres to yolk-shell and then to hollow spheres with the shell thickness of around 85 nm, close to the lateral size of LDH nanoflakes (Figure 13.8). The resulting hollow LDH microspheres yields largely enhanced activity as well as robust durability due to the significantly improved mass transport.

13.4.2 Coupling with Other Materials

Since the discovery of the ultraviolet (UV) light-induced photoelectrochemical water splitting on the TiO_2 surface in 1972,[57] research on photocatalysis based on metal oxide photocatalysts and their derivatives has never ended.[58,59] Considering the high adsorption capacity and the diversity of LDHs, the synergistic effect derived from the combination between LDH and TiO_2 or other photocatalysts has been extensively investigated.[23,60,61] It has been reported that by distributing TiO_6 units in an MTi-LDH (M = Ni, Zn, Mg), the composite materials displayed enhanced photocatalytic activity with a hydrogen production rate of 31.4 μmol h^{-1} as well as excellent recyclable performance.[61] The structural and morphological studies revealed that a high dispersion of TiO_6 octahedra in the LDH matrix was obtained by the formation of an M^{2+}-O-Ti network, rather different from the aggregation state of TiO_6 in the inorganic layered material $K_2Ti_4O_9$. Such a dispersion

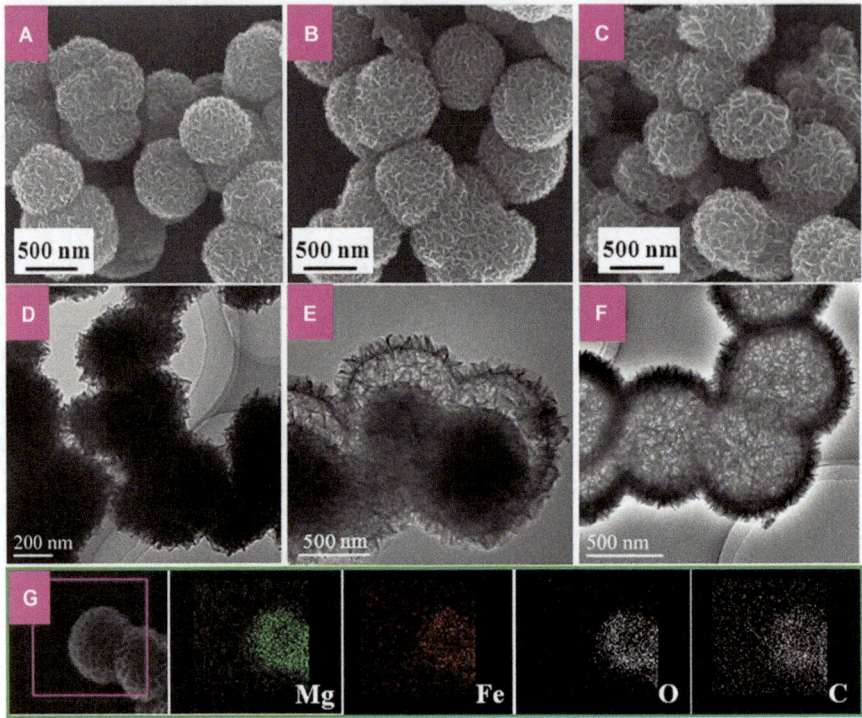

Figure 13.8 SEM and TEM images of MgFe-LDH microspheres with different inner architecture: (A) and (D) solid; (B) and (E) yolk-shell; (C) and (F) hollow; and (G) EDX mapping results of a single LDH hollow microsphere. Adapted with permission from ref. 56. Copyright 2013 John Wiley and Sons.

strategy for TiO_6 units within a 2D inorganic matrix decreased its bandgap to 2.1 eV and significantly depressed the electron–hole recombination process, which accounts for its superior water-splitting behaviour. Lu *et al.*[62] reported that by selectively reconstructing a Cu^{2+}, Mg^{2+}, Al^{3+}, Ti^{4+}-containing LDH precursor through calcination and rehydration process, anatase TiO_2 nanoparticles were found to be homogeneously distributed on the surface of the selectively reconstructed CuMgAl-LDH support. The composite showed superior photocatalytic properties to the single TiO_2 phase and the physical mixture of TiO_2 and LDH due to the presence of TiO_2/LDH heterojunction, which contributed efficient spatial separation between the photo-generated electrons and holes. The aggregation of TiO_2 nanoparticles was also suppressed by the 2D LDH framework.[63] By depositing TiO_2 on sodium dodecyl sulfate (SDS)-intercalated LDH, Huang and co-workers[64] prepared the composite SDS-LDHs/TiO_2 (Figure 13.9), which combined the high adsorption properties of SDS-LDH and the photocatalytic activity of TiO_2. The enrichment of dimethyl phthalate (DMP) onto the organic LDH composites and the external hydroxyl groups generated by TiO_2 produced a

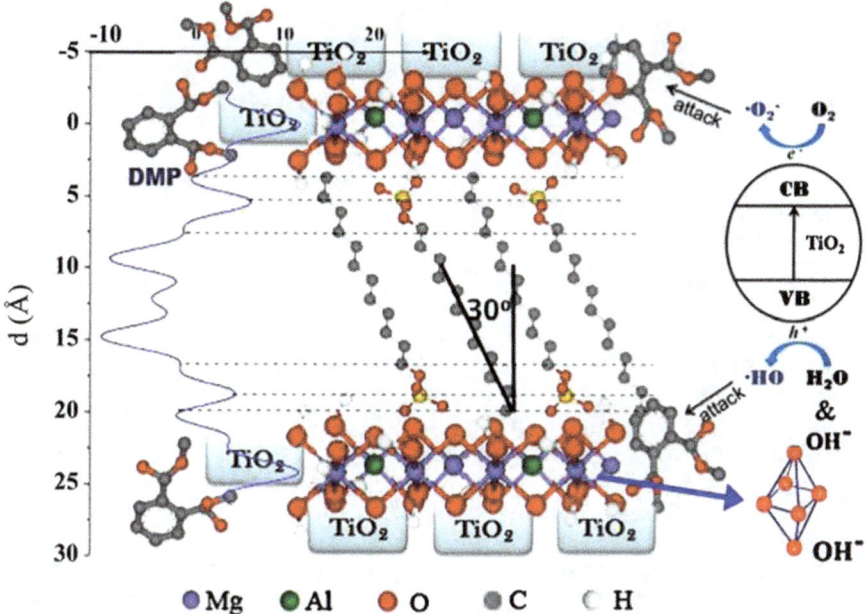

Figure 13.9 The simulated structure of the composite SDS-LDHs/TiO$_2$.
Reproduced from ref. 64, copyright (2013), with permission from
Elsevier.

synergistic effect leading to greatly enhanced degradation of DMP. The optimum amount of the LDH powders as a carrier for the enhancement of the
photocatalytic reactivity of SDS-LDHs/TiO$_2$ was estimated to be the mass
ratio of 1 : 1.

Other metal oxide such as ZnO, a semiconductor similar to TiO$_2$ photocatalyst,[65–67] has also been coupled with LDHs. The deposition of ZnO with
desired crystalline and nanostructure can be obtained through etching ZnAl-
LDH precursors or through reconstruction of calcined LDH materials. As
shown in Figure 13.10, ZnO with a higher percentage of exposed (0001)
facets was embedded on a hierarchical flower-like matrix using the *in situ*
topotactic transformation of an LDH precursor. The ZnO-embedded material, also named flower-like mixed metal oxides (F-MMO), showed enhanced photocatalytic activity under visible light irradiation when compared
with other ZnO nanorods and ZnO nanoplates with few exposed (0001)
facets.[68] It should be noted that although F-MMO was used as a photocatalyst, the active component should be a combination of ZnO and the
reconstructed LDH,[68,69] as the MMO would easily reconstruct to its original
LDH structure when dispersed in an aqueous solution due to the memory
effect of LDH.[70] Dvininov *et al.*[71] embedded the MgAl-LDH particles in SnO$_2$
domains by simple impregnation of the hydrotalcite with tin precursor
followed by hydrolysis and calcination, which improved the photocatalytic
activity of SnO$_2$ used for the degradation of methylene blue (MB).

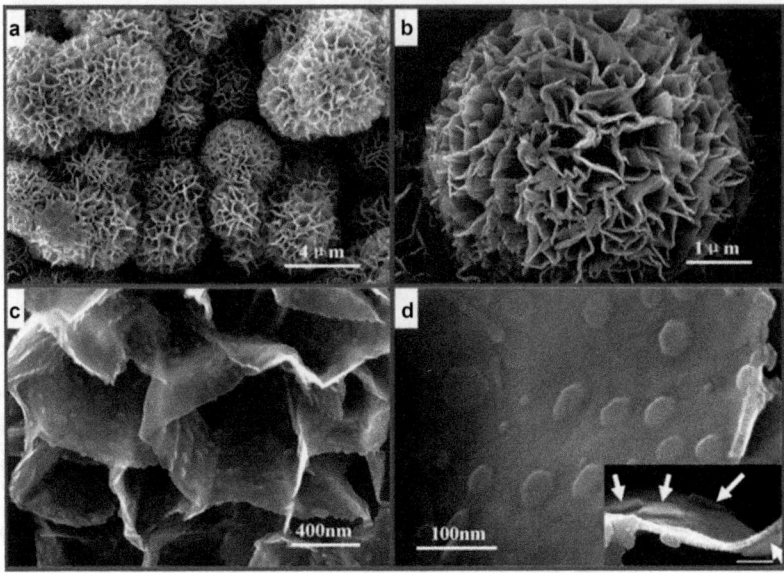

Figure 13.10 SEM images of the as-obtained F-MMO sample: (a) an overall view of the F-MMO microspheres; (b) a single F-MMO microsphere; (c) the nanoflakes of a F-MMO microsphere; (d) a portion of a F-MMO nanoflake with hexagonal ZnO nanoplatelets embedded on its surface (inset: the cross section of the F-MMO nanoflake). Reprinted from ref. 68.

The inactive MgAl-LDH particles in contact with SnO_2 act as a barrier layer for charge recombination and electron trap sites leading to a better charge separation. The combination between magnetic Fe_3O_4 nanoparticles and ZnCr-LDH not only improved the photocatalytic degradation of MB and methyl orange (MO), but also enhanced the separation and re-dispersion performance of LDH in aqueous solution.[22] The photocatalytic performance of CeO_2 deposited on MgAl-LDH is even far superior to the Degussa P25 photocatalyst for the degradation of phenol and 4-chlorophenol under UV irradiation.[72]

Using the layer-by-layer accumulation technique, CdS nanoparticles were densely immobilized on LDH sheets without aggregation into larger particles.[73] The 'quantum size effect' arising from mono-dispersed CdS broadened the bandgap of CdS, thus the absorbance and photoluminescence intensity of immobilized CdS particles were enlarged in LDH/CdS. LDH/CdS multilayers deposited on an F-doped SnO_2 (FTO) electrode behaved as an n-type semiconductor photoelectrode in an acetonitrile solution. Vectorial photo-induced electron transfer and energy transfer along the bandgap gradient from the interface between the LDH/CdS film and the electrolyte solution to the collecting FTO electrode enhanced the photocurrent generation in LDH/CdS multilayer films. By reacting Zn, Cd, Al-containing LDH with H_2S,[74] $Zn_xCd_{1-x}S$ nanoparticles have been successfully loaded onto

LDH surface. The ordered arrangement on an atomic level of the metal cations within the LDH layers led to a homogeneous array structure of the $Zn_xCd_{1-x}S$ solid, which suppressed the recombination of photo-excited electrons and holes, leading to improved photocatalytic activity. In addition, the absorption edge of $Zn_xCd_{1-x}S$ was monotonically shifted to the visible light region as the dissolved amount of Zn^{2+} ions into the CdS lattice decreased. The photocatalytic performance of the $Zn_xCd_{1-x}S$ increased with decreasing Zn mole fraction, and reached a maximum for a Zn mole fraction of 0.20.[75]

13.4.3   LDH Derivatives

Annealing treatment is the most common method to obtain LDH derivatives. The thermal stability of LDHs varies with their different constitutents.[76] In general, the thermal decomposition of LDH takes place in three steps: it starts from the removal of water molecules from surface, edge and interlayer space, then the elimination of hydroxyl groups from the brucite-like layers which is called dehydroxylation, followed by or overlapped with the decomposition of interlayer anions.[77] The first derivative of annealed LDH is the layered double oxide (LDO), which is the main product from the first two decomposition steps.[49,67,78,79] LDO has a memory effect which enables it to re-form the LDH structure after contacting with water and anions.[70] Increasing the calcination temperatures leads to the decomposition of LDO compounds with the formation of metastable phase at temperature 300–600 °C. The metastable mixed metal oxides (MMOs) normally have a porous structure and high thermal stability.[78] Calcination of LDHs above 600 °C is, generally, known to crystallize and yield AO and spinel-type oxides, AB_2O_4, which can resist water from LDH-structure re-forming.[23,35] These derived MMO materials have been reported to display outstanding magnetic, catalytic and textual properties.[79–81]

There are a large number of studies concerning the preparation of multi-cationic oxides or spinel-type oxides through the thermal decomposition of LDHs,[23] for example, zinc aluminium mixed metal oxide (ZnAl-MMO) nanostructures with prominent optical properties could be obtained by calcinating ZnAl-LDH nanoplates in air at 400–800 °C.[79,82] Calcination of LDHs has been an alternative to the traditional chemical and physical methods for the fabrication of a wide variety of MMO nanocomposite materials.[26] Due to the hierarchical structure along with a high specific surface area and wide pore size distribution, MMO derived from LDHs has been researched as a new type of highly effective photocatalysts,[83] or used as antireflection (AR) coating materials for photovoltaic applications due to their ability to enhance the transmittance of light.[32,84] Duan and co-workers[83] reported a hierarchical ZnAl-MMO framework obtained by calcination of the LDH precursor together with a biotemplated synthesis method. The formed polycrystal MMO demonstrated to be an effective and recyclable photocatalyst for the decomposition of dyes in water. Ahmed

et al.[85,86] reported that after high temperature annealing, the ZnAl-NO$_3$-LDH turned into ZnO and ZnAl$_2$O$_4$ in which the bandgap of ZnO and the oxygen vacancies in ZnO and ZnAl$_2$O$_4$ increased with the calcination temperatures. Cho *et al.*[87] also used LDH as a precursor to prepare a ZnO and ZnAl$_2$O$_4$ mixture. In their work, LDH was initially intercalated by terephthalate anions. After calcination in nitriding gas, the terephthalate collapsed down into C, leading to the production of MMO with C and N evenly distributed on the surface. The porous MMO demonstrated excellent performance in photocatalytic water oxidation reactions that was much higher than pure MMO and exhibited a 2.5-fold enhancement in the photocurrent density (0.053 mA) at 1.23 V *versus* reversible hydrogen electrode (RHE) as compared to N-doped MMO nanostructure photoanodes. Although the incident photon to current conversion efficiency (IPCE) of MMOs over the entire UV visible region was below 5%, the IPCE curve traced almost exactly the same as their UV-vis absorbance curves, indicating that all the photons (even with the longest wavelength) absorbed by the visible light absorbing sites created by the C and N co-doping were contributed to the photocatalytic water oxidation.

As ZnAl-LDH nanostructures did not absorb UV and visible light (Figure 13.11),[87] nanostructures prepared by calcination of LDH at 450 °C for 2 h in air demonstrated high absorbance in the UV region. When the calcination temperature increased from 450 °C to 650 °C, the UV absorbance spectrum of the nanostructures gradually shifted toward longer wavelengths.

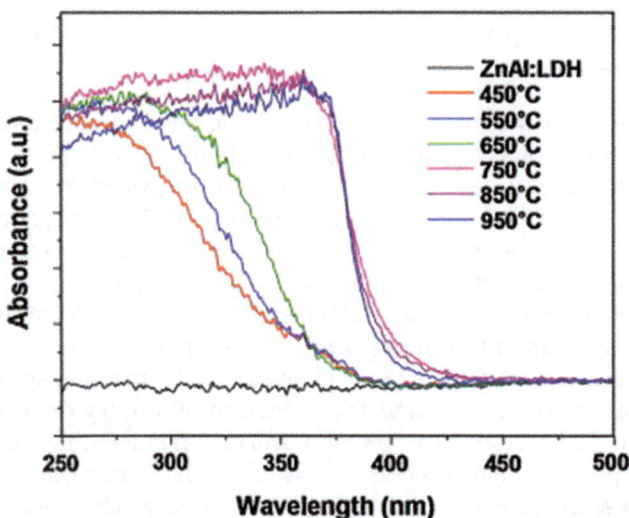

Figure 13.11 UV–vis diffuse reflectance spectra of ZnAl-LDH nanostructures and their nanostructures produced by calcination at various temperatures.
Reproduced from ref. 79, Copyright (2011), with permission from Elsevier.

Above 750 °C, the calcined MMO nanostructures showed strong absorption in the near ultraviolet region, which is the characteristic absorption of ZnO and $ZnAl_2O_4$.[79] The $ZnO/ZnAl_2O_4$ nanocomposites derived from the ZnAl-LDH precursors had superior photocatalytic performances to either single-phase ZnO or similar $ZnO/ZnAl_2O_4$ samples prepared by chemical co-precipitation or physical mixing methods. The heterojunction nanostructure and the strong coupling between ZnO and $ZnAl_2O_4$ phases were proposed to contribute the efficient spatial separation between the photo-generated electrons and holes, which concomitantly improve the photocatalytic activities.[26]

The spinel structure MM_2O_4, such as $ZnFe_2O_4$ and $ZnAl_2O_4$, derived from calcined LDH precursors have also been reported with good photocatalytic activity.[20] By dissolving a mixture of ZnO and $ZnFe_2O_4$ gained from the calcination of ZnFe-LDH at 500 °C in aqueous NaOH, pure $ZnFe_2O_4$ could be obtained with substantially increased surface area and pore volume. This type of $ZnFe_2O_4$ showed higher photocatalytic activity in the degradation of phenol than the corresponding ZnO and $ZnFe_2O_4$ mixture.[20] The MMO particles $Zn_{1-x}Mg_xO$, prepared by a polymer-based calcination–purification method,[81] effectively tuned the bandgap of ZnO in the metastable solid solution state with the photoluminescence changed from 2.12 eV to 2.32 eV. The enhanced photoluminescence in the visible region was due to the incorporation of magnesium ions on zinc ion lattice sites.

13.5 Photocatalytic Applications of LDH

13.5.1 LDH's Application in Photocatalytic Degradation of Organic Pollutants

Metal oxide nanostructures such as zinc oxide and titanium dioxide (TiO_2) are useful for photocatalytic, photoelectrochemical and photovoltaic processes. The energy levels for the conduction and valence bands and the electron affinity of ZnO are similar to those of TiO_2. However, both ZnO and TiO_2 are wide bandgap metal oxides, and they are only photocatalytically active under ultraviolet irradiation. Although bandgap narrowing can be achieved by either elevating the valence band maximum (VBM) or lowering the conduction band minimum (CBM) *via* doping or other techniques,[88] the ultrafine particles are difficult to separate from the reaction systems owing to their small particle sizes and the formation of a milky dispersion.[87] For other photocatalysts, such as CdS, photocorrosion is seemingly unavoidable during the photocatalytic processes.[89] As for LDHs, a large class of materials with stable structures and positively charged frameworks, have prospective applications in the photocatalysis area, which can be alternative for the metal oxide photocatalysts.

There are very many reports about the application of LDH and its derivatives for the photocatalytic degradation of non-biodegradable dyes and recalcitrant organic pollutants such as phenol and chlorophenols.[23,55]

Valente *et al.*[72] studied the photodegrading capabilities of CeO_2/MgAl-LDH composite under UV irradiation, and discovered that the degradation of phenol was superior to those obtained with the benchmark Degussa P25 TiO_2 photocatalyst under the same experimental conditions. They proposed that degradation of the contaminants may be taking place by two different mechanisms: ligand-to-metal charge-transfer (LMCT) and UV-generated electron–hole pairs. depending on the solution pH and the pK_a values of the contaminants. If the solution $pH < pK_a$, the acidic proton of the phenol substrate interacts with a surface basic group or is physisorbed *via* hydrogen bonding, and a charge-transfer complex is formed at defect sites near the CeO_2/LDH interface (Figure 13.12). Upon photon excitation, an electron could be transferred directly from the highest occupied molecular orbital (HOMO) of the ligand (phenol or 4-chlorophenol) to the catalyst, most likely to the 4f band of CeO_2. Then the electron was transferred to a suitable electron acceptor, O_2, to form radical species $O_2^{\bullet-}$, which in turn mineralized the contaminant phenol into small molecules or even into CO_2 and H_2O. However, if solution $pH > pK_a$, the interaction of deprotonated phenol

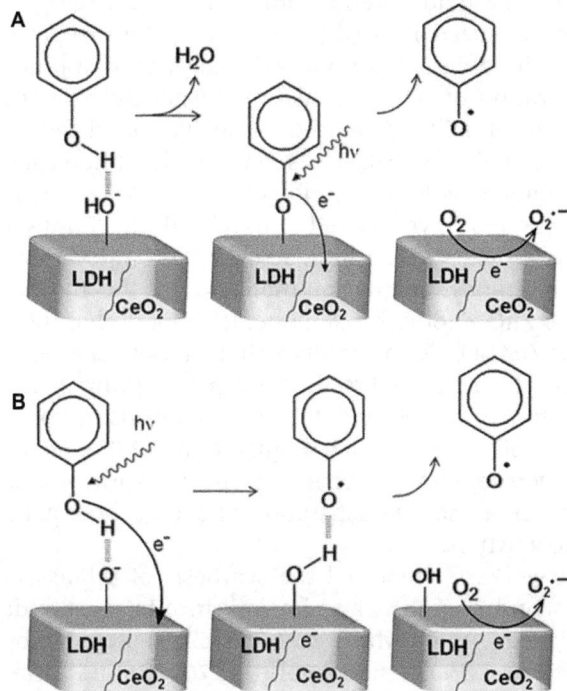

Figure 13.12   Proposal ligand-to-metal charge-transfer mechanism, initially responsible for the degradation of phenol; (A) inner-sphere complexation occurring on a Brønsted basic site, and (B) outer-sphere complexation occurring on a Lewis basic site.
Reprinted from ref. 72, Copyright (2011), with permission from Elsevier.

and the surface basic sites are hindered by electrostatic repulsion, the LMCT is also prohibited. Photocatalysis takes place by the better-known mechanism of UV-driven electron excitation from the valence band to the conduction band of CeO_2, followed by conversion of the electron–hole pair to oxidizing radical species (mainly $O_2^{\bullet-}$ and $^{\bullet}OH$) that in turn degrade the contaminant.

Chen *et al.*[23] synthesized highly efficient photo-responsive materials, titanium-containing MMO nanocomposites, from calcination of a NiTi-LDH and investigated their activities in degrading MB under UV and visible light irradiation. The materials had a high surface area and a broad adsorption range from visible to ultraviolet wavelength, enabling the photodegradation of MB to occur under UV and visible light irradiation. Pode and co-workers[60] used LDH as photocatalyst for the degradation of *p*-chlorophenol under UV light irradiation. The mineralization degree doubled with the addition of H_2O_2. Paratungstate combined with LDHs has also been reported to have high photocatalytic activities.[90] Paratungstate, a class of polyoxometalates (POM), has similar features with semiconductor.[91] By irradiating a $Mg_{12}Al_6(OH)_{36}$-$(W_7O_{24}) \cdot 4H_2O$ suspension in the near-UV area, trace aqueous organocholorine pesticide, hexachlorocyclohexane (HCH), was totally degraded and mineralized into CO_2 and HCl.[90] The disappearance of trace HCH was well fitted with Langmuir–Hinshelwood first-order kinetics, and the photogeneration of $^{\bullet}OH$ radicals was responsible for the degradation. Zn/Al/W(Mn) mixed oxides produced by calcination of the POM-containing LDH precursors in air at 600–700 °C showed higher photocatalytic activity in degradation of aqueous HCH solutions than the POM-LDH precursors.[92]

Xu and co-workers[93] synthesized hierarchical $ZnO/ZnAl_2O_4$ derived from ZnAl-LDH, and which showed high activities in photocatalytic degradation of both the anionic dye methylene orange (MO) and the cationic dye MB, superior to the ZnO photocatalyst alone. It is noted that the degradation of MO over $ZnO/ZnAl_2O_4$ is faster than that of MB (Figure 13.13). The enhancement may be attributed to the stronger coupling interaction between MO and the interface of ZnO and $ZnAl_2O_4$, as the adsorption of MO on $ZnO/ZnAl_2O_4$ is around 20.5%, much higher than 12.0% of MB. The homogeneously dispersed $ZnAl_2O_4$ phase in the network of ZnO also greatly contributed to the enhanced separation of charge–hole pairs and the subsequent photoactivities.

Zhu and co-workers[94] reported the synthesis of self-assembled carboxyl graphene (CG) and ZnAl-LDH and its derivative product, reduced CG/ZnAl-layered double oxide nanohybrid (rCG/ZnAl-LDO) prepared by high vacuum calcination of the CG/LDH nanohybrid at 700 °C, and tested their photocatalytic performance in degrading cationic MB and anionic dye orange G (OG) under visible light irradiation. The exfoliated LDH and CG solutions were prepared by using formamide as the solvent, and were mixed by electrostatically driven self-assembly and the formation of a layer-by-layer ordered nanohybrid (see Figure 13.14). The photocatalytic activity of the CG/LDH nanohybrid was not satisfactory, probably because of the low

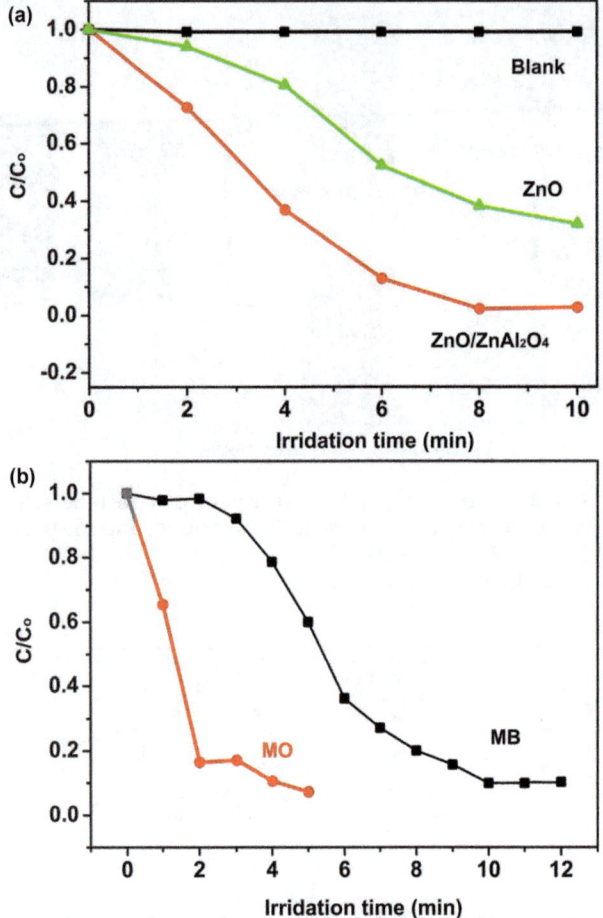

Figure 13.13 Photodegradation on (a) MB and (b) MO/MB mixture monitored as the normalized concentration *versus* irradiation time.
Reprinted from ref. 93, Copyright (2013), with permission from Elsevier.

photocatalytic activity of ZnAl-LDH. However, after calcination, rCG/LDO showed excellent photocatalytic properties. The degradation efficiency of MB by the rCG/LDO photocatalyst after 120 min was ∼100%, whereas efficiencies of about 57% and 39% were achieved by ZnAl-LDO and photolysis, respectively (Figure 13.15a). In the system of OG, as rCG/LDO had OG adsorption capacity, the degradation efficiency of OG by rCG/LDO after 90 min was 100%, much higher than that of pristine ZnAl-LDO (Figure 13.15b). The photoluminescence spectra (PL) intensity of ZnAl-LDO appeared to be obviously weaker after hybridization with rCG, suggesting the reduction of electron–hole recombination (Figure 13.15c). Moreover, the rCG/LDO showed a significant adsorption ability in the visible light region

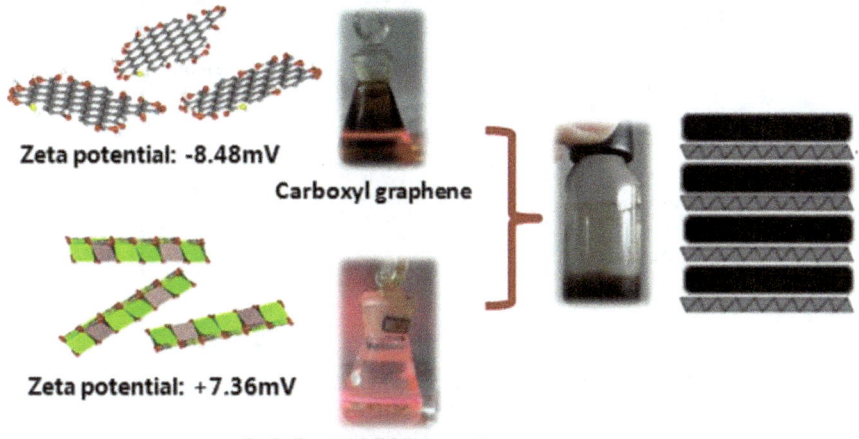

Figure 13.14   Schematic diagram of the process of electrostatic self-assembly be-
tween the negatively charged CG monolayer and the positively charged
ZnAl-LDH nanosheets.
Reproduced from ref. 94.

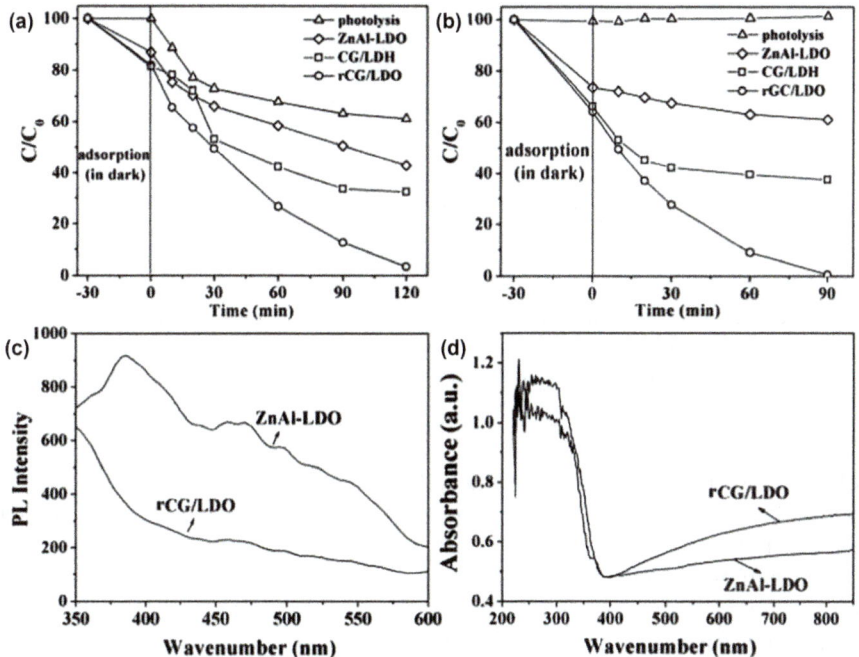

Figure 13.15   (a) Photocatalytic activity of the catalyst for MB decolorization;
(b) photocatalytic activity of the catalyst for OG decolorization; (c) PL
spectra of rCG/LDO and the pristine LDO and (d) UV-vis diffuse
reflectance spectra of rCG/LDO and the pristine LDO.
Reproduced from ref. 94.

(Figure 13.15d), contrary to the absorption edges observed in pristine ZnAl-LDO. Therefore, the hybridization of LDO with rCG plays a critical role in the photocatalytic reaction under visible light. The photocatalytic properties of the rCG/LDO nanohybrid exhibited only a 10% decrease for MB decolorization and 8% decrease for OG decolorization after 3 cycles of reuse. The high photostability was attributed to the effective physical contact and strong electronic coupling between graphene and ZnAl-LDO, and the excellent performance was attributed to the enhanced charge separation in ZnAl-LDO composed of both ZnO and $ZnAl_2O_4$ phases and between ZnAl-LDO and rCG, benefitting the spatial separation of electron–hole pairs and the production of hydroxyl radicals for dye degradations.

13.5.2 LDH's Application in Photocatalytic Reduction of Carbon Dioxide

Photocatalytic conversion of CO_2 is attractive from the viewpoint of sustainable energy production and greenhouse gas reduction. CO_2 can be adsorbed on a solid base, let alone LDH, whose anion layers are constantly intercalating with CO_3^{2-}. LDHs with an M^{2+}/M^{3+} ($M^{2+} = Mg^{2+}$, Zn^{2+}, Ni^{2+}; $M^{3+} = Al^{3+}$, Ga^{3+}, In^{3+}) ratio of 3 showed activity for the photocatalytic conversion of CO_2 to CO in water.[95] A copper-modified LDH was also reported to show activity for the photocatalytic conversion of CO_2 to methanol in the presence of H_2 gas,[96] and the methanol selectivity (26 mol%) obtained using ZnCuAl-LDH catalysts was improved to 68 mol% using ZnCuGa-LDH catalysts. In the interlayer space of LDH photocatalysts, CO_2 was suggested for the reaction with the hydroxy group bound to the Cu sites to form a hydrogen carbonate intermediate. Under UV-visible light, the Cu ions in the cationic layer facilitated charge separation utilizing the reduction/oxidation (redox) of Cu^{II}/Cu^{I}. Hydrogen carbonate species were gradually reduced to formic acid, formaldehyde, and finally to methanol utilizing the trapped photo-generated electrons as Cu^{I} ions. Therefore, the interlayer space of these LDH photocatalysts served as an active pocket for the reduction of CO_2 to methanol. An increase in the available reaction space would lead to enhanced photocatalytic activity. By substituting the interlayer CO_3^{2-} anions for $[Cu(OH)_4]^{2-}$, the methanol formation rates using ZnGa-LDH and ZnCuGa-LDH increased by a factor of 5.9 and 2.9, respectively.[97] The hydroxy groups that were bound to Cu sites were important, and the effects of the interlayer $[Cu(OH)_4]^{2-}$ were greater than those of the in-layer octahedral Cu sites because of its steric availability (accessibility) and semiconductivity (*e.g.* values of 3.0–4.2 eV).

13.5.3 LDH's Application in Photocatalytic Water Splitting

The search for suitable semiconductors as photocatalysts for the splitting of water into hydrogen by using solar energy is one of the primary missions

of material science, which has received considerable attention on account of the increasing number of global energy crises. It has been reported that MTi-LDHs (M = Ni, Zn, Mg) synthesized by the co-precipitation method showed remarkable photocatalytic performance for water splitting into hydrogen.[61] The H_2 generation rate was 31.4 µmol/h with 0.133 vol% lactic acid as the sacrificial electron donor, which was 18 times higher than that of $K_2Ti_4O_9$. Meng *et al.*[98] reported a 3D Ce^{III} polymer LDH which is photocatalytically active in H_2 evolution under UV irradiation. Silva *et al.*[49] synthesized a series of ZnTi-, ZnCe- and ZnCr-LDHs at different Zn/metal atomic ratios (from 4 : 2 to 4 : 0.25) and tested them for the visible light photocatalytic oxygen generation. The ZnCr-LDH with two adsorption bands in visible region at λ_{max} of 410 and 570 nm was found to be highly active, the quantum yields for oxygen generation were 60.9% and 12.2% at 410 and 570 nm, respectively. Besides, the efficiencies of the chromium layered double oxides for oxygen generation increased asymptotically with the Cr content. The overall efficiency of ZnCr-LDH for visible light oxygen generation was found to be 1.6 times higher than that of WO_3 under the same conditions. It has also been reported that hybridizing ZnCr-LDH nanoplates with graphene nanosheets or TiO_2 could dramatically increase their performance in photocatalytic O_2 generation.[8,99]

13.5.4   Other Applications

Photoelectrodes made from NiFeTi-LDH showed good photovoltaic performance.[100] By intercalating I_3^-/I^- into the LDH interlayer, the composite material can be applied for dye-sensitized solar cells (DSSC) as solid electrolyte.[101] Intimately mixed metal oxides derived from a calcined ZnTi-LDH precursor could serve as semiconductors for DSSC assembly,[67] and demonstrate high photovoltaic activity with the efficiency parameters $V_{oc} = 0.63$ V and $J_{sc} = 2.18$ mA cm^{-2}, comparable to an analogous photovoltaic cell prepared using TiO_2.

In addition, LDHs have been used as catalysts or catalyst supports,[102–105] and are of particular interest in therapeutic and pharmaceutical applications due to their low toxicity compared to other inorganic nanoparticles.[106] LDH nanosheets, like their bulk form, are attractive candidates as anion exchangers. They can be used as drug delivery capsules for controlled release of therapeutic, electroactive and photoactive materials,[7,107–109] and as polymer supports in organic synthesis.[35,110–113] Modification of LDHs by intercalating organic ions between lamellas can be used as a nanofiller to improve the flame retardancy and mechanical properties of the polymer matrix.[25,114–117] Certain kind of LDHs such as CoAl-LDH and NiAl-LDH can be used as electrode materials for supercapacitors due to their high activity for the faradic redox reaction, facilitating energy conversion and storage.[34] MgAl-LDHs and ZnAl-LDHs with chromophore molecules in the interlayers exhibited the prime requisites of pigments such as brilliant colours and insolubility,[118] which can be used for colouring cement and polymers.

13.6 Summary and Future Perspectives

LDHs and derivatives have been used in a wide range of photocatalytic applications. In this chapter, the most common LDH synthesis methods and the application of LDH and its derivatives such as MMO for photocatalysis have been summarized. As the techniques for the preparation of exfoliated LDH nanosheets becomes more and more mature,[9,19,119] it is anticipated that ultrathin and even transparent LDH single nanosheets will find their way into photocatalysis because of their unusual structural features, that is, ultimate two-dimensional anisotropy and the novel physical and chemical properties due to an extremely small thickness of around 1 nm. Furthermore, these nanosheets can be used as building blocks for the fabrication of a wide variety of functional nanostructured materials, such as the layer-by-layer assembly of nanosheets with appropriately charged counterparts through a wet process, which will also advance their potential applications in the near future.[19]

By taking advantage of the confined space between LDH layers, Duan *et al.*[120] prepared graphene nanosheets with good control over the number of layers within the two-dimensional galleries of LDHs. Wu *et al.*[16] prepared the water-dispersible composite material containing magnetite particles, graphene and LDHs, which demonstrated fast and efficient adsorption of arsenate from aqueous solution. The increasing reports on the combination of graphene and LDH forecast the trend that the combination of two different two-dimensional building blocks may lead to the formation of hierarchical composites that can take full advantages of each kind of material, which is also an effective way for the preparation of multifunctional materials with extraordinary properties.[121] It should be noticed that most of the properties of carbonaceous materials and LDHs are complementary. Therefore, the combination of graphene and LDHs into hierarchical nanocomposites is a promising method to integrate their particular properties together: graphene can provide good electrical conductivity and high mechanical strength, and LDHs can provide good chemical reactivity. LDHs/graphene is versatile in energy storage, catalysis and water treatment. It is also anticipated that the research on exfoliated LDH nanosheets and/or LDH-based composites will break new ground in LDH's photocatalytic applications in the near future by combining with other photocatalytic materials.

References

1. S. J. Yuan, Y. G. Li, Q. H. Zhang and H. Z. Wang, *Res. Chem. Intermediat.*, 2009, **35**, 685–692.
2. C. A. Kathleen and K. Athanasios, *Solid State Ionics*, 1988, **1**, 77–86.
3. S. J. Mills, A. G. Christy, J. M. R. Genin, T. Kameda and F. Colombo, *Mineral. Mag.*, 2012, **76**, 1289–1336.
4. F. Kovanda, E. Jindova, K. Lang, P. Kubat and Z. Sedlakova, *Appl. Clay Sci.*, 2010, **48**, 260–270.

5. Y. Wang and D. Zhang, *Mater. Res. Bull.*, 2011, **46**, 1963–1968.
6. N. Hao, Y. Yang, H. Wang, P. A. Webley and D. Zhao, *J. Colloid Interface Sci.*, 2010, **346**, 429–435.
7. R. Z. Ma and T. Sasaki, *Adv. Mater.*, 2010, **22**, 5082–5104.
8. J. L. Gunjakar, T. W. Kim, H. N. Kim, I. Y. Kim and S.-J. Hwang, *J. Am. Chem. Soc.*, 2011, **133**, 14998–15007.
9. Q. Wang and D. O'Hare, *Chem. Rev.*, 2012, **112**, 4124–4155.
10. A. Ekimov, A. L. Efros and A. Onushchenko, *Solid State Commun.*, 1985, **56**, 921–924.
11. F. Costa, M. Saphiannikova, U. Wagenknecht and G. Heinrich, in *Wax Crystal Control, Nanocomposites, Stimuli-Responsive Polymers*, Springer, Berlin Heidelberg, 2008, pp. 101–168.
12. J. Zhou, Z. P. Xu, S. Qiao, Q. Liu, Y. Xu and G. Qian, *J. Hazard. Mater.*, 2011, **189**, 586–594.
13. Y. F. Chao, P. C. Chen and S. L. Wang, *Appl. Clay Sci.*, 2008, **40**, 193–200.
14. R. Chitrakar, A. Sonoda, Y. Makita and T. Hirotsu, *Ind. Eng. Chem. Res.*, 2011, **50**, 9280–9285.
15. L. El Gaini, M. Lakraimi, E. Sebbar, A. Meghea and M. Bakasse, *J. Hazard. Mater.*, 2009, **161**, 627–632.
16. X. L. Wu, L. Wang, C. L. Chen, A. W. Xu and X. K. Wang, *J. Mater. Chem.*, 2011, **21**, 17353–17359.
17. J. S. Valente, F. Tzompantzi, J. Prince, J. G. H. Cortez and R. Gomez, *Appl. Catal., B*, 2009, **90**, 330–338.
18. V. Iliev, A. Ileva and L. Bilyarska, *J. Mol. Catal. A: Chem.*, 1997, **126**, 99–108.
19. L. Li, R. Ma, Y. Ebina, N. Iyi and T. Sasaki, *Chem. Mater.*, 2005, **17**, 4386–4391.
20. W. Q. Meng, F. Li, D. G. Evans and X. Duan, *J. Porous Mater.*, 2004, **11**, 97–105.
21. Z. P. Xu, J. Zhang, M. O. Adebajo, H. Zhang and C. Zhou, *Appl. Clay Sci.*, 2011, **53**, 139–150.
22. D. Chen, Y. Li, J. Zhang, J. Z. Zhou, Y. Guo and H. Liu, *Chem. Eng. J.*, 2012, **185**, 120–126.
23. X. Shu, J. He, D. Chen and Y. Wang, *J. Phys. Chem. C*, 2008, **112**, 4151–4158.
24. M. Intissar, J. C. Jumas, J. P. Besse and F. Leroux, *Chem. Mater.*, 2003, **15**, 4625–4632.
25. D. Y. Wang, F. R. Costa, A. Vyalikh, A. Leuteritz, U. Scheler, D. Jehnichen, U. Wagenknecht, L. Häussler and G. Heinrich, *Chem. Mater.*, 2009, **21**, 4490–4497.
26. X. F. Zhao, L. Wang, X. Xu, X. D. Lei, S. L. Xu and F. Z. Zhang, *Aiche J.*, 2012, **58**, 573–582.
27. O. Saber and H. Tagaya, *J. Porous Mater.*, 2003, **10**, 83–91.
28. Z. An, S. Lu, L. W. Zhao and J. He, *Langmuir*, 2011, **27**, 12745–12750.
29. D. Chen, X. Wang, T. Liu, X. Wang and J. Li, *ACS Appl. Mater. Interfaces*, 2010, **2**, 2005–2011.

30. H. M. He, H. L. Kang, S. L. Ma, Y. X. Bai and X. J. Yang, *J. Colloid Interface Sci.*, 2010, **343**, 225–231.
31. Y. H. Chuang, Y. M. Tzou, M. K. Wang, C. H. Liu and P. N. Chiang, *Ind. Eng. Chem. Res.*, 2008, **47**, 3813–3819.
32. A. Węgrzyn, A. Rafalska-Łasocha, D. Majda, R. Dziembaj and H. Papp, *J. Therm. Anal. Calorim.*, 2010, **99**, 443–457.
33. L. L. Wang, B. Li, X. C. Zhang, C. X. Chen and F. Zhang, *Appl. Clay Sci.*, 2012, **56**, 110–119.
34. R. Ma, J. Liang, X. Liu and T. Sasaki, *J. Am. Chem. Soc.*, 2012, **134**, 19915–19921.
35. F. Labajos, V. Rives and M. Ulibarri, *J. Mater. Sci.*, 1992, **27**, 1546–1552.
36. Z. P. Xu, G. S. Stevenson, C. Q. Lu, G. Qing, P. F. Bartlett and P. P. Gray, *J. Am. Chem. Soc.*, 2006, **128**, 36–37.
37. E. M. del Campo, J. S. Valente, T. Pavon, R. Romero, A. Mantilla and R. Natividad, *Ind. Eng. Chem. Res.*, 2011, **50**, 11544–11552.
38. A. Bankauskaite and K. Baltakys, *Sci. Sintering*, 2011, **43**, 261–275.
39. K. Dutta, S. Das and A. Pramanik, *J. Colloid Interface Sci.*, 2012, **366**, 28–36.
40. J. Liu, Y. Li, X. Huang, G. Li and Z. Li, *Adv. Funct. Mater.*, 2008, **18**, 1448–1458.
41. E. Hosono, S. Fujihara, I. Honma and H. Zhou, *Adv. Mater.*, 2005, **17**, 2091–2094.
42. X. D. Lei, Z. Lu, X. X. Guo and F. Z. Zhang, *Ind. Eng. Chem. Res.*, 2012, **51**, 1275–1280.
43. J. C. Dupin, H. Martinez, C. Guimon, E. Dumitriu and I. Fechete, *Appl. Clay Sci.*, 2004, **27**, 95–106.
44. S. Cho, S. Kim, E. Oh, S. H. Jung and K. H. Lee, *CrystEngComm*, 2009, **11**, 1650–1657.
45. A. I. Khan and D. O'Hare, *J. Mater. Chem.*, 2002, **12**, 3191–3198.
46. X. Shu, W. H. Zhang, J. He, F. X. Gao and Y. X. Zhu, *Solid State Sci.*, 2006, **8**, 634–639.
47. O. Saber, *J. Mater. Sci.*, 2007, **42**, 9905–9912.
48. Z. G. Xiong and Y. M. Xu, *Chem. Mater.*, 2007, **19**, 1452–1458.
49. C. G. Silva, Y. Bouizi, V. Fornés and H. García, *J. Am. Chem. Soc.*, 2009, **131**, 13833–13839.
50. N. Thi Dung, D. Tichit, B. Huong Chiche and B. Coq, *Appl. Catal., A*, 1998, **169**, 179–187.
51. C. H. Zhou, J. N. Beltramini, C. X. Lin, Z. P. Xu, G. Q. M. Lu and A. Tanksale, *Catal. Sci. Technol.*, 2011, **1**, 111–122.
52. A. Mantilla, F. Tzompantzi, J. Fernandez, J. Diaz Góngora, G. Mendoza and R. Gomez, *Catal. Today*, 2009, **148**, 119–123.
53. Y. F. Chen, S. H. Zhou, F. Li, F. Li and Y. W. Chen, *J. Lumin.*, 2011, **131**, 701–704.
54. C. C. Yang, S. Y. Chen and S. Y. Cheng, *Powder Technol.*, 2004, **148**, 3–6.

55. L. Mohapatra, K. Parida and M. Satpathy, *J. Phys. Chem. C*, 2012, **116**, 13063–13070.
56. M. Shao, F. Ning, J. Zhao, M. Wei, D. G. Evans and X. Duan, *Adv. Funct. Mater.*, 2013, **23**, 3513–3518.
57. A. Fujishima and K. Honda, *Nature*, 1972, **238**, 37–38.
58. Y. H. Guo and C. W. Hu, *J. Mol. Catal. A: Chem.*, 2007, **262**, 136–148.
59. Z. G. Xiong and X. S. Zhao, *J. Am. Chem. Soc.*, 2012, **134**, 5754–5757.
60. R. Pode, L. Cocheci, E. Popovici, E. M. Seftel and V. Pode, *Rev. Chim.*, 2008, **59**, 898–901.
61. Y. F. Zhao, P. Y. Chen, B. S. Zhang, D. S. Su, S. T. Zhang, L. Tian, J. Lu, Z. X. Li, X. Z. Cao, B. Y. Wang, M. Wei, D. G. Evans and X. Duan, *Chem. Eng. J.*, 2012, **18**, 11949–11958.
62. R. J. Lu, X. Xu, J. P. Chang, Y. Zhu, S. L. Xu and F. Z. Zhang, *Appl. Catal., B*, 2012, **111**, 389–396.
63. S. Pausova, J. Krysa, J. Jirkovsky, G. Mailhot and V. Prevot, *Environ. Sci. Pollut. Res.*, 2012, **19**, 3709–3718.
64. Z. Huang, P. Wu, Y. Lu, X. Wang, N. Zhu and Z. Dang, *J. Hazard. Mater.*, 2013, **246–247**, 70–78.
65. G. Q. Wan, D. X. Li, C. F. Li, J. Xu and W. G. Hou, *Chin. Chem. Lett.*, 2012, **23**, 1415–1418.
66. S. Cho, S. Kim, J. W. Jang, S. H. Jung, E. Oh, B. R. Lee and K. H. Lee, *J. Phys. Chem. C*, 2009, **113**, 10452–10458.
67. L. Teruel, Y. Bouizi, P. Atienzar, V. Fornes and H. Garcia, *Energy Environ. Sci.*, 2010, **3**, 154–159.
68. S. He, S. T. Zhang, J. Lu, Y. F. Zhao, J. Ma, M. Wei, D. G. Evans and X. Duan, *Chem. Commun.*, 2011, **47**, 10797–10799.
69. E. M. Seftel, E. Popovici, M. Mertens, E. A. Stefaniak, R. Van Grieken, P. Cool and E. F. Vansant, *Appl. Catal., B*, 2008, **84**, 699–705.
70. F. Tzompantzi, A. Mantilla, F. Banuelos, J. L. Fernandez and R. Gomez, *Top. Catal.*, 2011, **54**, 257–263.
71. E. Dvininov, M. Ignat, P. Barvinschi, M. A. Smithers and E. Popovici, *J. Hazard. Mater.*, 2010, **177**, 150–158.
72. J. S. Valente, F. Tzompantzi and J. Prince, *Appl. Catal., B*, 2011, **102**, 276–285.
73. T. Kameyama, K. Okazaki, K. Takagi and T. Torimoto, *Phys. Chem. Chem. Phys.*, 2009, **11**, 5369–5376.
74. X. Xu, R. J. Lu, X. F. Zhao, Y. Zhu, S. L. Xu and F. Z. Zhang, *Appl. Catal., B*, 2012, **125**, 11–20.
75. X. Xu, R. J. Lu, X. F. Zhao, S. L. Xu, X. D. Lei, F. Z. Zhang and D. G. Evans, *Appl. Catal., B*, 2011, **102**, 147–156.
76. M. Wei, X. Y. Xu, J. He, Q. Yuan, G. Y. Rao, D. G. Evans, M. Pu and L. Yang, *J. Phys. Chem. Solids*, 2006, **67**, 1469–1476.
77. J. Zhang, Y. F. Xu, G. Qian, Z. P. Xu, C. Chen and Q. Liu, *J. Phys. Chem. C*, 2010, **114**, 10768–10774.
78. J. Wang, J. D. Zhou, Z. S. Li, Y. He, S. S. Lin, Q. Liu, M. L. Zhang and Z. H. Jiang, *J. Solid State Chem.*, 2010, **183**, 2511–2515.

79. S. Cho and K. H. Lee, *J. Alloys Compd.*, 2011, **509**, 8770–8778.
80. X. F. Zhao, F. Z. Zhang, S. L. Xu, D. G. Evans and X. Duan, *Chem. Mater.*, 2010, **22**, 3933–3942.
81. G. Q. Lu, I. Lieberwirth and G. Wegner, *J. Am. Chem. Soc.*, 2006, **128**, 15445–15450.
82. M. S. Hadnadev-Kostic, T. J. Vulic, D. B. Zoric and R. P. Marinkovic-Neducin, *Chem. Ind. Chem. Eng. Q.*, 2012, **18**, 295–303.
83. Y. F. Zhao, M. Wei, J. Lu, Z. L. Wang and X. Duan, *ACS Nano*, 2009, **3**, 4009–4016.
84. J. B. Han, Y. B. Dou, M. Wei, D. G. Evans and X. Duan, *Chem. Eng. J.*, 2011, **169**, 371–378.
85. A. A. A. Ahmed, Z. A. Talib and M. Z. bin Hussein, *Appl. Clay Sci.*, 2012, **56**, 68–76.
86. A. A. A. Ahmed, Z. A. Talib, M. Z. bin Hussein and A. Zakaria, *J. Alloys Compd.*, 2012, **539**, 154–160.
87. S. Cho, J. W. Jang, K. J. Kong, E. S. Kim, K. H. Lee and J. S. Lee, *Adv. Funct. Mater.*, 2012, **23**, 2348–2356.
88. N. Serpone and A. V. Emeline, *J. Phys. Chem. Lett.*, 2012, **3**, 673–677.
89. A. L. Linsebigler, G. Lu and J. T. Yates Jr, *Chem. Rev.*, 1995, **95**, 735–758.
90. Y. H. Guo, D. F. Li, C. W. Hu, Y. H. Wang, E. B. Wang, Y. C. Zhou and S. H. Feng, *Appl. Catal., B*, 2001, **30**, 337–349.
91. Y. H. Guo and C. W. Hu, *J. Cluster Sci.*, 2003, **14**, 505–526.
92. Y. H. Guo, D. F. Li, C. W. Hu, E. Wang, Y. C. Zou, H. Ding and S. H. Feng, *Microporous Mesoporous Mater.*, 2002, **56**, 153–162.
93. R. Huo, Y. Kuang, Z. Zhao, F. Zhang and S. Xu, *J. Colloid Interface Sci.*, 2013, **407**, 17–21.
94. Z. J. Huang, P. X. Wu, B. N. Gong, Y. P. Fang and N. W. Zhu, *J. Mater. Chem. A*, 2014, **2**, 5534–5540.
95. K. Teramura, S. Iguchi, Y. Mizuno, T. Shishido and T. Tanaka, *Angew. Chem., Int. Ed. Engl.*, 2012, **51**, 8008–8011.
96. N. Ahmed, Y. Shibata, T. Taniguchi and Y. Izumi, *J. Catal.*, 2011, **279**, 123–135.
97. N. Ahmed, M. Morikawa and Y. Izumi, *Catal. Today*, 2012, **185**, 263–269.
98. B. F. Meng, W. S. You, X. F. Sun, F. Zhang and M. Y. Liu, *Inorg. Chem. Commun.*, 2011, **14**, 35–37.
99. J. L. Gunjakar, I. Y. Kim, J. M. Lee, N. S. Lee and S. J. Hwang, *Energy Environ. Sci.*, 2013, **6**, 1008–1017.
100. X. M. Ji, M. L. Li, Y. X. Zhao, Y. B. Wei and Q. H. Xu, *Solid State Sci.*, 2009, **11**, 1170–1175.
101. M. Bastianini, D. Costenaro, C. Bisio, L. Marchese, U. Costantino, R. Vivani and M. Nocchetti, *Inorg. Chem.*, 2012, **51**, 2560–2568.
102. T. J. Vulic, A. F. K. Reitzmann and K. Lazar, *Chem. Eng. J.*, 2012, **207**, 913–922.
103. L. Y. Liu, M. Pu, L. Yang, D. Q. Li, D. G. Evans and J. He, *Mater. Chem. Phys.*, 2007, **106**, 422–427.

104. W. Kagunya, Z. Hassan and W. Jones, *Inorg. Chem.*, 1996, **35**, 5970–5974.

105. S. Sankaranarayanan, C. A. Antonyraj and S. Kannan, *Bioresour. Technol.*, 2012, **109**, 57–62.

106. L. A. Utracki, M. Sepehr and E. Boccaleri, *Polym. Adv. Technol.*, 2007, **18**, 1–37.

107. G. Choi, J. H. Lee, Y. J. Oh, Y. Bin Choy, M. C. Park, H. C. Chang and J. H. Choy, *Int. J. Pharm.*, 2010, **402**, 117–122.

108. S. J. Choi and J. H. Choy, *Nanomedicine*, 2011, **6**, 803–814.

109. Y. Yasin, N. M. Ismail, M. Z. Hussein and N. Aminudin, *J. Biomed. Nanotechnol.*, 2011, 7, 486–488.

110. G. A. Wang, C. C. Wang and C. Y. Chen, *Polymer*, 2005, **46**, 5065–5074.

111. T. Nogueira, R. Botan, F. Wypych and L. Lona, *J. Appl. Polym. Sci.*, 2012, **124**, 1764–1770.

112. S. Martinez-Gallegos, M. Herrero and V. Rives, *J. Appl. Polym. Sci.*, 2008, **109**, 1388–1394.

113. Z. Matusinovic, M. Rogosic and J. Sipusic, *Polym. Degrad. Stab.*, 2009, **94**, 95–101.

114. C. Manzi-Nshuti, D. Y. Wang, J. M. Hossenlopp and C. A. Wilkie, *J. Mater. Chem.*, 2008, **18**, 3091–3102.

115. G. B. Huang, A. A. Zhuo, L. Q. Wang and X. Wang, *Mater. Chem. Phys.*, 2011, **130**, 714–720.

116. F. Costa, B. Satapathy, U. Wagenknecht, R. Weidisch and G. Heinrich, *Eur. Polym. J.*, 2006, **42**, 2140–2152.

117. S. L. Xu, Z. R. Chen, B. W. Zhang, J. H. Yu, F. Z. Zhang and D. G. Evans, *Chem. Eng. J.*, 2009, **155**, 881–885.

118. J. Bauer, P. Behrens, M. Speckbacher and H. Langhals, *Adv. Funct. Mater.*, 2003, **13**, 241–248.

119. J. Demel, J. Plestil, P. Bezdicka, P. Janda, M. Klementova and K. Lang, *J. Colloid Interface Sci.*, 2011, **360**, 532–539.

120. J. Sun, H. Liu, X. Chen, D. G. Evans, W. Yang and X. Duan, *Chem. Commun.*, 2012, **48**, 8126–8128.

121. M. Q. Zhao, Q. Zhang, J. Q. Huang and F. Wei, *Adv. Funct. Mater.*, 2012, **22**, 675–694.

CHAPTER 14

Health and Environmental Aspects of Green Photo-active Nanomaterials

H. HAYNES AND R. ASMATULU*

Department of Mechanical Engineering, Wichita State University, 1845 Fairmount, Wichita, KS 67260, USA
*Email: ramazan.asmatulu@wichita.edu

14.1 Introduction

Nanomaterials remain novel and exciting. Despite several decades of concentrated research in universities and companies around the world, new applications continue to be uncovered and new information continues to emerge. Along with the enthusiasm they express for the potential of nano-materials, researchers tend to express caution about the role that nano-technology will play in the production of future goods and energy. Most of the data available on nanomaterials is gathered from experiments, which show variable results even under necessarily controlled conditions. Laboratory tests can give an indication of the obstacles and hazards of full-scale nanomaterials production, which is also a controlled, though much larger, process. It is more challenging to extrapolate what impact nano-materials will have if released to the environment because of the complexity of the world outside the laboratory. If exponential growth in nanomaterials production occurs as the industry continues to develop,[1] any side effects of nanotechnology will be magnified over time. Some artificial nanomaterials in active use have already been detected in the environment.[2] As

RSC Green Chemistry No. 42
Green Photo-active Nanomaterials: Sustainable Energy and Environmental Remediation
Edited by Nurxat Nuraje, Ramazan Asmatulu and Guido Mul
© The Royal Society of Chemistry 2016
Published by the Royal Society of Chemistry, www.rsc.org

nanotechnology, and nanomaterials, spread further, research is required to limit the potential negative impacts on health and the environment.

Nanotechnology and the environment are often discussed together. Some researchers focus on the environmental benefits that nanomaterials can provide, whether through green energy generation, or by addressing other environmental issues. The nanomaterials explored for these applications are often photoactive, meaning that they absorb light energy. One of the most common compounds in this arena is titanium dioxide (titania, TiO_2), a versatile product that has been in continuous production for more than 80 years.[3] TiO_2 has variable crystalline structure, existing as anatase, rutile, and brookite crystals.[4,5] It is valued for its ability to absorb ultraviolet light,[3] which allows it to break down organic material.[1,5] Zinc oxide (ZnO) has similar properties to TiO_2 with regard to photocatalysis and photo-oxidation.[5,6] Fullerenes are composed of carbon atoms arranged in a hexagonal lattice structure. Common varieties are graphene, buckyballs (C_{60}), single-walled carbon nanotubes (SWCNTs), and multi-walled carbon nanotubes (MWCNTs). TiO_2,[7-9] ZnO,[9,10] and fullerenes[11] have all been used to create solar cells. TiO_2, ZnO,[12] and composites of CNTs and TiO_2[13] have also been investigated for their photocatalytic abilities. CNTs in a research setting have even been applied in fuel cell catalysts as a platinum replacement.[14] These materials represent only a fraction of the nanoproducts that may be applied to green technology, but they are the front runners in the field. In the sections that follow, this chapter will touch on some of the research associated with these materials, and some of the consequences that direct and environmental exposure to them may have on organisms and biological systems.

14.2   Model Making: Mass, Energy, Risk and Research in Manufacturing

To estimate the impact that large-scale nanomaterial production will have on the world, various models have been proposed. All are estimates because there is not enough data to make definitive calculations, but they can suggest the relative merits or risks of processes and policies. They can also indicate where information is lacking,[15,17] and where future research should be directed.[16] Energy, cost, and toxicity all play roles in decisions made about nanomaterials production.[16] The importance of these characteristics varies between producers, consumers, and environmentalists.[16,18] Estimated values for these factors can be combined in multi-criteria decision analysis (MCDA).[15-17] Computer models programmed with these criteria can, through iterative processes, allow the model to be refined.[17] Linkov *et al.* applied this analysis technique to four SWCNT production methods: chemical vapor deposition (CVD), arc discharge, laser vaporization, and high pressure carbon monoxide (HiPCO). They estimated which processes would be preferable from different viewpoints, and the results appear in

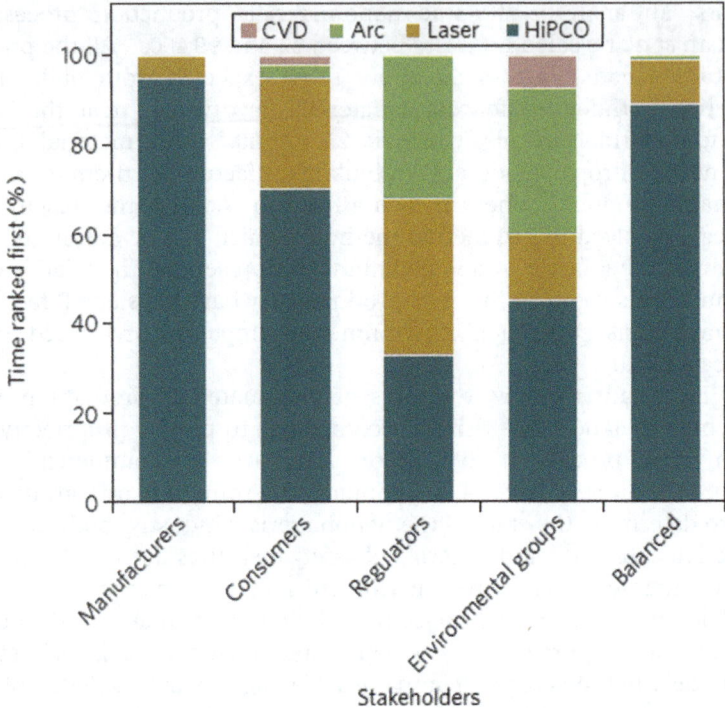

Figure 14.1   Relative merits of SWCNT production methods from five viewpoints. (Source: Adapted from Linkov *et al.*[16])

Figure 14.1. Figure 14.1 shows that there is a general preference for HiPCO, and a general opposition to CVD, results that could influence a manufacturer's choice of production method.[16] Even in relatively well-understood processes, there can be significant variation in the data inputs, which leads to variability in the results.[19] For a more detailed and accurate analysis of the impact of nanomaterials on the environment, it would be necessary to improve the consistency of information, and to include factors such as reactivity, size, bioavailability, and bioaccumulation in the analysis.[17,19]

In examining nanotechnology for 'green' purposes, the energy and resources used to produce the nanomaterials are among the most important factors. If a proposed material or energy-harvesting device requires an unsustainable resource input, then it is not an improvement over existing technologies. One measure of losses in production is the *E* factor of a material, which is the ratio of the mass of the chemicals used in a process, with the exceptions of water and combustion gas by-products,[20] to the mass of the end product.[21] In an ideal reaction this would be a one to one ratio, or an E-factor equal to 1, but in reality unreacted precursors and rinsing chemicals increase the total process mass.[20,21] For bulk chemical materials, the E-factor can range from 1 to 5 for large-scale production.[21] Nanomaterials, which are produced in laboratory batches, have much higher

E-factors; an analysis of some nanomaterials production processes by Eckleman *et al.* revealed E-factors between 22 and 99 400.[20] All the processes they studied had E-factors far above those expected from bulk material production,[20] though a few nanomaterials have values near the E-factor range for pharmaceuticals, which is 25 to 100.[21] Nanomaterial E-factors would need to drop considerably in bulk manufacture for them to become a sustainable product.[20] There is also a need to consider the toxicity of the chemicals involved in production methods, which can be accounted for by multiplying the E-factor by a scaled number representing the relative hazard of the materials used.[21] Thus, two products may have the same E-factor on a mass-only basis, but their environmental impacts vary based on the chemicals used.

It is important to know the risks of nanomaterials production as well as the mass balance. The drive for companies to protect proprietary information and processes complicates analyses of commercial manufacturing.[1,22] Surveys have been conducted by some research groups in an effort to determine the state of the nanomaterials industry, both in terms of production capacity[22] and in terms of safety measures used.[23] Helland *et al.* asked companies whether they performed risk assessments of their nanomaterials products, and the majority of their 40 respondents indicated that they did not.[24] Approximately 6 years later, Engeman *et al.* surveyed 78 nanoparticle manufacturing companies to gauge their knowledge of nanomaterial safety and attitudes toward industry regulation. Some respondents were uncertain about the level of risk inherent in their products, and a majority indicated that they lacked information needed for safe nanotechnology practices.[23] Robichaud *et al.* performed an analysis to estimate how much risk, as defined by insurance industry standards, is associated with the chemicals and methods used in nanotechnology production. They scaled five laboratory production techniques to commercial manufacturing levels, and considered the toxicity and likely emissions of the most hazardous precursor chemicals. Data drawn from an insurance industry database was applied to determine the risk of incidents, normal operations, and latent contamination. Among the processes examined were methods of producing C_{60}, SWCNTs, and TiO_2. Robichaud *et al.* estimated that these production methods had equal or lower risk during normal operations compared to the production of high-density plastics or petroleum refining.[25]

Additional research will be needed to inform industry risk-avoidance decisions, but the steps required to address the new hazards posed by nanomaterials can be challenging and expensive. A majority of companies surveyed by Engeman *et al.* felt that they would be better able to regulate safety than an external agency, but a few expressed concern that in the absence of regulation some companies would not move to improve safety practices.[23] An incentive for promoting safety in the absence of regulation is to avoid liability in the case of harm caused by nanomaterials, but in a legal proceeding it may be challenging for a plaintiff to prove that nanomaterials or the company that produced them are at fault.[18] Choi *et al.* examined

nanomaterials toxicity testing in a scenario where nanomaterial manufacturers are required to pay for and conduct testing of their own products. Their estimates for the time required to test the known nanomaterials, and the associated costs, varied wildly depending on nanomaterial toxicities and the budget portions that companies dedicate to the research. The most optimistic estimate suggested that testing could be completed in under a year, and their most conservative testing estimate required more than 50 years.[26] Toxicity testing is only a portion of the research that manufacturers will need to conduct. Methods of streamlining production and reducing waste will be needed to make large-scale production efficient,[16,25] and these concerns will limit the resources available for determining health and environmental consequences.[16] New safety equipment, as well as clean-up and waste disposal processes, for dealing with nanomaterials also drives cost.[23] Even when safety systems have been put in place, there is concern that as nanomaterials become commonplace, workers may lose perspective on the dangers they pose, and allow safety standards to lapse.[18,23]

Modeling the way in which nanoparticles may be released during manufacturing and product use provides a starting point for determining health and environmental impacts. Life cycle analysis (LCA) of products containing nanomaterials and their end-of-life disposal can help to determine the exposure potential.[15,19] A thorough environmental analysis would require knowledge of the concentration of nanomaterials and rate of release in wastes.[27] With nanoparticle production rates low compared to bulk materials,[1] and applications diverse, these analyses require theoretical constructs. Gottschalk *et al.* proposed means by which TiO_2, ZnO, and CNTs are likely to enter the environment by creating a flow model based on production methods and waste disposal. They considered the movement of nanomaterials between 'compartments,' such as soil, water, and sewage, to estimate the future concentration of nanomaterials in the environment. Their analyses showed a relatively low release of nanomaterials directly to the air, and higher release rates to sewer systems and landfills. TiO_2, ZnO, and some CNT nanomaterials are likely to be sent to a sewage treatment plant based on this model. From there, they can either remain in the solid sewage (sludge) or be released with treated water.[28] There is also potential for latent contamination of the manufacturing site during plant operation.[25]

14.3 Health Impacts of Exposure

Some exposure of workers during large-scale nanomaterial production is inevitable, even without unusual release incidents. It is necessary to determine the degree of exposure that will cause health impacts, and what those impacts will be. The toxicities of photo-active nanomaterials are among the most studied in nanotechnology because of their varied applications, both in energy and in other fields. The risks of exposure are greatest to workers in the nanotechnology field, but others would also be vulnerable following environmental release of nanomaterials during consumer product use.[1]

Nanomaterials are more reactive, and potentially more damaging, than bulk materials of the same composition because of their high surface area to volume ratio.[1,5,29] Testing with a variety of organisms to determine the impact of nanoparticle characteristics on toxicity has yielded inconsistent results, but some generalizations can be made. Smaller nanomaterials tend to be more toxic than larger nanomaterials.[30,31] Fibrous or rod-like nano-particles of any composition tend to be more hazardous than spherical or agglomerated nanoparticles.[32,33] For TiO_2, crystalline phase may influence degree or mode of toxicity,[4,32,34] but both rutile and anatase nanoparticles can cause toxic reactions.[31,32,34] As nanoparticle concentration increases, toxicity tends to increase,[32,35] unless particles aggregate, which tends to reduce toxicity.[36] Some tests indicate that the size of the nanoparticles has less effect on toxicity than the material itself.[37] Nanoparticle surface charge can impact toxicity, and may prevent contact with cells if they share a like charge.[31] The variables that affect nanoparticle toxicity continue to be refined, but impacts remain difficult to predict.

The means by which nanomaterials can cause cellular damage have been widely explored. Perhaps the most investigated avenue is their production of reactive oxygen species (ROS), which include hydrogen peroxide (H_2O_2) and singlet oxygen ($^1O_2$).[38–41] Noted free radical types of ROS, which have at least one unpaired electron, include superoxide ($O_2^{\bullet-}$), hydroxyl ($OH^{\bullet}$),[38–41] and hydrogen dioxide ($^{\bullet}HO_2$).[5] A certain volume of ROS are produced by natural biological processes.[38,39,42] As part of their cellular respiration function, mitochondria move electrons through the electron transport chain, and it is possible for electrons to 'leak' to the surrounding environment, generating some ROS.[41,42] The body provides antioxidant chemicals to defend against excess ROS,[38,43,44] but an imbalance causes oxidative stress,[41,44,45] which can be damaging to cells.[43] The development of ROS can be accelerated by outside catalysts, including transition metals,[38,46] and some transition metal oxides[5] such as Fe_2O_3.[47] ROS can be produced when photo-active com-pounds absorb energy and then transfer it to oxygen or water molecules.[5,38] Light exposure may increase ROS production by some nanomaterials,[29] but the absence of light does not preclude ROS generation[31] or toxicity.[4,5] A depiction of the reactions that may be initiated by the presence of a TiO_2 molecule appears in Figure 14.2.

Biological molecules can be transformed in a chain reaction initiated by ROS,[5,38] which may pull an atom or charge from a molecule, transforming the original molecule into a new free radical.[38] This radicalized biomolecule then pulls charges or atoms from another biomolecule, passing the trans-formation forward. Lipid peroxidation is one form of this chain reaction.[38] It can occur when a hydrogen atom in a fatty acid is removed by interaction with a reactive molecule or ion,[38] such as a hydroxyl radical.[46] The modified lipid may react again with other ROS or it may pull hydrogen from another molecule to fill the void, which passes the deficiency.[38] Polyunsaturated fatty acids, which are widely found in the central nervous system,[49] are a sus-ceptible target of lipid peroxidation.[38,49] Inside cells, ROS can cause damage

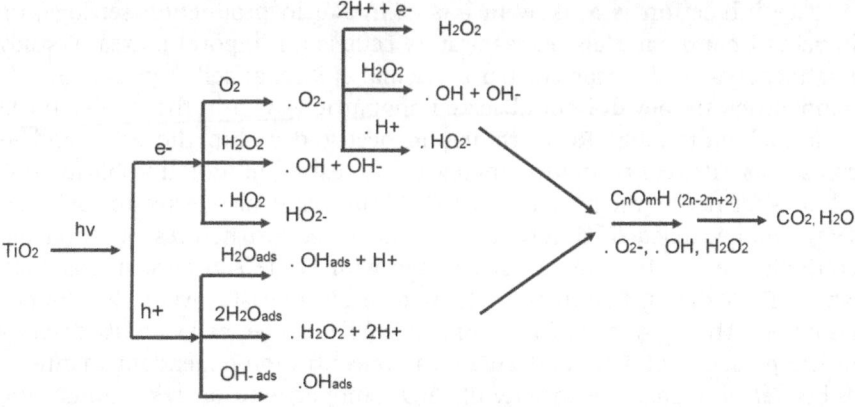

Figure 14.2 Illustration of chemical reactions that can occur between a metal oxide catalyst and molecules of oxygen and water.

to other vital structures, including mitochondria themselves.[40] DNA is also susceptible to damage by ROS reactions.[41,42,45] Molecules of DNA can be shortened or broken by ROS exposure (reviewed in ref. 42 and 45), and fragments from damaged chromosomes can be carried forward through cell reproduction.[50] DNA may develop adducts or mutations, possible precursors to cancer.[47] See ref. 42 for a review. Genetic variation may cause some individuals to be more likely to develop cancer than others when exposed to the same DNA stressors.[51] Lipid peroxidation is associated with accelerated plaque deposition in the brains of mice, and may be related to the development of Alzheimer's disease.[49] Other conditions thought to be related to the generation of ROS include metabolic syndrome,[44] muscle loss,[52] aging, diabetes, arthritis, and other neurological disorders. Reviewed in ref. 40, 42, and 44.

Elevated levels of antioxidants have been generated by cells in the presence of nanoparticles, indicating that the materials may be causing cytotoxic reactions and eliciting an immune response.[53] The outside addition of antioxidants can reduce toxicity in some nanoparticle exposure,[34,54] but not in all.[34] High levels of antioxidants are not necessarily associated with disease prevention, and can aggravate some conditions.[43] Nanoparticle toxicity can vary between cell types,[47,53] and impacts need not be limited to cell death. For example, nanomaterial-related DNA damage[4] could have a significant impact on the life of the affected organism. Further, nanomaterials known to produce ROS can show variable toxicity,[29,31,48] so other factors must influence the degree of cellular damage. It is however common in nanomaterial toxicity tests to look for signs of ROS and cellular distress. Some experimental results indicating nanomaterial toxicity at a cellular level are mentioned below. They may use nanoparticle concentrations on the order of milligrams per liter of liquid medium, but exact testing conditions vary.

Since inhalation is a likely means of intake in production settings, the impact of nanomaterials on respiratory cells is an important area of study. Bhattacharya *et al.* observed that bronchial epithelial cells can take in TiO_2 nanoparticles. They did not observe nanoparticles within these cells' nuclei or mitochondria, but ROS production occurred within the cells, and increased as TiO_2 concentration increased. In testing of lung fibroblasts, DNA adducts formed in the presence of TiO_2.[47] Gurr *et al.* also tested the effects of TiO_2 nanoparticles of different sizes and crystal structures on bronchial epithelial cells, observing DNA damage with some samples. In instances where DNA damage occurred, elevated levels of H_2O_2 were also found.[4] Hsaio and Huang observed that lung epithelial cancer cell viability declined in the presence of TiO_2 and ZnO in a concentration-dependent manner.[32] Akhtar *et al.* tested the toxicity of ZnO nanoparticles on two human lung cell lines, and observed that at high doses cell viability was reduced by half in 24 hours.[6]

Transdermal nanomaterials exposure is also possible. Shukla *et al.* examined the effects of TiO_2 nanoparticles on human skin cells. They found that cells took in nanoparticles 30 to 100 nm in diameter, and that some of the particles entered the cell nucleus. Damage to DNA within cells was detected, and it was noted that the higher the concentration of nanomaterials to which the cells were exposed, the higher the volume of ROS found in them. Lipid peroxidation was also observed.[55] Braydich-Stolle *et al.* exposed mouse skin cells to various sizes and crystallinities of TiO_2. They observed nanoparticles in mitochondria and membrane damage.[34] Alarifi *et al.* observed ROS and lipid peroxidation in human melanoma cells after exposure to ZnO. They also noted DNA damage.[56]

If nanomaterials pass through the organs initially exposed to them, they may be transported to other locations in the body, encountering a myriad of other cell types. Ghosh *et al.* observed evidence of mitochondrial DNA damage in lymphocytes exposed to TiO_2 at lower concentrations, but less of an effect at higher concentration exposure, possibly due to particle agglomeration.[50] Long *et al.* studied the impact of commercially produced TiO_2 nanoparticles on brain microglia cells. Nanoparticles were taken in by the cells, and they observed evidence of superoxide molecules in mitochondrial membranes. Some mitochondria near the particles showed deformation after 18 hours, even though nanoparticles aggregated over time to form non-nanoscale clusters. Cells continued to function in spite of the damage.[46] Akhtar *et al.* found that ZnO nanoparticles reduced the viability of liver cancer cells. They observed DNA fragmentation in some of the cells, as well as the presence of ROS and lipid peroxides.[6] Sayes *et al.* exposed cells to C_{60}, and observed cellular membrane damage in skin, liver cancer, and nerve cells that was consistent with lipid peroxidation.[53] In tests with liver cells and TiO_2 nanoparticles, Yoo *et al.* found evidence of ROS and toxicity.[54] In a potentially damaging interaction that need not be related to ROS, some carbon nanomaterials may reduce the ability of macrophages, cells in the immune system, to engulf and dispose of materials dangerous to the

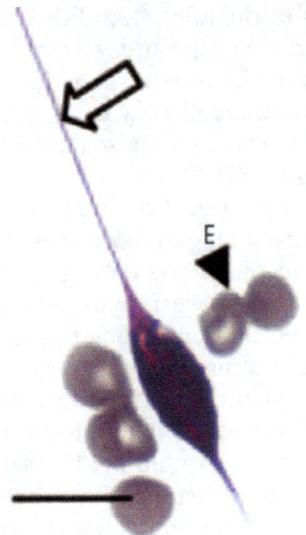

Figure 14.3   Cell trying to engulf a carbon nanotube, and unable to do so.
(Source: Adapted from Poland *et al.*[33])

body.[57,58] MWCNTs, which Poland *et al.* found to cause inflammation and granulomas reminiscent of asbestos exposure when injected into mice, can form fibers too long to be surrounded by macrophages.[33] An example of this phenomenon can be seen in Figure 14.3. Macrophages that are unable to engulf fibers can produce chemicals that trigger continuous inflammation in surrounding tissue.[59]

Many *in vivo* nanomaterials toxicity tests have been conducted with rodents. Inhalation testing of nanomaterials has been conducted by a variety of labs, and the inhalation of carbon fullerenes has received particular scrutiny. This may be due to the known health effects of carbon black, an amorphous form of carbon often used in toxicity tests as a positive control, which induces toxic effects for comparison with experimental results.[59] It has been noted that the inhalation effects of carbon nanotubes (CNTs) compared to equivalent quantities of microscopic carbon are more severe.[58,60] This may be in part because CNTs are fibers, rather than particles, and can bundle together into large fibers or agglomerations.[58,60,61] SWCNTs have been shown to be more toxic than MWCNTs in a variety of tests.[57,62] Lam *et al.* implanted SWCNTs in the tracheae of mice and observed that granulomas appeared in mice exposed to high doses of the material within 7 days, and that lung lesions tended to be larger after 90 days. Some of mice's lungs were visibly deformed after testing.[60] Shvedova *et al.* conducted inhalation testing on mice with SWCNTs in solution at quantities representing 1 month of acceptable levels of exposure to larger carbon particles based on OSHA standards. They found that SWCNTs caused notable inflammation in lung tissue that gradually reduced. Seven days after exposure, they observed

granulomas in the lungs of the mice.[58] Additionally, the researchers found that lung tissue thickened after exposure, was more pronounced with higher exposure, and was greatest when the test ended at 60 days. The thickening was not restricted to lung tissue directly in contact with nanoparticles, but rather was widespread.[58] Warheit et al. also observed granulomas in rat lung tissue after exposure to SWCNTs.[61]

Other nanoparticles have also been shown to cause lung damage. G. Oberdörster et al. observed lung inflammation in rats following exposure to TiO_2 nanoparticles, and that some damage persisted for at least a year after exposure.[30] Cho et al. exposed rats to ZnO nanoparticles through inhalation, and also observed thickened lung tissue.[36] As noted by Hext et al. in a review of inhalation testing with mice, rats, and hamsters, different species have different particulate sensitivities. Rats in particular may be more sensitive to inhaled nanodusts than mice or hamsters,[3] so results may vary from what human exposure would be. Inhalation testing by Z. Li et al. with ApoE mice tested whether aspirated SWCNTs could cause heart and blood vessel damage. They found mitochondrial DNA damage in aortic tissue after exposure to SWCNTs, and an increase in plaque formation in the aorta.[63] These conditions accord with the plaque formation[49] and DNA damage associated with ROS. Reviewed in ref. 41, 42, and 45.

In addition to transdermal and inhalation routes, ingestion is a potential nanomaterial exposure route. Cho et al. studied the consumption of TiO_2 and ZnO nanoparticles by rats. They found that very little TiO_2 dissolved in the digestive track, and they saw minimal change in Ti accumulation in organs as compared to a control group. Most of the ingested TiO_2 appeared to be excreted rather than retained. Conversely, they noted that ZnO rapidly dissolved in stomach acid, releasing zinc ions. At high exposure doses, rats were found to have lower body mass, and to have higher zinc levels than controls in their livers and kidneys.[64] Injection of nanomaterials has been used to study their impact on rodents without relying on the transfer of nanomaterials from natural points of exposure. The injection of CNTs by Poland et al. and the observed abdominal granulomas has been mentioned.[33] Toxicity and immune response need not be the only variables examined through this testing. Han et al. tested the impact of injected ZnO nanoparticles on rat learning behavior. They found that rats injected with ZnO nanomaterials were slower to learn the solution to a maze than control animals, and that when the solution changed they were slower to adapt than control rats.[65]

In considering the possible health implications of nanomaterials, it is important to note that fullerenes[60,63] and titanium dioxide (reviewed in ref. 3) are nearly insoluble in the body, which means that the potential exists for them to cause health effects at a period of time removed from the initial exposure if they are retained. It is possible that small nanomaterials are more difficult to remove from the body by natural clearing processes, perhaps because they can penetrate different tissues than larger particles.[30] Researchers in both Europe and the United States have studied the medical

histories of workers exposed to titanium dioxide to determine if there are observable long-term health impacts from being present during the production of this material.[3] It is known that nanoparticles are more reactive then bulk versions of the same material, so it is important to note that TiO_2 has been produced at nanoscale for some time,[22] with 200–300 nm particles used for pigments and 10–20 nm sized particles used for sunscreen.[3] Studies of workers exposed to TiO_2 did not show unusual disease trends that could be attributed primarily to TiO_2. However, it cannot be determined with any certainty how much nanomaterial these workers encountered in the course of their employment, or what safety standards were in place to protect them.[3] As noted above, and in contrast to fullerenes and TiO_2, ZnO nanoparticles may dissolve *in vivo*, releasing metal ions that still have cell and tissue damaging potential.[36,56] In acute exposure, Zn^{2+} ions can be lethal to rats.[36]

A few studies have considered toxicity related to surface functionalization, in which other molecules are bonded to the surfaces of nanoparticles to change the way in which they interact with their surroundings. This can include the conversion of particles from hydrophobic to hydrophilic,[66] and from harmful to inert in some circumstances. Nair *et al.* studied the interactions of ZnO nanomaterials, capped with starch or polyethylene glycol (PEG), with osteoblast cancer cells and two bacterial strains. They found that a starch coating reduced toxicity compared to PEG coatings, but that both were toxic to the osteoblast cells tested.[67] Yin *et al.* functionalized C_{60}, and larger C_{82} molecules surrounding gadolinium, by adding carboxyl or hydroxyl groups to their surfaces. In experiments with cells exposed to H_2O_2, the addition of these modified particles reduced damage to mitochondria due to oxidative stress.[40] Brunet *et al.* also observed a mildly beneficial effect of hydroxylated C_{60} under incandescent light.[48] Xia *et al.* observed no toxicity of hydroxylated C_{60}, but did observe ROS generation.[29] On the other hand, Brunet *et al.* observed that surface modification with poly(*N*-vinylpyrrolidone) (PVP) increased the toxicity of C_{60} compared to their unmodified samples.[48] Chen and Jafvert observed ROS production in the presence of illuminated SWCNTs that were functionalized with carboxyl groups, but which had potentially been altered by subsequent purification.[66] Aside from toxicity, surface modification may affect the way in which nanomaterials aggregate in blood vessels.[68]

In addition to functionalized coatings, a natural surface layer can accumulate on nanomaterials. When nanoparticles enter an organic medium, they attract and are attracted to biological materials.[69–71] A corona can develop around nanoparticles, encasing them in proteins, fats, and carbohydrates.[70,72] This can develop within 20 seconds of interaction with nanoparticles, but the molecules surrounding the nanoparticle change with time until the corona develops equilibrium with the environment.[73] The number and types of molecules that attach to a nanoparticle vary with surface area even if the material is unchanged.[70,72] Molecules that have been found in nanoparticle coronas include immunoglobulin, lipoproteins, and

coagulation factors.[72] Experimentation by Hellstrand *et al.* with polymer nanoparticles in plasma showed that they could attract cholesterol molecules.[70] Some researchers noted that the addition of fetal bovine serum (FBS) to a solution containing settled nanoparticles caused them to disperse and remain stable in solution, and therefore be more mobile than they had been in the absence of the added protein.[55,74] One side effect of this encapsulation is that nanoparticles at the center of a lipoprotein cluster will be perceived differently by cells and the immune system.[70] Efforts are underway to develop mathematical models to investigate how nanoparticle coronas will develop depending on the environment around them.[73]

14.4 Environmental Release and Impacts

The nanomaterials discussed thus far are among the most studied particles in nanotechnology because they have a wide variety of uses both inside and outside the energy field. TiO_2 is used in paint,[2,3] cosmetics,[3,75] sunscreen,[3,55] and food.[3] ZnO is also a component in paint and cosmetics.[22] CNTs are a potential component of composite materials,[22] and have even been tested in fertilizers.[76] Nanoparticles that leave a manufacturing setting in finished products may be exposed in the future through wear. For example, Kaegi *et al.* studied house paint that used TiO_2 as a colorant and found that nanoparticles of titania were washed off the surface by rain, and that this process could be observed even in a house that had not been recently painted. Runoff may be absorbed into the soil or transferred to sewers. A nearby drainage area was found to contain nanoparticles similar to those observed washing off the painted surfaces.[2] The use of nanomaterials in all manner of products is expected to increase with time,[1,28] leading researchers to explore ways in which they will impact the environment.

Kiser *et al.* explored how TiO_2 nanomaterials are removed from sewage during treatment by sampling waste material at different stages during treatment in a wastewater reclamation plant, and then searching for titanium in the samples. Based on samples taken at different times of the day every day for one month, they found that an average of $79\% \pm 23\%$ of the titanium that entered the plant was removed from the water. An additional day of testing at another time of year increased the average to $82 \pm 21\%$. Further, they determined that larger nanoparticles were filtered more effectively than smaller particles.[77] Subsequent testing by Kiser *et al.* showed that TiO_2, C_{60}, and silver nanoparticles can all be adsorbed by biosolids. With sufficient biomass concentration, more than 80% of non-functionalized C_{60} particles were removed from solution. That such different nanomaterials all adsorb onto biomaterials may be due to the heterogeneous composition of biomaterials in nature and in wastewater treatment plants.[71] Sewage sludge can be removed from treatment plants and disposed of on land as a fertilizer,[77] which means that nanomaterials remaining in the sludge can be transferred to soil, where their concentration is expected to increase with time.[28]

Nanomaterials deposited on land will be a source of exposure for bacteria, plants, and animals. Nanoparticles can move through soil, and their progress appears to be accelerated by precipitation.[78] Collins *et al.* placed ZnO and copper nanoparticles on soil in pots exposed to the environment and measured the progress of the nanomaterials from the soil surface into the lower levels of the pots. They found that their nanoparticles began to break down after a week, gradually releasing metal ions that continued down into the soil.[78] Waalewijn-Kool *et al.* tested the transformations of zinc materials in moist soil over a one year period. They detected increasing levels of zinc in soil moisture after exposure to ZnO nanoparticles, which tapered down over the course of a year. The same test was performed with ZnO nanoparticles coated with triethoxyoctylsilane. Coated particles also dissolved, but more slowly than uncoated nanomaterials. Dissolution was observed most often at lower concentrations of nanoparticles in the soil; at high concentrations, release of ions was impeded.[79] This suggests that in addition to being exposed to nanoparticles, organisms may eventually be exposed to their constituent materials through chemical breakdown. Even when not a part of a nanoparticle, zinc can alter the balance of bacteria and fungi present in soil.[80]

It is known that some nanomaterials have antibacterial properties, and this has been explored as a positive attribute in some experiments. However, the unintentional release of these nanomaterials to the environment may negatively impact the health of soil. Tsuang *et al.* studied the bactericidal uses of TiO_2 for medical sterilization. In their study, TiO_2 illuminated with UV light killed five species of bacteria, including two common to soil and one anaerobic bacterium. One hour of exposure destroyed almost all of the exposed colonies.[81] Banerjee *et al.* observed that TiO_2 films illuminated with near-UV light inhibited bacterial growth while in a culture medium.[5] Simon-Deckers *et al.* tested nanomaterial toxicity on *E. coli* and *C. metallidurans*. They found that both TiO_2 and MWCNTs induced toxicity in *E. coli*, but that *C. metallidurans* was more resistant to nanomaterial damage.[31] Brunet *et al.* observed some phototoxicity of TiO_2 at high concentrations under a low wattage incandescent lamp, but found lower concentrations neutral or somewhat beneficial. They also observed no significant toxicity with C_{60} in solution.[48] Brayner *et al.* tested the impact of ZnO nanoparticles on *E. coli*. At low concentrations they found that ZnO improved bacterial growth, and that at higher concentrations it inhibited bacterial growth. They suggested that the bacteria could metabolize lower concentrations of zinc, but that higher concentrations overwhelmed them. Examination of bacteria in some tests showed damaged cell walls; cell walls are rich in lipid molecules.[82] Tong *et al.* studied the effects of C_{60} nanoparticles on soil bacteria, and found no significant change in the respiration or volume of microorganisms in a soil sample that would suggest that the nanoparticles were toxic in a 6 month time period. They did see a slight change in the type of fatty acids found in the soil, which may imply that long-term alterations to the microorganisms could occur.[83] Collins *et al.* observed some microbial toxicity in their soil study of ZnO and copper nanoparticles.[78]

   Soil deposition of nanomaterials will affect plants, and if deposition occurs in an agricultural area this could have an impact on food production
and safety. In plant cells, both mitochondria and chloroplasts are potential
sources of ROS.[84] Some research into the impact of TiO_2 nanoparticles on
hydroponic plants suggests that they can slow water uptake by blocking the
nano-sized pores of roots. However, tests in the same study on potted plants
suggested that, under otherwise healthy conditions, they could develop
with little inhibition over time despite nanoparticle exposure.[85] Song *et al.*
exposed tomato (*Lycopersicon esculentum*) seeds and plants to TiO_2. They did
not observe any changes to germination rates in seeds exposed to TiO_2
nanoparticles, though some metal was detected. Based on this outcome and
additional testing with corn seeds, they speculated that the seed coats prevented nanoparticles from entering the seeds. They did not observe any
significant negative impacts on the growth of potted tomato plants exposed
to TiO_2, but did find traces of Ti in the leaves, stems, roots, and fruit of the
plants. They also observed some evidence of oxidative stress in plants grown
in a greenhouse, but not those grown in a lab.[75] Ghosh *et al.* studied the
effects of TiO_2 nanoparticles on hydroponically exposed onion bulbs (*Allium
cepa*) and on potted tobacco plants (*Nicotiana tabacum*). Evidence of DNA
damage and lipid peroxidation were found in onion roots. It was noted that
the degree of DNA damage tapered off at higher nanoparticle concentrations, perhaps due to increased agglomeration. They also observed DNA
damage in tobacco leaves after plant exposure.[50]

   Priester *et al.* studied the impact of nanomaterials on soybeans, growing
plants in the presence of various doses of CeO_2 or ZnO nanoparticles. They
found that plants grew and produced pods, but there were side effects. The
ZnO-exposed plants tended to produce fewer leaves at the outset, and their
leaves and pods contained less moisture than normal. Plants grown with
CeO_2 had a fewer leaves, some plants were stunted, and for some the biomass of produced soybeans was lower than normal. Some ZnO-exposed
plants had larger roots systems. After higher exposure to CeO_2 there was
a sharp decline in the plants' ability to fix nitrogen, which fell by 80%.[86]
Nitrogen fixation is the symbiotic process by which bacteria hosted in
nodules on the roots of soybean plants convert unusable N_2 molecules into
ammonia, which the plant uses to produce organic compounds, including
amino acids. This process is essential to the production of beans, and a lack
of nitrogen fixation reduces crop yield without additional fertilizer.[87]
Priester *et al.* also detected the metal components of the tested oxides in
the soybean root systems, and in the case of ZnO, high concentrations of
zinc moved through the plant into the leaves and soybeans.[86]

   As organisms that ingest soil and decomposing matter to extract nutrients, and because they can absorb material though their skin,[88] earthworms
in sewage waste dumping areas are certain to be exposed to nanomaterials.
Hu *et al.* examined the impact of TiO_2 and ZnO nanoparticles on the species
Eisenia fetida.[89] This particular species is used in vermicomposting, a process in which solid organic waste is fed to earthworms to process it into

fertilizer.[90] The experiments by Hu *et al.* showed that ZnO and TiO$_2$ can be ingested by and can accumulate in earthworms. The volume of nanoparticles ingested increased with the volume of nanomaterials present in the soil. ZnO appeared to accumulate in the worms at a greater rate than the TiO$_2$, and worms were not able to expel all of the ingested nanomaterials after exposure. When exposed to TiO$_2$ and ZnO particle concentrations of 1 g/kg soil or more, DNA damage was found in the worms, though ZnO appeared to be more toxic than TiO$_2$. Imaging of cells in the earthworms' digestive tracts after exposure to nanomaterials also revealed deformed mitochondria.[89] L. Li *et al.* exposed *E. fetida* to ZnO nanoparticles in a 96 h toxicity test using agar medium. They observed toxicity, and noted zinc within cell organelles.[91] Bigorgne *et al.* tested for damage to coelomocytes, cells from the worm immune system, after *in vitro* exposure to low concentrations of TiO$_2$ nanoparticles during a 24 hour toxicity test. They found that the cells took in detectable amounts of TiO$_2$ in 4 hours, and that the quantity encapsulated increased with time. They noted that the cells seemed to take in the larger nanoparticle aggregates. In the time period of the test, they did not observe damage to the cell or mitochondria, but the cell immune response was triggered.[74]

While some nanomaterials are not acutely toxic to earthworms, even at relatively high levels of exposure, there are potential impacts of their long term presence in the environment. Li and Alvarez tested the impact of C$_{60}$ nanoparticle exposure on *E. fetida*, found that at very high concentrations (50 g C$_{60}$/kg of soil) the worms reproduced more slowly than in control tests. They also found that worms did not retreat from soil containing C$_{60}$ as they do when faced with some contaminants, and so endured longer exposure to the nanoparticles.[88] Schlich *et al.* tested the impact of manufactured titania nanoparticles on another species, *E. andrei*. They observed seasonal variance in the reproduction rates of control worms, but that one of the nanoparticles tested lessened the seasonal changes in reproduction, altering the natural cycle.[92] McShane *et al.* exposed two earthworm species to various concentrations of TiO$_2$ nanomaterials, and found no significant change in reproduction or growth rates, but found that at high TiO$_2$ nanoparticle concentrations, *E. andrei* showed avoidance of contaminated soil. They did not observe the same reaction to soil contaminated with micro-sized titania particles.[93] Åslund *et al.* studied *E. fetida* exposure to sublethal concentrations of TiO$_2$ over several months, and while no unusual outward signs of toxicity were observed, they noted changes in the worms' metabolisms.[94]

In spite of the high potential rate of removal of nanomaterials from waste fluids in treatment plants, some volume of particles is expected to flow out of treatment plants and into surrounding water.[28,71] Because of their small size, the particles are likely to remain in suspension for an extended period. Fluid dynamics calculations indicate that spherical TiO$_2$ molecules at the 100 nm size require 14 hours to settle 1 mm in water at standard conditions.[77] In nature, the organic and ionic composition of the body of water into which the particles are discharged will affect the settling rate.[95,96]

Keller *et al.* examined the settling rate of metal oxide nanoparticles in water samples including freshwater, wastewater, and seawater. They found that as the organic material content of the water increased, the settling rate slowed and the nanoparticles became more negatively charged. As the water's ion concentration increased, nanoparticles settled more quickly, and had more neutral electrical charges. Thus in this experiment, nanoparticles in seawater settled more rapidly than nanoparticles in freshwater.[96] Graphs of the differences in settling rates in Keller *et al.*'s experiments are shown in Figure 14.4.

Part of the higher settling rate of nanoparticles in seawater is attributed by the researchers to the rapid agglomeration of particles in that medium as compared to freshwater.[96] High ion concentration in a fluid promotes aggregation.[46] A variety of physical forces influence nanoparticle interaction, and their impact varies with fluid viscosity, flow rate, temperature, pH, particle size, concentrations, and the electrical charges of the interacting surfaces. Similarly charged particles will repel one another, slowing agglomeration.[95] Keller *et al.* found that their TiO_2 nanoparticles in seawater tended to cluster into groups more than a micrometer in diameter within an hour. ZnO nanoparticles also agglomerated more rapidly in seawater, but not as rapidly as titania. In freshwater the nanoparticles also aggregated, but remained in groups 300–400 nm in size.[96] Even in ultrapure water, some aggregation can occur.[74] Nanoparticles are more likely to aggregate if their sizes are non-uniform,[95] so if the primary particles present are all manufactured to the same standard, they will tend to agglomerate more slowly than heterogeneously sized particles. While mathematical models can give an indication of how spherical nanoparticles will settle and aggregate in different environments, any nanoparticles with inconsistent cross-sections, especially fibrous nanomaterials, have interactions that are too variable to predict with theoretical methods.[95] Settling and agglomeration rates for nanoparticles will determine exposure periods for aquatic organisms.[96]

Nanomaterials released to the environment will be altered by their surroundings. Lower concentrations of nanoparticles can degrade quickly, releasing metal ions more rapidly than a large volume of particles that aggregates.[91] If nanoparticles aggregate, they can form groups too large to be readily eaten by small filter feeders.[62] Aggregation also reduces the surface area of nanoparticle clusters relative to their size, making them less reactive as a group. However, this can cause them to break down at a slower rate than if they remained individual nanoparticles.[69,91] Also, nanomaterials can aggregate with materials in their surroundings, including organic matter,[69] which can reduce the rates at which particles coalesce with each other, keeping them nano-sized.[96] Any surface treatments applied to nanomaterials will affect the way in which they interact with the environment.[28,37,69,71,77] For example, functionalized fullerenes may be easier to disperse in water than unfunctionalized molecules.[66] Metal oxides can withstand some acids,[2] but as discussed previously, may break down over time, releasing metal ions and changing the environmental impact scenario

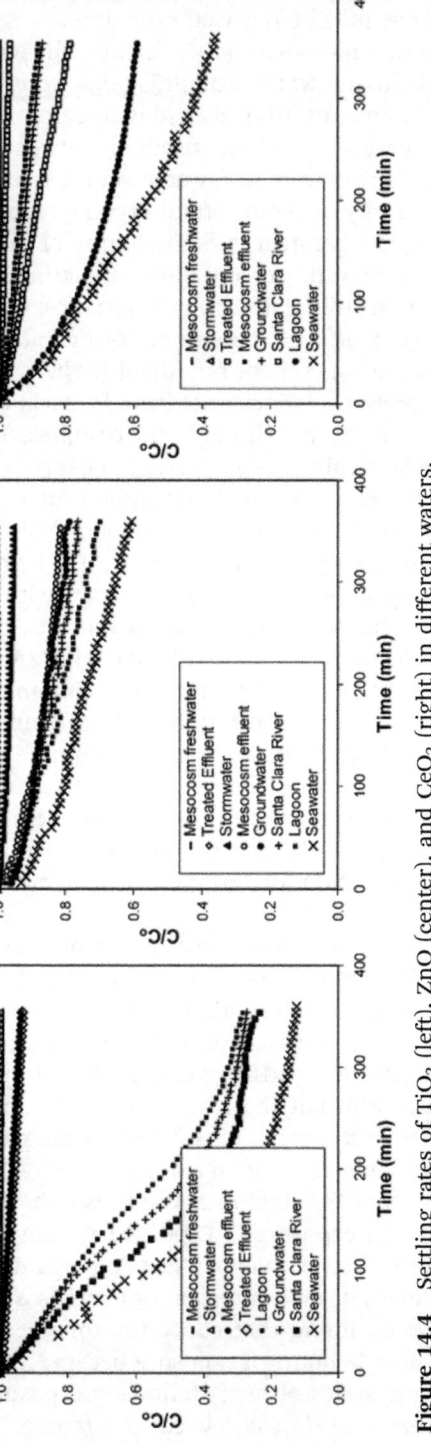

Figure 14.4 Settling rates of TiO_2 (left), ZnO (center), and CeO_2 (right) in different waters. (Source: Keller et al.[96])

from nanoparticle to heavy metal toxicity.[35,69,78] Nanomaterials aggregated to a binding substance may be released over time.[2]

It is also possible that nanomaterials will enter the water sources that are treated to produce drinking water. Abbott Chalew *et al.* analyzed the ability of current drinking water filtration methods to remove TiO_2, ZnO, and Ag nanoparticles. The coagulant, alum, used in conventional filtration was added to nanoparticle-contaminated water samples at the concentrations required to achieve clarity or remove total organic carbon (TOC) content in the water. After sample treatment, a high volume of TiO_2 (more than 90%) and Ag (80%) were removed, but 60–100% breakthrough of zinc was observed, depending on the initial water characteristics. Water samples filtered by membrane in a separate trial contained lower traces of Ag and Ti than coagulant filtration, but Zn content remained high.[97]

A number of researchers have explored the impacts of nanomaterials on aquatic organisms from the unicellular to the complex. Some of the bacterial toxicity noted at the beginning of this section is likely to be found in aquatic communities. Nyberg *et al.* studied the effects of C_{60} nanoparticles on samples of anaerobic bacteria in the presence of organic sludge to determine their impact on a microorganism community. They found no evidence of toxicity during testing periods of 89 to 154 days with low nanomaterials concentrations, and observed no significant changes to sample DNA profiles.[98] Lee and An tested the impact of TiO_2 and ZnO nanomaterials on algae (*P. subcapitata*). They found that both nanomaterials reduced the growth rate of the algae as concentration increased, and observed evidence of cell wall damage following ZnO exposure.[35] Multiple nanoparticle toxicity tests have been conducted using *Daphnia magna*. These crustaceans, referred to as daphnids, have been written into standard chemical analysis tests by organizations including the United States Environmental Protection Agency (USEPA) and the Organization for Economic Co-operation and Development (OECD).[99,100]

Zhu *et al.* exposed *Daphnia magna* to a variety of nanomaterials and their bulk counterparts, including TiO_2, ZnO, C_{60}, SWCNTs, and MWCNTs. Toxicity during a 48 hour test depended on the material and concentration. They observed that nanoparticles of ZnO caused the deaths of all the daphnids at a concentration of 10 mg/L. It took 14 times as many anatase TiO_2 nanoparticles to kill half the test animals. Among the carbon nanomaterials, SWCNTs were by far the most toxic, causing a 50% death rate at a concentration nearly one-tenth of that required of MWCNTs. For all the materials, percentages of the daphnids were also observed to be immobilized after nanomaterial exposure.[62] Exposure to nanoparticles at initially non-lethal doses can still have notable effects on organisms when the length of exposure is extended, or even after nanomaterials are removed from the environment. Zhu *et al.* found that concentrations of TiO_2 that were non-lethal to daphnids after 48 hours of exposure became lethal after 72 hours of exposure. In the same series of tests, following exposure to TiO_2 nanomaterials over the course of 21 days at low concentrations, daphnids produced

fewer offspring and grew at slower rates than control animals.[101] Lopes *et al.* found that after exposure to ZnO nanoparticles of various sizes, daphnids showed a lower feeding rate. During the 4 hours after their removal from nanoparticle contaminated environments, the daphnids were unable to recover to a control feeding rate, demonstrating the possibility of long-term effects due to exposure. During chronic testing, with exposure to ZnO nanoparticles over the course of 21 days, the researchers observed that the daphnids reproductive rates were also reduced; in some cases they were halved by exposure to less than 1 mg ZnO/L. The presence of ZnO also inhibited the animals' ability to molt.[37]

In the tests by Zhu *et al.*, analyses of the transparent animals showed that they had eaten nanoparticles in the surrounding water.[62,101] Some sank to the bottom of their containers, and others were prevented from moving by nanomaterials sticking to them.[62] Nanoparticles in solution are known to have an affinity for organic material,[2,71] and to deposit on surfaces in addition to aggregating in solution.[95] The presence of TiO_2 reduced the rate at which daphnids consumed algae, a typical food source. Also, daphnids that had consumed TiO_2 nanoparticles in non-lethal tests were not able to expel all of the nanoparticles after removal to a nanomaterial-free environment.[101] An image of a daphnid after ingesting TiO_2 nanoparticles is shown in Figure 14.5.

Aquatic exposure testing has been performed on fish, though this is a more complicated process than toxicity testing with daphnids. E. Oberdörster found that when C_{60} was added to their tanks, largemouth bass showed increased lipid peroxidation in brain tissue, indicating that nanomaterials had been absorbed by the fish and had caused a reaction.[102] Cedervall *et al.* observed altered metabolisms in fish that ingested polystyrene nanomaterials.[103]

Indirect impacts on toxicity have also been noted. X. Zhang *et al.* found that cadmium ions could be adsorbed onto TiO_2 nanoparticles in solution. Carp exposed to TiO_2 nanoparticles contaminated with cadmium showed higher concentrations of the metal than carp exposed only to cadmium.[104]

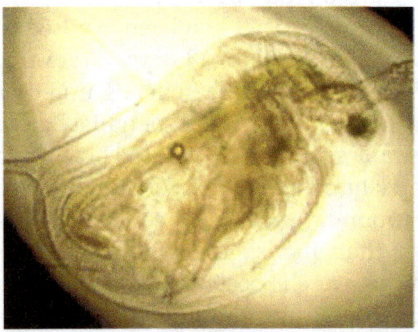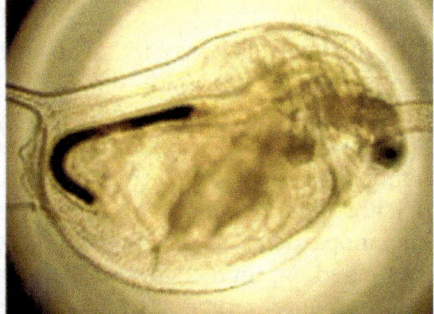

Figure 14.5   Left, a control daphnid; right, a daphnid that has ingested TiO_2 nanoparticles.
(Source: Zhu *et al.*[101])

14.5   Biomagnification

Priester *et al.* showed that the zinc from ZnO nanoparticles can travel through soybean plants from the soil and be deposited in the soybeans themselves,[86] Song *et al.* observed similar translocation of titanium in tomato plants exposed to TiO_2,[75] and it has been shown that filter feeding organisms can ingest whole nanomaterials.[62,89,101] That plants and filter feeding animals can take in and retain nanomaterials, as well as their constituent elements, raises the question of whether nanomaterials can be passed from them to higher trophic level consumers. Comparisons have been drawn between the potential spread of nanomaterials in the environment and the impact of past chemical releases, such as DDT.[102] In biomagnification, chemicals ingested by lower trophic level species can be concentrated as they are passed on to higher level predators. The predators may receive a harmful dose of a chemical that was originally present at minimally toxic levels in the environment.[105] DDT biomagnifications had a severe impact on hawks and eagles by limiting their ability to produce viable eggs.[106] Biomagnification is of particular concern in aquatic environments, where chemicals can spread freely. Microorganisms in the water, if not consumed, may sink to the bottom of their habitat once deceased, concentrating the ingested chemical there.[105] It is important to determine if nanomaterials have similar implications for the food chain. Chemicals noted in the past for biomagnifications tend to be soluble in lipids.[105,107] This is not characteristic of nanomaterials such as TiO_2, which are sometimes considered insoluble.[107] Bioaccumulation is affected by the digestion process and the composition of the organism's diet.[108] Organisms have systems in place to remove contaminants, so not all ingested contaminants will remain in the organism. And while predators may appear to contain higher concentrations of contaminants than their prey, their contamination may be due to ambient exposure rather than ingestion.[106] Further experimentation will be required to determine to what extent nanomaterials and their components can be transferred between consumers.

Some experiments along these lines have already been performed. Zhu *et al.* conducted an experiment in which daphnids that had been exposed to TiO_2 nanoparticles were fed to zebrafish (*Danio rerio*) for 14 days. Afterward, fish were fed an uncontaminated diet for 7 days. The researchers found that the TiO_2 levels in the fish increased for the first 5 days, and then plateaued. For comparison, the researchers exposed a separate group of zebrafish to nanoparticles by adding them to the tanks housing the fish for 14 days. The fish absorbed nanoparticles from the surrounding water, but concentrations of nanoparticles in the fish were lower than in fish that had eaten contaminated daphnids. However, the fish that were exposed to nanomaterials environmentally retained more nanomaterials than the fish that were exposed through food.[107] Cedervall *et al.* experimented with the transfer of polystyrene nanoparticles between trophic layers. They observed ingestion of 24 nm polystyrene nanoparticles by algae, which was then rinsed and fed to

daphnids. After algae consumption the daphnids were rinsed and fed to Crucian carp. The quantity of polystyrene nanoparticles ingested by the fish was not established, but significant changes in feeding behavior were observed following exposure. Tests showed alterations in the lipid balance of exposed fish compared to controls 22 days after the experiment commenced. Test fish moved slowly, and did not adapt to reductions in available food. The authors theorized that that these changes are due to the lipids pulled out of circulation in the fish to form particle coronas.[103]

These two examples show that nanomaterials may be transferred between trophic levels, but not necessarily that the concentration of nanomaterials will increase. With or without concentration, it does appear that the nanomaterials can have an impact on the higher level predators. It is also worth noting that lipid peroxides produced in a food source by ROS may be passed on to the consumer organism.[38] It is also likely that metal ions released by dissolved nanoparticles could be passed from a prey animal to a predator. This would mean that nanomaterials might have an impact on the health of higher trophic levels without any direct exposure.

14.6 Further Study

The impact of nanomaterials on the environment will depend on the degree of contamination. Humans are likely to encounter high doses through manufacturing accidents or through the use of products containing nanomaterials. For example, sunscreen, which may contain several hundred milligrams of nanoparticles per dose,[55] is an avenue of direct nanomaterial exposure. The concentration of nanomaterials in the environment will be lower than in manufacturing or in the commodities produced, but it is challenging to predict the eventual concentration. At high levels of exposure, the experiments mentioned above indicate that nanomaterials can have a negative impact. This is most evident in cells, where damage to cell structures have been observed, often in association with ROS.[4,6,55,56] Even without toxicity, their strong interactions with organic particles, to the point of adhering to bacteria and small aquatic animals, make nanomaterials potentially hazardous.[31,62] Once absorbed by an animal, nanoparticles can cause tissue damage that is slow to heal,[30,58,61] or has progressive after-effects.[60,63] They can also pull biomolecules from their environment,[70,73] which means that they can alter the balance of the living system around them. Exposure testing *in vivo* shows that in the absence of acute toxicity, nanomaterials can impact the metabolism, growth, behavior, and reproduction of animals.[65,92,94,101,103] Some nanomaterials dissolve, releasing their toxic constituent materials.[36,64,78,79] Even when they do not cause immediate harm, nanomaterials may have the potential to cause long-term, or multi-generation changes in organism populations.[89,94]

It has been suggested that because it is already a widely used material, the presence of TiO_2 in measurable concentrations will signal the locations where other nanomaterials are likely to accumulate.[77] Production of nanoscale TiO_2 is only expected to increase. It has been theorized that some

of the current bulk TiO_2 manufacturing may transition to nanoscale TiO_2, contributing to the increase in nanoparticle production along with the development of new applications.[1] If the scale of nanomaterial production is hundreds or thousands of tonnes per year,[1,22] local releases of nanomaterials could be a high volume in the absence of appropriate precautions. Essential to the safe use of nanomaterials will be communication. Information on nanomaterial safety and risks must be available for manufacturers to support safe nanomaterial production.[18,23] Closed production systems would reduce the potential for release of nanomaterials,[27] but the dangers of unreacted precursor chemicals should not be neglected. The impurities that may be included with fullerene nanoparticles – such as iron, nickel, and yttrium – can themselves cause toxic reactions.[57,59,60] Solvents used to suspend nanomaterials can also have toxic effects.[98] To purify the nanoparticles these traces must be rinsed away, which means that even if nanoparticles are not lost to the environment during this process, other toxic components may be released. Impurities can produce ROS in the same manner as nanoparticles,[59,66] meaning that the potential exists for environmental lipid peroxidation due to these by-products.

14.7 Conclusions

Nanotechnology remains a nebulous field despite the best efforts of researchers. The issue of nanomaterials in relation to the environment, health, and safety overlaps an exceptionally broad set of disciplines. Safety should be the first concern when exploring new technologies and materials, and all responsible organizations must consider the precautions necessary for proper nanotechnology use. In this brief chapter, information has been drawn from physics, chemistry, biology, mathematics, law, probability, economics, and food science. Collaboration is the only possible way to predict what the final impacts of the technology will be, and it should be encouraged whenever possible. Companies involved in this field must be prepared to carry out detailed life-cycle monitoring of those products containing nanomaterials, ensuring the safety of users and the environment.

Acknowledgements

The authors gratefully acknowledge the Kansas NSF EPSCoR (#R51243/700333), and Wichita State University for the financial and technical support of this work.

References

1. C. O. Robichaud *et al.*, *Environ. Sci. Technol.*, 2009, **43**(12).
2. R. Kaegi *et al.*, *Environ. Pollut.*, 2008, **156**, 233–239.
3. P. M. Hext *et al.*, *Ann. Occup. Hyg.*, 2005, **49**, 461–472.
4. J.-R. Gurr *et al.*, *Toxicology*, 2005, **213**, 66–73.

5. S. Banerjee *et al.*, *Curr. Sci.*, 2006, **90**(10), 1378–1383.
6. M. J. Akhtar *et al.*, *Int. J. Nanomed.*, 2012, 7, 845–857.
7. S. Ito *et al.*, *Thin Solid Films*, 2008, **516**, 4613–4619.
8. D. Cahen *et al.*, *J. Phys. Chem. B*, 2000, **104**, 2053–2059.
9. T. P. Chou *et al.*, *J. Phys. Chem. C*, 2007, **111**, 18804–18811.
10. M. Matsumura *et al.*, *Bull. Chem. Soc. Jpn.*, 1977, **50**(10), 2533–2537.
11. S. Barazzouk *et al.*, *J. Phys. Chem. B*, 2004, **108**, 17015–17018.
12. B. Dindar and S. İçli, *J. Photochem. Photobiol., A*, 2001, **140**(3), 263–268.
13. K. Woan *et al.*, *Adv. Mater.*, 2009, **21**, 2233–2239.
14. M. Zhang and L. Dai, *Nano Energy*, 2012, **1**, 514–517.
15. J. Shatkin, *J. Ind. Ecol.*, 2008, **12**(3), 278–281.
16. I. Linkov *et al.*, *Nat. Nanotechnol.*, 2011, **6**, 784–787.
17. T. Tervonen *et al.*, *J Nanopart Res.*, 2008, **11**(4), 757–766.
18. T. F. Malloy, *ACS Nano*, 2011, **5**(1), 5–12.
19. R. Hischier and T. Walser, *Sci. Total Environ.*, 2012, **425**, 271–282.
20. M. J. Eckelman *et al.*, *J. Ind. Ecol.*, 2008, **12**(3), 316–328.
21. R. A. Sheldon, *J. Chem. Technol. Biotechnol.*, 1997, **68**, 381–388.
22. F. Piccinno *et al.*, *J Nanopart Res.*, 2012, **14**(1109), Online.
23. C. D. Engeman *et al.*, *J Nanopart Res.*, 2012, **14**(749), Online.
24. A. Helland *et al.*, *Environ. Sci. Technol.*, 2008, **42**(2), 640–646.
25. C. O. Robichaud *et al.*, *Environ. Sci. Technol.*, 2005, **39**(22), 8985–8994.
26. J.-Y. Choi *et al.*, *Environ. Sci. Technol.*, 2009, **43**(9).
27. F. Gottschalk and B. Nowack, *J. Environ. Monit.*, 2011, **13**, 1145–1155.
28. F. Gottschalk *et al.*, *Environ. Sci. Technol.*, 2009, **43**(24), 9216–9222.
29. T. Xia *et al.*, *Nano Lett.*, 2006, **6**(8), 1794–1807.
30. G. Oberdörster *et al.*, *Environ. Health Perspect.*, 1994, **102**(suppl 5), 173–179.
31. A. Simon-Deckers *et al.*, *Environ. Sci. Technol.*, 2009, **43**(21), 8423–8429.
32. I.-L. Hsiao and Y.-J. Huang, *Sci. Total Environ.*, 2011, **409**(7), 1219–1228.
33. C. A. Poland *et al.*, *Nat. Nanotechnol.*, 2008, **3**, 423–428.
34. L. K. Braydich-Stolle *et al.*, *J Nanopart Res.*, 2009, **11**, 1361–1374.
35. W.-M. Lee and Y.-J. An, *Chemosphere*, 2013, **91**, 536–544.
36. W.-S. Cho *et al.*, *Part. Fibre Toxicol.*, 2011, **8**(27), Online.
37. S. Lopes *et al.*, *Environ. Toxicol. Chem.*, 2014, **33**(1), 190–198.
38. B. Halliwell and S. Chirico, *Am. J. Clin. Nutr.*, 1993, **57**, 715S–725S.
39. D. W. Morel *et al.*, *J. Lipid Res.*, 1983, **24**.
40. J.-J. Yin *et al.*, *Biomaterials*, 2009, **30**, 611–621.
41. J. F. Turrens, *J. Physiol.*, 2003, **552.2**, 335–344.
42. M. Valko *et al.*, *Mol. Cell. Biochem.*, 2004, **266**, 37–56.
43. National Center for Complementray and Alternative Medicine, National Insitutes of Health, U. S. Department of Health and Human Services, May 2010 (Updated Nov 2013).
44. T. Yokota *et al.*, *Diabetes Care*, 2013, **36**, 1341–1346.
45. T. J. Grahame and R. B. Schlesinger, *Part. Fibre Toxicol.*, 2012, **9**(21), Online.
46. T. C. Long *et al.*, *Environ. Sci. Technol.*, 2006, **40**(14), 4346–4352.

47. K. Bhattacharya *et al.*, *Part. Fibre Toxicol.*, 2009, **6**(17).
48. L. Brunet *et al.*, *Environ. Sci. Technol.*, 2009, **43**(12), 4355–4360.
49. D. Praticò *et al.*, *J. Neurosci.*, 2001, **21**(12), 4183–4187.
50. M. Ghosh *et al.*, *Chemosphere*, 2010, **81**, 1253–1262.
51. D. Ryberg *et al.*, *Carcinogenesis*, 1997, **18**(7), 1285–1289.
52. J. Wanagat *et al.*, *FASEB J.*, 2001, **15**, 322–332.
53. C. M. Sayes *et al.*, *Biomaterials*, 2005, **26**, 7587–7595.
54. K.-C. Yoo *et al.*, *Int. J. Nanomed.*, 2012, 7, 1203–1214.
55. R. K. Shukla *et al.*, *Toxicol. In Vitro*, 2011, **25**, 231–241.
56. S. Alarifi *et al.*, *Int. J. Nanomed.*, 2013, **8**, 983–993.
57. G. Jia *et al.*, *Environ. Sci. Technol.*, 2005, **39**(5), 1378–1383.
58. A. A. Shvedova *et al.*, *Am. J. Physiol.*, 2005, **289**, L698–L708.
59. K. Pulskamp *et al.*, *Toxicol. Lett.*, 2007, **168**, 58–74.
60. C.-W. Lam *et al.*, *Toxicol. Sci.*, 2004, **77**, 126–134.
61. D. B. Warheit *et al.*, *Toxicol. Sci.*, 2004, 77, 117–125.
62. X. Zhu *et al.*, *J Nanopart Res.*, 2009, **11**, 67–75.
63. Z. Li *et al.*, *Environ. Health Perspect.*, 2007, **115**, 377–382.
64. W.-S. Cho *et al.*, *Part. Fibre Toxicol.*, 2013, **10**(9), Online.
65. D. Han *et al.*, *Int. J. Nanomed.*, 2011, **6**, 1453–1461.
66. C.-Y. Chen and C. T. Jafvert, *Environ. Sci. Technol.*, 2010, **44**(17), 6674–6679.
67. S. Nair *et al.*, *J. Mater. Sci.: Mater. Med.*, 2009, **20**, S235–S241.
68. A. A. Clancy *et al.*, *Chem. Phys. Lett.*, 2010, **488**, 99–111.
69. G. V. Lowry *et al.*, *Environ. Sci. Technol.*, 2012, **46**, 6893–6899.
70. E. Hellstrand *et al.*, *FEBS J.*, 2009, **276**, 3372–3381.
71. M. A. Kiser *et al.*, *Water Res.*, 2010, **44**, 4105–4114.
72. M. Lundqvist *et al.*, *Proc. Natl. Acad. Sci.*, 2008, **105**(38), 14265–14270.
73. D. Dell'Orco *et al.*, *PLoS One*, 2010, **5**(6), Online.
74. E. Bigorgne *et al.*, *J Nanopart Res.*, 2012, **14:959**, Online.
75. U. Song *et al.*, *Ecotoxicol. Environ. Saf.*, 2013, **93**, 60–67.
76. M. Wu, *Adv. J. Food Sci. Technol.*, 2013, 5(4), 398–403.
77. M. A. Kiser *et al.*, *Environ. Sci. Technol.*, 2009, **43**(17), 6757–6763.
78. D. Collins *et al.*, *PLoS One*, 2012, 7(8), Online.
79. P. L. Waalewijn-Kool *et al.*, *Environ. Pollut.*, 2013, **178**, 59–64.
80. R. M. C. P. Rajapaksha *et al.*, *Appl. Environ. Microbiol.*, 2004, **70**(5), 2966–2973.
81. Y.-H. Tsuang *et al.*, *Artif. Organs*, 2008, **32**(2), 167–174.
82. R. Brayner *et al.*, *Nano Lett.*, 2006, **6**(4), 866–870.
83. Z. Tong *et al.*, *Environ. Sci. Technol.*, 2007, **41**(8), 2985–2991.
84. S. S. Sharma and K.-J. Dietz, *Trends Plant Sci.*, 2008, **14**(1), 43–50.
85. S. Asli and P. M. Neumann, *Plant, Cell Environ.*, 2009, **32**, 577–584.
86. J. H. Priester *et al.*, *Proc. Natl. Acad. Sci.*, 2012, **109**(37), E2451–E2456.
87. W. C. Lindemann and C. R. Glover. *Nitrogen Fixation by Legumes Guide A-129 (Revised May 2003 ed.)*, Cooperative Extension Service, College of Agriculture and Home Economics, New Mexico State University, http://aces.nmsu.edu/pubs/_a/A129/.

88. D. Li and P. J. J. Alvarez, *Environ. Toxicol. Chem.*, 2011, **30**(11), 2542–2545.

89. C. W. Hu *et al.*, *Soil Biol. Biochem.*, 2010, **42**, 586–591.

90. A. Aalok *et al.*, *J. Hum. Biol.*, 2008, **24**(1), 59–64.

91. L.-Z. Li *et al.*, *Environ Int*, 2011, **37**, 1098–1104.

92. K. Schlich *et al.*, *Environ. Sci. Eur.*, 2012, **24**(5), Online.

93. H. McShane *et al.*, *Environ. Toxicol. Chem.*, 2012, **31**(1), 184–193.

94. M. L. W. Åslund *et al.*, *Environ. Sci. Technol.*, 2012, **46**, 1111–1118.

95. A. R. Petosa *et al.*, *Environ. Sci. Technol.*, 2010, **44**(17), 6532–6549.

96. A. A. Keller *et al.*, *Environ. Sci. Technol.*, 2010, **44**(6).

97. T. E. Abbott Chalew *et al.*, *Environ. Health Perspect.*, 2013, **121**(10), 1161–1166.

98. L. Nyberg *et al.*, *Environ. Sci. Technol.*, 2008, **42**(6), 1938–1943.

99. USEPA, April 1996.

100. OECD/OCDE, April 13, 2004.

101. X. Zhu *et al.*, *Chemosphere*, 2010, **78**, 209–215.

102. E. Oberdörster, *Environ. Health Perspect.*, 2004, **112**(10), 1058–1062.

103. T. Cedervall *et al.*, *PLoS One*, 2012, 7(2), Online.

104. X. Zhang *et al.*, *Chemosphere*, 2007, **67**, 160–166.

105. B. T. Johnson and J. O. Kennedy, *Appl. Microbiol.*, 1973, **26**(1), 66–71.

106. J. S. Gray, *Mar. Pollut. Bull.*, 2002, **45**, 46–52.

107. X. Zhu *et al.*, *Chemosphere*, 2010, **79**, 928–933.

108. F. A. P. C. Gobas *et al.*, *Environ. Sci. Technol.*, 1999, **33**(1).

CHAPTER 15

Risk Assessments of Green Photo-active Nanomaterials

FARHANA ABEDIN,[a] MD. RAJIB ANWAR[b] AND
RAMAZAN ASMATULU*[b]

[a] Bioengineering Graduate Program, University of Kansas,
1530 W. 15th Street, Lawrence, KS 66045, USA; [b] Department of
Mechanical Engineering, Wichita State University, 1845 Fairmount,
Wichita, KS 67260, USA
*Email: ramazan.asmatulu@wichita.edu

15.1 Introduction

15.1.1 General Background

Nanomaterials have one or more of their external dimensions in the range of 1 to 100 nm, which allows them to have a very high surface-to-volume ratio. The increased surface area makes nanomaterials very reactive and able to induce properties that otherwise would have remained dormant.[1] The use of nanomaterials in various consumer goods, such as energy products, appliances, automobiles, home goods, personal care products (PCPs), windows, electronics, and computers, is thriving, and hence the manufacture of nanomaterials is on the rise. Photo-active nanomaterials, which absorb light, have gained attention in fields such as biomedical, paint, construction, energy, personal care, and wastewater treatment.[2-9]

Some examples of photo-active nanomaterials are metal oxide nanoparticles, such as titanium dioxide (TiO_2) and zinc oxide (ZnO), noble metal nanoparticles, such as gold and silver, quantum dots (QDs) (usually made

RSC Green Chemistry No. 42
Green Photo-active Nanomaterials: Sustainable Energy and Environmental Remediation
Edited by Nurxat Nuraje, Ramazan Asmatulu and Guido Mul
© The Royal Society of Chemistry 2016
Published by the Royal Society of Chemistry, www.rsc.org

from cadmium and selenium), carbon nanotubes, and fullerenes. The benefits of using photo-active nanoparticle in various fields are substantial. QDs exhibit enhanced quantum yields, wide absorption spectra, narrow tunable emissions, and resistance to photobleaching.[10] They also have the potential to be used in cellular labeling, *in vivo* cancer cell tracking, *in vitro* assay detection, and solar cells.[11-17] Metal oxide nanomaterials are used in paints, coatings, sunscreen and personal care products, antimicrobial applications, photocatalysts, solar cells, sensors, fuel cells, and electronics manufacturing.[18-26]

The growing market for nanomaterials has caused an increase in their production. The global production of TiO_2 nanomaterials in the year 2005 was approximately 0.002 megatons, which was worth about $70 million, and most were used in personal care products.[27] The commercial production of TiO_2 nanomaterials, which has grown to 0.005 megatons per year, is predicted to continue growing until 2025.[28] In 2007, the estimated global production was 0.02 megatons of metal oxide nanomaterials, 20 tons of metal nanomaterials, and 100 tons of carbon nanotubes (CNTs).[28] The increasing production and use of nanomaterials indicate that they may emerge as pollutants in the environment. The United States Government Accountability Office reported that products containing nanomaterials are anticipated to be approximately $2.6 trillion by 2015.[29] It is doubtful whether current testing methods are sufficient to assess the risk of nanomaterials to human health and environment.[29] It was found that the concentration of TiO_2 and ZnO nanomaterials in sludge-treated soil may rise at a rate of 42–89 $\mu g\,kg^{-1}$ per year and 1.6–3.3 $\mu g\,kg^{-1}$ per year, respectively.[30] The risks posed by these engineered nanomaterials to the environment as well as their exposure to humans are gaining increasing attention.

The fate of these nanomaterials in the environment is an important phenomenon that must be determined to fully understand their risks to human health and the environment. In the environment, nanomaterials may undergo physical, chemical, or biological transformation, which could alter their environmental risk.[7] Extensive research is needed to understand the fate, transportation, and transformation of these nanomaterials in the environment.[31] Figure 15.1 shows a schematic view of some of the possible sources and ways by which nanomaterials in commercial products may become released into the environment and exposed to humans.

15.1.2 Toxicity and Environmental Impacts

The most widely used nanomaterials that dominated the market in 2010 included silica, titania, alumina, iron, and zinc oxide.[32] It was estimated that 63–91% of globally engineered nanomaterials produced in 2010 would find their way to landfills, 8–28% to soils, 0.4–7% to water, and 0.1–1.5% to the atmosphere.[32] Nanoscale materials exhibit properties that could be different from the bulk materials, the difference is possibly due to the enhanced number of surface atoms.[33] Some physicochemical properties that

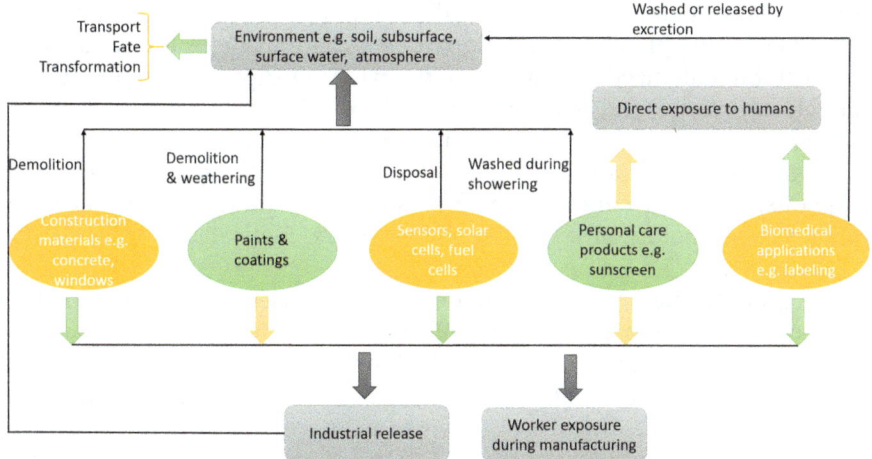

Figure 15.1 Schematic showing possible sources and methods of nanomaterial release into the environment and exposure to humans.

are distinct to nanomaterials and play an important role in nanomaterial research are enhanced magnetic properties, electrical and optical activity, and structural integrity.[34] Because of their nano-size, nanomaterials are able to translocate to biological and ecological regions that are not reachable by the bulk material.[35] The increased surface area of nanomaterials may enhance their absorbance of biomolecules or chemicals, which could cause negative impacts on human health.[35]

Nanomaterials can interact with biomolecules such as proteins.[36] Forces associated with nanomaterial–protein interactions include van der Waals force, hydrogen bonding, electrostatic force determined by surface charges, hydrophobic interactions, and pi–pi (π–π) stacking.[36] The interaction of nanomaterials with protein may alter the protein conformation, depending on concentration, size, and shape of the nanomaterials.[36] Those nanomaterials with a strong surface charge could lead to dispersion in a biosystem due to repulsion. However, in the biosystem, the surface charge and hydrophilicity of nanomaterials can also be altered by adsorption/desorption of other molecules.[36] In aquatic systems, factors that may impact the mobility, transport, fate, bioavailability, ecotoxicity, and reactivity include the following:[37]

- aggregation
- dissolution and weathering
- functionalization
- coating
- adsorption of dissolved species leading to transformation
- neoformation
- persistence
- interaction with other particles.

Some nanomaterial characteristics for toxicity assessment may include size distribution, surface area, reactivity, shape aggregation, chemical composition, and crystal structure.[35] The impact of nanomaterial exposure may vary with phases of its life cycle.[34] Hou *et al.* suggested the possibility of trophic transfer of engineered nanoparticles and biomagnification through the food chain, although further research is required to understand the impact of trophic transfer on higher-level animals such as mammals.[38] It was found that the trophic transfer factor for bacteria containing cadmium selenide (CdSe) quantum dots and ciliated protozoa was 5.4.[39] Some toxicity mechanisms for metal oxide nanomaterials include the generation of reactive oxygen species (ROS), release of metal ions, build-up of nanomaterials on membrane surfaces, and internalization of nanomaterials.[40] Some previous toxicity studies on certain photo-active nanomaterials are discussed below.

15.2 Photosensitive Nanomaterials

15.2.1 Quantum Dots

Quantum dots are 1–6 nm in size.[41] They are usually made of CdSe or cadmium telluride (CdTe), but elements from groups III/V and ternary semiconductors could also be used to make QDs.[41] They are synthesized as either a core–shell or core only, and are sometimes coated for their application.[41] The elements cadmium (Cd) and selenium (Se) are extremely toxic to organisms.[42] An environment-dependent transformation of QDs could elicit a cytotoxic effect. The *in vitro* toxicity of Cd-based QDs has been associated with the release of Cd^{2+}.[43,44] It was also observed that QDs could interfere with gene expression, and *in vitro* studies with CdTe QDs and bulk cadmium chloride ($CdCl_2$) showed that both induced similar transcription pattern changes for the human embryonic kidney 293 (HEK293) cell lines, indicating that free Cd^{2+} could be a major source of cytotoxicity.[45]

Chen *et al.* suggested that the localized high concentration of Cd^{2+} could arise at a certain organelle or other area within the cell due to irregular distribution of the QDs within the cells, which could explain the higher toxicity of QDs compared to the equivalent bulk $CdCl_2$.[45] Cadmium could cause mutation, elicit stress response, and interfere with the production of DNA, RNA, and protein.[42] Exposure of Cd-based QDs to an oxidative environment could cause the leaching of Cd^{2+}.[46] Although this could be avoided by protecting the core with a coating or stabilizing it by ligands, it has been found that even encapsulated QDs have shown to be toxic.[42] Moreover, the local pH or redox environment could lead to removal of ligands by protonation or photo-oxidation.[47] The toxicity of Cd-based QDs is not only confined to the leaching of Cd^{2+} but also due to the nanoscale size of the QDs.[45]

An *in vivo* study of thiol-capped CdTe QDs in hydras (fresh water polyps) showed that the QDs were more toxic than an equivalent amount of

cadmium chloride (CdCl$_2$).[48] This study was conducted using CdTe QDs as well as CdCl$_2$ salt solution, and measurement of the intracellular Cd concentration in the hydra polyps exhibited a lower concentration of Cd in those treated with CdTe QDs compared to those treated with CdCl$_2$ salt solution, indicating that the internalization routes for Cd^{2+} from CdCl$_2$ solution was different from that of the CdTe QDs.[48] Significant morphological alterations were observed for polyps exposed to CdTe QDs based on time and the concentration of QDs.[48] The authors of this study observed that a high dose of CdTe QDs, such as 25 nM, led to 35% mortality and inhibited the polyp head regeneration, whereas a lower concentration, such as 10 nM, led to inhibition to some extent during the first 48 hours, but recovery was observed later.[48] The reproductive rate of the hydra was also negatively impacted by the CdTe QDs, and the authors also reported nuclear damage of the organism, depending on time and concentration.[48]

Li and his colleagues showed the negative impacts of QDs on rat liver mitochondria.[49] They observed that QDs initiate a mitochondrial permeability transition (MPT), a transport mechanism by which the mitochondrial inner membrane permeability is suddenly enhanced to solutes.[49] During this time, detrimental proteins are released into the cytoplasm, inducing apoptosis.[50] QDs also impair mitochondrial respiration.[49] A study of the intracellular fate of COOH and NH$_2$ surface-coated QDs using murine macrophage-like cells indicated that they co-localize within the lysosome at about 2 hours of exposure.[51]

The QDs also co-localized either into or with the mitochondria, but the nanoparticle and cell type both could impact the co-localization of the nanoparticle with mitochondria.[51] Chen *et al.* suggested that the toxicity threshold of QDs could vary with cell type.[45] Irradiation can cause QDs to generate reactive oxygen species causing the cells to endure oxidative stress.[52] Reactive oxygen species (ROS) can be detrimental to cellular proteins, lipids, and DNA.[44] QDs also generate ROS when exposed to cells in the absence of light, and the mechanism of ROS production without light is still not well understood.[44] It is possible that the presence of oxygen in an aqueous or culture media is associated with the ROS formation by QDs in the absence of light.[44] QDs have shown to change the morphology of mitochondria, making them short and round, when exposed to a breast cancer cell line.[44] The above discussions indicate that QDs could have a detrimental impact on human health and aquatic life if they end up in the water system and environment.

Kušić *et al.* investigated the impact of QDs on pollutants such as phenol and toluene for simulated wastewater treatment by ultraviolet (UV) photolysis.[53] It was observed that QDs catalyzed the degradation of phenol and toluene, which could enhance the toxicity of the pollutants.[53] Therefore, it is possible that QDs can transform organic wastes, thus producing more toxic by-products. The presence of dissolved oxygen in a water system and sunlight can enhance the generation of ROS radicals by QDs, which may be

responsible for the degradation of the organic matter, rendering them toxic and directly impacting the aquatic life. Hence, QDs may alter the degradation pathway for organic pollutants.[53]

15.2.2 Metal Oxide Nanomaterials

Photo-excited TiO_2 and ZnO generate dissimilar electron–hole pairs, which renders them stable when compared with other 3d transition metal oxide semiconductor series.[54] These electron–hole pairs can initiate a redox reaction and trigger the generation of reactive oxygen species such as H_2O_2, $^{\bullet}O_2^{-}$, and hydroxyl radicals ($^{\bullet}OH$).[54] Excess ROS and $^{\bullet}OH$ have been associated with enhanced oxidative stress on cells. The reactions involved in the production of ROS and $^{\bullet}OH$ due to photoexcitation of TiO_2 are shown in Figure 15.2. The ROS and hydroxyl radicals can trigger a radical chain response by reacting with available biomolecules and generating different radicals. The ROS can be reduced to the more reactive hydroxyl radicals catalyzed by metal ions.[55–57] ROS causes enhanced damage to mitochondrial DNA (mtDNA) when compared to nuclear DNA.[55] ROS can hinder protein function and accelerate apoptosis.[58,59] The reactive hydroxyl radicals ($^{\bullet}OH$) will undergo reactions with biomolecules such as lipids, DNA, and proteins, almost at the site of generation, thus hindering enzymatic functions and causing cell and DNA damage.[56,57]

The hydroxyl radicals react with DNA bases or the deoxyribosyl backbone of DNA, causing a breakdown of the strands and damage to the bases.[56] The hydroxyl radicals have a longer lifetime of about 10^{-7} s and react in a diffusion-controlled manner.[60] TiO_2 has three distinct crystalline structures: rutile, anatase, and brookite.[54] TiO_2 nanoparticles have been associated with cytotoxicity and genotoxicity in cultured human cells,[61] and Uchino *et al.* linked the cytotoxicity of UVA-irradiated TiO_2 on Chinese hamster ovary (CHO) cells with the generation of hydroxyl radicals.[62] The formation of hydroxyl radicals is dependent on the crystal size and type of the TiO_2, and

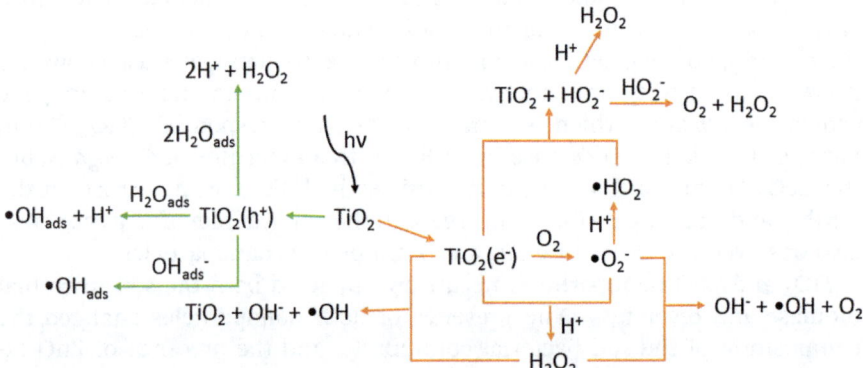

Figure 15.2   Generation of radicals due to photoexcitation of TiO_2.

the anatase form generated hydroxyl radicals in higher concentration upon irradiation with ultraviolet radiation (UVA) compared to the rutile form.[62] Ma *et al.* suggested that the risk assessment of TiO_2 nanomaterials should consider the environmental conditions since the photocatalytic activity generating ROS is impacted by the wavelength of the solar radiation.[63] UVA radiation is mostly responsible for the production of ROS by nano-TiO_2.[63]

A cytotoxicity study of TiO_2 nanoparticles on the human bronchial epithelial cell line has indicated that TiO_2 nanoparticles induced an apoptotic process and enhanced the expression of inflammation-related genes.[64] It was also observed in this study that nanoparticles aggregated near the perinuclear region.[64] TiO_2 nanoparticles were found to be less toxic than single-walled carbon nanotubes in trout, and it was observed that exposure of this fish to sub-lethal doses of TiO_2 nanoparticles led to respiratory distress, oxidative stress, and a decrease in Na^+K^+-ATPase activity in their gills and intestines.[65] Miller *et al.* demonstrated that TiO_2 nanoparticles are toxic to marine phytoplankton, which are the dominant primary producers of carbon and a major participant in the global carbon cycle.[60] This study indicated that the $^\bullet OH$ production rate in seawater was 10–20 times higher in the presence of TiO_2 nanoparticles irradiated with low-intensity UVA compared to its natural generation in freshwater.[60] A significant decrease in the growth rate of three species of phytoplankton (*Isochrysis golbana*, *Thalassiosira pseudonana*, and *Dunaliella tertiolecta*) was observed when they were exposed to TiO_2 nanoparticles in the presence of UVA.[60] Past investigation has shown that the lethal concentration (LC50) and lowest-observable-effect concentration (LOEC) of filtered TiO_2 nanoparticles for *Daphnia magna* organisms was 5.5 ppm and 2.0 ppm, respectively.[66]

C_{60} fullerenes exhibit a higher level of toxicity to *Daphnia magna* compared to TiO_2.[66] Free nanoparticles tend to aggregate, which may cause them to become trapped by sedimentation,[67] and this can impact the soil microbial biomass. One important factor that governs the aggregation of TiO_2 nanoparticle is the pH, and it was found that the aggregate size increased when the pH approached the point of zero charge.[67] The toxicity of nanoparticles is dependent on size, and smaller-sized nanoparticles elicited greater toxicity.[68,69] A study on the exposure of soil nematodes, *Caenorhabditis elegans* (*C. elegans*), to TiO_2 and cerium dioxide (CeO_2) nanoparticles showed a decrease in fertility and survival parameters, and an increase in gene expression *cyp35a2*, which is a measure of stress response.[68] TiO_2 (7 nm) nanoparticles led to a decrease in both growth and fertility of *C. elegans*, but the CeO_2 (15 nm and 45 nm) nanoparticles had significant impact on the fertility and not on growth.[68] A decrease in the survival rate of *C. elegans* was also observed for 15 nm of CeO_2 and 7 nm of TiO_2 nanoparticles.[68]

TiO_2 and ZnO nanoparticles negatively impacted both the soil microbial biomass and diversity.[70] The presence of both nanoparticles changed the composition of the soil bacterial community, and the presence of ZnO exhibited higher toxicity compared to the presence of TiO_2 nanoparticles in the same concentration.[70] Several authors reported a positive impact of TiO_2 on

the germination of seeds and root growth.[71,72] The positive effect on seed germination could be associated with the antimicrobial property of the anatase TiO_2.[71] TiO_2 nanoparticles can form large aggregates and trap algal cells, thus reducing the availability of light and negatively impacting their growth.[71] The aggregation of nanoparticles is dependent upon the surface chemistry of nanoparticles and water chemistry.[73] Some of the factors that may impact the aggregation of TiO_2 nanoparticles are ionic strength of the solution, presence of monovalent and divalent ions, natural organic matter (NOM), and pH.[73] TiO_2 forms large aggregates at an isoelectric point (IEP), and the aggregation of nanoparticles may determine their fate and transport within the aquatic environment.[73]

TiO_2 nanomaterials are being used in paints and exterior coatings as whitening pigments. Kaegi *et al.* has shown that considerable TiO_2 nano-materials are present in the run-off and storm water run-off from urban areas.[74] The eventual destination of these TiO_2 particles from run-off is surface water, and it has been shown that these particles either exist as single particles or agglomerates bound to organic matter.[74] Another source of exposure of TiO_2 nanomaterials to humans is personal care products such as toothpaste, select sunscreen, shampoo, deodorant, and shaving cream.[22] However, TiO_2 particles in neat sunscreen were found to be inactive when exposed to the Sun because of the inert alumina or silica coatings.[8] The ZnO particle-containing sunscreen exhibited ROS production when exposed to UVA.[8] Nonetheless, nanomaterials in personal care products may end up in the sewage system and find their way into treated effluent or biosolids.[22]

TiO_2 is also used as an additive in food, and nearly 36% of food grade TiO_2 particles were found to be less than 100 nm in at least one dimension.[22] Figure 15.3 shows the size distribution of food grade TiO_2 particles. A large fraction of TiO_2 in commercial products was seen to accumulate in biosolids of wastewater treatment plants, whereas a small fraction gathered in the liquid effluent, indicating that the possible fate of the majority of TiO_2 could be agricultural land, landfills, oceans, *etc.*[75] The size of TiO_x observed in the wastewater treatment plant effluent was in the range of a few hundred nanometers, and some were less than 100 nm.[75]

The cytotoxic effects of ZnO nanoparticles have been associated with its dissolution, releasing Zn^{2+} and causing ROS production. Xia *et al.*, using a phagocytic cell line (RAW 264.7) and a human bronchial epithelial cell line (BEAS 2B), showed that ZnO nanoparticles elicited toxicity in both cells, whereas TiO_2 nanoparticles were inert in the absence of illumination, and the CeO_2 nanoparticles exhibited cytoprotective effects against secondary oxidative stress.[76] The ZnO aggregates shrink in size as the Zn dissolves when exposed to cell culture medium, and serum proteins and lipids could play a positive role in the dissolution process.[76] Both dissolution of the ZnO nanoparticles releasing Zn^{2+} and the uptake of ZnO particle remnants were associated with the toxicity effects on the RAW 264.7 and BEAS 2B cells.[76] Xia *et al.* suggested that ZnO remnants could undergo intracellular dissolution, thus releasing Zn^{2+}, which could lead to organellar clumping, oxidative

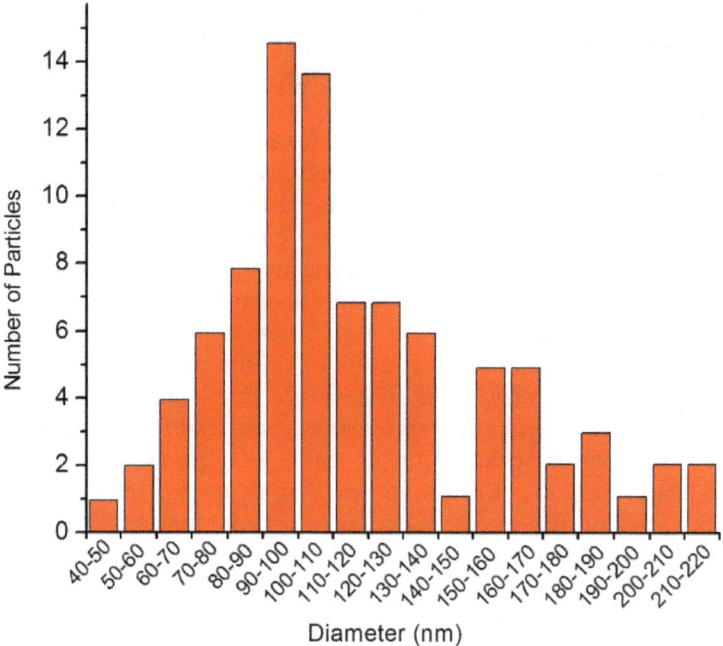

Figure 15.3 Size distribution of food grade TiO_2.

stress injury, intracellular Ca^{2+} release, mitochondrial depolarization, cytokine release, and cytotoxicity.[76] The ZnO nanoparticles were also observed to induce the generation of ROS (H_2O_2 and O^{2-}) in RAW 246.7 cells, and it was suggested that the ROS could be produced as the result of interaction with the ZnO particle surface and also arise from biological species such as damaged mitochondria.[76] Aggregation and dissolution of ZnO nanoparticles in hatchery water (E3 medium) were also reported by Bai *et al.*[77]

The suspension of ZnO nanoparticles in E3 medium consisted of small floating aggregates, dissolved zinc (Zn^{2+}, $Zn(OH)^+$, $Zn(OH)_2$), and large aggregates that settled at the bottom.[77] Exposure of ZnO nanoparticles decelerated the hatching of zebrafish embryos, which was linked to hypoxia and disturbance of the hatching enzyme; they also caused a reduction in larvae length and tail malformation.[77] Bai *et al.* suggested that the toxicity of ZnO nanoparticles could be partially due to the release of dissolved zinc and blocking of the pore canals of chorion by large aggregates, which leads to a reduction in the oxygen supply (hypoxia).[77] This study also indicated permeation of the small aggregates into the zebrafish eggs through the pore canals of chorion.[77]

An antimicrobial inhibition study on *E. coli* and *B. subtilis* for three different nanoparticles (SiO_2, TiO_2, and ZnO) showed that toxicity increased in this order: $SiO_2 < TiO_2 < ZnO$.[23] The enhanced antimicrobial activity by TiO_2 when illuminated, rather than in the dark, indicates that ROS is one of the factors contributing to the toxicity of TiO_2 nanomaterials, although there are

other factors since they also exhibit growth inhibition in the dark to a lesser extent.[23] The antimicrobial activity of ZnO and SiO$_2$ nanoparticles were not impacted by the presence of light.[23] ZnO nanoparticles exhibited near-complete growth inhibition for *B. subtilis* in the dark, but very little inhibition was observed for *E. coli*,[23] which indicated that dissolution of Zn or some undetermined mechanism could be involved in the toxicity. This study also indicated that cell physiology, metabolism, or degree of contact could also impact the toxicity exhibited by the nanoparticles.[23] Adam *et al.* also suggested that cytotoxicity mechanisms in the dark could include the formation of organic radicals in the dark or interference of membrane integrity by nanomaterials.[23] Franklin *et al.* reported that the toxicity of ZnO nanoparticles is due, in part, to dissolution.[78] In this study, a comparable toxicity was observed for the freshwater alga *Pseudokirchneriella subcapitata* when exposed to ZnO nanoparticles, bulk ZnO, and ZnCl$_2$.[78]

Ma *et al.* investigated the phototoxicity of ZnO nanoparticles in the free-living nematode *Caenorhabditis elegans* and showed that the size of the primary ZnO particles was an important factor in toxicity since ZnO nanoparticles exhibited higher toxicity to the nematode compared to bulk ZnO.[79] For ZnO nanoparticles, the mortality of the nematode was observed to be higher under natural sunlight compared to ambient artificial laboratory light, and no mortality was observed in the dark, irrespective of the duration of exposure to the ZnO nanoparticles and bulk ZnO,[79] thus indicating that the generation of ROS played a major role in causing toxicity for this case. Natural sunlight consists of 8–9% UVA, which can efficiently trigger ZnO to produce ROS, but the ambient artificial laboratory light contains negligible UVA.[79] Enhanced lipid peroxidation when exposed to ZnO nanoparticles and bulk ZnO under natural sunlight also indicated ROS generation by the ZnO.[79] The higher toxicity of the ZnO nanoparticles could be attributed to the availability of the enhanced surface area for ROS production.[79]

It was found that the transport of Zn/ZnO nanoparticles in the sandy soil loam is limited, depending on the ionic strength.[80] It was suggested that factors such as physicochemical binding or attachment between charged soil surfaces could impact the retention of nanoparticles within soil.[80] Aggregation of the ZnO nanoparticles increased with ionic strength, and the aggregates could be retained within by a physical straining mechanism.[80] Clay minerals in the soil could provide binding sites for the nanoparticles.[80] The retention of nanoparticles within soil could lead to the uptake of these particles by plants, which can adversely impact them. It was reported that zinc increased in the roots and shoots of corn plants in sandy loam soil that was treated with ZnO nanoparticles, and there was an uptake of ZnO aggregates in low concentration, indicating that only small aggregates were able to pass through the cell wall.[80] The transport of ZnO nanoparticles in the corn roots could be by an apoplastic pathway, which would allow them to reach the endodermis, and also by a symplastic pathway, which would lead them to the vascular cylinder.[80]

It was observed that organic soil matter could influence the motion of CeO_2 nanoparticles.[81] The concentration of Ce in the roots of corn plants grown in organic soil with uncoated and coated CeO_2 nanoparticles was higher compared to plants grown in unenriched soil, thus indicating that humic acid in the soil could play a role in attaching CeO_2 nanoparticles to corn roots.[81] The uptake of alginate-coated CeO_2 nanoparticles by corn roots was observed to be lower than the uncoated nanoparticles, indicating that the surface coating of CeO_2 nanoparticles could impact their uptake.[81] It was suggested that the coated CeO_2 nanoparticles attach to the colloidal soil clay with a heterogeneous charge distribution, which could adversely impact their availability for uptake by the corn roots.[81]

The soil type and CeO_2 surface charge could govern the translocation of the Ce from roots to shoots.[81] A study of wheat and pumpkin plants in a hydroponic culture exposed to uncoated CeO_2 nanoparticles showed that the nanoparticles did not trigger any toxic response and translocation of CeO_2 to pumpkin shoots, but not to wheat shoots.[83] The translocation of nano-particles to the shoots of plants may cause these nanoparticles to enter the food chain and impose adverse effects on humans and animals through biomagnification. Although CeO_2 nanoparticles could produce ROS, it was also found that they could act as an antioxidant in RAW 264.7 and BEAS-2B cells when the cells were exposed to an external source of oxidative stress.[76] An auto-regenerative radical scavenging process, shown in Figure 15.4, could be associated with its cytoprotective behavior. CeO_2 was also found to cause apoptosis in human BEAS-2B cells, which has been associated with en-hanced ROS production and oxidative stress.[82,84,85] The aggregation of

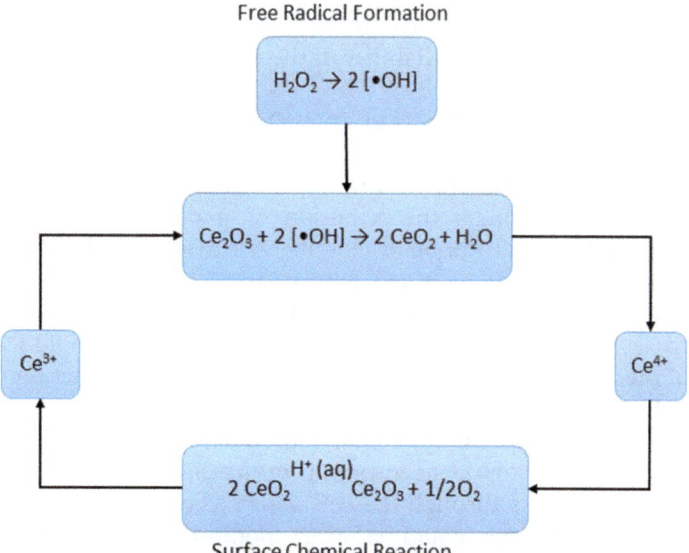

Figure 15.4   Radical scavenging process of nano CeO_2.

nanomaterials in the natural system will have a significant impact on the fate, transport, and toxicity in the environment.

Silicon dioxide (SiO_2) nanoparticles are less toxic compared to TiO_2 and ZnO, and they have shown cytotoxicity in various cell lines and aquatic organisms such as algae. SiO_2 nanoparticles cause oxidative stress in human bronchial cells (Beas-2B).[86] In this study, the toxicity of two types of silica nanoparticles were investigated – porous and fumed. It was found that the porous silica nanoparticles were more toxic than the fumed nanoparticles.[86] The silica nanoparticles were able to penetrate into cells and become localized around the nucleus.[86] The porous silica nanoparticles aggregated to a size of approximately 20 nm, whereas the aggregated size of the fumed nanoparticles was 400 nm.[86] Fujiwara *et al.* showed toxicity of 5 nm, 26 nm, and 78 nm silica nanoparticles in alga, *Chlorella kessleri*.[87] They observed that the smaller nanoparticles elicited more toxicity than the larger ones, and the toxicity to the alga was not impacted by the presence of light.[87] The exposure of silica nanoparticles to alga hindered cell division, thus leading to coagulation of cells and also decreased chlorophyll color.[87]

Silica nanoparticles could also elicit a pro-inflammatory response in mice when intraperitoneally treated with amorphous silica nanoparticles.[88] Silica nanoparticles in mice activated macrophages since their synthesis of nitric oxide (NO) increased.[88] It was observed that there was increased mRNA expression of inflammatory related genes within macrophages of the mice and a higher level of pro-inflammatory cytokines, indicating that the silica nanoparticles could cause inflammation.[88] Treatment of the RAW 264.7 cell line with silica nanoparticles led to oxidative stress since an increased production of ROS was observed.[88] These cells also exhibited an increased level of NO, and the silica nanoparticles negatively impacted cell viability, depending on the time and concentration, by apoptotic pathway.[88]

15.2.3 Metallic Nanomaterials

Silver nanoparticles are being used in commercial products because of their antibacterial activity. Therefore, there is a concern that silver nanoparticles may become released into the soil and aquatic environment. Silver nanoparticles accumulated within the cells of liver tissue of zebrafish exposed to a silver nanoparticle solution without containing any silver ions at the time of treatment.[89] The silver nanoparticles led to morphological changes of liver tissue, such as disruption in the hepatic cell cord.[89] Increased metallothionein-2 expression in the liver tissue of zebrafish exposed to silver nanoparticles indicated that the silver nanoparticles released silver ions that were cytotoxic.[89] The silver nanoparticles also caused oxidative stress, DNA damage, and apoptosis of liver tissue in zebrafish.[89]

It was found that silver nanoparticles were toxic to nitrifying bacteria.[90] Silver nanoparticles less than 5 nm in size hindered nitrification and microbial growth, indicating that smaller silver nanoparticles are more toxic to the bacteria.[90] In this investigation it was demonstrated that silver

nanoparticles, silver chloride, and silver ions all caused an increase in intracellular generation of ROS, and it was suggested that the toxicity of silver nanoparticles could be associated with the intracellular ROS-related apoptotic process.[90] The inhibition of the nitrification process correlated with the intracellular ROS production for the individual forms of silver.[90] Choi *et al.* also indicated that the silver nanoparticles were more toxic than the silver ions.[90] The authors of this study suggested that silver nanoparticles less than 5 nm could easily penetrate the cell membrane when compared to the charged silver ions, and the high surface area of the nanoparticles allows them to interact with the nitrifying cell membrane, thus inhibiting the nitrification process.[90] The authors indicated that other parameters could also contribute to the toxicity of silver nanoparticles since the correlations were different for various forms of silver.[90]

Meyer *et al.* investigated the toxicity of three types of coated silver nanoparticles – polyvinylpyrrolidone (PVP)-coated of the size 21 ± 17 nm (PVP$_S$), 75 ± 21 nm (PVP$_L$), and citrate-coated of the size 7 ± 11 nm (CIT$_{10}$) on the nematode *Caenorhabditis elegans* (*C. elegans*).[91] It was observed that in a potassium ion medium (high ionic strength), the silver nanoparticles aggregated to 1–1.6 μm, whereas in their stock suspension, the PVP-coated silver nanoparticles aggregated to approximately 140 nm particles, which settled down, but the citrate-coated nanoparticles were stable in suspension.[91] In this study, cellular internalization of the three types of silver nanoparticles by *C. elegans* was observed.[91] Internalization of the citrate-coated nanoparticles frequently prevented the laying of eggs of the *C. elegans*, which caused internal development of the embryo, and it was found that the citrate-coated silver nanoparticles were able to transfer to the fertilized eggs.[91] Exposure of all three types of silver nanoparticles negatively impacted the growth at concentrations of 50 mg L^{-1}, and it was found that the toxicity of the supernatants of the PVP-coated silver nanoparticles to the nematode was similar to the corresponding nanoparticles, but the supernatants of the citrate-coated silver nanoparticles exhibited no toxicity to the nematode.[91] Meyer *et al.* found that the toxicity of the PVP-coated silver nanoparticles could be associated with the dissolution of the silver, and that the citrate-coated silver nanoparticles could be associated with the internalization of the particles.[91] The type of coating also played an important role in the release of silver ions from the silver nanoparticles, since it was found that PVP-coated silver nanoparticles released more silver ions than the citrate-coated nanoparticles.[91,92] Smaller silver nanoparticles are more susceptible to dissolution than larger ones, and it was proposed that increased dissolution of the silver nanoparticles could be associated with desorption of the silver ions from the surface of the nanoparticles when exposed to river and lake water.[92]

An *in vivo* mice model study of bioaccumulation and cytotoxicity of gold nanoparticles ($\sim$ 12.5 nm) administered repeatedly on different tissues at varying concentrations indicated that these nanoparticles had no cytotoxic effect.[93] Studies on *in vitro* retina tissue models have demonstrated that even

a low concentration of 20–80 nm gold nanoparticles have an unwanted neurotoxic effect, especially on photoreceptors.[94] Although gold nanoparticles have been shown to be inert and non-cytotoxic, they can become cytotoxic upon surface functionalization. A study of epithelial cell viability of the human lung upon exposure to chitosan-functionalized gold nanoparticles showed that the resulting surface charge of functionalization can make the nanoparticles cytotoxic. However, the extent of these cytotoxic effects are dependent on the cell type.[95]

15.3 Factors Affecting Risk Assessment of Nanomaterials

In order to understand the risks posed by various nanomaterials to the environment and human health, extensive research on their fate, transport, and life cycle, as well as ecological and human health impacts are necessary.[31] Thomas *et al.* suggested that the following factors need to be considered when studying the fate, transport, and life cycle of engineered nanomaterials:[31]

- transformation
- mobility
- binding and leaching
- mechanism of transition from one media to another.

Life cycle analysis (LCA) may also reveal vital information regarding risk assessment of some of the engineered nanomaterials currently used worldwide:[31]

- collecting data on engineered nanomaterials
- development of standardized methodologies for LCA
- comparison between bulk and nanomaterials
- determination of specific nanomaterial emissions
- determination of environmental impacts of nanomaterials during their product life cycles.

The exposure assessment of nanomaterials in consumer products is an important factor for understanding the risks they pose to human health.[96] Exposure will be minimum if the nanomaterial is incorporated in the product such that there is a very low possibility of its emission.[31] Exposure of nanomaterials in consumer products to humans could occur through skin contact, ingestion, or inhalation.[31] Aggregation of nanomaterials can result in alterations in their behavior and exposure to humans.[31]

Currently, particulates that are assessed for emissions fall in the size range of <10 μm to <2.5 μm.[97] Although their effects on human health and the environment are being assessed, impacts specific to the nano size are

still deficient.[97] Hischier and Walser have suggested that it is vital to consider all new nano-related functionalities and to identify life cycle inventory analysis of widely used nanomaterials along with their production pathways.[97] For the life cycle inventory analysis, knowledge of indoor nanoparticle emissions during production and their release into air, water, and soil needs to be collected. The data gathered during this process must include details such as particle number, size distribution, surface charges, and elemental compositions.[97] For life cycle impact assessment, sufficient factors related to nanoparticle emissions to air, water, and soil need to be established.[97] With increasing nanotechnology products, more research is necessary in order to gain a thorough understanding of risks posed by nanomaterials to humans and the environment.

Workers are most vulnerable to maximum exposure to nanomaterials. According to Kuempel *et al.* there are still gaps in the area of exposure controls of nanomaterials.[98] Exposure to nanomaterials in the workplace could be minimized through use of personnel protective equipment and standard engineering controls such as exhaust ventilation and benchtop enclosures.[98] It was found that air-curtain fume hood design is effective against airborne nanomaterial release.[98] For risk assessment, the first step will be to identify the type of hazard and available approaches to contain it.[98] The following steps include estimation of its severity, workers' exposure, dose-based biological impact, and finally integration of the previous information to aid in risk management.[98] Although current testing methods are considered to be adequate for assessing toxicity of nanomaterials, additional factors such as dose metric, target tissue, and physicochemical properties should be considered.[98] For nanomaterials surface area dose or material volume correlates well with toxicity instead of mass dose metric.[98,99] As discussed in Section 15.2, nanomaterials are able to access regions within cells which are not reachable by larger sized materials. Moreover, studies have shown that nanomaterials are capable of translocating from one organ to the other although there is no definitive information regarding the translocation pathway nanomaterials follow in humans.[98] As discussed in Section 15.2, the physicochemical properties of nanomaterials relating their toxicity are different from those for larger particles which should be accounted for in the risk assessment. High uncertainty in the data used for risk assessment would indicate the need for more research to acquire accurate data.[98] Although there are toxicology data on certain nanomaterials, there is still lack of standardized test methods which makes it difficult to compare the existing data for risk assessment.[98] Differences in methods and assumptions have led to varying occupational exposure limits (OELs) for the same nanomaterial.[98] Since the *in vivo* toxicity studies are done using animal models, translation of the animal model data to understand impact on humans could result in uncertainty which should also be considered in risk assessment.[99] Kuempel and his colleagues suggested that a more sensitive, definite and quantitative approach is necessary for evaluation of workers' exposure.[98] According to them there is still room for improving risk

communication tools and development of specific criteria to classify nanomaterials based on their level of toxicity.[98] For consumer and environment exposure, data on consumer uses, product deterioration, disposal, and nanomaterial release and transport are necessary.[99] The concentration of engineered nanomaterials in the environment is below the detection limit for the majority of the test methods and the environment also contains naturally occurring nanomaterials.[100] This could make it difficult to assess current environmental exposure of engineered nanomaterials. Bhatt *et al.* provided some directions towards achieving selective detection of engineered nanomaterials.[100] There is a possibility that the development of standard risk assessment protocols can be slower than the pace at which nanomaterials will dominate the market.[34] According to Fan *et al.* it is necessary to focus on nanomaterial hazard and exposure and follow its use.[34] One of the task of the National Nanotechnology Initiative (NNI) under US federal government is to oversee the risks posed by engineered nanomaterials during manufacture and use.[34] Currently there are no specific regulations for nanomaterials.[34,35] It may be necessary to impose new laws for risk management of nanomaterials.[34] There are still questions whether nanomaterials fall under 'new materials' or whether it is possible to include them under regulations for the same chemical components in macroscale.[35] Risk assessment of nanomaterials is complex because the chemicals which are non-toxic in the macroscale may become toxic in the nanoscale.[35]

15.4 Conclusions

Studies in the past have shown that the photo-active nanomaterials could be toxic to various cell lines, terrestrial organisms, and aquatic organisms and hence their accumulation in the environment is a major concern now. The toxicity of nanomaterials could be attributed to various factors, such as release of metal ions, ability to produce/induce ROS generation, and penetration/accumulation within cells due to their nanoscale size and shape. Various means by which exposure and release of nanomaterials could occur are personal care products; biomedical applications; disposal of sensors and solar cells; industrial waste release; demolition and weathering of construction materials, paints and coatings; and worker exposure during manufacturing. Accumulation of nanomaterials within plants and animals may cause them to enter the food chain and, through trophic transfer, undergo more toxic biomagnification. In the environment the nanomaterials may undergo aggregation, dissolution, functionalization, interaction with and adsorption of other species. These may directly impact the bioavailability, fate and transport of the nanomaterials. For risk assessment of nanomaterials, it is vital to consider the impact of nanoscale-specific properties. It is vital to identify as many nano-related properties as possible. Thorough knowledge on the transformation, transport, fate, and emission of various nanomaterials is required for analyzing their impact on the environment.

Acknowledgements

The authors gratefully acknowledge the Kansas NSF EPSCoR (#R51243/700333), and Wichita State University for the financial and technical support of this work.

References

1. M. H. Kathawala, S. Xiong, M. Richards, K. W. Ng, S. George and S. C. J. Loo, *Small*, 2013, **9**, 1504–1520.
2. S. Liang, D. T. Pierce, C. Amiot and X. Zhao, *Synth. React. Inorg., Met.-Org., Nano-Met. Chem.*, 2005, **35**, 661–668.
3. S. Sortino, *J. Mater. Chem.*, 2012, **22**, 301–318.
4. C. A. Strassert, M. Otter, R. Q. Albuquerque, A. Höne, Y. Vida, B. Maier and L. De Cola, *Angew. Chem., Int. Ed.*, 2009, **48**, 7928–7931.
5. A. T. Saber, K. A. Jensen, N. R. Jacobsen, R. Birkedal, L. Mikkelsen, P. Møller, S. Loft, H. Wallin and U. Vogel, *Nanotoxicology*, 2012, **6**, 453–471.
6. L. Han, P. Wang and S. Dong, *Nanoscale*, 2012, **4**, 5814–5825.
7. J. Lee, S. Mahendra and P. J. J. Alvarez, *ACS Nano*, 2010, **4**, 3580–3590.
8. Z. A. Lewicka, W. W. Yu, B. L. Oliva, E. Q. Contreras and V. L. Colvin, *J. Photochem. Photobiol., A*, 2013, **263**, 24–33.
9. P. Xu, G. M. Zeng, D. L. Huang, C. L. Feng, S. Hu, M. H. Zhao, C. Lai, Z. Wei, C. Huang, G. X. Xie and Z. F. Liu, *Sci. Total Environ.*, 2012, **424**, 1–10.
10. J. Li and J.-J. Zhu, *Analyst*, 2013, **138**, 2506–2515.
11. F. Chen and D. Gerion, *Nano Lett.*, 2004, **4**, 1827–1832.
12. J. K. Jaiswal, H. Mattoussi, J. M. Mauro and S. M. Simon, *Nat. Biotechnol.*, 2003, **21**, 47–51.
13. X. Gao, Y. Cui, R. M. Levenson, L. W. K. Chung and S. Nie, *Nat. Biotechnol.*, 2004, **22**, 969–976.
14. Y. Xiao and P. E. Barker, *Nucleic Acids Res.*, 2004, **32**, e28.
15. Y. Zhang and T.-H. Wang, *Theranostics*, 2012, **2**, 631–654.
16. F. Hetsch, X. Xu, H. Wang, S. V. Kershaw and A. L. Rogach, *J. Phys. Chem. Lett.*, 2011, **2**, 1879–1887.
17. E. Goldman, I. Medintz and H. Mattoussi, *Anal. Bioanal. Chem.*, 2006, **384**, 560–563.
18. S. K. Hau, H.-L. Yip, N. S. Baek, J. Zou, K. O'Malley and A. K.-Y. Jen, *Appl. Phys. Lett.*, 2008, **92**, 253301.
19. K. D. Benkstein, N. Kopidakis, J. van de Lagemaat and A. J. Frank, *J. Phys. Chem. B*, 2003, **107**, 7759–7767.
20. K. R. Raghupathi, R. T. Koodali and A. C. Manna, *Langmuir*, 2011, **27**, 4020–4028.
21. A. A. Keller, H. Wang, D. Zhou, H. S. Lenihan, G. Cherr, B. J. Cardinale, R. Miller and Z. Ji, *Environ. Sci. Technol.*, 2010, **44**, 1962–1967.

22. A. Weir, P. Westerhoff, L. Fabricius, K. Hristovski and N. von Goetz, *Environ. Sci. Technol.*, 2012, **46**, 2242–2250.

23. L. K. Adams, D. Y. Lyon and P. J. J. Alvarez, *Water Res.*, 2006, **40**, 3527–3532.

24. N. S. Allen, M. Edge, A. Ortega, C. M. Liauw, J. Stratton and R. B. McIntyre, *Polym. Degrad. Stab.*, 2002, **78**, 467–478.

25. D.-H. Lim, W.-D. Lee, D.-H. Choi, H.-H. Kwon and H.-I. Lee, *Electrochem. Commun.*, 2008, **10**, 592–596.

26. S. Saha, S. K. Arya, S. P. Singh, K. Sreenivas, B. D. Malhotra and V. Gupta, *Biosens. Bioelectron.*, 2009, **24**, 2040–2045.

27. EPA Nanomaterial Case Studies: Nanoscale titanium dioxide in water treatment and in topical sunscreen. National Centre for Environmental Assessment Office of Research and Development U.S. Environmental Protection Agency, Research Triangle Park, NC, EPA A/600/R-09/057, 1–222, 2010.

28. R. Landsiedel, L. Ma-Hock, A. Kroll, D. Hahn, J. Schnekenburger, K. Wiench and W. Wohlleben, *Adv. Mater.*, 2010, **22**, 2601–2627.

29. United States Government Accountability Office Report on Nanotechnology, International Journal of Occupational and Environmental Health, 16, 525–539, 2010, http://www.maneyonline.com/doi/abs/10.1179/107735210799159932.

30. F. Gottschalk, T. Sonderer, R. W. Scholz and B. Nowack, *Environ. Sci. Technol.*, 2009, **43**, 9216–9222.

31. K. Thomas, P. Aguar, H. Kawasaki, J. Morris, J. Nakanishi and N. Savage, *Toxicol. Sci.*, 2006, **92**, 23–32.

32. A. Keller, S. McFerran, A. Lazareva and S. Suh, *J. Nanopart. Res.*, 2013, **15**, 1–17.

33. Y. Ju-Nam and J. R. Lead, *Sci. Total Environ.*, 2008, **400**, 396–414.

34. A. M. Fan and G. Alexeeff, *J. Nanosci. Nanotechnol.*, 2010, **10**, 8646–8657.

35. M. R. Gwinn and L. Tran, *Nanomed. Nanobiotechnol.*, 2010, **2**, 130–137.

36. X.-L. Chang, S.-T. Yang and G. Xing, *J. Biomed. Nanotechnol.*, 2014, **10**, 2828–2851.

37. M. Delay and F. Frimmel, *Anal. Bioanal. Chem.*, 2012, **402**, 583–592.

38. W.-C. Hou, P. Westerhoff and J. D. Posner, *Environ. Sci.: Processes Impacts*, 2013, **15**, 103–122.

39. R. Werlin, J. H. Priester, R. E. Mielke, S. Kramer, S. Jackson, P. K. Stoimenov, G. D. Stucky, G. N. Cherr, E. Orias and P. A. Holden, *Nat. Nanotechnol.*, 2011, **6**, 65–71.

40. A. B. Djurišić, Y. H. Leung, A. M. C. Ng, X. Y. Xu, P. K. H. Lee, N. Degger and R. S. S. Wu, *Small*, 2015, **11**, 26–44.

41. U. Resch-Genger, M. Grabolle, S. Cavaliere-Jaricot, R. Nitschke and T. Nann, *Nat. Methods*, 2008, **5**, 763–775.

42. M. Bottrill and M. Green, *Chem. Commun.*, 2011, **47**, 7039–7050.

43. K. G. Li, J. T. Chen, S. S. Bai, X. Wen, S. Y. Song, Q. Yu, J. Li and Y. Q. Wang, *Toxicol. In Vitro*, 2009, **23**, 1007–1013.

44. J. Lovrić, S. J. Cho, F. M. Winnik and D. Maysinger, *Chem. Biol.*, 2005, **12**, 1227–1234.
45. N. Chen, Y. He, Y. Su, X. Li, Q. Huang, H. Wang, X. Zhang, R. Tai and C. Fan, *Biomaterials*, 2012, **33**, 1238–1244.
46. C. Kirchner, T. Liedl, S. Kudera, T. Pellegrino, A. Muñoz Javier, H. E. Gaub, S. Stölzle, N. Fertig and W. J. Parak, *Nano Lett.*, 2005, **5**, 331–338.
47. J. Aldana, N. Lavelle, Y. Wang and X. Peng, *J. Am. Chem. Soc.*, 2005, **127**, 2496–2504.
48. A. Ambrosone, L. Mattera, V. Marchesano, A. Quarta, A. S. Susha, A. Tino, A. L. Rogach and C. Tortiglione, *Biomaterials*, 2012, **33**, 1991–2000.
49. J. Li, Y. Zhang, Q. Xiao, F. Tian, X. Liu, R. Li, G. Zhao, F. Jiang and Y. Liu, *J. Hazard. Mater.*, 2011, **194**, 440–444.
50. N. Zamzami and G. Kroemer, *Nat. Rev. Mol. Cell Biol.*, 2001, **2**, 67–71.
51. M. J. D. Clift, C. Brandenberger, B. Rothen-Rutishauser, D. M. Brown and V. Stone, *Toxicology*, 2011, **286**, 58–68.
52. A. C. S. Samia, X. Chen and C. Burda, *J. Am. Chem. Soc.*, 2003, **125**, 15736–15737.
53. H. Kušić and D. Leszczynska, *Chemosphere*, 2012, **89**, 900–906.
54. S. Banerjee, J. Gopal, P. Muraleedharan, A. K. Tyagi and B. Rai, *Curr. Sci.*, 2006, **90**, 1378–1383.
55. F. M. Yakes and B. Van Houten, *Proc. Natl. Acad. Sci.*, 1997, **94**, 514–519.
56. M. Valko, M. Izakovic, M. Mazur, C. Rhodes and J. Telser, *Mol. Cell. Biochem.*, 2004, **266**, 37–56.
57. K. Hensley, K. A. Robinson, S. P. Gabbita, S. Salsman and R. A. Floyd, *Free Radical Biol. Med.*, 2000, **28**, 1456–1462.
58. C. Fleury, B. Mignotte and J.-L. Vayssière, *Biochimie*, 2002, **84**, 131–141.
59. M. L. Circu and T. Y. Aw, *Free Radical Biol. Med.*, 2010, **48**, 749–762.
60. R. J. Miller, S. Bennett, A. A. Keller, S. Pease and H. S. Lenihan, *Plos One*, 2012, 7, DOI: 10.1371/journal.pone.0030321.
61. J. J. Wang, B. J. S. Sanderson and H. Wang, *Mutat. Res., Genet. Toxicol. Environ. Mutagen.*, 2007, **628**, 99–106.
62. T. Uchino, H. Tokunaga, M. Ando and H. Utsumi, *Toxicol. In Vitro*, 2002, **16**, 629–635.
63. H. Ma, A. Brennan and S. A. Diamond, *Environ. Toxicol. Chem.*, 2012, **31**, 2099–2107.
64. E.-J. Park, J. Yi, K.-H. Chung, D.-Y. Ryu, J. Choi and K. Park, *Toxicol. Lett.*, 2008, **180**, 222–229.
65. G. Federici, B. J. Shaw and R. D. Handy, *Aquat. Toxicol.*, 2007, **84**, 415–430.
66. S. B. Lovern and R. Klaper, *Environ. Toxicol. Chem.*, 2006, **25**, 1132–1137.
67. V. K. Sharma, *J. Environ. Sci. Health*, 2009, **44**, 1485–1495.
68. J.-Y. Roh, Y.-K. Park, K. Park and J. Choi, *Environ. Toxicol. Pharmacol.*, 2010, **29**, 167–172.

69. G. Oberdorster, E. Oberdorster and J. Oberdorster, *Environ. Health Perspect.*, 2005, **113**, 823–839.
70. Y. Ge, J. P. Schimel and P. A. Holden, *Environ. Sci. Technol.*, 2011, **45**, 1659–1664.
71. L. Clément, C. Hurel and N. Marmier, *Chemosphere*, 2013, **90**, 1083–1090.
72. L. Zheng, F. Hong, S. Lu and C. Liu, *Biol. Trace Elem. Res.*, 2005, **104**, 83–91.
73. S. Ottofuelling, F. Von Der Kammer and T. Hofmann, *Environ. Sci. Technol.*, 2011, **45**, 10045–10052.
74. R. Kaegi, A. Ulrich, B. Sinnet, R. Vonbank, A. Wichser, S. Zuleeg, H. Simmler, S. Brunner, H. Vonmont, M. Burkhardt and M. Boller, *Environ. Pollut.*, 2008, **156**, 233–239.
75. M. A. Kiser, P. Westerhoff, T. Benn, Y. Wang, J. Pérez-Rivera and K. Hristovski, *Environ. Sci. Technol.*, 2009, **43**, 6757–6763.
76. T. Xia, M. Kovochich, M. Liong, L. Mädler, B. Gilbert, H. Shi, J. I. Yeh, J. I. Zink and A. E. Nel, *ACS Nano*, 2008, **2**, 2121–2134.
77. W. Bai, Z. Zhang, W. Tian, X. He, Y. Ma, Y. Zhao and Z. Chai, *J. Nanopart. Res.*, 2010, **12**, 1645–1654.
78. N. M. Franklin, N. J. Rogers, S. C. Apte, G. E. Batley, G. E. Gadd and P. S. Casey, *Environ. Sci. Technol.*, 2007, **41**, 8484–8490.
79. H. Ma, N. J. Kabengi, P. M. Bertsch, J. M. Unrine, T. C. Glenn and P. L. Williams, *Environ. Pollut.*, 2011, **159**, 1473–1480.
80. L. Zhao, J. R. Peralta-Videa, M. Ren, A. Varela-Ramirez, C. Li, J. A. Hernandez-Viezcas, R. J. Aguilera and J. L. Gardea-Torresdey, *Chem. Eng. J.*, 2012, **184**, 1–8.
81. L. Zhao, J. R. Peralta-Videa, A. Varela-Ramirez, H. Castillo-Michel, C. Li, J. Zhang, R. J. Aguilera, A. A. Keller and J. L. Gardea-Torresdey, *J. Hazard. Mater.*, 2012, **225–226**, 131–138.
82. M. Das, S. Patil, N. Bhargava, J.-F. Kang, L. M. Riedel, S. Seal and J. J. Hickman, *Biomaterials*, 2007, **28**, 1918–1925.
83. F. Schwabe, R. Schulin, L. K. Limbach, W. Stark, D. Bürge and B. Nowack, *Chemosphere*, 2013, **91**, 512–520.
84. H.-J. Eom and J. Choi, *Toxicol. Lett.*, 2009, **187**, 77–83.
85. E.-J. Park, J. Choi, Y.-K. Park and K. Park, *Toxicology*, 2008, **245**, 90–100.
86. H.-J. Eom and J. Choi, *Toxicol. In Vitro*, 2009, **23**, 1326–1332.
87. K. Fujiwara, H. Suematsu, E. Kiyomiya, M. Aoki, M. Sato and N. Moritoki, *J. Environ. Sci. Health, Part A: Toxic/Hazard. Subst. Environ. Eng.*, 2008, **43**, 1167–1173.
88. E.-J. Park and K. Park, *Toxicol. Lett.*, 2009, **184**, 18–25.
89. J. E. Choi, S. Kim, J. H. Ahn, P. Youn, J. S. Kang, K. Park, J. Yi and D.-Y. Ryu, *Aquat. Toxicol.*, 2010, **100**, 151–159.
90. O. Choi and Z. Hu, *Environ. Sci. Technol.*, 2008, **42**, 4583–4588.
91. J. N. Meyer, C. A. Lord, X. Y. Yang, E. A. Turner, A. R. Badireddy, S. M. Marinakos, A. Chilkoti, M. R. Wiesner and M. Auffan, *Aquat. Toxicol.*, 2010, **100**, 140–150.

92. J. Dobias and R. Bernier-Latmani, *Environ. Sci. Technol.*, 2013, **47**, 4140–4146.

93. C. Lasagna-Reeves, D. Gonzalez-Romero, M. A. Barria, I. Olmedo, A. Clos, V. M. Sadagopa Ramanujam, A. Urayama, L. Vergara, M. J. Kogan and C. Soto, *Biochem. Biophys. Res. Commun.*, 2010, **393**, 649–655.

94. E. Söderstjerna, P. Bauer, T. Cedervall, H. Abdshill, F. Johansson and U. E. Johansson, *PLoS ONE*, 2014, **9**, e105359.

95. P. Schlinkert, E. Casals, M. Boyles, U. Tischler, E. Hornig, N. Tran, J. Y. Zhao, M. Himly, M. Riediker, G. J. Oostingh, V. Puntes and A. Duschl, *J. Nanobiotechnol.*, 2015, **13**, DOI: 10.1186/s12951-014-0062-4.

96. T. Thomas, K. Thomas, N. Sadrieh, N. Savage, P. Adair and R. Bronaugh, *Toxicol. Sci.*, 2006, **91**, 14–19.

97. R. Hischier and T. Walser, *Sci. Total Environ.*, 2012, **425**, 271–282.

98. E. D. Kuempel, C. L. Geraci and P. A. Schulte, *Ann. Occup. Hyg.*, 2012, **56**, 491–505.

99. J. S. Tsuji, A. D. Maynard, P. C. Howard, J. T. James, C.-w. Lam, D. B. Warheit and A. B. Santamaria, *Toxicol. Sci.*, 2006, **89**, 42–50.

100. I. Bhatt and B. N. Tripathi, *Chemosphere*, 2011, **82**, 308–317.

CHAPTER 16

Energy Harvesting from Solar Energy Using Nanoscale Pyroelectric Effects

ARMANJ HASANYAN,[a] RAMAZAN ASMATULU*[b] AND DAVRESH J. HASANYAN[a]

[a] Department of Engineering Science and Mechanics, Virginia Tech, Blacksburg, Virginia 24061, USA; [b] Department of Mechanical Engineering, Wichita State University, Wichita, Kansas 67260, USA
*Email: ramazan.asmatulu@wichita.edu

16.1 Introduction

In the last decade, piezoelectric materials have found widespread application in sensors, actuators, loud speakers, *etc.* because of their ability to convert electrical energy to mechanical, thermal, and magnetic energy, and vice versa. This has led to an accumulation of research on the development of piezoelectric-based energy harvesting devices as power generators in a variety of portable and low-power consumption devices. The process of extracting energy from the surrounding environment is termed 'energy harvesting.' Energy harvesting, which originated from the windmill and water wheel, is widely being considered as a low-maintenance solution for a wide variety of applications.

Note that energy-harvesting techniques are numerous.[2–10,14–16,18–22] Photovoltaic-solar energy is directly converted into electrical energy using polarized solar cells (semiconductor devices). A mechanical (vibration), electrostatic method, and relative movement between electrically charged

RSC Green Chemistry No. 42
Green Photo-active Nanomaterials: Sustainable Energy and Environmental Remediation
Edited by Nurxat Nuraje, Ramazan Asmatulu and Guido Mul
© The Royal Society of Chemistry 2016
Published by the Royal Society of Chemistry, www.rsc.org

capacitor planes can be also utilized. The work against the electrostatic force between the plates provides the harvested energy in an electromagnetic method. Electromagnetic induction arising from the relative motion (rotation or linear) between a magnetic flux and a conductor is used, and piezoelectric method-active materials are employed to generate the energy when mechanically stressed.[1–5,8,10]

Another attractive area for harvesting energy is from thermal sources. Thermal energy (a temperature gradient) is converted into electrical energy using, for example, Seebeck's effect.[7,20] Thermal energy (temperature variation) is also converted *via* the pyroelectric effect.[6,14,19,22]

It must be mentioned that when using a thermoelectric module, which has a limited temperature gradient due to the limited heat exchange (Seebeck's effect), a maximum efficiency of $\sim 3\%$–4% can be expected. However, in contrast, a pyroelectric device may reach an efficiency up to 50%.[1,19]

Vibration energy can be converted into electrical energy through piezoelectric, electromagnetic, and capacitive transducers. Among them, piezoelectric vibration-to-electricity converters have received much attention, because they have high electromechanical coupling and no external voltage source requirement, and are particularly attractive for use in micro-electro-mechanical systems (MEMS).[10,11,16,18] Paradiso and Starner[15] have researched a recent commercial wristwatch that uses thermoelectric modules to generate enough power to run the clock's mechanical components. The thermoelectric modules in the clock work by using the thermal gradient produced through body heat. The pyroelectric effect is another possibility for converting heat into electricity. Several authors[6,14,19,22] have proposed pyroelectric energy harvesting using materials such as PZT-5A, PMN-PT, PVDF, and thin films. They have concluded that with a higher pyroelectric coefficient, more power is generated. Namli and Taya[14] proposed a thermal energy harvester with a piezo-shape memory alloy (SMA) composite. They studied the electro-elastic coupling of the piezoelectric structure combined with the thermal response of the SMA.

For more details about energy-harvesting methods, challenges, and thermal sources, the reader is referred to several other works.[1,7,14,19,22] A piezoelectric energy harvester in an infinite degrees-of-freedom system is often modeled as a mass + spring + damper + piezo structure (as a lumped model) together with an energy storage system. This approach is simple but cannot capture all phenomena specific to the distributed system. We model an energy harvester as a distributed system and, in particular, show that the effect of boundary conditions on the energy harvesting coefficient can have a significant influence.

The work presented in this chapter deals with modeling the harvesting from mechanical, electrical (piezoelectric), and thermal (pyroelectric) bimorph-type structures. Also, the influence of boundary conditions, in plain conservative follower force, vibration frequency, and material properties on a bimorph energy harvester is analyzed while under a thermal-electrical field. The pyroelectric and thermal expansion coefficients are also

considered. The proposed model for the bimorph can be extended for more general cases. An example is to composites made of magneto-thermo-electro-elastic SMAs.

16.2   Model and Constitutive Equations

Considered here is a thermally active thermo-piezoelectric bilayer structure of length $2L$ and thickness $H = h_p + h_m$, where h_p is the thickness of the piezoelectric layer, and h_m is the thickness of the elastic layer. The system of coordinates is chosen in such a way that the x_1 axis is directed along the neutral line, the x_2 axis is directed across the width, and the x_3 axis is orthogonal to both of them. For simplicity, the structure is assumed to be a two-dimensional beam, where the field functions depend only on the space coordinates x_1 and x_3. We also consider a piezoelectric layer that is poled in the x_3 direction, as shown in Figure 16.1. Furthermore, we assume the following:

- that the material of each layer is linearly elastic
- that the strains and displacements are small
- that the length of the composite is much larger than its total thickness $(L \gg H)$
- that the thermal field distribution is constant across each layer
- that Bernoulli's (Kirchhoff's) hypothesis is valid for both layers.

The displacements in x_1 and x_3 directions are given as

$$\begin{cases} u_1(x_1, x_3) = u(x_1) - x_3 \dfrac{\partial w}{\partial x_1} \\ u_3(x_1, x_3) = w(x_1) \end{cases} \tag{16.1}$$

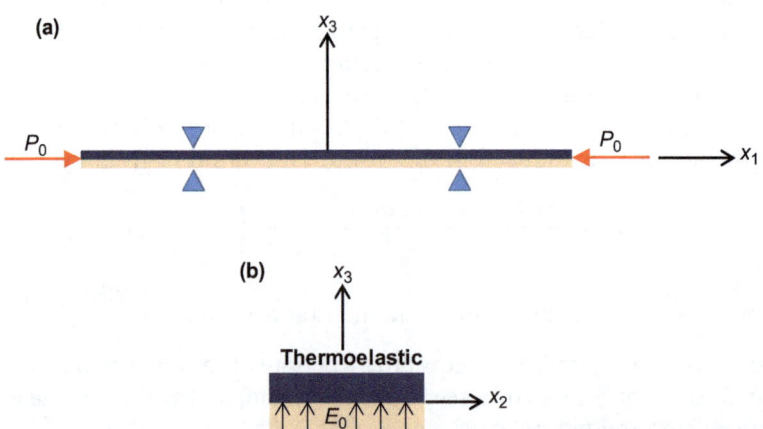

Figure 16.1   (a) Piezoelectric-thermo-elastic bilayer under thermal field θ and a conservative compressive follower force P_0. (b) Bi-layer cross section.

Based on the above assumptions, the equations of motion and Maxwell's electro-magneto-static equations for the thermo-elastic and thermo-electro-elastic layers are written as:[9,12,13,17]

$$T^{(k)}_{ij,i} = \rho^{(k)} \frac{\partial^2 u^{(k)}_j}{\partial t^2} \tag{16.2}$$

$$D^{(k)}_{i,i} = 0 \tag{16.3a}$$

$$e_{ijm}E^{(k)}_{j,m} = 0, \quad (k=1,2) \tag{16.3b}$$

where $F_{,i} = \dfrac{\partial F}{\partial x_i}$, T_{ij} is the stress tensor, ρ is the density, D_i is the electric displacement, E_i is the electric field, and e_{ijm} is the permutation index. The superscript 'k' is used to denote the layer, with $k=1$ indicating the thermo-piezoelectric layer and $k=2$ indicating the thermo-elastic layer.

The constitutive equations are written as

$$\begin{cases} S^{(1)}_i = s^{(1)}_{ij}T^{(1)}_j + d^{(1)}_{ji}E^{(1)}_j + \alpha^{(1)}_i \vartheta \\ D^{(1)}_i = d^{(1)}_{ij}T^{(1)}_j + \varepsilon^{(1)}_{ij}E^{(1)}_j + p^{(1)}_i \vartheta \end{cases} \tag{16.4}$$

for the thermo-piezoelectric layer, and

$$S^{(2)}_i = s^{(2)}_{ij}T^{(2)}_j + \alpha^{(2)}_i \vartheta \tag{16.5}$$

for the thermo-elastic layer. In these equations, S_i is the strain vector, T_i is the stress vector, ϑ is the thermal field across the two layers, s_{ij} is the compliance matrices of the piezoelectric and pure elastic media, d_{ij} is the piezoelectric coefficient, ε_{ij} is the dielectric permittivity, α_i is the thermal expansion coefficient, and p_i is the pyroelectric coefficient. The two-index stress T_{ij} and one-index stress T_i are related as $T_{11}=T_1$, $T_{22}=T_2$, $T_{12}=T_6$ and others (for details see ref. 9, 12, 13, and 17).

Within the scope of Bernoulli's (Kirchhoff's) hypothesis of beam theory, only the strain S_1 is induced in the beam. This strain is given by:

$$S_1 = \frac{\partial u_1(x_1,x_3)}{\partial x_1} = \frac{\partial u(x_1)}{\partial x_1} - x_3 \frac{\partial^2 w}{\partial x_1^2} = \varepsilon - x_3 \kappa \tag{16.6}$$

where $\varepsilon = \dfrac{\partial u(x_1)}{\partial x_1}$ is a strain along the neutral axis, and $\kappa = \dfrac{\partial^2 W(x_1,t)}{\partial x_1^2}$ is the bending of the neutral axis. Eqn (16.6) denotes the linear behavior of the strain S_1 over the entire cross section of the beam, and x_3 defines the vertical distance from the neutral axis.

Next, the boundary conditions for the electrical quantities are provided. If there are no electrodes on the surface of the beam and if the layer on these surfaces is in contact with a non-conductive medium (*i.e.*, insulating glue,

a vacuum, or air), then the component of the electric induction vector $D_1^{(1)}$ normal to these surfaces is equal to zero:

$$D_1^{(1)} = 0 \qquad (16.7)$$

For the electrical field, the following boundary conditions should be satisfied:

$$D_3^{(k)}\Big|_{x_3=z_k} = D_3^{(k+1)}\Big|_{x_3=z_k}$$

$$E_1^{(k)}\Big|_{x_3=z_k} = E_1^{(k+1)}\Big|_{x_3=z_k} \qquad (16.8)$$

where $(k=0, 1, 2)$ and $x_3 = z_k$ is used to denote the location of the interface surfaces, as shown in Figure 16.1. Later, we will assume that the surrounding air is a vacuum. If the electrodes are in a closed-circuit condition with a known complex conductivity $Y = Y_0 + iY_1$, then:[9,17]

$$I = \iint \frac{dD_3^{(1)}}{dt} d\Gamma = 2VY \qquad (16.9)$$

where Γ is the surface over the electrodes, V is an applied voltage, and I is the magnitude of the current. If the electrodes are in an open-circuit condition, then

$$I = \iint \frac{dD_3^{(1)}}{dt} d\Gamma = 0 \qquad (16.10)$$

For the mechanical load on the surface of the beam,

$$T_5^{(2)}\Big|_{x_3=z_2} = q_1^+$$

$$T_5^{(1)}\Big|_{x_3=z_0} = q_1^-$$

$$T_3^{(2)}\Big|_{x_3=z_2} = q_3^+$$

$$T_3^{(1)}\Big|_{x_3=z_0} = q_3^- \qquad (16.11)$$

where q_i^+ and q_i^- are the forces applied at $x_3 = z_2$ and $x_3 = z_0$. The boundary conditions on the composite edges are provided in the discussion of the vibration of a bilayer composite. In order to construct a theory of beams, some additional assumptions regarding the electrical quantities must be made. As in the theory of piezoelectric shells and plates, the assumed hypotheses depend on the electrical conditions on the surfaces of the composite layers. For the piezoelectric layers, the electric field component $E_3^{(1)}(x_1, x_3, t)$ will be assumed not to be a function of the coordinates x_1 and x_3, i.e.,

$$E_3^{(1)}(x_1, x_3, t) = E_0(t) \qquad (16.12)$$

16.3　Tangential Force and Bending Moment

Using the above constitutive relations (eqn (16.4) and (16.5)) and representation (16.6), we express the induced stresses in the layers of various phases as

$$T_1^{(1)} = \frac{1}{s_{11}^{(1)}} (\varepsilon - x_3\kappa - d_{31}^{(1)}E_3^{(1)} - \alpha_1^{(1)}\vartheta) \tag{16.13}$$

for the thermo-piezoelectric layer, and

$$T_1^{(2)} = \frac{1}{s_{11}^{(2)}} (\varepsilon - x_3\kappa - \alpha_1^{(2)}\vartheta) \tag{16.14}$$

for the thermo-elastic layer. By integrating the stress over the thickness, we
obtain the resultant tangential force T_1 as

$$T_1 = \sum_{k=1}^{2} \int_{z_{k-1}}^{z_k} T_1^{(k)}(x_1, x_3, t)dx_3 = A\varepsilon - B\kappa - A_{01}E_0 - A_\vartheta\vartheta \tag{16.15}$$

where z_0, z_1, and z_2 are locations of layered surfaces from the mid-plane (in
our case it could be neutral plane) with $z_1 - z_0 = h_{\mathrm{p}}$, $z_2 - z_1 = h_{\mathrm{m}}$;

$$B = \frac{z_1^2 - z_0^2}{2s_{11}^{(1)}} + \frac{z_2^2 - z_1^2}{2s_{11}^{(2)}} \tag{16.16a}$$

$$A_{01} = \frac{d_{31}^{(1)}h_{\mathrm{p}}}{s_{11}^{(1)}} \tag{16.16b}$$

$$A_\vartheta = \frac{\alpha_1^{(1)}h_{\mathrm{p}}}{s_{11}^{(1)}} + \frac{\alpha_1^{(2)}h_{\mathrm{m}}}{s_{11}^{(2)}} \tag{16.16c}$$

The bending moment M_1 is calculated according to

$$M_1 = \sum_{k=1}^{2} \int_{z_{k-1}}^{z_k} x_3 T_1^{(k)}(x_1, x_3, t)dx_3 = B\varepsilon - D\kappa - C_1E_0 - C_\vartheta\vartheta \tag{16.17}$$

where

$$D = \frac{h_{\mathrm{p}}^3}{3s_{11}^{(1)}} + \frac{h_{\mathrm{m}}^3}{3s_{11}^{(2)}} \tag{16.18a}$$

$$C_1 = -\frac{d_{31}^{(1)}h_{\mathrm{p}}^2}{2s_{11}^{(1)}} \tag{16.18b}$$

$$C_\vartheta = \frac{\alpha_1^{(2)}h_{\mathrm{m}}^2}{2s_{11}^{(2)}} - \frac{\alpha_1^{(1)}h_{\mathrm{p}}^2}{2s_{11}^{(1)}} \tag{16.18c}$$

In the context of the above simplification, the second equation in eqn (16.4) is used to result in

$$D_3 = \int_{-h_p}^{0} D_3^{(1)} dx_3 = C_2 E_0 - C_3 \kappa - P_\vartheta \vartheta \tag{16.19}$$

where

$$C_2 = \varepsilon_{33}^{(1)} \left(1 - K_1^2\right) h_p \tag{16.20a}$$

$$C_3 = -\frac{\varepsilon_{33}^{(1)}}{2} h_p^2 r_1 \tag{16.20b}$$

$$K_1^2 = \frac{\left(d_{31}^{(1)}\right)^2}{\varepsilon_{33}^{(1)} s_{11}^{(1)}} \tag{16.20c}$$

$$r_1 = \frac{d_{31}^{(1)}}{\varepsilon_{33}^{(1)} s_{11}^{(1)}} \tag{16.20d}$$

$$P_\vartheta = \left(\frac{\alpha_1^{(1)} d_{31}^{(1)}}{s_{11}^{(1)}} - p_3^{(1)}\right) h_p \tag{16.20e}$$

We then combine eqn (16.15), (16.17), and (16.19), and write

$$
\begin{cases}
A\varepsilon - Bk - A_{01}E_0 - A_\vartheta \vartheta = T_1 & \text{(16.21a)} \\
B\varepsilon - Dk - C_1 E_0 - C_\vartheta \vartheta = M_1 & \text{(16.21b)} \\
C_2 E_0 - C_3 k - P_\vartheta \vartheta = D_3 & \text{(16.21c)}
\end{cases}
$$

The unknown function $W(x_1, t)$ should be determined using eqn (16.2), Maxwell's eqn (16.3), and the boundary conditions on the composite edges $x_1 = \pm L$.

Note that from eqn (16.21), we can see that if the coefficient $B \neq 0$, then the bending term κ produces a tension T_1, and vice versa. These two modes can be decoupled only if $B = 0$. However, the coefficient B is always zero for symmetrically laminated composites. For beam-layer structures, the electro-elasticity relations (eqn (16.21)) can be simplified by choosing a position of a system of coordinates so that the coefficient $B = 0$, from which we can determine, for example, z_0 location of system of coordinates from the bottom surface of the beam. For the pure elastic case, such a choice gives us a location of the neutral line.[9,17] In addition, if we consider a low-frequency

vibration at $B=0$; then the longitudinal and transversal motions of beam-layer can be fully decoupled.

16.4   Equations of Motion of Bilayer Thermo-electro-elastic Composites

In beam theory, the equations of motion are obtained by integrating the three-dimensional equations of motion (eqn (16.2) and (16.3)) over the beam thickness. We write the equations of motion in the following form:

$$\frac{\partial T_1}{\partial x_1} + X_1 = \rho \frac{\partial^2 \mathbf{u}}{\partial t^2} - \tilde{\rho} \frac{\partial^3 \mathbf{w}}{\partial x_1 \partial t^2} \tag{16.22a}$$

$$\frac{\partial Q}{\partial x_1} + X_3 = \rho \frac{\partial^2 \mathbf{w}}{\partial t^2} \tag{16.22b}$$

$$Q = \frac{\partial M_1}{\partial x_1} - \tilde{\rho} \frac{\partial^2 \mathbf{u}}{\partial t^2} - \tilde{\tilde{\rho}} \frac{\partial^3 \mathbf{w}}{\partial x_1 \partial t^2} \tag{16.22c}$$

where $\rho = h_P \rho_P + h_m \rho_m$, $\tilde{\rho} = \frac{\rho_1}{2}\left(z_1^2 - z_0^2\right) + \frac{\rho_2}{2}\left(z_2^2 - z_1^2\right)$, $\tilde{\tilde{\rho}} = \frac{\rho_1}{3}\left(z_1^3 - z_0^3\right) + \frac{\rho_2}{3}\left(z_2^3 - z_1^3\right)$, $X_1 = q_1^+ - q_1^-$, q_1^+ and q_1^- are applied shear stresses to the top and bottom of the composite, respectively, $X_3 = q_3^+ - q_3^-$, and q_3^+ and q_3^- are the applied normal stresses to the top and the bottom of the composite, respectively.

The total charge Q on each electrode connected to the generator circuit is obtained by integrating the induction D_3 from eqn (16.9) and (16.19) over the entire surface of the electrodes Γ. Then, the conduction current is calculated as

$$I = \iint \frac{\mathrm{d}D_3^{(1)}}{\mathrm{d}t} \mathrm{d}\Gamma = 2VY = -\frac{\partial Q}{\partial t} = \frac{\partial}{\partial t}\left(\Gamma C_2 E_0 - \Gamma P_9 \vartheta - C_3 \iint \frac{\partial^2 w}{\partial x_1^2} \mathrm{d}\Gamma\right) \tag{16.23}$$

16.5   Problem Formulation for Beam with Arbitrary Support Locations

We assume that the beam occupies the interval $-L \leq x_1 \leq L$ and is fixed at arbitrary points $x_1 = \pm c$. This beam is subjected to a tangential follower force P_0 at the free ends $x_1 = \pm L$ (see Figure 16.1). It should be noted that a cantilever beam is obtained for $c = 0$, and a simply supported beam is obtained for $c = L$.

As we mention above, by choosing a position of a system of coordinates so that the coefficient $B=0$ and considering a low-frequency type of motion, then the bending equation can be decoupled from the longitudinal motion, and from eqn (16.22) we can get:[9,11–13,17]

$$-D\frac{\partial^4 W}{\partial x_1^4} - P_0\frac{\partial^2 W}{\partial x_1^2} = \rho\frac{\partial^2 W}{\partial t^2} \tag{16.24}$$

where P_0 is a follower tangential force. Next, we are interested in the bilayer's pure bending harmonic motion, *i.e.*, $(W(x_1,t), E_0(t), \vartheta(t)) = (w(x_1), E, \theta)e^{i\omega t}$, where ω is the circular frequency of motion. The displacements are denoted as

$$w_1(x_1) = w(x_1) \quad \text{if } c \leq x_1 \leq L \tag{16.25a}$$

$$w_3(x_1) = w(x_1) \quad \text{if } -L \leq x_1 \leq -c \tag{16.25b}$$

$$w_2(x_1) = w(x_1) \quad \text{if } -c \leq x_1 \leq c \tag{16.25c}$$

We solve eqn (16.24) with the boundary conditions at $x_1 = \pm L$ and continuity conditions at $x_1 = \pm c$. The boundary conditions at $x_1 = \pm L$ are written as

$$M_1 = -D\frac{d^2 w}{dx_1^2} - C_1 E - C_9 \theta = 0 \tag{16.26}$$

$$N = \frac{dM_1}{dx_1} = 0$$

The continuity conditions yield

$$w_1(c) = w_2(c) = 0 \tag{16.27a}$$

$$w_1(-c) = w_2(-c) = 0 \tag{16.27b}$$

$$\frac{dw_1(c)}{dx_1} = \frac{dw_2(c)}{dx_1} \tag{16.27c}$$

$$\frac{dw_3(-c)}{dx_1} = \frac{dw_2(-c)}{dx_1} \tag{16.27d}$$

$$\frac{d^2 w_1(c)}{dx_1^2} = \frac{d^2 w_2(c)}{dx_1^2} \tag{16.27e}$$

$$\frac{d^2 w_3(-c)}{dx_1^2} = \frac{d^2 w_2(-c)}{dx_1^2} \tag{16.27f}$$

For simplicity, we will consider the symmetric problem. In this case, $w_3(x_1) = w_1(-x_1)$, $w_2(x_1) = w_2(-x_1)$, and the conditions at $x_1 = -c$ and $x_1 = -L$ (eqn (16.27b), (16.27d), and (16.27f)) are replaced by the following symmetry conditions:

$$\frac{dw_2(0)}{dx_1} = 0 \tag{16.28}$$

$$\frac{d^3w_2(0)}{dx_1^3} = 0$$

Note that the boundary value problem (eqn (16.24)–(16.28)), without thermo-electric properties was discussed by V. Gnuni in 2006. Next, the following non-dimensional parameters are introduced:

$$x = x_1/L,$$

$$\alpha = c/L,$$

$$\lambda = \frac{3P_0 L^2 s_{11}^{(1)}}{2h_p^3},$$

$$\Omega^4 = \frac{3\omega^2 L^4 s_{11}^{(1)} \rho_p}{h_p^2},$$

$$p = \left(\left(\left(\frac{\lambda s}{s+h^3} \right)^2 + \frac{\Omega^4 s(1+h\sigma)}{s+h^3} \right)^{1/2} - \frac{\lambda s}{s+h^3} \right)^{1/2}, \text{ and}$$

$$q = \left(\left(\left(\frac{\lambda s}{s+h^3} \right)^2 + \frac{\Omega^4 s(1+h\sigma)}{s+h^3} \right)^{1/2} + \frac{\lambda s}{s+h^3} \right)^{1/2},$$

where $s = s_{11}^{(2)}/s_{11}^{(1)}$,

$$h = h_m/h_p, \text{ and}$$

$$\sigma = \rho_m/\rho_p.$$

The solution of eqn (16.24) is given by

$$w_k(x) = a_{1+4(k-1)} \cosh(px) + a_{2+4(k-1)} \sinh(px) + a_{3+4(k-1)} \cos(qx)$$
$$+ a_{4+4(k-1)} \sin(qx) \tag{16.29}$$

where $a_{1+4(k-1)}$, $a_{2+4(k-1)}$, $a_{3+4(k-1)}$, and $a_{4+4(k-1)}$ $(k=1, 2)$ are unknown coefficients to be determined from boundary conditions (eqn (16.25)–(16.28)).

Using boundary conditions (eqn (16.25)–(16.28)) for unknown coefficients a_{1+4k}, a_{2+4k}, a_{3+4k}, and a_{4+4k} ($k = 0, 1$), the following system of linear algebraic equations is obtained:

$$\hat{A} \cdot \vec{X}^T = \frac{L^2}{D} C_1 E \vec{X}_{01}^T + \frac{L^2}{D} C_9 \theta \vec{X}_{01}^T \tag{16.30}$$

where $\vec{X}^T$ and $\vec{X}_{01}^T$ are transposes of

$$\vec{X} = (a_1, a_2, a_3, a_4, a_5, a_6, a_7, a_8) \tag{16.31}$$

$$\vec{X}_{01} = (0, 0, 0, 0, 0, 0, -1, 0)$$

We also let

$$\hat{A} = \begin{pmatrix} 0 & 0 & 0 & 0 & 0 & p & 0 & q \\ 0 & 0 & 0 & 0 & 0 & p^3 & 0 & -q^3 \\ \cosh p\alpha & \sinh p\alpha & \cos q\alpha & \sin q\alpha & 0 & 0 & 0 & 0 \\ 0 & 0 & 0 & 0 & \cosh p\alpha & \sinh p\alpha & \cos q\alpha & \sin q\alpha \\ p\sinh p\alpha & p\cosh p\alpha & -q\sin q\alpha & q\cos q\alpha & -p\sinh p\alpha & -p\cosh p\alpha & q\sin q\alpha & -q\cos q\alpha \\ p^2\cosh p\alpha & p^2\sinh p\alpha & -q^2\cos q\alpha & -q^2\sin q\alpha & -p^2\cosh p\alpha & -p^2\sinh p\alpha & q^2\cos q\alpha & q^2\sin q\alpha \\ p^2\cosh p & p^2\sinh p & -q^2\cos q & -q^2\sin q & 0 & 0 & 0 & 0 \\ p^3\sinh p & p^3\cosh p & q^3\sin q & -q^3\cos q & 0 & 0 & 0 & 0 \end{pmatrix} \tag{16.32}$$

Assuming $\Delta = \det(\hat{A}) \neq 0$ from eqn (16.30), we find all unknown coefficients $\vec{X} = (a_1, a_2, a_3, a_4, a_5, a_6, a_7, a_8)$, i.e.,

$$\vec{X}^T = \hat{C} \cdot \frac{1}{\Delta} [\tilde{E} \vec{X}_{01}^T] \tag{16.33}$$

where the matrix $\hat{C}$ has elements $\hat{C} = (C_{i,j})$ equal to $\hat{C} = (C_{i,j}) = \hat{A}^{-1}$ $\det(\hat{A}) = \hat{A}^{-1}\Delta$, and $\tilde{E} = \frac{L^2}{D} C_1 E + \frac{L^2}{D} C_9 \theta$. The solution (eqn (16.29)) is then rewritten in the following form

$$w_1(x) = -\frac{v_1}{\Delta} \tilde{E} \tag{16.34}$$

$$w_2(x) = -\frac{v_2}{\Delta} \tilde{E} \tag{16.35}$$

The detailed expressions for the coefficients in eqn (16.34) and (16.35) are presented in Appendix A.

Having the solution for eqn (16.29) or eqn (16.34) and (16.35), we determine the conversion and energy-harvesting coefficients for piezoelectric-thermo-elastic bimorphs. Under thermodynamic equilibrium, the internal energy density of an infinitesimally small volume element in the piezoelectric material is given by

$$U^p(x_1, x_3, t) = \frac{1}{2} S_1^{(1)} T_1^{(1)} + \frac{1}{2} E_3^{(1)} D_3^{(1)} \tag{16.36}$$

Substitution of eqn (16.6) into eqn (16.36), and using eqn (16.4), we obtain

$$U^{\mathrm{p}}(x_1, x_3, t) = \frac{1}{2s_{11}^{(1)}} (x_3 \kappa)^2 + \frac{\alpha_1^{(1)}}{2s_{11}^{(1)}} x_3 \kappa \, \vartheta + \frac{C_2}{2h_{\mathrm{p}}} E_0^2 - \frac{P_\vartheta}{2h_{\mathrm{p}}} E_0 \vartheta \qquad (16.37)$$

where $\kappa = \dfrac{\partial^2 W}{\partial x_1^2}$, and C_2 and P_ϑ are defined in eqn (16.20a) and (16.20e). The energy for the thermo-elastic layer is given by

$$U^{\mathrm{m}}(x_1, x_3, t) = \frac{1}{2} S_1^{(2)} T_1^{(2)} = \frac{1}{2s_{11}^{(2)}} (x_3 \kappa)^2 + \frac{\alpha_1^{(2)}}{2s_{11}^{(2)}} x_3 \kappa \vartheta \qquad (16.38)$$

Once we determine the internal energy density of each layer, the total energy of the bilayer bender is obtained by volume integration. Assuming that the width of the structure is unity, we write

$$U(t) = \int_{-L}^{L} \int_{0}^{h_{\mathrm{p}}} U^{\mathrm{p}} \mathrm{d}x_3 \mathrm{d}x_1 + \int_{-L}^{L} \int_{-h_{\mathrm{m}}}^{0} U^{\mathrm{m}} \mathrm{d}x_3 \mathrm{d}x_1 \qquad (16.39)$$

Using the symmetry conditions $w_3(x_1) = w_1(-x_1)$, $w_2(x_1) = w_2(-x_1)$, and non-dimensional parameters $x = x_1/L$, $\alpha = c/L$, eqn (16.39) results in

$$U(t) = C_2 L E_0^2 - P_\vartheta L E_0 \vartheta + \frac{DL}{2} \int_{-1}^{1} k^2 \mathrm{d}x - \frac{C_\vartheta 9 L}{2} \int_{-1}^{1} k \mathrm{d}x \qquad (16.40)$$

or using eqn (16.34) and (16.35), we write

$$U(t) = \gamma_1 E_0^2 + \gamma_2 \vartheta^2 + \gamma_3 E_0 \vartheta \qquad (16.41)$$

where γ_i, $(i = 1, 2, 3)$ are presented in Appendix B. By treating the electrical and the coupled terms as $U(t) = QV_0$, where $Q(t)$ is the charge and $V_0(t)$ is the voltage, we derive the generated charge by substitution into eqn (16.41). An expression for the electrical field in terms of voltage $E_0(t) = V_0(t)/h_{\mathrm{p}}$ is obtained. This is differentiated with respect to V_0, where h_{p} is the distance between the top and bottom surfaces of the electrodes in the piezoelectric layer, as shown in Figure 16.1. The result is

$$Q(t) = \frac{\partial U(t)}{\partial V_0} = \frac{2\gamma_1}{h_{\mathrm{p}}^2} V_0(t) + \frac{\gamma_3}{h_{\mathrm{p}}} \vartheta(t) = C_{\mathrm{v}} V_0(t) + Q_\vartheta \vartheta(t) \qquad (16.42)$$

where

$$C_{\mathrm{v}} = \frac{2\gamma_1}{h_{\mathrm{p}}^2} = \frac{2}{h_{\mathrm{p}}^2} \left(C_2 L + \frac{C_1^2 L}{D\hat{\Delta}^2} R_1 \right) \qquad (16.43a)$$

$$Q_\vartheta = \frac{\gamma_3}{h_{\mathrm{p}}} = -\frac{P_\vartheta L}{h_{\mathrm{p}}} + \frac{2C_1 C_\vartheta L}{h_{\mathrm{p}} D\hat{\Delta}^2} R_1 + \frac{C_1 C_\vartheta L}{h_{\mathrm{p}} D\hat{\Delta}} R_2 \qquad (16.43b)$$

The values of R_1 and R_2 are presented in Appendix B. Using eqn (16.43), we determine the amplitude of the generated charge as

$$Q_{\text{gen}}^V = C_v V = \frac{2}{h_p^2}\left(C_2 L + \frac{C_1^2 L}{D\hat{\Delta}^2}R_1\right)V \qquad (16.44\text{a})$$

from the applied voltage $V_0(t) = Ve^{i\omega t}$ and

$$Q_{\text{gen}}^\vartheta = Q_\vartheta \theta = \left(-\frac{P_\vartheta L}{h_p} + \frac{2C_1 C_\vartheta L}{h_p D\hat{\Delta}^2}R_1 + \frac{C_1 C_\vartheta L}{h_p D\hat{\Delta}}R_2\right)\theta \qquad (16.44\text{b})$$

from the thermal gradient $\vartheta = \theta e^{i\omega t}$.

Recognizing that $Q_{\text{gen}}^V = C_v V$, the capacitance is

$$C_v = \frac{2}{h_p^2}\left(C_2 L + \frac{C_1^2 L}{D\hat{\Delta}^2}R_1\right) \qquad (16.45)$$

The generated voltage amplitude from ϑ is

$$V_{\text{gen}}^\vartheta = \frac{Q_{\text{gen}}^\vartheta}{C_v} \qquad (16.46)$$

The generated electrical energy amplitude from ϑ is

$$U_{\text{gen}}^\vartheta = \frac{1}{2}Q_{\text{gen}}^\vartheta V_{\text{gen}}^\vartheta \qquad (16.47)$$

16.6 Discussions and Numerical Results

The non-dimensional thermal energy harvesting coefficient is finally derived in the following form:

$$\tilde{Q}_{\text{gen}}^\vartheta = \frac{-s_{11}^{(1)}Q_{\text{gen}}^\vartheta}{\alpha_1^{(1)}d_{31}^{(1)}L\theta} = (1-\tilde{p}) + \frac{3}{4}\frac{(\alpha_T h^2 - s)}{(s+h^3)}\left(\frac{2R_1}{\hat{\Delta}^2} + \frac{2R_4}{\hat{\Delta}}\right) \qquad (16.48)$$

where $\alpha_T = \alpha_1^{(2)}/\alpha_1^{(1)}$, $\tilde{p} = (p_3^{(1)}s_{11}^{(1)})/(\alpha_1^{(1)}d_{31}^{(1)})$, $h = h_m/h_p$, and $s = s_{11}^{(2)}/s_{11}^{(1)}$.

We can state that eqn (16.48) is derived for the first time in such a general form. In particular, for a static case, assuming also $\alpha_T = 0$, eqn (16.48) coincides with counterpart derived in ref. 14. The material properties shown in Table 16.1 will be considered during the numerical simulation of eqn (16.48). It should be noted that the derivation in this work can be extended to the magneto-thermo-electro-elastic shape memory alloy composites. For simplicity, the two materials for SMA can be also PZT-5A with an aluminum substrate (this case is discussed in ref. 14).

Using the following properties, the critical bucking load is $\lambda_{\text{cr}} \geq 1.65$. This is determined using the formula for a static Euler's column:

$$P_{\text{cr}} = \frac{\pi^2 D}{4L^2} \qquad (16.49)$$

Table 16.1 Material properties used in numerical simulation.

Material property	PZT-5A	Aluminum
Compliance $(10^{-12}\ m^2\ N^{-1})$	16.4	14.3
Density $(kg\ m^{-3})$	7750	2700
Height (mm)	0.75	0.5
Thermal expansion $(10^{-6}\ K^{-1})$	2	22
Pyroelectric coefficient $(10^{-6}\ Cm^{-2}\ K^{-1})$	238	—
Piezoelectric coefficient $(10^{-12}\ C\ N^{-1})$	-171	—
Dielectric permittivity $(10^{-12}\ F\ m^{-1})$	15051.8	—

The value of the follower force will be taken below the critical value.

The effect of varying the vibrational frequency on the generation of charge is shown in Figure 16.2. Parameters s, h, σ, α_T, and $\tilde{p}$ are chosen according to Table 16.1. The regions that peak to infinity represent the vibrational frequency being equal to the natural frequency. In Figure 16.2a, the support location has a strong effect on the charge coefficient. Figure 16.2b–e show the effect of varying the value of the follower force. When $\lambda > 0$, the follower force is in compression, and when $\lambda < 0$, it is in tension. From these plots, λ has the strongest effect on the system when the vibration frequency is below the first natural frequency. When the system is static $(\Omega = 0)$, as shown in Figure 16.3, by including the follower force, the energy-harvesting coefficient can be improved. When the magnitude of the compressive follower force increases, the strongest effect is at higher values of α. For $\alpha = 1.0$, the charge coefficient approaches to infinity monotonically as the compressive follower force approaches λ_{cr}. Contrary to conventional uses of bimorphs, where a compressive follower force is applied, a tensile follower force is also considered. In Figure 16.2d and e, or Figure 16.3, for higher values of α, the tensile follower force decreases the energy harvesting coefficient, but for lower values of α, it can improve the energy harvester. For a cantilever beam, the energy harvesting coefficient can be improved to $\tilde{Q}_{gen}^{\vartheta} = 28$ for $\lambda = -1.6$ from $\tilde{Q}_{gen}^{\vartheta} = 20$ for $\lambda = 0$. This is 40% increase of $\tilde{Q}_{gen}^{\vartheta}$.

An energy harvester can also be improved by increasing the probability of capturing resonance in environments where the vibrational frequency fluctuates. For example, in Figure 16.4, the first natural frequency of Figure 16.2b and d is considered. In Figure 16.4a, a compressive load is applied, and the width of the peak, which will be called the bandwidth, is widest for $\alpha = 1$ as compared to $\alpha = 0$ and $\alpha = 0.5$. Conversely, in Figure 16.4b, when there is tensile load, the bandwidth is widest for $\alpha = 0$ and $\alpha = 0.5$, as compared to $\alpha = 1$. Because of this, when designing an energy harvester, it can be optimized for environments where the frequency fluctuates. The formulation in this paper can also be a guild to experimenters who try to capture resonance with energy harvesters.

Next, the effect of the nondimensional energy harvester's properties – h, s, and ρ – on the generation of charge is studied. The remaining material properties L, ϑ, α_T, $p_3^{(1)}$, and $d_{31}^{(1)}$ affect Q_{gen}^{ϑ} linearly from our formulation of

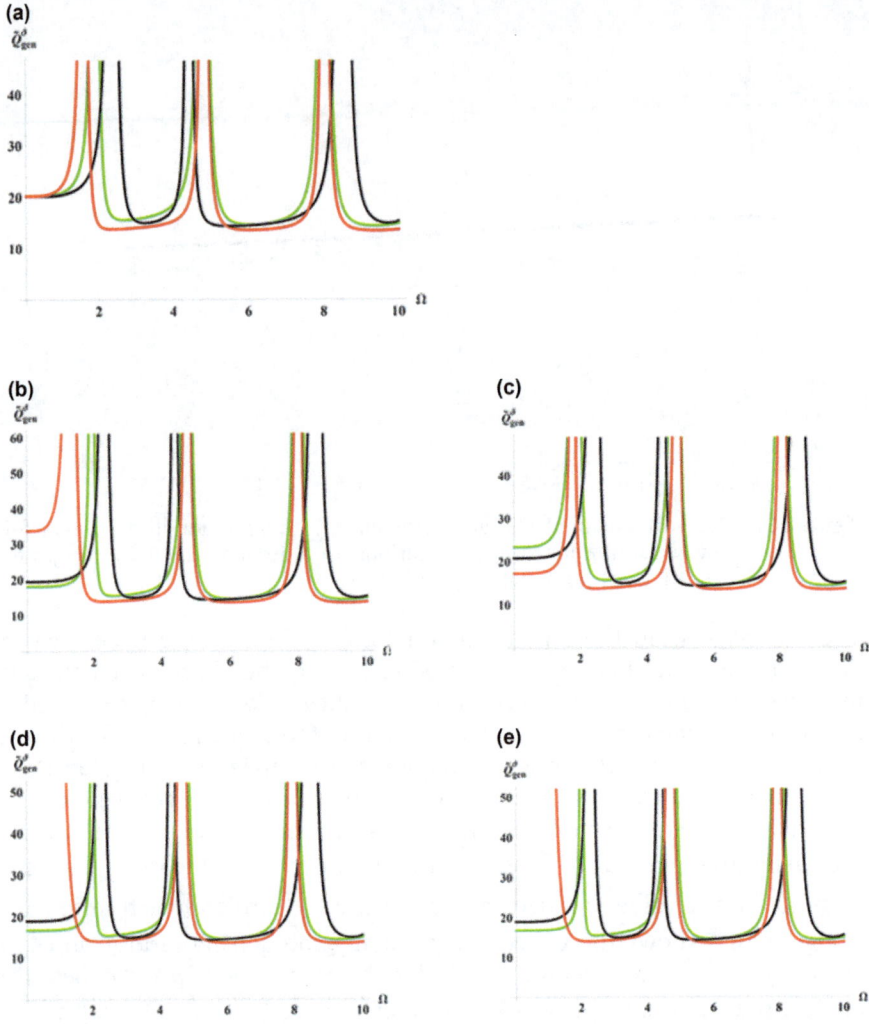

Figure 16.2 Non-dimensional charge coefficient $\tilde{Q}^{\vartheta}_{gen}$ *versus* non-dimensional frequency Ω for support locations $\alpha = 0$ (green), $\alpha = 0.5$ (black), and $\alpha = 1.0$ (red) for follower force (a) $\lambda = 0$, (b) $\lambda = 0.8$, (c) $\lambda = 1.60$, (d) $\lambda = -0.8$, and (e) $\lambda = -1.60$.

eqn (16.44b), and they will not be considered. The variation of these properties represents changing the material properties while the piezoelectric properties are held constant. When plotting the changes of one of these parameters, the remaining parameters are set to the values of Table 16.1. In Figure 16.5, the effect of varying the volume fraction h is shown when $\Omega = 0$. In Figure 16.5a, when $\lambda = 0$, the support location has no effect on the generation of charge as h is changed. At $h = 1.3$, the energy harvesting coefficient approaches a maximum value and increasing h to infinity, the energy harvesting coefficient converges to a constant value. These characteristics are

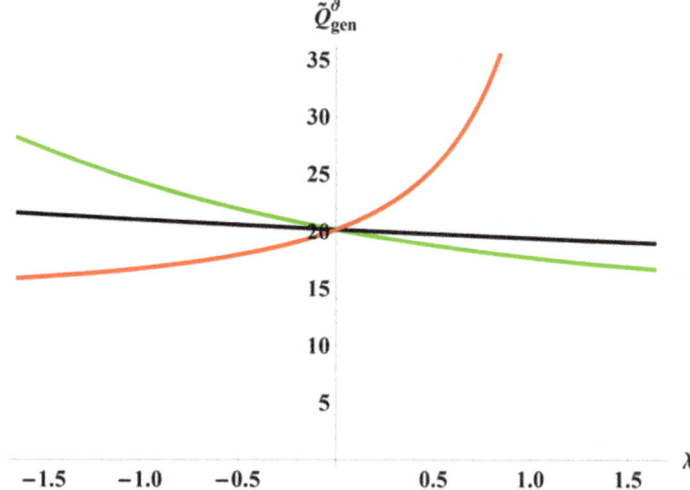

Figure 16.3 Non-dimensional charge coefficient $\tilde{Q}^{\vartheta}_{\text{gen}}$ *versus* non-dimensional follower force λ for support locations $\alpha = 0$ (green), $\alpha = 0.5$ (black), and $\alpha = 1.0$ (red) for a static system $\Omega = 0$.

also observed when there is a follower force included, as can be seen in Figure 16.5c–e. Similar to the discussion earlier, the charge coefficient is improved in the presence of a compressive follower force when $\alpha = 1$, and it is improved in the presence of a tensile follower force when $\alpha = 0$ and $\alpha = 0.5$.

Figure 16.6 shows the effect of varying the compliance ratio s on $\tilde{Q}^{\vartheta}_{\text{gen}}$. From these results, it is determined that decreasing the stiffness of the material relative to the piezoelectric stiffness, the energy harvester improves. Increasing the material stiffness to infinity causes the value of $\tilde{Q}^{\vartheta}_{\text{gen}}$ to converge to a finite value, but decreasing it causes $\tilde{Q}^{\vartheta}_{\text{gen}}$ to approach infinity.

Figure 16.7 shows the effect of the varying the density ratio σ on $\tilde{Q}^{\vartheta}_{\text{gen}}$, where σ only appears as a product of Ω^4 in the expression of p and q. Because of this, the variation of σ will have no effect on $\tilde{Q}^{\vartheta}_{\text{gen}}$ when $\Omega = 0$. Thus, a dynamic system at vibrating frequencies $\Omega = 1.5$ and $\Omega = 3.5$ is considered in Figure 16.7. These represent the frequency before the first natural frequency and between the first and second natural frequency, respectively, for the values in Table 16.1. Varying the densities of the piezoelectric and the substrate does change the values of the natural frequencies. This explains why at some values of σ, $\tilde{Q}^{\vartheta}_{\text{gen}}$ is infinity. This is because resonance effects are observed.

16.7 Conclusions

An analytical analysis of a bimorph's thermal energy harvesting coefficient has been performed. The analysis also takes into account pyroelectric and thermal expansion effects. The most general analytical expression for the

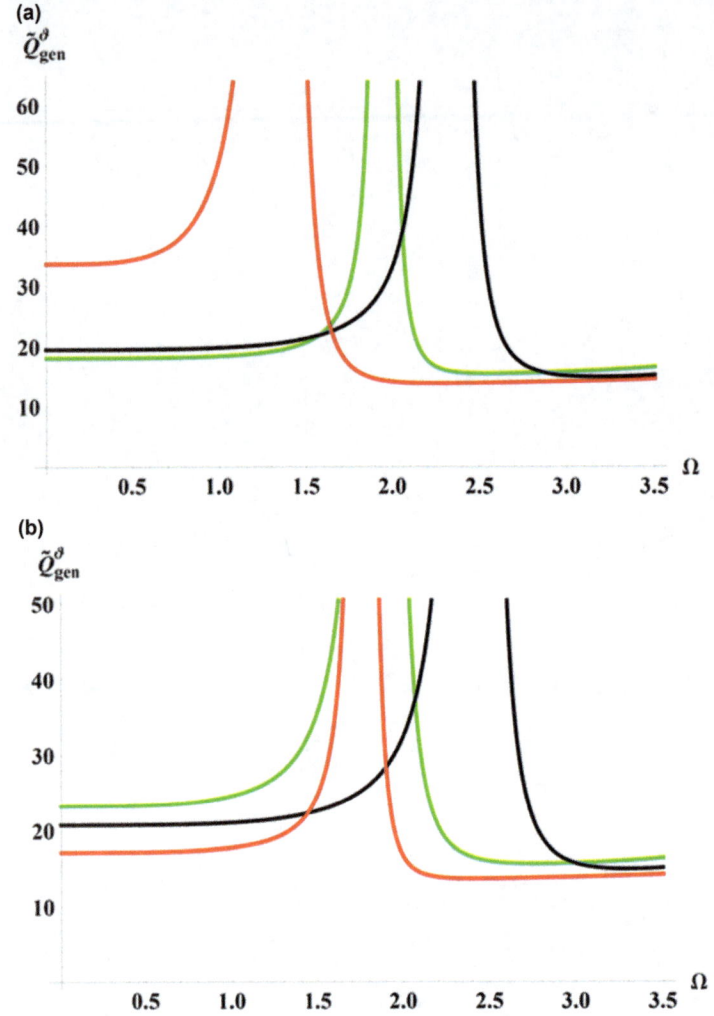

Figure 16.4 Display of the first natural frequency's bandwidth of Figure 16.2b and d for support locations $\alpha = 0$ (green), $\alpha = 0.5$ (black), and $\alpha = 1.0$ (red) for follower force (a) $\lambda = 0.8$ and (b) $\lambda = -0.8$.

energy conversation coefficients are presented for a bilayer. These coefficients were derived for more general situations, when mechanical, electrical, and thermal fields are present. We derived coefficients (transformation coefficients) for sensing, actuating, and energy harvesting. As a particular case, the analytical expressions for energy harvesting coefficients due to pyroelectric and thermal expansion effects were obtained. The influences of support location, material properties, and a conservative follower force were analyzed in order to optimize the energy harvester. The numerical simulation of the thermal energy harvesting coefficient showed that support

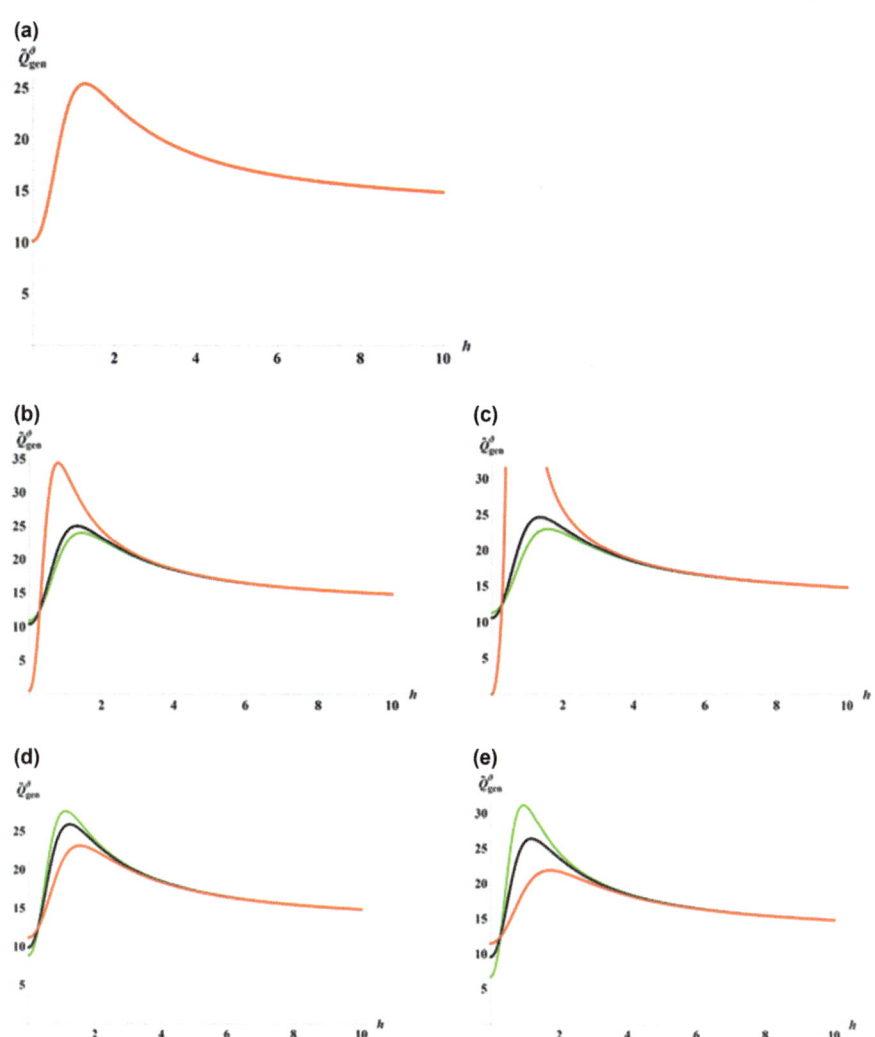

Figure 16.5 Non-dimensional charge coefficient $\tilde{Q}^9_{\mathrm{gen}}$ *versus* volume fraction h for a static case $\Omega = 0$ at support locations $\alpha = 0$ (green), $\alpha = 0.5$ (black), $\alpha = 1.0$ (red) for follower force (a) $\lambda = 0$, (b) $\lambda = 0.8$, (c) $\lambda = 1.6$, (d) $\lambda = -0.8$, and (e) $\lambda = -1.6$.

location strongly influences the generated charge. The follower force was seen to influence the energy harvester the most when the vibration frequency was below the first resonance frequency. Contrary to convention, where a compressive in-plane follower force is applied, a tensile follower force was shown to also improve the energy harvester. In addition, by varying the support location and follower force, the bandwidth, or the range of vibrating frequency over which resonance effects are observed, can be made wider so that the probability of the energy harvester operating at resonance frequency can be increased. This is beneficial for designs that operate in

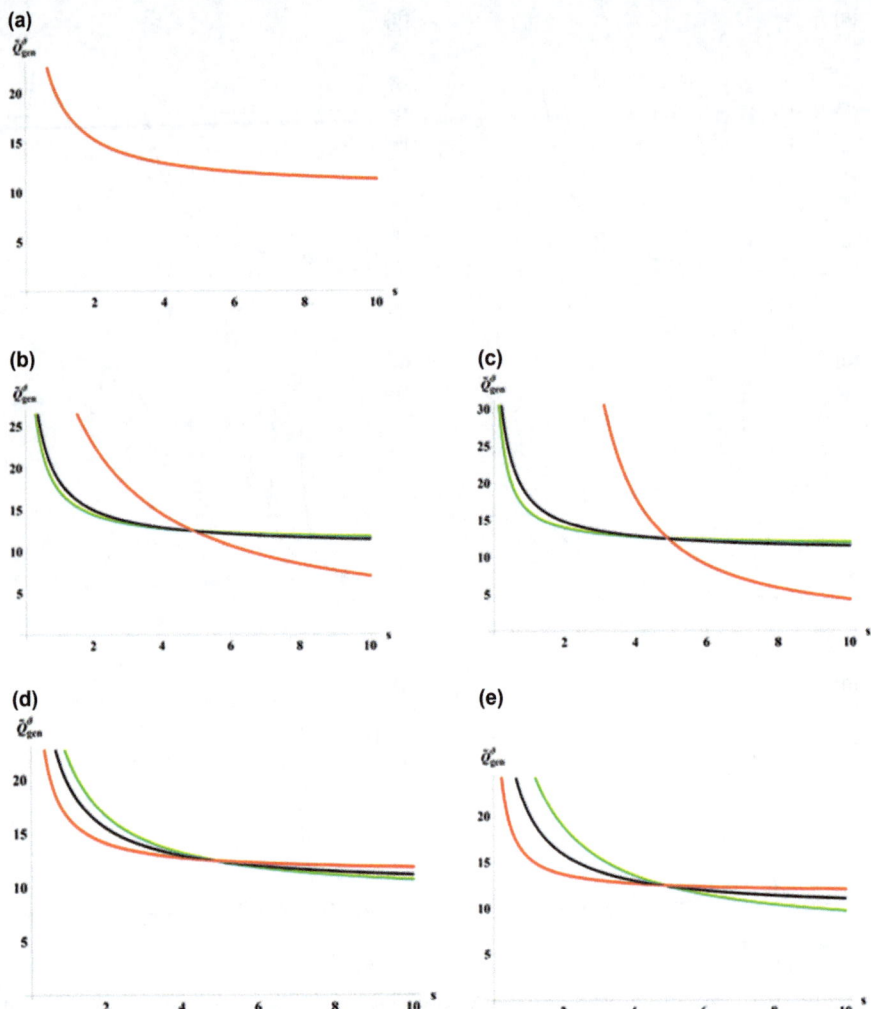

Figure 16.6   Non-dimensional charge coefficient $\tilde{Q}^{\vartheta}_{\mathrm{gen}}$ *versus* compliance ratio s for a static case $\Omega = 0$ at support locations $\alpha = 0$ (green), $\alpha = 0.5$ (black), $\alpha = 1.0$ (red) for follower force (a) $\lambda = 0$, (b) $\lambda = 0.8$, (c) $\lambda = 1.6$, (d) $\lambda = -0.8$, and (e) $\lambda = -1.6$.

environments where the vibration frequency fluctuates over a certain frequency range.

The effect of volume fraction and compliance ratio was studied for a static system. It was shown that at a certain volume fraction, the energy harvester is optimized. The compliance ratio improves the energy harvester by taking it close to zero. The density ratio does not affect the charge coefficient directly. By changing the densities of the two layers of the bimorph, the natural frequency of the system is changed, and this indirectly influences the charge coefficient. It was also shown that the density ratio has no effect on a static

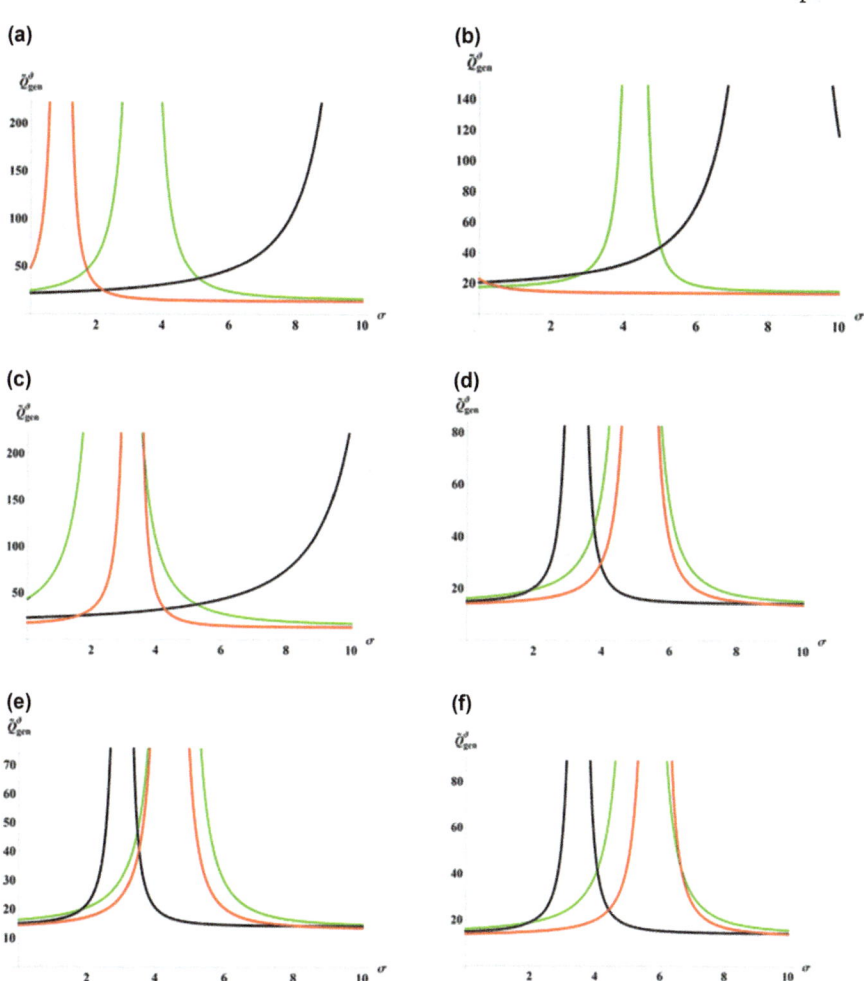

Figure 16.7 Non-dimensional charge coefficient $\tilde{Q}^9_{\text{gen}}$ *versus* density ratio σ for support locations $\alpha = 0$ (green), $\alpha = 0.5$ (black), $\alpha = 1.0$ (red) and (a) $\lambda = 0$, $\Omega = 1.5$ (b) $\lambda = 1.6$, $\Omega = 1.5$, (c) $\lambda = -1.6$, $\Omega = 1.5$, (d) $\lambda = 0$, $\Omega = 3.5$, (e) $\lambda = 1.6$, $\Omega = 3.5$, and (f) $\lambda = -1.6$, $\Omega = 3.5$.

system, as expected, since it appears as a coefficient of the vibrational frequency in the equilibrium equations.

16.8   Appendix A

$$v_1 = \omega_1 p^3 \cosh px + \omega_2 q^3 \frac{\cosh px}{\cosh p(1-\alpha)} + \omega_3 p^3 + \omega_4 \frac{q^3}{\cosh p(1-\alpha)},$$

$$v_2 = \omega_5 p^3 \cosh px + \omega_6 pq^2 \frac{\cosh px}{\cosh p(1-\alpha)} + \omega_7 p^3 + \omega_8 \frac{pq^2}{\cosh p(1-\alpha)},$$

$$\hat{\Delta} = p^3 q^2 \left(\frac{p^2}{q^2 \cosh p(1-\alpha)} + \frac{q^2}{p^2 \cosh p(1-\alpha)} + T \right),$$

$$T = 2\cos q(1-\alpha) - \frac{p \tanh p\alpha + q \tan q\alpha}{q} \sin q(1-\alpha) + \frac{q}{p}\tanh p(1-\alpha)$$

$$\times \left(\sin q(1-\alpha) + \frac{p \tanh p\alpha + q \tan q\alpha}{q} \cos q(1-\alpha) - \frac{p^2}{q^2}\sin q(1-\alpha) \right)$$

$$\omega_1 = 1 - \tanh p(1-\alpha)\tanh px,$$

$$\omega_2 = \frac{p}{q}\cos q(1-\alpha)$$

$$+ \left(\sin q(1-\alpha) + \frac{p \tanh p\alpha + q \tan q\alpha}{q} \cos q(1-\alpha) \right) \tanh px,$$

$$\omega_3 = -\cos qx + \frac{p \tanh p\alpha + q \tan q\alpha}{q}\sin qx + \frac{p}{q}\sin qx \tanh p(1-\alpha),$$

$$\omega_4 = \frac{p}{q}\cos q(1-\alpha)\left(-\cos qx + \frac{p \tanh p\alpha + q \tan q\alpha}{q}\sin qx \right)$$

$$- \frac{p}{q}\sin qx \left(\sin q(1-\alpha) + \frac{p \tanh p\alpha + q \tan q\alpha}{q}\cos q(1-\alpha) \right),$$

$$\omega_5 = 1 + \tanh p\alpha \tanh px,$$

$$\omega_6 = \cos q(1-\alpha)(1 + \tanh p\alpha \tanh px),$$

$$\omega_7 = \cos qx + \tan q\alpha \sin qx,$$

$$\omega_8 = \cos q(1-\alpha)(-\cos qx + \tan q\alpha \sin qx).$$

16.9   Appendix B

$$R_1 = \int_{-\alpha}^{0} \left(\frac{d^2 v_2}{dx^2} \right)^2 dx + \int_{0}^{1-\alpha} \left(\frac{d^2 v_1}{dx^2} \right)^2 dx,$$

$$R_2 = \int_{-\alpha}^{0} \frac{d^2 v_2}{dx^2} dx + \int_{0}^{1-\alpha} \frac{d^2 v_1}{dx^2} dx,$$

$$\gamma_1 = C_2 L + \frac{C_1^2 L}{D \hat{\Delta}^2} R_1,$$

$$\gamma_2 = \frac{LC_\vartheta^2}{D \hat{\Delta}^2} R_1 + \frac{LC_\vartheta^2}{D \hat{\Delta}} R_4,$$

$$\gamma_3 = - P_\vartheta L + \frac{2C_1 C_\vartheta L}{D \hat{\Delta}^2} R_1 + \frac{C_1 C_\vartheta L}{D \hat{\Delta}} R_4$$

References

1. S. R. Anton and H. A. Sodano, *Smart Mater. Struct.*, 2007, **16**(3), R1–R21.
2. S. Bauer, *IEEE Trans. Dielectr. Electr. Insul.*, 2006, **13**(5), 953–962.
3. S. P. Beeby, R. N. Torah, M. J. Tudor, P. Glynne-Jones, T. ODonnell, C. R. Saha and S. Roy, *J. Micromech. Microeng.*, 2007, **17**, 1257–1265.
4. L.-C. J. Blystad, E. Halvorsen and S. Husa (2008). Simulation of a MEMS Piezoelectric Energy Harvester Including Power Conditioning and Mechanical Stoppers. Technical Digest, PowerMEMS 2008, Sendai, Japan, November 2008, 237–240.
5. L.-C. J. Blystad, E. Halvorsen and S. Husa, *IEEE Trans. Ultrason., Ferroelectr., Freq. Control*, 2010, **57**(4), 908–919.
6. V. A. Borisenok, A. S. Koshelev and E. Z. Novitsky, *Bull. Russ. Acad. Sci.: Phys.*, 1996, **60**(10), 1660–1662.
7. F. J. Di Salvo, *Science*, 1999, **285**, 703–706.
8. A. Erturk and D. J. Inman, *Smart Mater. Struct.*, 2008, **17**, 065016.
9. D. Hasanyan, J. Gao, Y. Wang, R. Viswan, M. Li, Y. Shen, J. Li and D. Viehland, *J. Appl. Phys.*, 2012, **112**, 013908.
10. Y. B. Jeon, R. Sood, J. H. Jeong and S. G. Kim, *Sens. Actuators, A*, 2005, **122**, 16–22.
11. S. Kim, W. W. Clark and Q. M. Wang, *J. Intell. Mater. Syst. Struct.*, 2005, **16**, 847–854.
12. M. Krommer and H. Irschik, *Acta Mech.*, 2000, **141**, 51–69.
13. L. Librescu, D. Hasanyan, Z. Qin and D. R. Ambur, *J. Therm. Stresses*, 2003, **26**(11–12).
14. O. C. Namli and M. Taya, *J. Appl. Mech.*, 2011, **78**.
15. J. A. Paradiso and T. Starner, *IEEE*, 2005, **4**, 18–27.
16. G. Poulin, E. Sarraute and F. Costa, *Sens. Actuators, A*, 2004, **116**, 461–471.
17. N. N. Rogacheva, *J. Appl. Math. Mech.*, 2010, **74**, 1009–1027.
18. S. Roundy, D. Steingart, L. Frechette, P. Wright and J. Rabaey, *Lect. Notes Comput. Sci.*, 2004, **2920**, 1–17.
19. G. Sebald, D. Guyomar and A. Agbossou, *Smart Mater. Struct.*, 2009, **18**, 125006.

20. H. A. Sodano, G. E. Simmers, R. Dereux and D. J. Inman, *J. Intell. Mater. Syst. Struct.*, 2007, **18**, 3–10.
21. M. Ujihara, G. P. Carman and D. G. Lee, *Appl. Phys. Lett.*, 2007, **91**, 093508.
22. J. Xie, X. P. Mane, C. W. Green, K. M. Mossi and K. K. Leang, *J. Intell. Mater. Syst. Struct.*, 2009, **21**, 243–249.

Subject Index